ATTITUDES AREN'T FREE
THINKING DEEPLY ABOUT DIVERSITY
IN THE US ARMED FORCES

JAMES E. PARCO
DAVID A. LEVY

Air University Press
Maxwell Air Force Base, Alabama

February 2010

Published by Books Express Publishing
Copyright © Books Express, 2011
ISBN 978-1-780392-01-1

Books Express publications are available from all good retail
and online booksellers. For publishing proposals and direct
ordering please contact us at: info@books-express.com

Contents

CONTENTS

★ ★ ★ ★ ★ ⭐ ★ ★ ★ ★ ★

If you have an important point to make, don't try to be subtle or clever. Use a pile driver. Hit the point once. Then come back and hit it again. Then hit it a third time—a tremendous whack.

—Winston Churchill

Board of Reviewers

Associate Editors

Dr. Amit Gupta – Air War College
Maj Melinda Moreau – School of Advanced Air and Space Studies
Dr. Fred R. Blass – Florida State University

Dr. Mike Allsep – Air Command and Staff College
Dr. Filomeno Arenas – Squadron Officer College
Lt Col Scott Cook – Air Command and Staff College
Dr. Cyndy Cycyota – US Air Force Academy
Maj Joel England, JD – US Air Force
Dr. Barry Fagin – US Air Force Academy
Dr. John Farrell – Squadron Officer College
Dr. Claudia Ferrante – US Air Force Academy
Dr. Steve Fraser – Florida Gulf Coast University
Lt Col Andy Gaydon – Air Command and Staff College
Lt Col Kent Halverson – US Air Force Academy
CDR Kevin Haney, retired – US Naval Academy
CDR Dan Heidt – Air Command and Staff College
Lt Col Sharon Heilmann – US Air Force Academy
MAJ William D. Hemphill – US Military Academy
Dr. Kurt Heppard – US Air Force Academy
Lt Col Tyson Hummel – US Air Force
Dr. J. T. LaSaine – Air Command and Staff College
Lt Col Robert Lass – Air Command and Staff College
Col Walter Leach – US Army War College
Dr. Patricia Maggard – Squadron Officer College
Lt Col Gaylon McAlpine, retired – Air Command and Staff College
Mr. Robert Mishev – McKinsey & Co.
Col Brett Morris – Air Command and Staff College
Maj Amy Nesbitt – US Air Force Academy
Lt Col Rick Rogers – Air Command and Staff College
Mr. Michael J. C. Roth – President Emeritus, USAA Investment Mgt. Co.
Col Joe Sanders – US Air Force Academy
1st Lt Anthony Stinton – US Air Force
Dr. John Terino – Air Command and Staff College
Dr. Jonathon Zartman – Air Command and Staff College

Foreword

Men and women in the uniforms of our armed services share many things in common, but none more important than their love of country and pride of being an American Soldier, Sailor, Airman, or Marine. Although we may wear different insignia, perform different jobs, and observe different customs, when it comes to the job of defending our nation, the diversity of our missions becomes a fundamental source of our great power.

Make no mistake, diversity is a hallmark of the modern US armed forces, not just in terms of our mission elements, but also within the composition of every unit. However, the journey to get where we are today didn't happen overnight, and reflecting back on our history, we overcame many challenges along the way. If we've learned anything about warfare from our earliest experiences back in 1776, end strengths, budget allocations, or technological innovation is never enough to fully maximize our military capabilities in war or peace. The cornerstone of any military success story starts with great ideas championed by principled leaders. Dialoguing with others who challenge our thinking and frames of reference is essential. By creating environments which are conducive to intellectual discourse, we can further our critical thinking skills and bolster our adaptive capacity to constantly see the world from different perspectives—to learn—to understand. As we move forward, we must constantly strive to think deeply, because ideas matter.

Attitudes Aren't Free: Thinking Deeply about Diversity in the US Armed Forces is an innovative approach to foster the much-needed discussion on social issues within the military domain and the first volume of its kind. As you venture into the pages that follow, you will encounter a variety of essays on critical topics that are near and dear to our hearts. The volume has been edited in such a manner to give readers an opportunity to engage in a variety of debates framed by some of the leading voices in the field. Each of the experts shares his or her perspectives and suggests ways ahead based on what he or she believes to be best.

Plurality of perspective is the essence of diversity. It can become a great source of strength when used to seek a deeper understanding of contentious issues. But for diversity to add value, it is critical that each of us opens our mind and makes a sincere effort to understand the perspectives of others. There is never a guarantee that people with different perspectives will come to agreement, nor should there be. Freedom of conscience is one of the oldest American traditions, and it exemplifies the ideals for which we fight. However, what matters most is that we remain sufficiently open-minded, listen to each other's perspectives, and, when appropriate, share our own. Our goal should always be to seek a greater understanding and a mutual respect of our fellow Americans—particularly those who have sworn to give their lives in the defense of our country. By embracing the diversity of perspective that brought the United States of America into being 234 years ago, we will ensure its thriving future in the decades to come.

William T. Lord
Lieutenant General, USAF
January 2010

Preface

The *Janus-faced nature* of the US armed forces requires military leaders to prepare highly effective forces to pursue missions dictated by civilian leadership while at the same time conforming to the values of the society which pays their way. This relationship sometimes creates dilemmas, and none more profound than the recent trends which have resulted in conflict between societal values and those espoused within military culture.

Since World War II, the US military has emerged as an iconic example of diversity. In nearly every unit across the armed services, you will find men and women from every race, religion, and creed serving side by side in the defense of our nation. But the diversity evident in today's military isn't the result of a deliberate strategy to create an inclusive organizational culture as much as the result of an emergent strategy where the integration of minority groups has been resisted at every turn. Instead, the military has periodically been directed to make changes at the direction of its civilian leadership to ensure the composition of the armed forces is reflective of the larger society.

In 1948, President Truman decisively ended racial segregation in the military by executive order. Although racial equality was achieved with the stroke of a pen, the integration of women across the roles of military service proved to be more complicated and continued to lag for several more decades. While gender integration was one of the most hotly contested social issues in the twentieth century, Congress eventually took the lead in the mid-1970s, integrating women through appointments to military academies. Still, it would be two decades before women received equal opportunity in select combat roles.

Determining the "right" social policy governing qualifications for military service in the United States has been one of the most contentious topics and promises to dominate the political stage for the foreseeable future. As America continues its wars in Iraq and Afghanistan, one needs to look no further than the ongoing debates regarding religious expression in the military or the merits of allowing openly gay people to serve to appreciate the challenges military leaders face. On one hand, these issues are largely irrelevant for the members of the military. Civilian leadership dictates policy and military leaders enforce it in the ranks. Once clear standards are established and communicated down the chain of command, compliance is not an option. However, unlike base speed limits or established duty hours, policies governing social issues can cause emotions to run very high.

Effective leaders understand the necessity to foster inclusive environments based on trust and respect to maximize unit performance. To ensure all members feel like an equal part of the team, they must be given opportunities to share their thoughts and opinions at appropriate times. In doing so, leaders can help their units develop an appreciation for the diversity of perspective within their ranks. Increased understanding doesn't necessarily equate to agreement with others, but rather a shared respect that as American citizens, all are entitled to their opinions and in turn, they should respect the opinions of others.

Attitudes Aren't Free: Thinking Deeply about Diversity in the US Armed Forces emerged from a vision to collect essays from the brightest voices of experts across the range of contentious social issues to catalyze productive discussions between military members of all ranks and services. Forty-nine experts contributed to the following 29 chapters writing on the primary themes of religious expression, homosexuality, gender, race, and ethics. Chapters appearing in this volume passed the scrutiny of a double-blind peer-review by one or more referees from the board of reviewers, listed following the table of contents. The chapters are largely written in a colloquial, intellectual op-ed fashion and capture a "snapshot" of the current discussions regarding a particular topic of interest to uniformed personnel, policy makers, and senior leaders. Each section seeks to frame the spectrum of perspectives captured within the current debates and lines of argument.

Authors were specifically asked not to address all sides of the issue, but rather to produce a well-reasoned argument explaining why they believe their well-known position on an issue is in the best interests of the military members and make specific recommendations about how best to address the policy issues from their perspective.

The volume is arranged in four primary sections by theme (Religious Expression, Homosexuality, Race and Gender, and Social Policy Perspectives 2010). Within each section, readers will find multiple chapters—each embracing a different perspective surrounding the section's theme. Thus, because of the unbalanced nature of many of the individual chapters, it is critically important that readers focus on the entire spectrum of perspectives presented within a section to ensure they have the context necessary to frame any single perspective.

Through the chapters presented, we hope to have captured the full range of prevailing perspectives which are most likely to lead to thoughtful and productive discussions. Despite how strongly a military member might hold an opinion, it is likely that someone else in uniform holds a different view, and that's okay. Diversity of opinion has been the hallmark of the United States since its dramatic birth in 1776 and has continued unfettered through today where we now have developed the most innovative and effective military the world has ever known. Thus, it is imperative that we continue to reflect upon the diversity of ideas about how best to formulate the "right" social policy to ensure our service members can most effectively execute their missions.

Acknowledgments

This book would not have been possible without the exceptional efforts of more than 100 individuals who were directly involved with bringing it to fruition. Our sincere thanks go out to all of them.

First and foremost, we'd like to express our gratitude to the contributing authors. We are indebted to them for their willingness to collaborate with us on this project. Each of them put a great deal of time and effort into developing a chapter that was not only creative and thought-provoking, but also passed the scrutiny of a double-blind peer review process. Second, we would like to acknowledge members of our board of reviewers who read one or more essays and helped us decide which essays would eventually make their way into the volume as chapters and which ones wouldn't. This book emerged as a superior publication to what it otherwise would have been without their candid, expert review and advice. Specifically, we would like to individually recognize Dr. Amit Gupta from the Air War College, Dr. Randy Blass from Florida State University, and Maj Melinda Moreau from the School of Advanced Air and Space Studies. As our associate editors, they reviewed the entire manuscript and made recommendations about how best to present the chapters. Dr. Gupta also contributed to the appendix.

There are also many people behind the scenes who put an inordinate amount of time into reading every word of the book and ensuring the highest standards of quality for the printed publication. A great deal of credit for the cover goes to Daniel Armstrong of Air University Press for his exceptional skill as a graphic artist and designer building on the initial cover design concept by Ghislain Viau of Creative Publishing Design in California. The typesetters, Vivian O'Neal and Ann Bailey, worked wonders with the layout operating on a very compressed timeline, which was greatly appreciated. Thanks also to Diane Clark for handling all of the background administrative burdens for the project to include the distribution of thousands of preordered copies of the book. We were equally fortunate to have six outstanding editors from Air University Press contribute to this volume: Belinda Bazinet, Sandi Davis, Tammi Long, Andrew Thayer, Catherine Parker, and Demorah Hayes. Everyone did a tremendous job of identifying the innumerable errors and making the book as close to error-free as humanly possible. We'd also be remiss in not specifically recognizing and publicly thanking Ms. Demorah Hayes. As our primary point of contact at AU Press, she went above and beyond to ensure not only that the quality of the book was to AU Press standards, but more importantly, that her team worked with the alacrity necessary to get this project to press in under six months time, which in the publishing world, is a truly remarkable feat. More importantly, she has been one of the nicest people with whom we ever had the pleasure of working.

Finally, we'd like to thank the senior leadership who made this project possible. Lt Gen Bill Lord originally sponsored an idea several years ago to begin teaching a graduate seminar at Air University on social policy perspectives and has remained a staunch advocate of fostering a new venue for the critical exchange of ideas ever since. We'd also like to thank Dr. Dan Mortensen and Gen John Shaud, retired, of the Air Force Research Institute, who agreed to sponsor this manuscript as an AU Press book. Thanks to their faith in the quality of this volume, we are able to make it available to men and women serving in the Department of Defense free of charge. Last but not least, we'd also like to thank Gen Bill Looney, retired, not only for his support and endorsement of this book, but more importantly for his guidance, wisdom, and leadership as a previous commander of the Air Force's Air Education and Training Command.

All said, despite that credit is to be shared with so many, all errors and omissions are entirely our own.

James E. Parco   *David A. Levy*
Montgomery, Alabama   *Colorado Springs, Colorado*
January 2010

★★★★★ ★ ★★★★★

A PERSPECTIVE FROM
THE COMMANDER IN CHIEF

Barack H. Obama
44th President of the United States of America
10 December 2009

Reprinted transcript of the president's acceptance speech at the Nobel Peace Prize ceremony in Oslo, Norway

---

Your Majesties, Your Royal Highnesses, distinguished members of the Norwegian Nobel Committee, citizens of America, and citizens of the world: I receive this honor with deep gratitude and great humility. It is an award that speaks to our highest aspirations—that for all the cruelty and hardship of our world, we are not mere prisoners of fate. Our actions matter, and can bend history in the direction of justice.

And yet I would be remiss if I did not acknowledge the considerable controversy that your generous decision has generated. In part, this is because I am at the beginning, and not the end, of my labors on the world stage. Compared to some of the giants of history who've received this prize—Schweitzer and King; Marshall and Mandela—my accomplishments are slight. And then there are the men and women around the world who have been jailed and beaten in the pursuit of justice; those who toil in humanitarian organizations to relieve suffering; the unrecognized millions whose quiet acts of courage and compassion inspire even the most hardened cynics. I cannot argue with those who find these men and women—some known, some obscure to all but those they help—to be far more deserving of this honor than I.

But perhaps the most profound issue surrounding my receipt of this prize is the fact that I am the Commander-in-Chief of the military of a nation in the midst of two wars. One of these wars is winding down. The other is a conflict that America did not seek; one in which we are joined by 42 other countries—including Norway—in an effort to defend ourselves and all nations from further attacks.

Still, we are at war, and I'm responsible for the deployment of thousands of young Americans to battle in a distant land. Some will kill, and some will be killed. And so I come here with an acute sense of the costs of armed conflict—

1

filled with difficult questions about the relationship between war and peace, and our effort to replace one with the other.

Now these questions are not new. War, in one form or another, appeared with the first man. At the dawn of history, its morality was not questioned; it was simply a fact, like drought or disease—the manner in which tribes and then civilizations sought power and settled their differences.

And over time, as codes of law sought to control violence within groups, so did philosophers and clerics and statesmen seek to regulate the destructive power of war. The concept of a "just war" emerged, suggesting that war is justified only when certain conditions were met: if it is waged as a last resort or in self-defense; if the force used is proportional; and if, whenever possible, civilians are spared from violence.

Of course, we know that for most of history, this concept of "just war" was rarely observed. The capacity of human beings to think up new ways to kill one another proved inexhaustible, as did our capacity to exempt from mercy those who look different or pray to a different God. Wars between armies gave way to wars between nations—total wars in which the distinction between combatant and civilian became blurred. In the span of 30 years, such carnage would twice engulf this continent. And while it's hard to conceive of a cause more just than the defeat of the Third Reich and the Axis powers, World War II was a conflict in which the total number of civilians who died exceeded the number of soldiers who perished.

In the wake of such destruction, and with the advent of the nuclear age, it became clear to victor and vanquished alike that the world needed institutions to prevent another world war. And so, a quarter century after the United States Senate rejected the League of Nations—an idea for which Woodrow Wilson received this prize—America led the world in constructing an architecture to keep the peace: a Marshall Plan and a United Nations, mechanisms to govern the waging of war, treaties to protect human rights, prevent genocide, restrict the most dangerous weapons.

In many ways, these efforts succeeded. Yes, terrible wars have been fought, and atrocities committed. But there has been no Third World War. The Cold War ended with jubilant crowds dismantling a wall. Commerce has stitched much of the world together. Billions have been lifted from poverty. The ideals of liberty and self-determination, equality and the rule of law have haltingly advanced. We are the heirs of the fortitude and foresight of generations past, and it is a legacy for which my own country is rightfully proud.

And yet, a decade into a new century, this old architecture is buckling under the weight of new threats. The world may no longer shudder at the prospect of war between two nuclear superpowers, but proliferation may increase the risk of catastrophe. Terrorism has long been a tactic, but modern technology allows a few small men with outsized rage to murder innocents on a horrific scale.

Moreover, wars between nations have increasingly given way to wars within nations. The resurgence of ethnic or sectarian conflicts; the growth of secessionist movements, insurgencies, and failed states—all these things have increasingly trapped civilians in unending chaos. In today's wars, many more civilians are killed than soldiers; the seeds of future conflict are sown, economies are wrecked, civil societies torn asunder, refugees amassed, children scarred.

I do not bring with me today a definitive solution to the problems of war. What I do know is that meeting these challenges will require the same vision, hard work, and persistence of those men and women who acted so boldly decades ago. And it will require us to think in new ways about the notions of just war and the imperatives of a just peace.

We must begin by acknowledging the hard truth: We will not eradicate violent conflict in our lifetimes. There will be times when nations—acting individually or in concert—will find the use of force not only necessary but morally justified.

I make this statement mindful of what Martin Luther King Jr. said in this same ceremony years ago: "Violence never brings permanent peace. It solves no social problem: it merely creates new and more complicated ones." As someone who stands here as a direct consequence of Dr. King's life work, I am living testimony to the moral force of non-violence. I know there's nothing weak—nothing passive—nothing naïve—in the creed and lives of Gandhi and King.

But as a head of state sworn to protect and defend my nation, I cannot be guided by their examples alone. I face the world as it is, and cannot stand idle in the face of threats to the American people. For make no mistake: Evil does exist in the world. A non-violent movement could not have halted Hitler's armies. Negotiations cannot convince al Qaeda's leaders to lay down their arms. To say that force may sometimes be necessary is not a call to cynicism—it is a recognition of history; the imperfections of man and the limits of reason.

I raise this point, I begin with this point because in many countries there is a deep ambivalence about military action today, no matter what the cause. And at times, this is joined by a reflexive suspicion of America, the world's sole military superpower.

But the world must remember that it was not simply international institutions—not just treaties and declarations—that brought stability to a post–World War II world. Whatever mistakes we have made, the plain fact is this: The United States of America has helped underwrite global security for more than six decades with the blood of our citizens and the strength of our arms. The service and sacrifice of our men and women in uniform has promoted peace and prosperity from Germany to Korea, and enabled democracy to take hold in places like the Balkans. We have borne this burden not because we seek to impose our will. We have done so out of enlightened self-interest—because we seek a better future for our children and grandchildren, and we believe that their lives will be better if others' children and grandchildren can live in freedom and prosperity.

So yes, the instruments of war do have a role to play in preserving the peace. And yet this truth must coexist with another—that no matter how justified, war promises human tragedy. The soldier's courage and sacrifice is full of glory, expressing devotion to country, to cause, to comrades in arms. But war itself is never glorious, and we must never trumpet it as such.

So part of our challenge is reconciling these two seemingly irreconcilable truths—that war is sometimes necessary, and war at some level is an expression of human folly. Concretely, we must direct our effort to the task that President Kennedy called for long ago. "Let us focus," he said, "on a more practical, more attainable peace, based not on a sudden revolution in human nature but on a gradual evolution in human institutions." A gradual evolution of human institutions.

What might this evolution look like? What might these practical steps be?

To begin with, I believe that all nations—strong and weak alike—must adhere to standards that govern the use of force. I—like any head of state—reserve the right to act unilaterally if necessary to defend my nation. Nevertheless, I am convinced that adhering to standards, international standards, strengthens those who do, and isolates and weakens those who don't.

The world rallied around America after the 9/11 attacks, and continues to support our efforts in Afghanistan, because of the horror of those senseless attacks and the recognized principle of self-defense. Likewise, the world recognized the need to confront Saddam Hussein when he invaded Kuwait—a consensus that sent a clear message to all about the cost of aggression.

Furthermore, America—in fact, no nation—can insist that others follow the rules of the road if we refuse to follow them ourselves. For when we don't, our actions appear arbitrary and undercut the legitimacy of future interventions, no matter how justified.

And this becomes particularly important when the purpose of military action extends beyond self-defense or the defense of one nation against an aggressor. More and more, we all confront difficult questions about how to prevent the slaughter of civilians by their own government, or to stop a civil war whose violence and suffering can engulf an entire region.

I believe that force can be justified on humanitarian grounds, as it was in the Balkans, or in other places that have been scarred by war. Inaction tears at our conscience and can lead to more costly intervention later. That's why all responsible nations must embrace the role that militaries with a clear mandate can play to keep the peace.

America's commitment to global security will never waver. But in a world in which threats are more diffuse, and missions more complex, America cannot act alone. America alone cannot secure the peace. This is true in Afghanistan. This is true in failed states like Somalia, where terrorism and piracy is joined by famine and human suffering. And sadly, it will continue to be true in unstable regions for years to come.

The leaders and soldiers of NATO countries, and other friends and allies, demonstrate this truth through the capacity and courage they've shown in Afghanistan. But in many countries, there is a disconnect between the efforts of those who serve and the ambivalence of the broader public. I understand why war is not popular, but I also know this: The belief that peace is desirable is rarely enough to achieve it. Peace requires responsibility. Peace entails sacrifice. That's why NATO continues to be indispensable. That's why we must strengthen U.N. and regional peacekeeping, and not leave the task to a few countries.

That's why we honor those who return home from peacekeeping and training abroad to Oslo and Rome; to Ottawa and Sydney; to Dhaka and Kigali—we honor them not as makers of war, but of wagers—but as wagers of peace.

Let me make one final point about the use of force. Even as we make difficult decisions about going to war, we must also think clearly about how we fight it. The Nobel Committee recognized this truth in awarding its first prize for peace to Henry Dunant—the founder of the Red Cross, and a driving force behind the Geneva Conventions.

Where force is necessary, we have a moral and strategic interest in binding ourselves to certain rules of conduct. And even as we confront a vicious adversary that abides by no rules, I believe the United States of America must remain a standard bearer in the conduct of war. That is what makes us different from those whom we fight. That is a source of our strength. That is why I prohibited torture. That is why I ordered the prison at Guantanamo Bay closed. And that is why I have reaffirmed America's commitment to abide by the Geneva Conventions. We lose ourselves when we compromise the very ideals that we fight to defend. And we honor—we honor those ideals by upholding them not when it's easy, but when it is hard.

I have spoken at some length to the question that must weigh on our minds and our hearts as we choose to wage war. But let me now turn to our effort to avoid such tragic choices, and speak of three ways that we can build a just and lasting peace.

First, in dealing with those nations that break rules and laws, I believe that we must develop alternatives to violence that are tough enough to actually change behavior—for if we want a lasting peace, then the words of the international community must mean something. Those regimes that break the rules must be held accountable. Sanctions must exact a real price. Intransigence must be met with increased pressure—and such pressure exists only when the world stands together as one.

One urgent example is the effort to prevent the spread of nuclear weapons, and to seek a world without them. In the middle of the last century, nations agreed to be bound by a treaty whose bargain is clear: All will have access to peaceful nuclear power; those without nuclear weapons will forsake them; and those with nuclear weapons will work towards disarmament. I am committed to upholding this treaty. It is a centerpiece of my foreign policy. And I'm working with President Medvedev to reduce America and Russia's nuclear stockpiles.

But it is also incumbent upon all of us to insist that nations like Iran and North Korea do not game the system. Those who claim to respect international law cannot avert their eyes when those laws are flouted. Those who care for their own security cannot ignore the danger of an arms race in the Middle East or East Asia. Those who seek peace cannot stand idly by as nations arm themselves for nuclear war.

The same principle applies to those who violate international laws by brutalizing their own people. When there is genocide in Darfur, systematic rape in Congo, repression in Burma—there must be consequences. Yes, there will be engagement; yes, there will be diplomacy—but there must be consequences when those things fail. And the closer we stand together, the less likely we will be faced with the choice between armed intervention and complicity in oppression.

This brings me to a second point—the nature of the peace that we seek. For peace is not merely the absence of visible conflict. Only a just peace based on the inherent rights and dignity of every individual can truly be lasting.

It was this insight that drove drafters of the Universal Declaration of Human Rights after the Second World War. In the wake of devastation, they recognized that if human rights are not protected, peace is a hollow promise.

And yet too often, these words are ignored. For some countries, the failure to uphold human rights is excused by the false suggestion that these are somehow Western principles, foreign to local cultures or stages of a nation's development. And within America, there has long been a tension between those who describe themselves as realists or idealists—a tension that suggests a stark choice between the narrow pursuit of interests or an endless campaign to impose our values around the world.

I reject these choices. I believe that peace is unstable where citizens are denied the right to speak freely or worship as they please; choose their own leaders or assemble without fear. Pent-up grievances fester, and the suppression of tribal and religious identity can lead to violence. We also know that the opposite is true. Only when Europe became free did it finally find peace. America has never fought a war against a democracy, and our closest friends are governments that protect the rights of their citizens. No matter how callously defined, neither America's interests—nor the world's—are served by the denial of human aspirations.

So even as we respect the unique culture and traditions of different countries, America will always be a voice for those aspirations that are universal. We will bear witness to the quiet dignity of reformers like Aung Sang Suu Kyi; to the bravery of Zimbabweans who cast their ballots in the face of beatings; to the hundreds of thousands who have marched silently through the streets of Iran. It is telling that the leaders of these governments fear the aspirations of their own people more than the power of any other nation. And it is the responsibility of all free people and free nations to make clear that these movements—these movements of hope and history—they have us on their side.

Let me also say this: The promotion of human rights cannot be about exhortation alone. At times, it must be coupled with painstaking diplomacy. I know that engagement with repressive regimes lacks the satisfying purity of indignation. But I also know that sanctions without outreach—condemnation without discussion—can carry forward only a crippling status quo. No repressive regime can move down a new path unless it has the choice of an open door.

In light of the Cultural Revolution's horrors, Nixon's meeting with Mao appeared inexcusable—and yet it surely helped set China on a path where millions of its citizens have been lifted from poverty and connected to open societies. Pope John Paul's engagement with Poland created space not just for the Catholic Church, but for labor leaders like Lech Walesa. Ronald Reagan's ef-

forts on arms control and embrace of perestroika not only improved relations
with the Soviet Union, but empowered dissidents throughout Eastern Europe.
There's no simple formula here. But we must try as best we can to balance isolation and engagement, pressure and incentives, so that human rights and dignity
are advanced over time.

Third, a just peace includes not only civil and political rights—it must encompass economic security and opportunity. For true peace is not just freedom
from fear, but freedom from want.

It is undoubtedly true that development rarely takes root without security;
it is also true that security does not exist where human beings do not have access to enough food, or clean water, or the medicine and shelter they need to
survive. It does not exist where children can't aspire to a decent education or a
job that supports a family. The absence of hope can rot a society from within.

And that's why helping farmers feed their own people—or nations educate
their children and care for the sick—is not mere charity. It's also why the world
must come together to confront climate change. There is little scientific dispute
that if we do nothing, we will face more drought, more famine, more mass
displacement—all of which will fuel more conflict for decades. For this reason,
it is not merely scientists and environmental activists who call for swift and
forceful action—it's military leaders in my own country and others who understand our common security hangs in the balance.

Agreements among nations. Strong institutions. Support for human rights.
Investments in development. All these are vital ingredients in bringing about the
evolution that President Kennedy spoke about. And yet, I do not believe that we
will have the will, the determination, the staying power, to complete this work
without something more—and that's the continued expansion of our moral
imagination; an insistence that there's something irreducible that we all share.

As the world grows smaller, you might think it would be easier for human
beings to recognize how similar we are; to understand that we're all basically
seeking the same things; that we all hope for the chance to live out our lives with
some measure of happiness and fulfillment for ourselves and our families.

And yet somehow, given the dizzying pace of globalization, the cultural
leveling of modernity, it perhaps comes as no surprise that people fear the loss
of what they cherish in their particular identities—their race, their tribe, and
perhaps most powerfully their religion. In some places, this fear has led to conflict. At times, it even feels like we're moving backwards. We see it in the Middle East, as the conflict between Arabs and Jews seems to harden. We see it in
nations that are torn asunder by tribal lines.

And most dangerously, we see it in the way that religion is used to justify the
murder of innocents by those who have distorted and defiled the great religion
of Islam, and who attacked my country from Afghanistan. These extremists are
not the first to kill in the name of God; the cruelties of the Crusades are amply
recorded. But they remind us that no Holy War can ever be a just war. For if you
truly believe that you are carrying out divine will, then there is no need for restraint—no need to spare the pregnant mother, or the medic, or the Red Cross

worker, or even a person of one's own faith. Such a warped view of religion is not just incompatible with the concept of peace, but I believe it's incompatible with the very purpose of faith—for the one rule that lies at the heart of every major religion is that we do unto others as we would have them do unto us.

Adhering to this law of love has always been the core struggle of human nature. For we are fallible. We make mistakes, and fall victim to the temptations of pride, and power, and sometimes evil. Even those of us with the best of intentions will at times fail to right the wrongs before us.

But we do not have to think that human nature is perfect for us to still believe that the human condition can be perfected. We do not have to live in an idealized world to still reach for those ideals that will make it a better place. The non-violence practiced by men like Gandhi and King may not have been practical or possible in every circumstance, but the love that they preached—their fundamental faith in human progress—that must always be the North Star that guides us on our journey.

For if we lose that faith—if we dismiss it as silly or naïve; if we divorce it from the decisions that we make on issues of war and peace—then we lose what's best about humanity. We lose our sense of possibility. We lose our moral compass.

Like generations have before us, we must reject that future. As Dr. King said at this occasion so many years ago, "I refuse to accept despair as the final response to the ambiguities of history. I refuse to accept the idea that the 'isness' of man's present condition makes him morally incapable of reaching up for the eternal 'oughtness' that forever confronts him."

Let us reach for the world that ought to be—that spark of the divine that still stirs within each of our souls.

Somewhere today, in the here and now, in the world as it is, a soldier sees he's outgunned, but stands firm to keep the peace. Somewhere today, in this world, a young protestor awaits the brutality of her government, but has the courage to march on. Somewhere today, a mother facing punishing poverty still takes the time to teach her child, scrapes together what few coins she has to send that child to school—because she believes that a cruel world still has a place for that child's dreams.

Let us live by their example. We can acknowledge that oppression will always be with us, and still strive for justice. We can admit the intractability of depravation, and still strive for dignity. Clear-eyed, we can understand that there will be war, and still strive for peace. We can do that—for that is the story of human progress; that's the hope of all the world; and at this moment of challenge, that must be our work here on Earth.

RELIGIOUS EXPRESSION

★ ★ ★ ★ ★ ★ ★ ★ ★ ★ ★ ★ ★ ★ ★ ★ ★ ★ ★ ★

Jefferson and Madison did not bequeath us a "Christian nation." The United States has never had an established church, and our Constitution grants no special preference to Christianity.

—Rev. Barry W. Lynn

We must have ten thousand Christian missionaries and a million bibles to complete the occupation of this land [Japan].

—Gen Douglas MacArthur

Indeed, protecting free exercise of religion is particularly important in the Armed Services because it is a key component in developing and strengthening the Warrior Ethos, an indispensible factor in fighting and winning our Nation's wars.

—Dr. Jay Alan Sekulow and Robert W. Ash

It has been suggested, that [the military chaplaincy] has a tendency to introduce religious disputes into the Army, which above all things should be avoided, and in many instances would compel men to a mode of Worship which they do not profess.

—Gen George Washington

RELIGIOUS EXPRESSION

The Founding Fathers understood all too clearly the lessons gleaned from our European ancestors when government officially supports a particular faith tradition. Fortunately, these lessons weren't lost on the architects of the US Constitution. During the earliest days of the American Revolution, spirited discussions over the diversity of religious beliefs often dominated the political debates, just as they do today. Regardless of the side people found themselves on back then, everyone could agree upon the notion that religion and government must be allowed to freely exist in parallel without unnecessarily becoming entangled in each other's domain. Because when they do, beware.

Since those early days of American history, much has changed. But when it comes to the value placed on the freedom of religious expression and the risk of mixing it with official government capacities, the discussions are as spirited as they've ever been. What seemed to work for Jefferson, Madison, Franklin, and Adams was the emergence of a shared perspective by way of open, candid, and honest discussion about their individual views. By developing a common understanding of one another and establishing clear policy guidance as to how people and government should properly behave, the Founders hoped to create a workable system where all constituencies would get what they needed, but not necessarily what they wanted. On one hand, it worked very well. Now more than 230 years later, we are effectively functioning as a society of the diversely religious and nonreligious under the same basic legal structure. Nevertheless, the same tension which emerged from our nation's beginning is ever present today and is unlikely to diminish anytime soon.

Over the past six years, there has been a very public resurgence of complaints levied against military commanders' excessive religious expression in official capacities, which may give the appearance of official government endorsement. Others have come to the defense of the accused, claiming that actions restricting the religious speech of military commanders are repressive and unconstitutional. So who is right?

The point of this section is not to convince anyone of an opinion which differs from the one he or she already has on the subject. Quite frankly, it is almost unthinkable that anyone is undecided when it comes to his or her thoughts on religious beliefs. However, to paraphrase the great American writer Mark Twain, *"It isn't what we don't know that gets us in trouble, it's the things we know for sure that just ain't so."* What seems to be true regarding this ongoing debate is the danger of legislating Truth. As soon as anyone claims exclusivity over Truth, regardless of perspective, look for the shrapnel to fly. Thus, the goal of *Attitudes Aren't Free* is to provide readers with a collection of essays that span the spectrum of opinions and offer the perspectives of some of the most credible, outspoken, and impassioned experts on the subject. These authors were specifically asked to avoid giving a balanced perspective of the religious expression issue, but rather to clearly articulate their particular perspective and make specific policy recommendations which they believe would be in the best interests of our men and women in uniform. By subjecting ourselves to thoughtful, articulate experts who hold a wide diversity of opinions, perhaps we may find some common ground to continue moving forward in the evolution of the "great American experiment" just as the Founding Fathers did so long ago.

The six chapters that follow in the Religious Expression section are authored by some of the most prominent people engaged in the current debates over religious expression in the US military and are briefly described below.

Rev. Barry W. Lynn, executive director of Americans United for Separation of Church and State, provides a historical account of the roots of religious liberty in the United States through an examination of the doctrine of separation of church and state and its application to the military chaplaincy. Lynn recommends a formal end to government sponsorship of evangelistic rallies, a formal end to the quasi-official status of evangelical groups, and a reform of the chaplaincy from its current state.

Gordon James Klingenschmitt, a former US Navy chaplain, chronicles the 2009 seizure of Bibles printed in Pashto and Dari languages from American service members serving in Afghanistan and describes the difference between evangelism and proselytization. Using his own example of being court-martialed for worshiping in public in uniform, he raises the question about the line between constitutionally protected free speech and "totalitarian suppression." He recommends that military members should never be forced to pray to the government's "nonsectarian" god and warns of how censorship of religious expression can threaten our long-term security.

In an award-winning essay, **Maj Paula Grant**, a USAF staff judge advocate, provides a legal perspective on the First Amendment and analyzes its establishment, free exercise, and free speech clauses. Comparing the two sets of guidelines issued by the USAF in 2005 and 2006, Major Grant recommends against public prayers at command functions and coercive evangelizing or proselytizing, and advocates no official endorsement of religion. More specifically, she argues that DOD should issue enhanced religious guidelines with clear behavioral implications.

Lt Col Jim Parco and **Dr. Barry Fagin**, former colleagues at the US Air Force Academy, reprint an article originally published in the June 2007 issue of *The Humanist*, which provides an account of religious issues that emerged at the US Air Force Academy from 2003 to 2007. They propose that officers who take command voluntarily affirm the "Oath of Equal Character" to eliminate any misperceptions that their religious beliefs, or lack thereof, would ever be used as a basis to judge a subordinate's moral character.

In "Against All Enemies, Foreign and Domestic," the senior research director of the Military Religious Freedom Foundation, **Chris Rodda**, provides a variety of cases in which, she argues, the rise of evangelical Christian influence in the US armed forces over the past 15 years has created a system conducive to the coercion of junior military members by superiors who don't share their particular religious beliefs. She recommends equating "evangelism" and "proselytization" in General Order 1B and calls for military leaders to merely enforce the laws and regulations already in place.

Finally, **Dr. Jay Alan Sekulow** and **Robert W. Ash** from the American Center of Law and Justice provide a constitutional perspective with respect to free exercise of religion. They argue strict "church-state separation" has never been required in the United States and is not required now. They recommend robust tolerance training and an increased trust in military leaders to know how best to train their troops. Furthermore, they advocate that courts and civilian society should defer to the experience of military commanders and not second-guess their judgment merely because it is inconsistent with expectations within civilian society.

★★★★★ ★ ★★★★★

Editorial note: In each of the four sections of this book, it is imperative for readers to *consider each essay in the context of the entire section in which it appears*. Taking any one chapter out of context could present a biased view and potentially confound the goal of this volume. During our earliest days, Jefferson and Adams recognized that the success of the United States as a federated republic depended upon open, honest, and transparent discussions—even with those with whom we might disagree. Since then, nothing has changed. Only through open dialogue within the ranks can we ensure a better understanding of others who see the world differently. But there comes a point when leaders have to take a hard look at the evidence and enforce the rules, despite their own personal convictions, which may conflict. To paraphrase author Ayn Rand, leaders have the power to choose, but no power to escape the necessity of choice.   *J.P. & D.L.*

CHAPTER 1

RELIGION IN THE MILITARY
FINDING THE PROPER BALANCE

Barry W. Lynn

Questions about the proper role of religion in the US military have intensified in recent years. Allegations have been made of favoritism toward evangelical Christianity. At the same time, some service members from minority faiths or who have no religious interest have claimed that their rights are not respected.

The men and women serving in our nation's armed forces are entitled to the same right of religious freedom as members of the general public. However, military service does present unique circumstances and concerns that are not present outside of the military context. In most walks of civilian life, for example, it would be unacceptable for the government to use tax funds to provide chaplains and pay their salaries. In civilian life, religious leaders and houses of worship are supported by voluntary donations, not government support.

The military context is different. Service members are usually stationed away from their homes and may even be sent to foreign lands. These individuals are not able to worship at their home congregations and may in fact be many miles (or even an ocean) away from any congregation they would recognize or feel comfortable attending. Some accommodation must be made for them.

The military chaplaincy was created to provide for this need. Chaplains are charged with an unusual mission that has few, if any, exact parallels in the civilian world: providing for the religious needs of a variety of individuals, including a wide array of Christian faiths and non-Christian beliefs as well. While chaplains are not expected to personally provide every religious service or ritual requested, they may be asked to facilitate others' worship by supplying materials or religious texts or arranging places where believers can meet.

The military's hierarchical nature also presents unique challenges for religious-liberty issues. In the civilian world, federal laws prohibit religious discrimination and provide some measure of protection to employees from unwanted proselytism. It is difficult to replicate this structure in the military context. The hierarchical nature of the chain of command and the military's need to stress discipline can make it difficult for a subordinate to feel entirely comfortable challenging a superior. Thus, any allegation of religious pressure down the chain of command requires heightened scrutiny.

Americans United for Separation of Church and State has been following the issue of religion in the military closely since 2005, when reports reached the organization of improper proselytization, religious coercion, and religious discrimination at the Air Force Academy in Colorado Springs. At that time, Americans United examined many of the complaints and prepared a report, which was later included as part of an official government investigation of the problems at the Academy.[1]

Americans United's interest in the issue did not end with the issuance of that report. Americans United has continued to work alongside the Military Religious Freedom Foundation to make sure that religious liberty is respected in the military.

The Roots of Religious Liberty

Members of the armed services are responsible for protecting American freedoms. Unfortunately, it's quite possible that some of them don't know the origin of some of those freedoms. During the debate over religious liberty at the Air Force Academy, several national organizations attacked the stands taken by Americans United and the Military Religious Freedom Foundation. Some claimed that Americans United and the Military Religious Freedom Foundation are hostile to religion and even that separation of church and state is not a valid constitutional concept.[2]

The First Amendment guarantees five core freedoms: religion, speech, press, assembly, and petition. In the case of freedom of religion, the core right is expressed in just 16 words: "Congress shall make no law respecting an establishment of religion or prohibiting the free exercise thereof."

Originally a prohibition on Congress, the First Amendment—and indeed other portions of the Bill of Rights—has now been extended to the states through the Fourteenth Amendment. The words of the religion clauses encompass two key concepts: The government will not make laws that foster an "establishment" of religion (or give any or all religions special preference), and the government will protect the right to engage in religious activities.

This is the genesis of the separation of church and state. Note that the First Amendment does not say that the government will not create an official church, as existed in Great Britain and many other nations at the time the amendment was drafted. Rather, it bars laws "respecting an establishment of religion." The Founders wanted something stronger than a mere ban on a national church, and their words have been interpreted to mean that government will not make laws that advance religion or interfere in theological matters.

At the time the First Amendment was drafted, many states had established churches. Some required people to pay church taxes. Thomas Jefferson and James Madison were great opponents of this system and worked together (aided by dissenting clergy) to end the established church in Virginia and pass a law guaranteeing religious liberty for everyone—Christian and non-Christian.

This law, the Virginia Statute for Religious Freedom, is considered by many scholars to have been a model for the First Amendment. Although Jefferson was in France when the Bill of Rights was written, his influence is felt through his collaboration and correspondence with Madison, who was in many ways Jefferson's protégé. Jefferson, for example, wrote the Virginia Statute for Religious Freedom, but it was Madison who pushed it through the legislature and made it law.

Jefferson and Madison had nearly identical views on religious freedom. Both saw coercion and state sponsorship of religion as a great evil. In this thinking, they were probably influenced by the many centuries of religious warfare and bloodshed that had plagued Europe, as both men were keen students of history.

Yet Jefferson and Madison were not hostile to religion. Evidence for this is found in the great outpouring of support they received from religious leaders. Many members of the clergy were weary of government's attempts to control religion and eagerly endorsed the efforts of Jefferson and Madison to sever the tie.

Jefferson and Madison did not bequeath us a "Christian nation." The United States has never had an established church, and our Constitution grants no special preference to Christianity. In fact, there is only one reference to religion in the Constitution proper: Article VI bans "religious tests" for federal office. The Constitution contains no mention of God.

Again, none of this was done out of hostility toward religion. In fact, the Founders believed that decoupling church and state would lead to a great flowering of religious freedom and diversity in America. Time has proven them right. Some scholars have estimated the number of distinct religious denominations in the country to be as high as 2,000, while people who say they have no religion account for a growing percentage of the population.

The phrase "separation of church and state" was used by both Jefferson and Madison to describe the First Amendment.[3] Madison, one of the primary authors of the First Amendment, is considered authoritative on this matter. As president, he vetoed attempts to give churches federal support and even expressed reservations about issuing proclamations calling for days of prayer and fasting. (Jefferson did not issue them at all.)

Madison also had concerns about chaplains both in Congress and in the military.[4] Madison worried that any entanglement between religion and government would be to the detriment of both institutions.

Despite Madison's concerns, the constitutionality of the military chaplaincy does not appear to be in doubt. A challenge to the chaplaincy on establishment clause grounds was launched in 1979 by two students at Harvard Law School. The case, *Katcoff v. Marsh*, eventually reached the Second US Circuit Court of Appeals, which ruled that the chaplaincy is constitutional, since its primary objective is to ensure the free exercise of religion. (The issue did not reach the Supreme Court, but this is not surprising since the vast majority of cases filed in federal court never get that far.)[5]

The Second Circuit held that the chaplaincy is necessary since service personnel are often sent overseas, sometimes to far-flung places, where they might not have a house of worship to join. The court did not address the issue of the chaplaincy at domestic military bases, many of which are served by nearby communities with a wide variety of houses of worship. Broadly speaking, the court affirmed the idea that the chaplaincy's permissible purpose is to facilitate the free exercise of religion in circumstances where the military has put people in a situation that otherwise burdens their ability to engage in their religious freedom.

Challenges and Coercion

As we examine this history, we can see its application to the challenges American society faces today, in both military and nonmilitary contexts. One is diversity. Religious diversity flourishes in an atmosphere of tolerance and respect. Our First Amendment, and its attendant church-state wall, have fostered that atmosphere. Yet Americans United found some resistance to these concepts among cadets and staff when it examined the problems at the Air Force Academy.

Another challenge revolves around questions of sharing faith and allegations of proselytism. In civilian contexts, individuals are free to share their faith and invite others to explore it. Indeed, many Christians consider efforts to spread their faith part of the "Great Commission" handed down to them by Christianity's founders. But in hierarchical structures, efforts to share faith are sometimes perceived as unwanted and unwelcome forms of coercion. Concerns are often expressed that efforts to resist such coercion could affect job advancement.

Again, Jefferson and Madison provide some guidance. Jefferson and Madison believed there should be no state-sponsored coercion in religious matters. Thus, in the military context, there must be no sponsorship of events or actions designed to convert someone or to promote certain faiths over others. Interpersonal relations can be guided by commonsense rules: One invitation to attend church may be acceptable. Repeated invitations after no interest has been expressed or implications that acceptance of such invitations is the key to advancement/promotion are not welcome and may in fact be unlawful.

In short, we can say that America's doctrine of church-state separation contains three central concepts:

No coercion in religious matters: Individuals must be free to embrace or reject any faith. People have the right to change their minds about religion. The decisions people make about religion—which group to join or whether to join any—are private and are no business of the government.

No one should be expected to support a religion against his or her will: Support for religion—financial, physical, and emotional—must be voluntary. No American should be taxed to pay for the faith of another. All religious groups must be supported through voluntary channels.

Religious liberty encompasses all religions: Americans may join any number of religious groups. In the eyes of the law, all religions are equal. Larger groups do not have more rights than smaller ones. No group was meant to have favored status or a special relationship with the government.

Recommendations for the Military

How should these concepts be applied in the armed services? Americans United believes adoption of the following set of principles would help the military deal more effectively with potential religious liberty concerns. Please note that some of the concepts may reflect current military policies or regulations. The problem is, they are apparently not being enforced with vigor or seem occasionally to be ignored entirely. This must change.

End all sponsorship or other direct support of evangelistic rallies or events designed to persuade service personnel to adopt a certain set of religious beliefs. No branch of the government, including the military, should sponsor an evangelistic event. This includes rallies featuring proselytizing preachers, "Christian rock" bands, revivals, seminars that are in reality covers for evangelism, and similar events. It is not the job of the military to coerce service personnel to adopt new religious beliefs, discard the ones they have, or affiliate with a religious body. The military is required to accommodate the religious needs of its soldiers. This is a far cry from promoting religion.

Reform the chaplaincy. We must return the chaplaincy to its traditional role. Chaplains should be facilitators of religious worship, not promoters of their own faiths. A chaplain's role is to assist soldiers in discharging their religious duties. In some cases, this may involve leading a religious service, but in others, a more passive role might be played. Chaplains must be willing to work with and respect a variety of religious believers. Those who wish to engage in the elevation of one religion over all others or in proselytizing activities on behalf of their own faiths are not good candidates for the chaplaincy. (Obviously, a chaplain has the right to discuss his or her faith if approached and asked about it, but spreading a particular version of religion should not be viewed as the chief goal of the job.)

The armed services might consider moving back toward rules that were in place in the 1980s that roughly proportioned chaplains according to the religious demographics of the military as a whole. Currently, the chaplaincy seems to be heavily tilted toward evangelical/fundamentalist denominations. Members of these denominations often represent traditions that place a premium on recruitment of new members and the aggressive spreading of their particular interpretation of the gospel. They view service members as a "mission field" and consider it their calling to proselytize on behalf of their own faith.

This view is at odds with the traditional view of the chaplain. Individuals who adhere to this perspective will probably make poor chaplains, since their main goal is to win souls for their own religion, not assist individual soldiers with whatever religious needs they may have. These aggressive forms of proselytism are bound to increase friction and interfaith tension in the military. This runs counter to the stated goals of the armed services. Thus, there is nothing wrong with the military leadership acknowledging this fact and screening potential chaplains to determine their views on these issues. Those whose main goal is proselytism should be rejected for service.

In 2008, an Army chaplain from the Unitarian Universalist tradition, writing on a personal blog, reflected on his decision to serve in the armed forces. In doing so, the chaplain provided a succinct description of what a chaplain's job should be:

> My primary duty as a military chaplain is to insure that all of the soldiers under my care are given the necessary time, space, materials, and freedom to practice their religion. It is not to proselytize, to convert people to my faith, or to hinder those who hold a faith other than my own. It is to insure that I help soldiers to explore and connect deeper with the religious faith they are called to, be it Christianity, Buddhism, Judaism, Islam, Atheism, Humanism, Paganism, Wicca, Hinduism, or anything else.[6]

This paragraph should be required reading for any individual interested in entering the military chaplaincy. Anyone who is not willing to respect these principles should consider alternative employment.

View skeptically extra-legal claims by conservative religious and legal organizations. Some conservative groups claim that chaplains have a religious-freedom right under the First Amendment to proselytize. This assertion is unlikely to stand up in court. In the 1980s, a chaplain at a Veterans Administration hospital sued under Title VII after he had been told to stop proselytizing by his superiors.

The Seventh US Circuit Court of Appeals ruled that the hospital had the authority to curb the chaplain's actions. Although much of the opinion deals with this aspect of the dispute, one section did state that the Veterans Administration may also have the duty, under the establishment clause, to crack down on proselytism. Observed the court in *Baz v. Walters*:

> The V.A. provides a chaplain service so that veterans confined to its medical facilities might have the opportunity to participate in worship services, obtain pastoral counseling and engage in other religious activities if they so desire. If there were not a chaplaincy program, veterans might have to choose between accepting the medical treatment to which their military service has entitled them and going elsewhere in order to freely exercise their chosen religion. This itself might create a free exercise problem. (The First Amendment "obligates Congress, upon creating an Army to make religion available to soldiers who have been moved by the Army to areas of the world where religion of their own denominations is not available to them.") But, at the same time, the V.A. must ensure that the existence of the chaplaincy does not create establishment clause problems. Unleashing a government-paid chaplain who sees his primary role as proselytizing upon a captive audience of patients could do exactly that. The V.A. has established rules and regulations to ensure that those patients who do not wish to entertain a chaplain's ministry need not be exposed to it. Far from defining its own institutional theology, the medical and religious staffs at Danville are merely attempting to walk a fine constitutional line while safeguarding the health and well-being of the patients.[7]

Remind chaplains of the distinction between public and private events. It is to be expected that there would be a difference between a private funeral service for a fallen soldier and a public event, such as an induction or graduation ceremony. A private funeral will reflect the religious views of the deceased and will feature prayers, worship, and liturgy that come from that tradition. It should also be done in consultation with the family members of the deceased and reflect their wishes, not the chaplain's.

A public event is different. The audience will consist of service personnel of many religious traditions (as well as those who hold no particular beliefs). Sectarian prayer, proselytism, and other denomination-specific practices are inappropriate at such events. If there must be prayers at public events, they should be nonsectarian. Furthermore, the military should adopt policies stating that any chaplain who takes advantage of a public event to proselytize or promote his or her specific faith should be corrected.[8]

It is fair to point out there is some debate over what constitutes "nonsectarian" prayer. People may differ on whether terms such as "Lord" or "God the Father" are appropriate. That discussion will continue, but as it does, it should be noted that there is a consensus on what types of prayers are not nonsectarian. Those that end "in Jesus' name" or reference specific tenets of a certain faith do not meet that standard.

The chief of chaplains for each respective branch should take the lead in ensuring that public events do not become occasions for proselytism. Manuals and other materials distributed to chaplains should stress this issue, if they do not already do so. Furthermore, there must be enforcement. A chaplain who knowingly and repeatedly violates these standards should be disciplined in the same manner any other officer would be for disregarding orders or violating policy.

Religious coercion along the chain of command should be banned. High-ranking officers should be reminded that there is to be no religious coercion or pressure through the chain of command. Officers should never show preferential treatment to coreligionists, pressure subordinates to join their faith, or imply to subordinates that adopting different religious beliefs will be advantageous. Those who do should be held accountable through the proper channels. How, where, and when someone worships should be a private matter. An individual's religious preference should have no bearing on performance reviews, promotion, or duty assignments.

The military should strive to instill a healthy respect for religious diversity in all of its officers. This issue can be discussed at an appropriate point during officer training. The logical place for such a discussion is alongside instruction about racial diversity and what constitutes sexual harassment. The military strives for a workplace that does not tolerate racial discrimination or sex-based discrimination. Likewise, it should not tolerate any form of religious discrimination (or its converse, preference based on shared religious beliefs). Existing policies that cover racial and sex-based discrimination can most likely be modified to address religious issues.

Military leaders must understand that a good soldier can hold a variety of beliefs or nonbeliefs. Men and women of many different backgrounds feel called upon to serve their country through the armed services. Many are Christians, but others represent non-Christian traditions or nontheistic approaches to life.

Unfortunately, the mind-set of some military leaders seems grounded in simplistic "God-and-country" rhetoric—that is, their belief is that one cannot be an effective soldier unless one has also adopted, at the very least, some form of religious belief. At its most extreme, this view manifests itself as "Christian soldier" rhetoric—the belief that the most effective soldiers are those who view their work as an evangelistic mission or those who loudly proclaim to have a personal relationship with Jesus Christ.

These are dangerous views, and they are fallacious as well. There are indeed atheists in foxholes. They have always been there and always will.[9] Effective soldiers come from many different religious and nonreligious traditions, just as they come from different racial and socioeconomic backgrounds.

Service personnel should have a better understanding of our rights and their origins. Members of the military are expected to defend American rights and freedoms, possibly sacrificing their lives for them. It is difficult to expect anyone to do such things if these freedoms remain abstractions or are shrouded in obscurity.

As part of their regular classroom training, military personnel should be told about the rise of religious freedom in America. They should be told how our nation came to be home to so many different religious beliefs and taught about the role separation between church and state played in securing these liberties. It is not safe to assume that this information is common knowledge among the American people. Public-opinion polls show that is not the case.

Soldiers should be taught to respect religious differences. It should be made clear to them that respecting someone else's religious choice in no way de-emphasizes their own. A soldier can truly believe that his chosen faith is "right" and "true," while still respecting a fellow soldier's decision to affiliate with another faith group. There should be zero tolerance for those who fail to respect the views of others or who engage in activities such as slurs or assault due to religious differences.

In recent years, some public schools have adopted curriculum materials designed to instill respect for religious pluralism. This material could easily be adapted for military use. Such materials are not designed to change anyone's religious views; rather, they stress the point that people can believe deeply in their own faith tradition while still respecting the equal rights of others and working toward common civic goals.

End the quasi-official status of evangelical groups. Several evangelical organizations seek to convert members of the military. This is their right, but they must do so outside of official channels. These organizations have no right to use the power and prestige of the military to spread religious messages.

In 2006, a group called Christian Embassy came under scrutiny after it released a video that included interviews with several high-ranking military officers at the Pentagon. The Department of Defense's inspector general later concluded that seven officers acted inappropriately by appearing in the film, which Christian Embassy used for fund-raising.

It was also reported that the group had free access to the Pentagon. In fact, during the investigation, some officers told the inspector general that they believed Christian Embassy had some type of permission or authority to be in the Pentagon. The office recommended that the organization's special access to the Pentagon be discontinued.

Following the Christian Embassy flap, other reports surfaced of close relationships between branches of the military and evangelistic organizations. In May 2007, Americans United and the Military Religious Freedom Foundation protested Army and Air Force sponsorship of an evangelistic rally at Stone Mountain Park in Georgia. The event was sponsored by Task Force Patriot USA, a group that says it exists for "the purpose of sharing the fullness of life in Jesus Christ with all U.S. military, military veterans and families." Military cosponsorship of the event was subsequently dropped.[10]

Branches of the military should cease working with these organizations. The military should not sponsor evangelistic events or even promote them. Doing so may imply that the military endorses a certain brand of Christianity. The military, as an arm of the government, may not endorse any form of religion. Enforcement of this basic constitutional principle must come from the highest sources and filter down the chain of command.

Conclusion

The First Amendment guarantees all Americans religious-freedom rights. At the same time, the unique demands of military service place special controls and regulations on religious free exercise that might not survive in other contexts.

The military may, for example, place restrictions on efforts by service personnel to proselytize the native population of Muslim nations or ban such activity outright. Such regulations have been promulgated and are in effect in both Iraq and Afghanistan. The belief is that efforts to convert Iraqis and Afghans from Islam to conservative Christianity reinforce the perception that the United States is engaged in a type of "religious war," which could disrupt efforts to bring stability and effective governance to both nations.

We believe that the military may exercise control over and curb activities by chaplains and other military personnel, since those persons are acting as official agents of the government. Such controls would likely not survive in a civilian context, nor would they be desirable. In the military, they are both needed and required.

Likewise, it is not unreasonable to expect officers in a hierarchical chain of command to refrain from religious coercion or from closely aligning themselves, in an official capacity, with certain religious groups at the expense of others. A theologically diverse military that reflects the makeup of the nation at large is in our country's best interest.

These regulations are not designed to stifle religious freedom. To the contrary, requiring chaplains to be respectful of all faiths and to refrain from engaging in heavy-handed forms of proselytism protects religious liberty.

Demanding respect for religious pluralism along the chain of command is not just reasonable, it is vital.

At all times, any soldier is free to explore the faith of his or her choosing. But that act must be voluntarily initiated and free of even the hint of coercion. A military whose chaplaincy or hierarchy is too closely aligned with one narrow expression of Christianity and sees its goals partly in theological terms ("saving souls," "winning converts for Jesus," "advancing the kingdom," etc.) is bound to eventually fail to meet its objectives and end up doing a disservice to the very people it is pledged to support—the American public.

Notes

1. The Air Force report can be viewed at http://www.foxnews.com/projects/pdf/HQ_Review _Group_Report.pdf.

2. This historical review is based on several sources, primarily Isaac Kramnick and R. Laurence Moore, *The Godless Constitution: The Case against Religious Freedom* (New York: W. W. Norton, 1997); Leo Pfeffer, *Church, State, and Freedom* (Boston: Beacon Press, 1953); and John M. Swomley, *Religious Liberty and the Secular State: The Constitutional Context* (Buffalo, NY: Prometheus Books, 1987).

3. Jefferson's famous letter to the Danbury Baptists, which contains the "wall of separation between church and state" reference, can be read at http://www.au.org/resources/history/old-docs/ jeffersons-letter-to-the.pdf.

4. See Madison's "Detached Memoranda," at http://press-pubs.uchicago.edu/founders/ documents/amendI_religions64.html.

5. See full text of the ruling at http://www.usafa.edu/isme/ISME07/Rosen07.html.

6. Collected from http://celestiallands.org/wayside/?p=62.

7. The full opinion may be read at http://openjurist.org/782/f2d/701/baz-v-n-walters-sh-d.

8. The US Supreme Court has endorsed this standard for prayers held before meetings of government bodies. See *Marsh v. Chambers*, http://caselaw.lp.findlaw.com/scripts/getcase.pl?court =US&vol=463&invol=783.

9. The Military Association of Atheists and Freethinkers maintains a list of active-duty and retired service personnel who identify as nontheistic. It can be viewed at www.maaf.info/expaif .html.

10. See http://www.au.org/media/church-and-state/archives/2007/07/military-backs-a.html.

About the Author

The Reverend Barry W. Lynn is the executive director of Americans United for Separation of Church and State, a Washington, DC-based organization dedicated to the preservation of the Constitution's religious liberty provisions. In addition to his work as a long-time activist and lawyer in the civil liberties field, Lynn is an ordained minister in the United Church of Christ, offering him a unique perspective on church-state issues. A member of the Washington, DC, and US Supreme Court bars, Lynn earned his law degree from Georgetown University Law Center in 1978. In addition, he received his theology degree from Boston University School of Theology in 1973. Lynn is the author of *Piety & Politics: The Right-Wing Assault on Religious Freedom* and a coauthor of *First Freedom First: A Citizen's Guide to Protecting Religious Liberty and the Separation of Church and State.*

CHAPTER 2

Burning Bibles and Censoring Prayers
Is That Defending Our Constitution?

Gordon James Klingenschmitt

Introduction

The principle of religious freedom is perhaps the most original American ideal. Many of our forefathers sought to escape religious persecution and create a democracy, which categorically rejects totalitarian forms of government that either enforce one religion upon an entire society (as most Islamic states do) or forbid any forms of religious expression (as most Communist states did). Finding a balanced approach to "democratic diversity" defeats totalitarianism and permits individuals to express their own faith. It is a challenging but worthwhile endeavor, especially within the ranks of the military.

The American military has served as a crucible for recent debate about freedom of religious expression and the private rights of our Soldiers, Sailors, Airmen, Marines—even chaplains—to publicly express their own private faith. Two recent "headline" case studies merit careful consideration: the rights of our Soldiers to receive, own, and even distribute Christian Bibles in their unofficial off-duty capacity and the rights of military chaplains of all faiths to pray diverse Christian, Jewish, Buddhist, Muslim, or other prayers in public ceremonies.

Case Study No. 1:
Pentagon Defends Burning Christian Bibles

A Pentagon spokesman under the Obama administration recently acknowledged the seizure and destruction of privately owned Bibles from American Soldiers serving in Afghanistan. The Bibles had been printed in the local Pashto and Dari languages and sent by private donors (in 2008 under the Bush administration) to American Christian Soldiers and chaplains for distribution to American troops on overseas military bases during optionally attended Christian worship services.[1]

If the Bibles hadn't been recently seized and destroyed, our troops could have *legally* given the Bibles as gifts during off-duty time to Afghani citizens who welcome our troops in their homes, as an expression of American gratitude for Afghani hospitality, thereby promoting the democratic ideals of freedom of religion and freedom of the press.

On 4 May 2009, the Muslim-controlled al-Jazeera television network released year-old video footage[2] of the subject Christian Bibles, given to American Soldiers inside the chapel on the Bagram Air Base while listening to a chaplain whose sermon encouraged outreach and personal evangelism. Media fireworks ensued. The American values of freedom of religion, press, assembly, and speech offended some Afghani extremist Muslim groups[3] and angered a small group of American atheists.[4] They immediately demanded the chaplain be punished for "proselytizing" because he simply repeated Jesus' words in church, quoting Matthew 28:19, which instructs Christians to "go ye therefore and teach all nations."

When questioned about the authenticity of the al-Jazeera video, US Army Col Greg Julian admitted the al-Jazeera reporting was biased against the American Christians: "Most of this is taken out of context . . . this is irresponsible and inappropriate journalism."[5] Can you imagine the media outrage if we had burned the Koran instead of the Bible? The hypocrisy of the contrast between al-Jazeera's celebration at our burning of Christian Bibles at Bagram (a story ignored by mainstream American media) and their horrified allegations of "torture" when we allegedly mistreated the Koran at Guantanamo Bay (a story covered widely by American mainstream media)[6] is breathtaking, in my opinion.

To avoid further public controversy, American military personnel caved in to pressure from atheist and Muslim extremist groups and confiscated troops' privately owned Bibles amid concern they would be used to try to convert Afghans.[7] "Military rules forbid troops of any religion from proselytizing while deployed there," United States Central Command (USCENTCOM) spokesman Lt Col Mark Wright told CNN.[8] The constitutionality of any such "rules" remains untested by American courts.

The new Afghani constitution clearly protects freedom of the press and freedom of religion among foreigners. No current Afghani laws were broken by our troops, and no charges were filed against any commander or Soldier. Al-Jazeera filmmaker Brian Hughes admitted the Bibles could have been useful in helping Soldiers learn the Pashto and Dari languages of the Afghan people.

Without compensation to their owners, the privately owned Bibles were confiscated and destroyed. US military spokesman Maj Jennifer Willis told Reuters reporters, "I can now confirm that the Bibles shown on al-Jazeera's clip were, in fact, collected by the chaplains and later destroyed. They were never distributed." Her careful use of the words "later" and "now" indicates the Bibles were destroyed in August 2009, only after and because of the recently publicized video. But regardless of when the Bibles were destroyed, the Obama administration now defends their destruction as required by policy. "The translated Bibles were never distributed as far as we know," Willis continued, "because the Soldier understood that if he distributed them, he would be in violation of General Order No. 1, and he would be subject to punishment."[9]

"Punishment" for exercising freedom of the press? How can the free printing of books be both constitutionally protected and punishable, simultaneously? And what information printed inside that book is so dangerous that it must be banned?[10] Why has our Pentagon suddenly adopted a Communist ideology that publicly defends the burning of the Bible as a banned book? (By contrast, pornography remains rampant on our military bases; ask any G.I.) Further discussion of USCENTCOM's General Order No. 1 and its defensibility under American law will follow, but first, let's examine Afghanistan's new religious law.

Examining the development of the new Afghani constitutional law and its dramatic social conflict with its predecessor law of the totalitarian Taliban regime will help us understand how to support the new democracy and defeat the remnants of the old regime. A short case study concerning Abdul Rahman will highlight this old-law versus new-law conflict and perhaps inform our role as outsiders involved in that continuing struggle to assist those seeking a truly democratic future.

The Afghani Constitution

The supposed USCENTCOM military rules that "require" burning Bibles appear to conflict not only with the First Amendment of the US Constitution,[11] but also with the new Afghani constitution ratified in September 2005, which specifically protects everybody's freedom to print and distribute Christian Bibles. The Afghani constitution states in Article 34: "Freedom of expression shall be inviolable. Every Afghan shall have the right to express thoughts through speech, writing, illustrations as well as other means in accordance with provisions of this constitution. Every Afghan shall have the right, according to provisions of law, to print and publish on subjects without prior submission to state authorities." The Afghani constitution also protects practitioners of other faiths, including Americans practicing the Christian faith, by stating that followers of other religions are "free to exercise their faith and perform their religious rites." Article 7 emphasizes that all legislation must be defined "in accordance with international human rights laws."[12] This latter phrase may relate to United Nations declarations designed to protect free religious expression for all faiths, again including Christianity.

Yet these apparently democratic phrases of their constitution may conflict with totalitarian phrases of their constitution, such as "the religion of the state of the Islamic Republic of Afghanistan is the sacred religion of Islam," and "no law can be contrary to the beliefs and provisions of the sacred religion of Islam."[13] In matters upon which the constitution of Afghanistan remains silent, it recognizes and delegates jurisprudence (among six other tribal schools) to the Hanafi tradition, which extremists believe *mandates the death penalty for apostasy from Islam.* Hence some legal tension exists between the subordinate but extremist Hanafi tradition and the moderate but overarching Afghani constitution.

Abdul Rahman

When Afghani citizen Abdul Rahman was arrested in February 2006 for the "crime" of apostasy against Islam because he converted to Christianity, the Afghan courts had their first opportunity to test the possibility of conflicts between the Afghani constitution, which may have protected his rights, and the subordinate Hanafi tradition, which may have mandated his death. Three trial judges disputed his fate, one of whom, Judge Ansarullah Mawlafizada, stated, *"the Prophet Muhammad has said several times that those who convert from Islam should be killed* if they refuse to come back. Islam is a religion of peace, tolerance, kindness and integrity. That is why we have told him if he regrets what he did, then we will forgive him"* (emphasis added).[14] Clearly this ideology reminds us of the "Taliban mentality," which enslaved their women under burkas and flew planes into the World Trade Center and Pentagon. So long as this ideology remains uncontested, both America and freedom remain in grave danger.

Another Muslim extremist, Maulavi Habibullah, told more than a thousand clerics and young people gathered in Kabul that "Afghanistan does not have any obligations under international laws. *The prophet says, when somebody changes religion, he must be killed*" (emphasis added).[15] When facing a possible death sentence, Abdul Rahman held firm to his convictions: "They want to sentence me to death and I accept it. . . . I am a Christian, which means I believe in the Trinity. . . . I believe in Jesus Christ."[16] Ultimately the Afghani courts avoided ruling on the merits of their own constitutional dilemma. Instead, they declared Rahman insane and deported him to Italy. This occurred in the face of great international pressure, including a protest from the Bush administration.

An Honest Question for Totalitarians

The preceding background lays the foundation for the following deeper question about long-term American security. People have argued that by burning our own troops' Bibles, we enhance the short-term security of our forces. If we hadn't destroyed the Bibles, our military might be seen as Christian crusaders. In principle, I almost agree. Our troops should never be tasked to promote one faith over another. But shouldn't we still promote democracy and freedom of individual religious choice over totalitarian suppression?

In the days of Ronald Reagan, this was not debatable. Totalitarian regimes that banned free presses were openly declared as evil. We Cold War warriors cannot soon forget the Soviet barricades that banned guns, drugs, and Bibles as "equally dangerous" illegal contraband. But in the long term, one must wonder if our own military has recoiled under pressure from atheists or Muslim extremists.

It's Not Proselytizing, It's Evangelism

The 4 May 2009 al-Jazeera video proves the chaplain properly explained USCENTCOM's General Order No. 1, which prohibits "proselytizing of any religion, faith or practice"—that is, forcible religious conversions by threats or weapons. However, it does permit Soldiers of any religion to engage in non-threatening "evangelism"—that is, voluntary conversations about their faith. It also allows the giving of private gifts, including books, to Afghani citizens during off-duty hours in the Soldiers' unofficial capacity.

Before assuming "forced proselytizing" and "voluntary evangelism" mean the same things, you might consider that academic theologians[17] must necessarily draw some legal distinction between the terms for several reasons, the first of which appeals to our own Constitution.

How Would You Enforce General Order No. 1?

Imagine that you're a commander, and some young Soldier who is off duty, unarmed, and not in uniform (in his private, unofficial capacity) is invited to the home of an Afghani national. During dinner, he casually talks about his faith in Jesus Christ and gives the Afghani his own Bible as a gift. You counsel him, as you don't think it's wise, but he claims he has freedom to act as he did. Let's also assume he writes his congressman. Now you hesitate to reassign him since it might be considered an act of reprisal. So will you punish him if he does it again? Will you court-martial your Soldier for violating General Order No. 1? Let's suppose you do. How likely is it that his legal defense will involve the First Amendment, which both you and he have sworn to defend? Will you now oppose this? As the prosecuting JAG officer, will you argue that the First Amendment does not protect our own Soldiers' private rights to free religious expression and freedom of the press?

In doing so, not only would you risk violating your own oath, but you'd likely lose the argument in court, since the Supreme Court has ruled "military personnel do not forfeit their constitutional rights as a price of enlistment. Except as otherwise required by 'interests of the highest order,' soldiers as well as civilians are entitled to follow the dictates of their faiths."[18] It's not unthinkable that a commander will try to argue in court someday soon that "interests of the highest order" require enforcing totalitarian atheistic silence upon our off-duty troops, necessitating the seizure of their privately owned Bibles. Perhaps "interests of the highest order" require that we defeat the Taliban ideology, not embrace their anti-liberty values.

Bibles Cannot Proselytize All by Themselves

Christian scripture and the teachings of Jesus are actually embraced as one of five "Injil" holy books for Muslim teaching and praised 12 times in the Koran as compatible with Islam. Christian Bibles cannot proselytize converts, since the teachings of Islam praise those same Bibles as useful to the Muslim faith. Knee-jerk confiscation neglects the possibility that the Bibles were intended for distribution to Afghani Christians who love the scriptures, not Afghani terrorists who hate them. Bibles cannot proselytize those who already share that faith.

Just some friendly advice: You'll be better off saying "the Soldiers and chaplains didn't forcibly proselytize anybody using threats or weapons, so they didn't violate General Order No. 1 just by talking about their faith."

Policy Implications

Unless we draw a reasonable distinction between forced proselytizing and voluntary evangelism, General Order No. 1 will become unenforceable, and somebody will:

1. wrongly prosecute a Soldier for merely talking about his or her own private faith,

2. support the Taliban's totalitarian enforcement of the Hanafi tradition over the Afghani constitution,

3. become a domestic enemy of the US Constitution,

4. become a persecutor of the Christian faith,

5. oppose the very notions of religious liberty upon which our Founding Fathers established our nation,

6. turn our own military into a Taliban-controlled regime, and

7. take steps toward converting the rest of America into the Communist or Muslim-extremist enemy of religious liberty.

Why would anybody want to do that?

Let's just agree that "evangelism is not a crime" and "the Bible is not a banned book." Coincidentally, my nationally organized radio campaign and online petitioners put nearly 200,000 faxes into the US Congress during August 2009 stating exactly that. Scores of inquiries have begun flowing downhill, and the Bible-burning commanders will answer to Congress, when Bible-defending petitioners enjoy an 85 percent public approval rating on similar religious liberty issues. Pentagon promotions to flag rank are scrutinized by the same congressmen we're faxing. Nothing escapes the public eye. The transparent system designed by our Founding Fathers works very well.

Gen Douglas MacArthur Encouraged Evangelism

After World War II, to carry out the democratization of Japan, five-star General of the Army Douglas MacArthur brought Christian leaders to the country to meet with Emperor Hirohito and encouraged mass distribution of Bibles to the population. MacArthur later stated to a visiting American churchman, "We must have ten thousand Christian missionaries and a million bibles to complete the occupation of this land."[19]

In 2009, instead of confronting Muslim and atheist enemies of religious liberty, Adm Mike Mullen, chairman of the Joint Chiefs of Staff, now appears to support the destruction of the Soldiers' privately owned Bibles, stating during a Pentagon briefing that the military's position is that it will never "push any specific religion." He did not address the possibility that by seizing and burning privately owned Bibles the Obama administration may now endorse enforcing atheistic silence upon our troops against their private rights. How far has liberty fallen in 60 short years?

Case Study No. 2:
Praying in Public in Jesus' Name

Having considered the first case study about the wisdom or lack thereof in burning the privately owned Bibles of American Soldiers, let us transition to our second case study involving the suppression of free speech by chaplains who pray publicly "in Jesus name." To better understand the context which led to this suppression, we should consider some historical background, starting with the complaints of atheists who oppose freedom of speech by military chaplains.

Mikey Weinstein stands as a leading voice of opposition to religious expression by Christians in the military. For example, he demanded the chaplain who distributed the Bibles in the al-Jazeera video be court-martialed for encouraging voluntary evangelism in church, despite the Pentagon's claim that the same chaplain somehow assisted in confiscating the Bibles. Mr. Weinstein's group, the Military Religious Freedom Foundation, has repeatedly sued the military to silence our troops and chaplains, but they've not yet won any lawsuits.

Court-Martial the Army Chief of Chaplains?

The favorite new ploy of the Military Religious Freedom Foundation is to intimidate some troops into silence by issuing press releases demanding the court-martial of any Soldier or chaplain who talks publicly about his or her faith. For example, Weinstein recently demanded the Army court-martial their chief of chaplains, Maj Gen Douglas Carver, because Chaplain Carver issued a proclamation calling for a day of prayer and fasting for our troops.[20] The chief of chaplains' proclamation called for voluntary prayer by chaplains of all diverse faiths to act "in keeping with your religious traditions," and support our troops who face difficult traumas, pressures, and temptations toward suicide. Carver consulted with two senior Jewish chaplains before issuing the proclamation.

Weinstein claims that no chaplain can encourage prayer or Bible reading without violating what he pretends is a constitutional mandate separating church from state. "These inciteful actions are grossly offensive to not only Muslims in Afghanistan and across the world, but to all those who hold faith in the U.S. Constitution," said Weinstein of the troops' Bibles in Afghanistan.[21]

Courts Defend Christian Speech by Military Commanders

Since losing his 2005 lawsuit about "too much Jesus" at the Air Force Academy, Mr. Weinstein gained little policy traction in his quest to prohibit freedom of expression by military Christians. US district judge James A. Parker in Albuquerque, New Mexico, in a 16-page decision, said the complainants:

> could not claim their First Amendment rights were violated since they no longer attended the Academy. Moreover, the group failed to give specific examples of which cadets were harmed, or when. Without that personal link or connection to future misconduct, plaintiffs have simply not shown that they will suffer an injury in fact that is both concrete and particularized and actual or imminent.[22] In 2005, an Air Force task force concluded there was no overt religious discrimination at the school but that some cadets and staff were insensitive.[23]

Other inspector-general complaints filed by Mr. Weinstein are routinely publicized by the *New York Times*.[24] But when they are resolved in favor of free Christian expression, the *New York Times* ironically fails to report the outcome. For example, on 30 March 2009 the Air Force dismissed Weinstein's complaint against Col Kimberly Toney, who was cleared of any wrongdoing after sending an inspirational email to 3,000 subordinates.[25]

Can Military Chaplains Pray Publicly in Jesus' Name?

In June 2009, the US Congress again began consideration of a bipartisan bill supporting military chaplains' rights, cosponsored by two North Carolina congressmen, Mike McIntyre (D-NC) and Walter Jones (R-NC), in the House Armed Services Committee. The pro-chaplain bill, HR 268, would simply guarantee military chaplains of all faiths the right to pray publicly according to the dictates of their conscience. "If called upon to lead a prayer outside of a religious service, a chaplain shall have the prerogative to close the prayer according to the dictates of the chaplain's own conscience," the new bill declares.[26]

The bill was deemed necessary due to recent legal attacks against the 1860 law, originally signed into military regulations by Pres. Abraham Lincoln, now codified in US Code Title 10 Section 6031: "An officer in the Chaplain Corps may conduct '*public worship*' according to the manner and forms of the church of which he is a member" (emphasis added). Chaplains who respect that old law have come under fire by liberals redefining the scope of "public worship."

Censoring Chaplains' Prayers

In 2006, the Air Force temporarily mollified atheist complainers by issuing restrictive guidelines,[27] and Navy secretary Donald C. Winter signed regulation SECNAVINST 1730.7C, *Religious Ministry within the Department of the Navy*, redefining public worship as only that which occurs safe inside Sunday chapel. Both Air Force and Navy guidelines at that time temporarily prohibited chaplains of Christian, Muslim, Buddhist, or Jewish faiths from using sectarian words like "Jesus," "Allah," "Buddha," or "Adonai" in their public prayers and mandated that all chaplains conform to pray neutered, government-sanitized "nonsectarian" prayers instead. The secular groups then demanded enforcement of those restrictions to proselytize the evangelicals to water down their prayers or be punished with government sword.

Public Worship versus Worshiping in Public

As a former US Navy evangelical chaplain, I took a stand by personally whistle-blowing and then violating those restrictions. Ironically my complaint appeared above the fold on the front page of the *Washington Times*[28] directly beside the story about Abdul Rahman's apostasy trial in Afghanistan. The following week, I dared to pray "in Jesus' name" in uniform[29] at a Thursday press conference outside the White House. Six months later, I was court-martialed for worshiping in public in uniform, since the Navy judge enforced SECNAVINST 1730.7C and ruled that public worship was protected, but worshiping in public was not.[30]

Restrictive Guidelines Rescinded in 2006

Together, my friends and I fought back and won on Capitol Hill. When 300,000 petitioners, 75 congressmen, 35 pro-family groups, and 85 percent of polled voters expressed outrage on my behalf,[31] Pres. George Bush signed the conference report to the 2007 Defense Authorization Act passed by a pro-faith Congress, who rescinded the exact same Navy policy enforced against me and also rescinded the restrictive Air Force guidelines, restoring "public worship" to its original broader meaning.

Since 2007, all military chaplains have been free again[32] to pray publicly in Jesus' name (or however they wish) everywhere, seven days a week, even in uniform, and even at public ceremonies outside of chapel. This victory for religious freedom, which was not grandfathered back to my case, cost me my career and pension.[33]

Policy Implications

Since Congress rescinded prayer content restrictions in 2006:

1. Free speech, freedom of religious expression, and diversity rights are now extended to chaplains and troops equally.

2. No longer are the words "Jesus" or "Christ" or "savior" banned as illegal speech.

3. Chaplains are free to pray in Jesus' name even in public, even in front of non-Christians, so long as members of other faiths are granted the same liberty, taking turns.

4. Commanders who attempt to silence or censor their chaplains' or troops' religious speech no longer enjoy policy protection; they do so at their own risk.

5. Commanders who attempt to enforce a universalist, nonsectarian brand of watered-down faith upon all the troops stand in danger of violating both the establishment clause and the free exercise clause of the First Amendment.

6. Commanders who censor the content of religious speech may reduce the short-term complaints of easily offended listeners, but they risk the long-term goals of protecting American ideals of free speech and free religious expression.

7. Commanders can better educate Soldiers of all faiths on the ideals of tolerance when they set a personal example of tolerance toward faiths they may not personally share.

What the Supreme Court Said

Regarding censorship of public prayer content, the US Supreme Court ruled in a 1991 case, *Lee v. Weisman*, that the government cannot censor the content of anyone's prayer:

> The government may not establish an official or civic religion as a means of avoiding the establishment of a religion with more specific creeds. . . . The State's role did not end with the decision to include a prayer and with the choice of clergyman. Principal Lee provided Rabbi Gutterman with a copy of the "Guidelines for Civic Occasions" and advised him that his prayers should be nonsectarian. Through these means, the principal directed and controlled the content of the prayers. Even if the only sanction for ignoring the instructions were that the rabbi would not be invited back, we think no religious representative who valued his or her continued reputation and effectiveness in the community would incur the State's displeasure in this regard. It is a cornerstone principle of our Establishment Clause jurisprudence that it is no part of the business of government to compose official prayers for any group of the American people to recite as a part of a religious program carried on by government, Engel v. Vitale, (1962), and that is what the school officials attempted to do.

The First Amendment, as interpreted by our Supreme Court, protects everybody's right to compose their own diverse prayer content, whenever invited to pray. When this school prayer dictum is properly applied as precedent for all legislative or military prayers, then no government, especially the federal government, can censor or regulate the content of military chaplains' public prayers or enforce its preferred nonsectarian religion over a sectarian view.

Two Conflicting Appeals Court Rulings

Others may disagree, and the uncertainty about the question is compounded by two conflicting 2008 court rulings by the Fourth and Eleventh Circuit Courts of Appeals. Respectively, *Turner v. Fredericksburg* ruled governments are able (but not required) to mandate nonsectarian prayer speech in government forums, but *Pelphrey v. Cobb County* ruled governments cannot censor anybody's prayers, ever.[34] The tendency of local governments and commanders to make up their own conflicting policies has led Congress to reexamine the issue. Mr. Jones and Mr. McIntyre's bill currently lies on the table in the House Armed Services Committee.[35]

Local Battles over Chaplains' Rights

Since Congress rescinded nationwide restrictions, other local restrictions have emerged. One Air Force commander in Oklahoma issued a basewide ban on Jesus prayers by chaplains, only to have that ban lifted by the next commander. A handful of evangelical Army chaplains reported religious harassment and were not promoted by hostile commanders. Rutherford Institute attorney Art Schulcz, suing the Navy on behalf of 65 evangelical chaplains, lost a DC Appeals Court decision which granted Catholic priests retention well beyond the mandatory retirement age for Protestants.[36]

Last summer the American Civil Liberties Union (ACLU) threatened to sue the US Naval Academy to stop their noon prayer tradition, ongoing since 1845. But the Departments of Justice and the Navy defended the students' right to pray, led by rotating chaplains of diverse faiths: Muslim, Jewish, Catholic, Protestant, and Buddhist.[37]

Conclusion

As partly stated in another journal article recommended for further reading,[38] please consider four concluding personal thoughts:

1. I believe in good order and discipline. Soldiers, Sailors, Airmen, Marines—even chaplains—should march in formation, salute the flag, and obey lawful orders. But no military member should ever be punished, excluded, censored, or forced to pray to the government's nonsectarian god. Furthermore, military members should never have their books seized and burned by our own American government because the content of the book is too religious.

2. I believe in freedom. The commander's key to "order" isn't totalitarian suppression but equal access and equal opportunity for all.

3. Who's proselytizing whom? When easily offended folks disagreed with my prayers, or refused the Bibles I freely offered, they were never punished. But when I declined to pray to the government's nonsectarian, neutered version of god, I was punished with the full weight of the US government and so were our troops wrongly threatened with such punishment in Afghanistan. The enemies of religious liberty now proselytize the Christians with government sword, not vice versa.

4. Censorship of religious expression may enhance short-term security by mollifying those who are easily offended, but it hurts long-term security by teaching the complainers that we will do their bidding, that we willingly enable the enemies of religious liberty, and that we will promote intolerance and hatred of all things Christian or otherwise sectarian, which only encourages them to complain louder, since their complaints succeed in converting us without much fight for liberty. We fear their complaints so much that we begin to turn our own swords

upon ourselves. Instead we need courage to simply let them be offended and defend the uniquely American ideals of free speech, free press, and free religious expression.

When commanders enforce state atheism or abuse rank to universalize one preferred nonsectarian, "pluralistic" religion upon their entire unit, they're shoving their nonreligion down our throats, not vice versa.

The prophet Elijah said to the prophets of Baal, "You call on the name of your god [Baal], and I will call on the name of the name of the Lord [Jehovah], and the God who answers by fire, He is God." And all the people answered and said, "That is a good idea."[39]

Democratic diversity is a good idea. Totalitarian pluralism is a bad idea. Totalitarian atheism is even worse. Let all diverse people pray, publicly each to their own God, print their own books, and take turns sharing their own individualized faith ideas with people in the world around them. It's called religious freedom. And that is a *great* idea.

Notes

1. "Military Burns Bibles Sent to Afghanistan," United Press International, 20 May 2009, http://www.breitbart.com/article.php?id=upiUPI-20090520-093347-7043&show_article=1 (accessed 31 Mar 2009).

2. "US Troops Urged to Share Faith in Afghanistan - 04 May 09," *al-Jazeera* television video, www.youtube.com/watch?v=hVGmbzDLq5c (accessed 30 May 2009).

3. These groups apparently disregard their own constitution.

4. These individuals apparently disregard our Constitution.

5. "U.S. Burns Bibles in Afghanistan Row," *al-Jazeera*, 22 May 2009, http://english.aljazeera.net/news/middleeast/2009/05/200952017377106909.html (accessed 30 May 2009).

6. "U.S. Military Confirms Guantanamo Koran Abuse," *ABC News*, 4 June 2005, www.abc.net.au/news/newsitems/200506/s1384449.htm (accessed 30 May 2009).

7. The Bibles were ultimately destroyed by being burned.

8. "Military Burns Unsolicited Bibles Sent to Afghanistan," 19 May 2009, *CNN.com*, www.cnn.com/ 2009/WORLD/asiapcf/05/20/us.military.bibles.burned/ (accessed 30 May 2009).

9. "Soldiers in Afghanistan Given Bibles," *The Huffington Post*, 4 May 2009, www.huffingtonpost.com/2009/05/04/soldiers-in-afghanistan-g_n_195674.html (accessed 30 May 2009).

10. Hint: Read John 3:16.

11. All military members swear allegiance to defend the US Constitution against its foreign and domestic enemies.

12. The Constitution of Afghanistan 2004, www.afghanan.net/afghanistan/constitutions/constitution2004.htm (accessed 30 May 2009).

13. "Afghanistan," Persecution News, The Voice of the Martyrs, www.persecution.net/afghanistan.htm (accessed 30 May 2009).

14. Sanjoy Majumder, "Mood Hardens against Afghan Convert," *BBC News*, 24 March 2006, http://news.bbc.co.uk/2/hi/south_asia/4841334.stm (accessed 30 May 2009).

15. Abdul Waheed Wafa, "Preachers in Kabul Urge Execution of Convert to Christianity," *New York Times*, 25 March 2006, www.nytimes.com/2006/03/25/international/asia/25convert.html (accessed 30 May 2009).

16. Brian Goldstone, "Violence and the Profane: Islamism, Liberal Democracy, and the Limits of Secular Discipline," *Anthropological Quarterly* 80, no. 1 (Winter 2007): 207–35.

17. This is also true for lexicographers at USCENTCOM, or the Pentagon, or Congress.

18. Justice Blackmun, US Supreme Court, dissented overall but agreed (in this sentence) with the majority, *Goldman v. Weinberger* (No. 84-1097), 236 U.S.App.D.C. 248, 734 F.2d 1531, www .law.cornell.edu/supct/html/historics/USSC_CR_0475_0503_ZD1.html (accessed 30 May 2009).

19. Rodger R. Venzke, *Confidence in Battle, Inspiration in Peace: The United States Army Chaplaincy, 1945–1975* (Washington, DC: Office of the Chief of Chaplains, 1977), 24–25, as cited by Dr. John W. Brinsfield, "The Army Chaplaincy and World Religions," *The Army Chaplaincy Journal*, Winter–Spring 2009, 13.

20. John Hanna, "Group Suing Military Seeks Top Chaplain's Ouster," Associated Press, 8 April 2009, www.militaryreligiousfreedom.org/press-releases/ap_ouster.html (accessed 30 May 2009).

21. "Weinstein: Video Proves Proselytization Rampant at U.S. Military Bases," Military Religious Freedom Foundation press release, 4 May 2009, www.militaryreligiousfreedom.org/press -releases/mrff_video_proves.html (accessed 30 May 2009).

22. "Religious-Bias Suit against Air Force Academy Dismissed," Associated Press, 30 October 2006, www.firstamendmentcenter.org/news.aspx?id=17592 (accessed 30 May 2009).

23. Ibid.

24. Two examples are "Questions Raised Anew about Religion in Military," *New York Times*, 1 March 2009; and "Air Force Looks into 'Inspirational' Video," *New York Times*, 15 March 2009.

25. Steve Mraz, "AF Colonel Cleared of Promoting Religion," *Stars and Stripes*, 2 April 2009, www.military.com/news/article/April-2009/air-force-colonel-cleared-of-promoting-religion .html (accessed 30 May 2009).

26. Gordon James Klingenschmitt, "In Defense of Religious Liberty: Congress Ponders Bi-Partisan Bill," *The Washington Times*, 7 May 2009, A19.

27. Stewart Ain, "Jesus Barred from Air Force Invocations," *The Jewish Week*, 10 February 2006, www.persuade.tv/frenzy2/JewishWeek10Feb06.pdf (accessed 30 May 2009).

28. Eric Pfeiffer, "Navy Prayer Rule Ignites Debate," *The Washington Times*, 23 March 2006, http://persuade.tv/frenzy2/WashTimes23Mar06.pdf (accessed 31 May 2009).

29. Navy Uniform Regulation 6405 states: "Chaplains have the option of wearing their uniform when conducting worship services and during the performance of rites and rituals distinct to their faith groups."

30. Navy judge Lewis T. Booker's entire ruling can be reviewed at persuade.tv/frenzy6/ Klingordrule.pdf (accessed 30 May 2009).

31. Gordon James Klingenschmitt, "Crossing Swords: Let Us Pray," *Proceedings*, January 2007, 22, persuade.tv/Frenzy8/ProceedingsJan07.pdf (accessed 30 May 2009).

32. They are free on paper at least.

33. Bob Unruh, "Navy Dismisses Chaplain Who Prayed 'in Jesus' Name,'" *Worldnet Daily*, 12 January 2007, www.worldnetdaily.com/news/article.asp?ARTICLE_ID=53731 (accessed 30 May 2009).

34. The *Pelphrey v. Cobb County* decision may be read at www.ca11.uscourts.gov/opinions/ ops/200713611.pdf (accessed 30 May 2009).

35. Klingenschmitt, "In Defense of Religious Liberty."

36. Ibid.

37. Cong. Walter B. Jones, "Jones Condemns ACLU Threat to U.S. Naval Academy Prayer," press release, 26 June 2008, http://jones.house.gov/release.cfm?id=691 (accessed 30 May 2009).

38. Klingenschmitt, "Crossing Swords."

39. 1 Kings 18:24 (New American Standard Bible, 1995).

About the Author

Gordon James Klingenschmitt is a 1991 graduate of the United States Air Force Academy, holds graduate degrees in divinity and business, and is currently pursuing a PhD in theology from Regent University. After a decade of service to the USAF in missiles and intelligence, he cross-commissioned to the US Navy where he served as a chaplain. His programs aboard USS *Anzio* won six awards for community service, including best in Navy. When the secretary of the Navy mandated SECNAVINST 1730.7C, requiring chaplains to offer nonsectarian prayers outside of chapel or face punishment for disobeying orders, Chaplain Klingenschmitt deliberately prayed "in Jesus name" in uniform outside the White House in front of the national media and demanded his own court-martial for violating the policy. The Navy judge ruled "worshiping in public" as punishable as a misdemeanor crime, and the chaplain was convicted. Shortly thereafter he was honorably discharged, ending short his 16-year career. Later, SECNAVINST 1730.7C was rescinded, restoring military chaplains' right to pray without censorship. He may be contacted through his Web site: www.PrayInJesusName.org.

CHAPTER 3

THE NEED FOR (MORE) NEW GUIDANCE REGARDING RELIGIOUS EXPRESSION IN THE AIR FORCE

Paula M. Grant

Introduction

Over the past few years, the United States armed services have repeatedly found themselves in the news because of conflicts surrounding religious expression. These conflicts raise constitutional issues, as commanders and lawyers attempt to strike a balance between members' rights under three major First Amendment clauses—the establishment clause, the free exercise clause, and the free speech clause. In striking that balance, commanders and lawyers must currently sift through many layers of confusing guidance. The lack of clear, commander-friendly guidance on the issue of religious expression in the Air Force compels commanders to waste valuable mission time searching for answers.

This chapter will briefly review the history of recent conflicts surrounding religious expression in the military, explore the history of Supreme Court and congressional mandates on religious expression issues, and examine Department of Defense rules on religious accommodation and expression. In addition, this chapter will analyze both current and previously issued Air Force guidance on religious expression, including the 2005 "Interim Guidelines Concerning Free Exercise of Religion in the Air Force" and the 2006 "Revised Interim Guidelines Concerning Free Exercise of Religion in the Air Force." Finally, this chapter will suggest new guidance regarding religious expression for uniformed Air Force members which should be considered for implementation Department of Defense–wide. The suggested new guidance incorporates Supreme Court, congressional, and Department of Defense mandates, yet is clear enough for commanders to apply without the necessity for consultation with a lawyer-chaplain team.

Problem Background

The United States military has recently been forced to deal with several high-visibility religious issues, including those at the service academies, in basic

---

Major Grant wrote this essay while a student at Air Command & Staff College. It won the 2009 Dean's Leadership and Communication Studies Award.

training, at the Pentagon, and in deployed locations. Starting with the Air Force, in 2003 the Christian Leadership Ministries published an annual adver- tisement in the United States Air Force Academy (USAFA) base paper includ- ing the statement, "We believe that Jesus Christ is the only real hope for the world. If you would like to discuss Jesus, feel free to contact one of us!" The signatories included over 200 USAFA faculty and staff, including a majority of USAFA department heads.[1]

In 2004, Christian Embassy, a group established in 1975 to minister to mem- bers of Congress, ambassadors, presidential appointees, and Pentagon officials,[2] filmed a promotional video inside the Pentagon showing several generals and senior defense officials talking about the importance of religion in their jobs and lives. In 2007, the Department of Defense inspector general publicly released a report finding that senior Army and Air Force personnel violated the Joint Ethics Regulation when they participated in the video while in uniform and on active duty.[3]

On 28 April 2005, Americans United for the Separation of Church and State sent a multipage complaint to then–Secretary of Defense Donald Rums- feld, documenting what it called systematic and pervasive religious bias and intolerance at the highest levels of the USAFA command structure.[4] On 2 May 2005, the acting secretary of the Air Force directed a team investigation to as- sess the religious climate at USAFA.[5] Also in May 2005, Chaplain (Capt) Melinda Morton, assigned to USAFA, stated that the religious problem at USAFA "is pervasive."[6]

In June 2005, the Headquarters Review Group Concerning the Religious Climate at the US Air Force Academy (hereafter, Review Group) released its report. The report documented seven specific events of what appeared to be "questionable behavior" and referred those events for command follow-up. In addition, the Review Group identified nine findings regarding the overall climate and made nine recommendations. The first recommendation was to "develop policy guidelines for Air Force commanders and supervisors regarding religious expression."[7]

In July 2005, Chaplain (Brig Gen) Cecil Richardson, then the deputy chief of Air Force chaplains (a major general and chief of Air Force chaplains at the time of publication), stated in a *New York Times* interview, "We [chaplains] will not proselytize, but we reserve the right to evangelize the unchurched."[8] The distinction, he said, is that proselytizing is trying to convert someone in an ag- gressive way, while evangelizing is more "gently" sharing the gospel.[9]

On 6 October 2005, USAFA graduate Mikey Weinstein joined four 2004 USAFA graduates in suing the Air Force in federal district court, claiming that USAFA illegally imposed Christianity on cadets at USAFA. The case was dis- missed by the judge a year later, who ruled that the plaintiffs had graduated and were thus unable to prove any direct harm.[10]

On 25 October 2006, former Navy chaplain Gordon Klingenschmitt filed suit against the Navy in federal district court for, among other claims, violating his First Amendment rights by discouraging him from praying in the name of

Jesus.[11] While he was a Navy chaplain, Klingenschmitt's commander issued him a direct order which instructed him that he could only wear his uniform if conducting a bona fide religious service. Soon afterward, Klingenschmitt conducted a prayer vigil in uniform outside the White House, followed by a news conference to pressure Pres. George W. Bush to issue an executive order regarding military chaplains' right to pray as they wished. Klingenschmitt was subsequently court-martialed for failing to obey a direct order and was involuntarily separated from the Navy.[12]

The religious issues continued well into 2008. In February 2008, USAFA was criticized by Muslim and religious freedom organizations for playing host to and paying three speakers who critics say are evangelical Christians pretending to be alleged former Muslim terrorists.[13] On 5 March 2008, Army SPC Jeremy Hall, an atheist, and the Military Religious Freedom Foundation, headed by Weinstein, filed suit against Secretary of Defense Robert Gates and SPC Hall's commander, MAJ Freddy Welborn. The suit alleged that SPC Hall was denied his First Amendment right to be free of government-sponsored religious activity.[14] On 10 October 2008, the plaintiffs dropped the suit.[15]

In August 2008, the *Air Force Times* interviewed Chaplain Richardson. A reporter asked him to respond to a question about whether he was concerned that a Christian chaplain who was visited by a troubled Airman who wasn't interested in religion might steer the Airman toward Jesus. Chaplain Richardson's response, "Well, you know, sometimes Jesus is what they need. They're asking for it."[16]

On 24 September 2008, PVT Michael Handman, a Jewish Soldier attending basic training at Fort Benning, Georgia, suffered a beating at the hands of fellow Soldiers.[17] The Military Religious Freedom Foundation, in a 16 October 2008 letter to Secretary Gates, alleged that prior to the beating, PVT Handman was a victim of anti-Semitic actions by his drill sergeants.[18]

On 25 September 2008, the Military Religious Freedom Foundation and Army SPC Dustin Chalker, an atheist, filed suit against Secretary Gates. The suit alleges that the plaintiff was required to attend military functions or formations which included sectarian Christian prayers, thus violating his rights under the establishment clause of the First Amendment.[19]

On 16 January 2009, Col Kimberly Toney, commander of the 501st Combat Support Wing at Alconbury, England, sent an e-mail to all Airmen in the wing, inviting them to watch an attached video link highlighting an inspirational story.[20] The attached link was to a Catholic Web site which had posted a video about Nick Vujicic, a man who was born without arms or legs. In the video, Mr. Vujicic attributed his ability to deal with his disability to his faith in Jesus.[21] In addition, information on the Web site attacked Pres. Barack Obama's stance on abortion by depicting him in a Nazi uniform and calling him a "forerunner of the Antichrist."[22] Colonel Toney sent an apology by e-mail to all Airmen in the wing. The Third Air Force investigated the matter and concluded that Colonel Toney acted "inadvertently and unintentionally and did not willfully violate Air Force Policy."[23] The issues described above are merely the ones which received media attention and likely represent the tip of the iceberg with respect to religious conflict in today's multifaith military. In addition, the issues highlight the underlying tension between the First Amendment's two main clauses dealing with religious expression, the establishment clause, and the free exercise clause ("Congress shall make no law respecting an establishment of religion nor prohibiting the free exercise thereof").[24] The underlying tension exists because the establishment clause appears to limit religious expression while the free exercise clause appears to encourage it.

The Air Force first issued direct guidance on the exercise of religion in August 2005, when it issued "Interim Guidelines Concerning Free Exercise of Religion in the Air Force." That guidance was soon followed by "Revised Interim Guidelines Concerning Free Exercise of Religion in the Air Force" in February 2006. The guidance does not appear to have alleviated the confusion and misunderstanding surrounding the subject, as is evidenced, at a minimum, by Chaplain Richardson's August 2008 comments to the *Air Force Times*, discussed above. If the chief chaplain of the Air Force can't get it right, it is difficult to see how commanders in the field, trained in neither law nor religion, can be expected to pick their way through the legally confusing and emotionally charged topic, given the current state of Air Force guidance.

First Amendment Legal Framework

In formulating official guidance regarding military members' exercise of religion in the Air Force, one must address the rights inherent in two seemingly contradictory clauses in the First Amendment: the establishment clause ("Congress shall make no law respecting an establishment of religion")[25] and the free exercise clause ("nor prohibiting the free exercise thereof").[26] In addition, because the exercise of religion often involves both actual and symbolic speech, officials attempting to formulate guidance must also consider the rights of military members under the free speech clause of the First Amendment ("Congress shall make no law . . . abridging the freedom of speech").[27] Officials must possess a clear understanding of the legal framework enclosing the interplay of these three First Amendment clauses before they can formulate new religious guidance.

Establishment Clause

The establishment clause "mandates government neutrality between religion and religion, and between religion and non-religion."[28] Consequently, the government cannot act in a way which favors one religion over another, nor can it act in a way which favors religion over nonreligion. *Lemon v. Kurtzman*[29] remains the Supreme Court's most influential case on the establishment clause. In *Lemon*, the Supreme Court articulated what has become known as "the *Lemon* test," a standard against which to measure government action to determine if it is constitutional under the establishment clause.[30]

In 1971 three lawsuits—two from Rhode Island and one from Pennsylvania—were reviewed by the US Supreme Court. The plaintiffs in the lawsuits asserted that certain state monetary support of church-affiliated private schools violated the establishment clause. In holding that the Rhode Island and Pennsylvania systems violated the establishment clause, the Supreme Court articulated the *Lemon* test. For a statute to pass muster under the establishment clause (and therefore be held constitutional), all three prongs of the *Lemon* test must be satisfied:

1. The statute must have a secular legislative purpose (meaning a legitimate, nonreligious purpose as judged by an objective observer).[31]
2. The statute's principal or primary effect must neither advance nor inhibit religion (the statute must be "religion neutral").
3. The statute must not foster "an excessive government entanglement with religion," meaning that the government should not involve itself in the workings of a religion (or a religious organization) and vice versa.[32]

Most establishment clause cases which reach the Supreme Court have a problem with the second prong of the *Lemon* test, the "effects" prong. In many effects-prong cases, the government establishes a rule which appears neutral on its face, but its primary effect either aids or hinders religion. In *Lemon v. Kurtzman*, the issue was not the effects prong but the "entanglement" or third prong. The states of Rhode Island and Pennsylvania were careful to ensure that the money provided to the private schools was used only for secular purposes. To that end, the states set up extensive auditing systems to monitor the private schools' use of the state money. Ironically, because the states went to such lengths to ensure that the money was used only for nonreligious purposes (so the states would not be violating the Constitution by aiding religion), the states ended up involving themselves too intimately in the business of the religious schools. Consequently, the Supreme Court found excessive entanglement under the third prong of the *Lemon* test and held that the Pennsylvania and Rhode Island systems were in violation of the Constitution.

Lemon v. Kurtzman has not been overturned by the Supreme Court, but at times the Supreme Court has used alternate tests to determine constitutionality under the establishment clause. The coercion test is one alternative, under which the Court looks at whether the state has by its actions essentially forced someone to support or participate in religion.[33] In holding unconstitutional a rabbi-led prayer at a middle-school graduation ceremony, the Supreme Court applied the coercion test and stated that "at a minimum, the Constitution guarantees that government may not coerce anyone to support or participate in religion or its exercise."[34] Applying both the *Lemon* and coercion tests, the Supreme Court also struck down student-led prayer at a high-school football game.[35] When considering the concept of coercion, the Supreme Court would likely give more leeway to the government as students grow older and more mature; the older a student is, the less likely the Court is to find that he was coerced.

In 2003, the Fourth Circuit Court of Appeals applied the coercion test to "voluntary" prayer at the noon meal at the Virginia Military Institute (VMI). Cadets were required to stand quietly during the "voluntary" prayer and were not allowed to go about their business until it was over. In finding the prayer unconstitutional, the court reasoned that because of the strict military-type environment at VMI, any real voluntariness was taken out of the situation.[36] The Fourth Circuit made this holding in spite of the fact that VMI cadets are older and more mature than the middle-school or high-school students at issue in previous cases. While the decision of a court of appeals obviously doesn't carry the same weight as a Supreme Court decision, it is indicative of how the issue would be resolved, were it to reach the Supreme Court. In addition, the VMI decision also sheds some light on how a court would rule in similar situations in a military context.[37]

Another substitute test used by the courts has been called the endorsement test. In applying that test, the court will look to whether a reasonable observer, aware of the history of the conduct at question, would view the government action as endorsing religion.[38] The endorsement test is favored by some *Lemon* test critics because it is more common-sense based and far less formulaic than the *Lemon* test. *Lemon v. Kurtzman* remains good law, however. As a result, when analyzing a government action under the establishment clause, the *Lemon* test must be considered first, before looking at either the coercion or endorsement tests.

Free Exercise Clause

Counterbalancing the establishment clause is the free exercise clause ("nor prohibiting the free exercise thereof").[39] Sometimes government action, instead of appearing to "establish" religion, may unintentionally burden religion.[40] Just as the government doesn't actually have to "establish" a religion in the strict sense of the word to be guilty of violating the establishment clause, so too the government need not actually "prohibit" the exercise of religion to be guilty of violating the free exercise clause. Most of the free exercise clause cases involve government action which is not necessarily directed at religion but may limit someone's ability to practice his or her religion through laws which are "religion neutral." For example, a law which prohibits animal sacrifice, while generally applicable and not directed at practitioners of the Santeria religion, would nonetheless inhibit the Santeria practitioners' ability to sacrifice animals as part of their religious practices.[41]

The leading such religion-neutral case is *Employment Division, Department of Human Resources of Oregon v. Smith*.[42] Smith's religion required him to use peyote as part of church ceremonies, but an Oregon state statute proscribed the possession of peyote. Because of his use of peyote as part of his Native American church, he was fired from his job at a drug rehabilitation facility. Smith applied for unemployment compensation and was denied because he was fired for misconduct. He claimed that the denial of unemployment compensation violated the free exercise clause because it prevented him from freely exercising his religion.

Smith asserted that the court, in reviewing his case, should apply the most stringent review standard, known as strict scrutiny. To pass strict scrutiny, the government would have to show that the application of the statute was in furtherance of a compelling government interest and that the government used the least restrictive means of furthering that compelling government interest. Previous cases had applied this standard to determine the constitutionality of similar laws. The Supreme Court, in a decision joined by only five of the nine justices (a bare majority), refused to apply the strict scrutiny standard suggested by Smith and previously applied by other courts. Instead, the Supreme Court held that if a law is religion neutral and of general applicability (the law applies to everyone, not just religious practitioners), as long as it is otherwise (procedurally) valid, it passes muster under the free exercise clause.[43] Thus, the Supreme Court held that the government could pass a law or enact a practice which burdens someone's ability to practice religion, as long as that law or practice was not directed at the religious practitioner and the law or practice applied to everyone and not just the religious practitioner.

In direct response to the Supreme Court justices' refusal to apply the strict scrutiny standard to religion-neutral laws of general applicability which incidentally inhibit religious practitioners' ability to practice their religion (such as the Oregon statute criminalizing peyote possession), in 1993 Congress passed the Religious Freedom Restoration Act (RFRA).[44] RFRA prohibits the government from placing a substantial burden on a person's exercise of religion (even if the burden is a result of a rule of general applicability) unless the government can show that the application of the burden is in furtherance of a compelling governmental interest and that it is the least restrictive means of furthering that compelling governmental interest. Evidently unhappy with the result in *Smith*, Congress simply legislated the application of strict scrutiny to similar cases in the future.

The constitutionality of RFRA as it applies to the federal government was affirmed by the Supreme Court in 2006.[45] This is significant for the military, as part of the federal government, because RFRA applies to military actions which substantially burden a person's free exercise of religion. When courts review actions by the military which substantially burden a member's free exercise of religion, they will apply strict scrutiny. Consequently, any military policy or practice which substantially burdens a military member's free exercise of religion must be in furtherance of a compelling governmental interest and must be the least restrictive means of furthering that compelling governmental interest.

Free Speech Clause

The free speech clause prohibits the government from abridging the freedom of speech.[46] Protections under the First Amendment's free speech clause include religious speech.[47] Issues arise under the free exercise clause when a government action somehow limits religious *conduct*, while issues arise under the free speech clause when a government action somehow limits religious *speech*.

Government action limiting free speech can be one of two types, either content-based or content-neutral.[48] A content-based speech restriction is one which limits a particular type of message. For example, a law which prohibited anyone from stating "I am a Christian" or "I am a Muslim" would be a content-based speech restriction. By contrast, a content-neutral speech restriction doesn't limit the message but instead imposes what has been called a "time, place, or manner restriction."[49] For example, a law which mandates that protesters must conduct protests at least five feet away from city streets between the hours of 0800 and 2200 would be a permissible time, place, or manner restriction. Content-based speech restrictions are subject to strict scrutiny (see discussion above, under free exercise clause), while content-neutral restrictions are subject to a much lower degree of scrutiny. Content-neutral restrictions will be upheld as long as they are reasonable in light of the purpose served by the forum.[50]

Also factoring into the legal analysis is whether the speech is made in a public or nonpublic forum. A public forum is one which, by tradition or otherwise, has been used for public debate and assembly, such as public parks. All other areas, including military bases, are considered nonpublic forums. Speech may be regulated more closely in a nonpublic forum.[51]

It is well established that the military may regulate certain types of speech by its members which if made by civilians would be protected.[52] The decisive case on free speech in the military is *Parker v. Levy*.[53] In that case, an Army officer encouraged African-American Soldiers to refuse to serve in Vietnam and called Special Forces members liars, thieves, killers of peasants, and murderers of women and children.[54] CPT Howard Levy was convicted of conduct unbecoming an officer and a gentleman and of conduct prejudicial to good order and discipline in the armed forces. He appealed his conviction to the Supreme Court on the basis that his First Amendment rights had been violated. In upholding the conviction, the Supreme Court recognized that the military is separate from civilian society in some respects and that the demands of the military are such that, under certain circumstances, military members' free speech rights may be trumped by the needs of the military society and mission. In so holding, the Court stated, "While members of the military are not excluded from the protection granted by the First Amendment, the different character of the military community and of the military mission requires a different application of those protections."[55]

For example, the Uniform Code of Military Justice (UCMJ) prohibits officers from using contemptuous words against a long list of civilian officials[56] and prohibits members from using disrespectful language toward superiors.[57] While civilians are free to use contemptuous words against any number of elected officials, military members may not. Civilians may use disrespectful words against superiors at work without risking criminal prosecution, while military members may not. It is irrelevant for purposes of the UCMJ whether a member makes contemptuous or disrespectful speech while on or off duty and on or off base.[58] In addition, Articles 133 and 134 of the UCMJ prohibit conduct which is prejudicial to good order and discipline or is service discrediting and conduct which is unbecoming an officer. Conduct includes speech.[59] No religious speech is explicitly prohibited by the UCMJ, but it is conceivable that under the right factual circumstances, a member's speech, including religious speech, could potentially violate Articles 133 or 134.

Consider the following example: A dental officer is a born-again Christian who is required to bear witness as part of his religion. While patients are in his dental chair, he repeatedly proselytizes to them. Several patients have complained to him and to the dental unit commander, but the dentist refuses to stop proselytizing. The dentist brags to other members of the unit that God is a higher authority than the unit commander and that he will not stop bearing witness. After noticing a patient's Star of David necklace, he tells the patient (a dependent of a military member) that she is going to Hell unless she accepts Jesus into her heart. The patient complains to the post commander. Under the above circumstances, it is conceivable that the dentist could be charged with either of Articles 133 or 134 because his open defiance of the commander violates good order and discipline and is unbecoming an officer.

The Court of Appeals for the Armed Forces has held that the military may prohibit speech which "interferes with or prevents the orderly accomplishment of the mission or presents a clear danger to loyalty, discipline, mission or morale of the troops."[60] Courts have not yet ruled on this issue in the context of religious speech.

Courts analyzing religious speech by military members will also look to whether the speech is private or official. To determine whether speech is private or official, courts will most likely look at the totality of the circumstances, including the status of the speaker, the status of the listener, and the context and characteristics of the speech itself.[61]

Consider the following examples: At one end of the spectrum is religious speech made by one junior enlisted member to another at an off-base social establishment after duty hours. Neither Sailor in this example is in uniform, nor is either in a supervisory relationship toward the other. That speech is likely to be held purely private and enjoy the highest levels of protection under the free speech and the free exercise clauses.

An example at the other end of the religious speech spectrum is the post commander, in uniform at a mandatory commander's call during duty hours in the post theater, telling the entire battalion about his recent conversion to Islam. The post commander exhorts all present to recognize that worshipping Allah is the way to Heaven. That speech is likely to be held as official speech and would therefore trigger an analysis under the establishment clause. Under the establishment clause analysis, a court would likely apply the *Lemon* test and determine that under the "effects" prong, the base commander's action violated the establishment clause because the speech had the primary effect of establishing religion. The speech would also fail both the coercion test and the endorsement test, given the mandatory nature of the commander's call and the fact that a reasonable person attending the commander's call would see the religious speech as a government endorsement of religion. In the example above, the post commander has clearly violated the establishment clause.

The two religious speech situations described above are relatively easy; most religious issues which arise in today's armed forces are not nearly so clear cut and fall somewhere in the middle of the two extremes described. Consequently, commanders need clear guidance regarding religious speech and actions because it is on commanders that the burden to sort out religious issues regularly falls.

Applicable Department of Defense Regulation

Department of Defense Directive 1300.17, *Accommodation of Religious Practices within the Military Services*, applies to each of the services. It directs military commanders to consider the following factors when determining whether to grant a request for accommodation of religious practices:

1. The importance of military requirements in terms of individual and unit readiness, health and safety, discipline, morale, and cohesion.
2. The religious importance of the accommodation to the requester.
3. The cumulative impact of repeated accommodations of a similar nature.
4. Alternative means available to meet the requested accommodation.
5. Previous treatment of the same or similar requests, including treatment of similar requests made for other than religious reasons.[62]

This directive was promulgated in 1988 and has not been substantially altered since. The recently released 2009 version of the directive remains essentially the same as the 1988 version, in spite of the fact that the 1993 Religious Freedom Restoration Act (RFRA) changed the way that commanders should view requests for accommodation.[63] RFRA mandates that government policies which substantially burden someone's free exercise of religion must be in furtherance of a compelling governmental interest and that the burden must be the least restrictive means of furthering that compelling governmental interest. While the military will always have a compelling governmental interest in completing the military mission, courts will look closely at whether the burden placed on a member's ability to exercise his or her religion is the least restrictive means available. Consequently, RFRA compels commanders to grant religious accommodations when at all possible.

The Joint Ethics Regulation[64] is also applicable to each of the services. It states that "an employee shall not use or permit the use of his Government position or title or any authority associated with his public office in a manner that could reasonably be construed to imply that his agency or the Government sanctions or endorses his personal activities or those of another."[65] This provision prohibits military members from using their official positions to endorse private organizations, including religious organizations.

The Army, Navy, and Air Force all have service-specific regulations (or sections of regulations) dealing with religious accommodation.[66] In addition, all three services have published guidance on chaplain activities.[67] However, none

of the services currently has comprehensive guidance dealing with accommodation, ministry, and free exercise of religion issues.[68]

Guidelines Covering
Religious Expression in the Air Force

In June 2005 the Headquarters Review Group Concerning the Religious
Climate at the US Air Force Academy published a report detailing seven specific events of "questionable behavior" concerning religious expression and made
nine recommendations to the acting secretary of the Air Force.[69] Several of the
report's recommendations are specific to USAFA, but a number of others are
applicable Air Force–wide. For instance, the Review Group recommended that
the Air Force "develop policy guidelines for Air Force commanders and supervisors regarding religious expression . . . ; reemphasize the requirement for all
commanders to address issues of religious accommodation up front, when planning, scheduling and preparing operations; and develop guidance that integrates
the requirements for cultural awareness and respect across the learning continuum, as they apply to Airmen operating in Air Force units at home as well as
during operations abroad."[70]

In apparent response to the Review Group report, in August 2005 the Air
Force established "Interim Guidelines Concerning Free Exercise of Religion in
the Air Force."[71] The four-page "Interim Guidelines" addressed the "key areas"
of religious accommodation, public prayer outside of voluntary worship settings, individual sharing of religious faith in the military context, the chaplain
service, e-mail, and other communications as well as good order and discipline.
With respect to public prayer, the "Interim Guidelines" stated that it "should
not usually be included in official settings such as staff meetings, office meetings, classes or officially sanctioned activities such as sports events or practice
sessions,"[72] but allowed for exceptions such as mass casualties, imminent combat, and natural disaster.[73] The "Interim Guidelines" further advised that "a
brief, non-sectarian prayer may be included in non-routine military ceremonies
or events . . . such as a change-of-command or promotion . . . where the purpose
of the prayer is to add a heightened sense of seriousness or solemnity, not to
advance specific religious beliefs."[74]

The "Interim Guidelines" cautioned members that when sharing religious
faith, they must be "sensitive to the potential that personal expressions may
appear to be official expressions," especially when they involve superior/subor-
dinate relationships. The "Interim Guidelines" further noted that the "more
senior the individual, the more likely that personal expressions may appear to
be official expressions."[75]

The "Interim Guidelines" discussion of the chaplain service states, "Chaplains are commissioned to provide ministry to those of their own faiths, to facilitate ministry to those of other faiths, and to provide care for all service members, including those who claim no religious faith." The "Interim Guidelines"

further caution chaplains to "respect professional settings where mandatory participation may make expressions of religious faith inappropriate."[76]

Public and congressional response to the "Interim Guidelines" was immediate. Christian organizations interpreted the "Interim Guidelines" as a prohibition against chaplains mentioning the name of Jesus or evangelizing and began a national petition campaign urging President Bush to enact an executive order allowing military chaplains to pray according to their faiths.[77] Rep. Walter Jones (R-NC), along with approximately 70 other members of Congress, endorsed a 25 October 2005 letter to President Bush also urging an executive order.[78] In addition, a group representing hundreds of evangelical Christian chaplains threatened to remove its chaplains from the military unless chaplains were given more leeway in public prayers.[79]

In February 2006, only six months after issuing the "Interim Guidelines," the Air Force issued "Revised Interim Guidelines Concerning Free Exercise of Religion in the Air Force."[80] At a minimum, the timing of the release of the new guidelines suggests that the Air Force was well aware of the controversy the first set of guidelines had caused.

At only one page, the "Revised Interim Guidelines" are substantially shorter than the previous version, and many consider them a watered-down version of the "Interim Guidelines." Markedly absent from the "Revised Interim Guidelines" is any reference to religious coercion by supervisors. Instead, the "Revised Interim Guidelines" assure superiors that they "enjoy the same free exercise rights as other airmen."[81] In addition, the "Revised Interim Guidelines" state that the Air Force respects "the rights of chaplains to adhere to the tenets of their religious beliefs" and that chaplains "will not be required to participate in religious activities, including public prayer, inconsistent with their faiths."[82] It is unclear whether chaplains are free to exhort the name of Jesus in public prayers under the "Revised Interim Guidelines," but the removal of the term "nonsectarian" from the guidelines could not have been accidental.

Conservative Christian groups praised the "Revised Interim Guidelines." A senior official at Focus on the Family stated, "We hope these guidelines will bring an end to the frontal assault on the Air Force by secularists who would make the military a wasteland of relativism, where robust discussion of faith is impossible."[83] Representative Jones said the guidelines "are a step in the right direction."[84]

Mikey Weinstein criticized the new guidelines, calling them "a terrible disappointment and a colossal step backward."[85] The national director of the Anti-Defamation League also expressed disappointment, stating, "Taken as a whole, these revisions significantly undermine the much-needed steps the Air Force has already taken to address religious intolerance."[86] The executive director of Americans United for the Separation of Church and State condemned the new guidelines, stating that they "focus heavily on protecting the rights of chaplains while ignoring the rights of nonbelievers and minority faiths."[87]

The House of Representatives continued to have a keen interest in the issue of religious expression in the armed forces. During 2006, the National Defense Authorization Act made its way through the Senate and the House. While in the House, a group led by Representative Jones attached an amendment to the bill which stated, "Each Chaplain shall have the prerogative to pray according to the dictates of the Chaplain's own conscience, except as must be limited by military necessity, with any such limitation being imposed in the least restrictive manner feasible."[88] The Senate version included no such amendment.

While in committee discussing the differences between the House and Senate versions, Senator John Warner, chair of the Senate Armed Services Committee, suggested that the amendment should not be included in the final version of the bill because Congress would not have enough time [before the end of the congressional session] to fully debate and discuss the issue.[89] Senator Warner stated that he had talked to each of the head chaplains for the various services and that each opposed the inclusion of the amendment as worded.[90]

In an apparent compromise, the committee members agreed that the amendment would be excluded from the final version of the bill, but that the following language would be included in the report:

> The Secretary of Defense will hold in abeyance enforcement of the regulations newly promulgated by both the Air Force and Navy[91] until such time as the Congress has had an opportunity to hold its hearings, go through a deliberative process, and then decide whether it wishes to act by way of sending a conference report to the President for purposes of becoming the law of the land.[92]

Senator Warner recognized that the report language had no force of law on the services.[93]

President Bush did not issue an executive order regarding military chaplains, nor did Congress revisit the specific issue. The secretary of defense did not order either the Air Force or the Navy to rescind the 2006 regulations. In fact, Secretary of the Air Force Michael W. Wynne issued a memorandum on 21 November 2006 stating that "the Air Force intends to defer taking such further action on such guidance until there has been an opportunity for the Congress to hold such hearings [on religious guidelines]."[94] Consequently, the February 2006 "Revised Interim Guidelines" remain valid Air Force guidance.

Since 2005, when the Review Group report determined that in the Air Force there was "a lack of operational instructions that commanders and supervisors can use as they make decisions regarding appropriate exercise of religion in the workplace,"[95] the Air Force has produced two separate (and some say conflicting) versions of guidance regarding religious expression. However, the fact that religious conflict issues continue to arise may be evidence that commanders and supervisors are either unaware of the guidance or unclear how the guidance should be applied. While some commanders may have the luxury of being able to consult a lawyer or chaplain for every religious issue which arises, others may be unable to or disinclined to do so. The point is that commanders should have available to them clear guidance regarding religious expression in the Air Force which may be straightforwardly applied (by commanders, not lawyers or chaplains) to real situations. To date, the Air Force has not provided such guidance.

An issue which arises when attempting to formulate such guidance is that the more detailed the guidance, the more likely it is that it will run afoul of the *Lemon* test. Any guidelines promulgated by the Air Force will have to pass muster against the three prongs of the test (purpose, effects, and entanglement). The more government involvement, the more likely it is that a court would find, as the Supreme Court did in *Lemon v. Kurtzman*, that the government has entangled itself too much in religion.[96] Any new guidelines promulgated by the Air Force should be detailed enough for commanders and supervisors to follow, yet not so detailed as to risk violating the entanglement prong of the *Lemon* test. In addition, any new guidelines will obviously have to strike the appropriate balance between the establishment and free exercise clauses, while bearing in mind members' rights under the free speech clause.

Suggested New Guidance for Uniformed Air Force Members

Rather than the four-page 2005 "Interim Guidelines" or the single-page 2006 "Revised Interim Guidelines," the Air Force is in need of guidance which is short and clear and which commanders can readily apply to actual situations. To that end, the Air Force should adopt the following three rules regarding religious expression in the Air Force:

1. No public prayer at command functions. Command functions include both events which are actually mandatory (staff meetings, changes of command, graduation exercises at military schools, meals for trainees and cadets), as well as those which are *de facto* mandatory by nature of the military environment (retirements, dining-ins, military balls, awards ceremonies). The Supreme Court has long recognized the special nature of the military environment,[97] and an important feature of that special environment is the coercion inherent in superior-subordinate relationships.[98] While there is an argument to be made that public prayer should not be abolished at "voluntary" events such as dining-ins, in the military environment the pressure which often accompanies an "invitation" to attend such an event renders them compulsory in fact if not in name.

By eliminating public prayer at command functions, chaplains who feel pressured to pray in a nonsectarian manner against the tenets of their religion will be relieved of that conflict, and military members who do not wish to pray at all will not be forced to stand uncomfortably silent or risk the disapproval of both the majority and military superiors. A reasonable alternative to a chaplain-led public prayer is a chaplain-led moment of silence. During a moment of silence before command events, those who wish to pray may do so according to the tenets of the religion to which they belong, while those who do not are not forced to.

2. No coercive evangelizing or proselytizing.[99] While some have attempted to articulate a distinction between evangelizing and proselytizing, from the standpoint of the First Amendment, they are on equal footing. Both evangelizing and proselytizing are protected speech under the First Amendment to the same degree as other speech is protected. Because the military may restrict members' speech which "interferes with or prevents the orderly accomplishment of the mission or presents a clear danger to loyalty, discipline, mission or morale of the troops,"[100] evangelizing or proselytizing may be restricted when it is coercive. When either evangelizing or proselytizing is coercive, it interferes with the orderly accomplishment of the mission and is a clear danger to morale. Someone who is being pressured to listen to unsolicited gospel or to convert to an unwanted religion cannot possibly be accomplishing the mission in an orderly manner. Whether evangelizing or proselytizing is coercive is fact dependent, but examples of potentially coercive situations would be those involving a supervisor-subordinate relationship, those involving disparities in rank, grade, or position (especially if the senior member is a commander), or those involving repeated attempts when the listener has made it clear that the evangelizing or proselytizing is unwelcome.

3. No official endorsement of any particular religion or religion in general. This rule applies to both actions and speech, and the key to this rule is the descriptive term "official." This rule is also fact dependent, but certain generalities apply. For example, the more senior a member is, the more likely it is that he or she will be perceived as acting or speaking in an official capacity.[101] A member in a position of authority vis-à-vis another member (commander, supervisor, coach, instructor) is more likely to be perceived as acting or speaking in an official capacity than one who is on equal footing with another member. In addition, a member in uniform is more likely to be perceived as acting or speaking in an official capacity than one in civilian clothes. Finally, one who is involuntarily present during the religious action or speech is more likely to perceive that the action or speech is official in nature than one who is voluntarily present. The more captive the audience, the more likely it is that the audience will perceive the action or speech as official.[102]

The above three rules are short, easily understood, and capable of ready application to real-world situations. In spite of their brevity, however, the rules are comprehensive enough to cover virtually all situations involving issues of religious expression in the Air Force.[103] In addition, the rules strike an appropriate balance between the three First Amendment clauses and avoid running afoul of either Supreme Court decisions or RFRA.

Most importantly, limiting Air Force guidance to the three simple rules listed above places the decision-making authority with respect to religious expression issues more clearly where it belongs—with commanders. The simplicity of the rules themselves encourages commanders to apply those rules to factual situations, without the need to consult a lawyer-chaplain team for every religious expression issue which arises in a squadron. Commanders are, of course, free to consult with lawyers or chaplains as necessary, but the simplicity of the rules should limit the necessity for commanders to do so.

Conclusion

The recent high-visibility issues regarding religious expression in the military highlight the need for clear, commander-friendly guidance on the topic. While several layers of guidance currently exist, much of it can be confusing to commanders trained in neither the law nor theology. The emotionally charged nature of religious issues will almost guarantee that commanders will continue to have to deal with them in the future. In doing so, commanders should have available to them comprehensive yet easily applied rules, such as the three simple rules suggested above. Consequently, the Air Force specifically, and Department of Defense generally, should issue (more) new guidance on religious expression, modeled after the suggested new guidance provided in this chapter.

Notes

1. Laurie Goodstein, "Air Force Academy Staff Found Promoting Religion," *New York Times*, 23 June 2005.

2. Gordon Lubold, "Religion on Their Sleeves?" *Air Force Times*, 25 December 2006.

3. DOD, *Alleged Misconduct by DOD Officials Concerning Christian Embassy*, Inspector General Report No. H06L102270308, 20 July 2007.

4. US Air Force, *Report of the Headquarters Review Group Concerning the Religious Climate at the U.S. Air Force Academy*, 22 June 2005.

5. Ibid.

6. Goodstein, "Air Force Academy Staff Found Promoting Religion."

7. US Air Force, *Report of the Headquarters Review Group*, ii.

8. Laurie Goodstein, "Evangelicals Are a Growing Force in the Military Chaplain Corps," *New York Times*, 12 July 2005.

9. Ibid.

10. "Suit against Air Force Academy Dismissed," *Washington Times,* 26 October 2006.

11. *Klingenschmitt v. Winter*, in US District Court for the District of Columbia, filed 25 October 2006.

12. Alan Cooperman, "Navy Chaplain Guilty of Disobeying a Lawful Order," *Washington Post,* 15 September 2006.

13. Neil MacFarquar, "Speakers at Academy Said to Make False Claims," *New York Times*, 7 February 2008.

14. *Military Religious Freedom Foundation (MRFF) v. Gates*, in US District Court for the District of Kansas, filed 5 March 2008.

15. Ibid.

16. Patrick Winn, "Religion at Issue," *Air Force Times*, 11 August 2008.

17. "Debate Rages over Attack on Jewish Soldier," *JTA,* 13 October 2008.

18. MRFF Press Release, 16 October 2008.

19. *MRFF v. Gates.*

20. Geoff Ziezulewicz, "AF Colonel Accused of Imposing Religion," *Stars and Stripes*, 20 February 2009.

21. Eric Lichtblau, "Air Force Looks into 'Inspirational' Video," *New York Times,* 14 March 2009.

22. Ibid.

23. Steve Mraz, "Alconbury Colonel Cleared of Promoting Religion," *Stars and Stripes*, 2 April 2009.

24. US Constitution, Amendment 1, 1791.

25. Ibid.

26. Ibid.

27. Ibid.

28. *McCreary County v. ACLU*, in *Supreme Court Reporter*, vol. 125 (2005), 2733.

29. *Lemon v. Kurtzman*, in *United States Supreme Court Reports*, vol. 403 (1971), 602.

30. Ibid.

31. *McCreary County v. ACLU*, 2722.

32. *Lemon v. Kurtzman*, 602.

33. *Lee v. Weisman*, in *United States Supreme Court Reports*, vol. 505 (1992), 577 (nonsectarian prayer at high school graduation coercive).

34. Ibid., 587.

35. *Santa Fe Independent School District v. Doe*, in *United States Supreme Court Reports*, vol. 530 (2000), 290.

36. *Mellen v. Bunting*, in *Federal Reporter*, 3d series, vol. 327 (2003), 355.

37. This conclusion is of course based upon the Supreme Court's current makeup. As the makeup of the Court changes, so too may the likely outcome of establishment clause cases which reach the Court.

38. *Lynch v. Donnelly*, in *United States Supreme Court Reports*, vol. 465 (1984), 668.

39. US Constitution, Amendment 1, 1791.

40. Intentional burdening of religion won't be discussed here, as those are easy cases, and clearly unconstitutional.

41. Contrast with *Church of the Lukumi Babalu Aye, Inc. v. City of Hialeah*, in *United States Supreme Court Reports*, vol. 508 (1993), 520, in which the City of Hialeah, Florida, passed city ordinances specifically targeted at Santeria practitioners.

42. *Employment Division, Department of Human Resources of Oregon v. Smith*, in *United States Supreme Court Reports*, vol. 494 (1990), 872.

43. Ibid., 878.

44. 42 US Code 2000bb.

45. *Gonzales v. O Centro Espirita Beneficiente Uniao do Vegetal*, in *United States Supreme Court Reports*, vol. 546 (2000), 418.

46. US Constitution, Amendment 1, 1791.

47. *Widmar v. Vincent*, in *United States Supreme Court Reports*, vol. 454 (1981), 269.

48. See *Turner Broadcasting, Inc. v. Federal Communications Commission*, in *United States Supreme Court Reports*, vol. 512 (1994), 622.

49. *Clark v. Community for Creative Non-Violence*, in *United States Supreme Court Reports*, vol. 468(1984), 293.

50. *Cornelius v. NAACP*, in *United States Supreme Court Reports*, vol. 473 (1985), 806.

51. Ibid.

52. *Parker v. Levy*, in *United States Supreme Court Reports*, vol. 417 (1974), 733.

53. Ibid.

54. Ibid., 737.

55. Ibid., 758.

56. Uniform Code of Military Justice (UCMJ), Art. 88.

57. UCMJ. Arts. 89, 91.

58. Rules for Courts-Martial 202, 2008 edition.

59. *Parker v. Levy*, 733.

60. *United States v. Brown*, in *Military Justice Reports*, vol. 45 (1996), 389.

61. David E. Fitzkee and Linell Letendre, "Religion in the Military: Navigating the Channel between Religion Clauses," *Air Force Law Review* 59, no. 207, 38.

62. DOD Instruction 1300.17, *Accommodation of Religious Practices within the Military Services*, 10 February 2009, Enclosure, para. 1.

63. It is inconceivable that the 2009 version of the directive makes no mention of RFRA, in spite of RFRA's major impact on the way commanders must view requests for religious accommodation.

64. 5 Code of Federal Regulations (CFR) Section 2635.702(b).

65. Ibid.

66. Army Regulation (AR) 600-20, *Army Command Policy*, 2002; Secretary of the Navy Instruction (SECNAVINST) 1730.8A, *Accommodation of Religious Practices*, 31 December 1997, I; and Air Force Instruction (AFI) 36-2706, *Military Equal Opportunity Program*, 29 July 2004.

67. AR 165-1, *Chaplain Activities in the United States Army*, 25 March 2004; Naval Warfare Publications (NWP) 1-05, *Religious Ministry in the US Navy*, 2003; and Air Force Policy Directive (AFPD) 52-1, *Chaplain*, 2 October 2006.

68. To date, the most comprehensive and clear guidelines on religious expression in the federal workplace were promulgated in 1997, during the Clinton administration. Those guidelines specifically excluded uniformed military personnel.

69. US Air Force, *Report of the Headquarters Review Group*, ii.

70. Ibid.

71. US Air Force, "Interim Guidelines Concerning Free Exercise of Religion in the Air Force", 2005, available at http://www.usafa.af.mil/superintendent/pa/religious.cfm.

72. Ibid.

73. Ibid.

74. Ibid.

75. Ibid.

76. Ibid.

77. "ACLJ National Petition Tops 160,000," *Business Wire*, 14 December 2005.

78. Ibid.

79. Bryant Jordan, "Evangelicals Threaten to Remove Chaplains from Military," *Air Force Times*, 2 January 2006.

80. US Air Force, "Revised Interim Guidelines Concerning Free Exercise of Religion in the Air Force", 2006.

81. Ibid.

82. Ibid.

83. Erin Roach, "Air Force Religion Guidelines Garner Both Praise and Criticism," *Baptist Press*, 10 February 2006.

84. Ibid.

85. Ibid.

86. Ibid.

87. "Air Force Retreats from Religious Guidelines after Religious Right Push," *Church and State*, 1 March 2006.

88. Joint Conference, House and Senate Armed Services Committees, addressing prayer in the armed forces, 19 September 2006.

89. Ibid.

90. Ibid.

91. US Air Force, "Revised Interim Guidelines"; and SECNAVINST 1730.7C, *Religious Ministry within the Department of the Navy*, 21 February 2006.

92. Joint Conference, addressing prayer in the armed forces.

93. Ibid.

94. Michael W. Wynne, secretary of the Air Force, to ALMAJCOM-FOA-DRU/CC, memorandum, 21 November 2006.

95. US Air Force, "Report of the Headquarters Review Group", iii.

96. In fact, it is possible that some may argue that the August 2005 iteration of the "Interim Guidelines" contained enough detail to border on violating the entanglement prong of the *Lemon* test. The February 2006 iteration suffers from no such excess of detail; the length has been cut from four pages to one page, while critics and supporters alike are left wondering exactly what the guidance means.

97. See *Parker v. Levy*, 733.

98. See *United States v. Duga*, in *Military Justice Reports*, vol. 10 (1981), 206 for discussion of coercion in the context of the right against self-incrimination.

99. *Evangelize* means to preach the gospel or to convert to Christianity. *Proselytize* means to induce someone to convert to one's own religious faith. *American Heritage Dictionary*, 4th ed., 2006.

100. *United States v. Brown*, in *Military Justice Reports*, vol. 45 (1996), 395.

101. The old adage, "There's no such thing as a casual conversation with a general officer," comes to mind.

102. For example, a colonel discussing religious topics in the minutes before an official staff meeting begins would likely be perceived by a captain who is required to attend the meeting as speaking in an official capacity.

103. Accommodation of religious practices is not covered by the suggested rules, as that has been covered by DOD Instruction 1300.17.

Bibliography

Alleged Misconduct by DOD Officials Concerning Christian Embassy. Department of Defense Inspector General Report No. H06L102270308, 20 July 2007.

Cooperman, Alan. "Navy Chaplain Guilty of Disobeying a Lawful Order." *Washington Post*, 15 September 2006.

Congressional Record. *United States House of Representatives and Senate Armed Services Committees, Joint Conference*, 19 September 2006.

Department of Defense. *Alleged Misconduct by DOD Officials Concerning Christian Embassy*. Inspector General Report No. H H06L102270308, 20 July 2007.

Dobosh, William J., Jr. "Coercion in the Ranks: The Establishment Clause Implications of Chaplain-Led Prayers at Mandatory Army Events." *Wisconsin Law Review* 6 (2007): 1493–1562.

Employment Division, Department of Human Resources of Oregon v. Smith, in *United States Supreme Court Reports*. Vol. 494, 1990.

Fagin, Barry S., and James E. Parco. "The One True Religion in the Military." *The Humanist*, September–October 2007.

Fitzkee, David E., and Linell A. Letendre. "Religion in the Military: Navigating the Channel between Religion Clauses." *Air Force Law Review* 59 (2007): 1.

Goodstein, Laurie. "Air Force Academy Staff Found Promoting Religion." *New York Times*, 23 June 2005.

Jordan, Bryant. "Evangelicals Threaten to Remove Chaplains from Military." *Air Force Times*, 2 January 2006.

Lee v. Weisman, in *United States Supreme Court Reports*. Vol. 505, 1992.

Lemon v. Kurtzman, in *United States Supreme Court Reports*. Vol. 403, 1971.

Lichtblau, Eric. "Air Force Looks into 'Inspirational' Video." *New York Times*, 14 March 2009.

Lubold, Gordon. "Religion on Their Sleeves?" *Air Force Times*, 25 December 2006.

MacFarquar, Neil. "Speakers at Academy Said to Make False Claims." *New York Times*, 7 February 2008.

McCreary County v. ACLU, in *Supreme Court Reporter*. Vol. 125, 2005.

Military Religious Freedom Foundation, Joint MRFF/NAACP. To Secretary of Defense. Press release letter, 16 October 2008.

Military Religious Freedom Foundation and Specialist Dustin Chalker v. Secretary of Defense Gates, US District Court for the District of Kansas, filed 25 September 2008.

Military Religious Freedom Foundation and Specialist Jeremy Hall v. Secretary of Defense Gates, U.S. District Court for the District of Kansas, filed 5 March 2008.

Milonig, Lt Col William. *The Impact of Religious and Political Affiliation on Strategic Military Decisions and Policy Recommendations*, Research Report. Carlisle Barracks, PA: US Army War College, March 2006.

Mraz, Steve. "Alconbury Colonel Cleared of Promoting Religion." *Stars and Stripes*, 2 April 2009.

Neem, Johann. "Beyond the Wall: Reinterpreting Jefferson's Danbury Address." *Journal of the Early Republic* 27, no. 1 (Spring 2007): 139.

Parker v. Levy, in *United States Supreme Court Reports*. Vol. 417, 1974.

Roach, Erin. "Air Force Religion Guidelines Garner Both Praise and Criticism." *Baptist Press*, 10 February 2006.

Santa Fe Independent School District v. Doe, in *United States Supreme Court Reports*. Vol. 530, 2000.

Schweiker, Kenneth J. "Military Chaplains: Federally Funded Fanaticism and the United States Air Force Academy." *Rutgers Journal of Law and Religion* 8, no. 1 (Fall 2006).

US Air Force. "Interim Guidelines Concerning Free Exercise of Religion in the Air Force," August 2005.

US Air Force. *Report of the Headquarters Review Group Concerning the Religious Climate at the U.S. Air Force Academy*, 22 June 2005.

US Air Force. "Revised Interim Guidelines Concerning Free Exercise of Religion in the Air Force," 9 February 2006.

Wilcox, Clyde. "Laying up Treasures in Washington and in Heaven: The Christian Right and Evangelical Politics in the Twentieth Century and Beyond." *Magazine of History* 17, no. 2 (January 2003): 3.

Winn, Patrick. "Religion at Issue." *Air Force Times*, 11 August 2008.

Wynne, Michael. ALMAJCOM-FOA-DRU/CC Memorandum, 9 February 2006.

About the Author

Maj Paula M. Grant is the staff judge advocate at the 8th Fighter Wing, Kunsan AB, Republic of Korea. She has served in various legal positions at base, numbered air force, and major command levels, including area defense counsel and assistant professor of law at the United States Air Force Academy. She deployed to Kenya in 2000 and Afghanistan in 2006. Major Grant is a graduate of Air Command and Staff College, Wellesley College, and Northeastern University School of Law. She is a member of the Colorado and Massachusetts bars.

CHAPTER 4

THE ONE TRUE RELIGION
IN THE MILITARY

James E. Parco
Barry S. Fagin

Over the past several years, the US Air Force Academy (USAFA) has been under scrutiny for issues of religious tolerance that have caused many to wonder, What on earth is going on at that place? On the one hand, the same thing is happening at USAFA that's happening at colleges across the United States. Students are leaving home (many for the first time) and embarking on individual journeys of self-discovery, meeting new people from different backgrounds with different perspectives, and engaging with trained faculty who will strive to motivate each of them to discover life's truths for themselves. On the other hand, unit cohesion, morale, and the US Constitution have all been challenged at USAFA by a growing evangelical Christian community that espouses a duty to proselytize to non-Christians and to the "unchurched."

The media has done a fairly thorough job identifying cases of religiously intolerant behavior at USAFA and also on the military's response and official findings (examples also listed in accompanying timeline). In the popular press, Mikey Weinstein's 2006 book *With God on Our Side* offers a very personal and impassioned portrayal of the evolution of the Academy's evangelical climate. Our aim here isn't to retell the stories that brought us here, but rather to provide a larger context to help explain why these issues occurred and suggest appropriate action.

The Air Force Academy, located in Colorado Springs, Colorado, is quite similar to many other small colleges. With a student body of 4,300, there are approximately 530 faculty members, many with terminal degrees. The core curriculum requires 90-plus credit hours in the humanities, social sciences, engineering, and basic sciences. Students have the opportunity to select most of the majors available at any world-class institution of higher learning, and many of them are accredited by national scholarly associations.

But it isn't the similarities between the Academy and other colleges that help one to understand the genesis of problems, but rather the profound differences. Unlike other universities, military academies (West Point and Annapolis included) are part of the armed forces and so hire 100 percent of their students after graduation (many of whom stay on the job for the next 20 years). This places

---

Originally published in *The Humanist*, July 2007.

an additional responsibility on military academies to ensure that each admitted student is "acceptable" to work for and alongside other commissioned officers.

Additionally, students (cadets) at the academies are considered constantly "on duty" and thus live and work in the same environment. Although in most cases college students are free to do as they choose once they're off campus, cadets aren't. They have, at best, limited authority to criticize or speak their minds, and, typically, the only allowable place to address a grievance is through an individual's chain of command. But what if the grievance is within that chain of command? Other avenues such as the Office of the Inspector General or the local Military Equal Opportunity Office exist, but many cadets are unaware of them. And those who do know about them are often reticent to "complain."

Given the homogeneity among the military academies, one still wonders why the Air Force Academy has had publicly visible religious tolerance issues arise, whereas the US Military Academy (West Point) and US Naval Academy have not. Clearly the large evangelical presence in Colorado Springs is a contributing factor. Colorado Springs is home to Focus on the Family, The Navigators, New Life Church, and dozens of other evangelical Christian groups. Beyond these influences, a systems perspective is required to understand the underlying fundamental issues at the Academy.

In truth, USAFA is an amazing place. Located on some of the most beautiful real estate in Colorado, it attracts some of the most capable and dedicated staff (comprised of military officers, noncommissioned officers, and civilians) devoted to the development of recent high-school graduates into second lieutenants capable of serving in the Air Force. The Academy is well funded, and its institutional processes are well established. So how is it possible that there could be scandals of sexual harassment and religious intolerance there?

Part of the answer is simple but not obvious: structural instability. The Air Force embraces a culture of mobility, and for good reason. In today's security environment, it's essential that military forces be able to operate globally in joint operations and readily execute their missions. Thus, to ensure that the personnel base has a requisite variety of experiences, the human resources function routinely moves its personnel from place to place in the spirit of "professional development." Every two to four years, officers (primarily) move to new jobs in order to gain a broad base of experiences sufficient to readily adapt to complex and uncertain environments. The philosophy is that by having a wide range of experiences, the individual will be a more capable commander when reaching that point in his or her career. The Academy's military staff and faculty are included in this model of constant turnover.

The dilemma here is that USAFA is a developmental educational institution. Its focus is to transform the student population from kids to adults, from civilians to officers, from diverse backgrounds and perspectives to a single, shared philosophy. To do this, a high degree of expertise in the various mission elements of military training, academics, and athletics is required. But because the majority of personnel brought to the Academy are active-duty and noncommissioned officers from the line of the Air Force, very few to none of the new commanders, new faculty, or new staff have sufficient experience or expertise in the areas to which they are being assigned to be immediately effective. As an example, each year 50 percent of the commanders of the cadet squadrons are new, and none of them have ever been commanders before. Similarly, each year 25 percent or more of the faculty are new. The vast majority don't have terminal degrees in the teaching area assigned, and most have never been instructors before. The key USAFA staff positions over the past decade show a similar pattern of constant turnover. This means that the students, particularly those in the upper classes, tend to be the most experienced collective body at the institution.

Like at any school, intolerance, harassment, bigotry, cheating, and other bad behaviors exist. The Academy actively pursues a diverse student body from all over the country and recognizes that because each class brings with it many influences from varied environments, conflicts between students along their individual paths of development will occur. But sufficient structures should be in place to facilitate their learning.

One of the axioms of organizational theory is that "every system is perfectly designed to yield the behaviors observed." So when issues of harassment and intolerance arise, the cadets can't be blamed entirely. The organizational structure must be analyzed to make the necessary changes.

To the Academy's credit, it has always been transparent about conflicts that have arisen there. While the school has made some progress in this area, we submit it hasn't been enough. Scandals involving sexual harassment and religious intolerance resulted largely as an effect of a culture that had developed within the cadet wing. Regrettably, few officers, faculty, or staff were around long enough to understand that culture, identify its problems, and work to change the behaviors.

My God Is Bigger Than Your God

US military officers take an oath of allegiance to one thing—not to the president or to the nation generally, but to the US Constitution. And, as guaranteed by the Constitution, there is absolutely no requirement for members of the armed forces to be of a certain skin color; a certain gender; or affiliate with, practice, or submit to any religious or spiritual beliefs.

When someone puts on a military uniform, nothing changes with his or her personal or religious beliefs. However, when people submit to wearing that uniform, they are necessarily obliged to another set of values and beliefs—a "shared religion" if you will—and that religion is patriotism, whereby their bible is the Constitution, their cross the US flag.

This so-called religion is necessary to ensure the creation of a shared reality where everyone in the military unit is included and treated with respect. Every leader, commander, and supervisor must be mindful that diversity is one of the greatest strengths in an organization. Each individual must have the freedom to appropriately express his or her views without denigrating the views of others or making others uncomfortable in the practice of their own.

Like it or not, this is precisely the fine line the framers drew for us to walk by way of the First Amendment.

Some have challenged the Academy, alleging that their religious beliefs require them to testify to the truth of those beliefs and that to prevent such testimony would limit their freedom of religious expression. Prior to 2005, a recurrent example was an annual advertisement purchased by staff and faculty during the Christmas holiday season and published in the school (base) newspaper. The full-page advertisement included the words "We believe that Jesus Christ is the only real hope for the world. If you would like to discuss Jesus, feel free to contact one of us!" The ad then listed the names of over 200 faculty and staff of the Air Force Academy, including many senior leaders. Although it's doubtful that anyone meant for the advertisement to be anything other than a friendly holiday greeting, it ended up identifying the evangelical Christians in each organizational element. Once any form of organizational power is attached to a particular belief structure and this belief structure is promoted by organizational superiors, it becomes a basis for a discriminating environment. Since proselytizing is part of the evangelical Christian belief system, do those who subscribe to it have the right to proselytize?

The First Amendment tells us the answer is yes. However, it also instructs us that when there is a power differential between superior and subordinate (regardless of on- or off-duty status), there can be no forcible discussion of religious beliefs, as this could be perceived as an official government endorsement and promotion of a particular belief system. In today's military and political environment, it has never been so important to advocate for the rights of all within the military rank and file to believe as they choose without oppression by superiors. The Constitution is clear on this one—the government will neither entangle itself in nor endorse any religious beliefs. You always have the right to swing your fists (off duty), but remember, those rights stop at the tip of my nose.

The Unique Challenges Posed to Evangelical Christians in the Military

We can gain insight into the need for change by understanding the unique challenges evangelical Christians face in a military environment. On the one hand, members of the military live with the fact that they could be asked to surrender their lives at any moment. Those who see combat face life and death issues on a regular basis and are forced to grapple with the fundamental questions of existence in a way those they protect will never face. This means that for many in the military, if not most, religion is part and parcel of their original decision to serve, their loyalty to country and family, and their source of strength in times of great stress. While the shared military "religion" of patriotism and loyalty to the Constitution are the only common requirements for military service, it's unrealistic to expect the spiritual beliefs of soldiers to vanish once they put on a uniform. Indeed, the explicit enforcement of such a requirement prior to enlistment would likely cause the armed forces to shrink to unacceptable levels.

None of this is a problem for faith traditions that don't proselytize. However, for those in uniform who claim certainty regarding untestable claims and a religious obligation for others to share that certainty, tremendous problems arise. Consider the following set of religious beliefs:

1. One faith exclusively possesses the truth of an untestable claim, and all other faiths are false.

2. Eternal life is the reward for believers in the one true faith.

3. Eternal hell is certain for everyone else.

4. It is required to share this belief with others.

5. It is ultimately incompatible to associate with unbelievers.

The more of these principles a military leader accepts, the more he or she will find leadership challenges lurking around every corner. As you work your way down the list, you are faced with increasing social, moral, and especially constitutional quandaries.

If, for example, someone believes that his faith tradition makes people better human beings, who among his colleagues is he more likely to trust? It goes against everything we know about human nature, especially adolescent human nature, to assume that members of one evangelical faith tradition won't be disposed to prefer members of that same tradition. USAFA cadets of minority faiths have expressed exactly this concern with regard to both their daily lives and their future careers in the military. The military requires teamwork, trust, and equal confidence in everyone in uniform in order to do its job. Special treatment based on race, religious belief, or any other factor unrelated to performance is inimical to morale, is harmful to the unit, and jeopardizes the mission. On purely pragmatic grounds, we would argue that the impact of theological disputes on mission effectiveness is one of the most important principles that should guide the regulation of religious speech in the military.

What Is to Be Done?

To address the unique challenges presented by evangelism in the military, we propose changes in three areas: structure, demographics, and culture.

If the Air Force Academy is serious about canceling its membership in the "scandal-of-the-month" club, it must recognize that its responsibility for 4,300 18- to 24-year-olds who seek a college education makes it fundamentally different from other Air Force bases. Professional staff must have greater latitude to engage controversial topics, including but not limited to religion, in the best traditions of Western intellectual inquiry. Staff should also remain at the Academy long enough to accumulate the necessary expertise to mentor young people, to understand appropriate guidelines for religion in the military, and to enforce them from positions of credibility and expertise. Accordingly, we propose that the superintendent (the highest ranking USAFA official and a three-star general) should serve a minimum of six years, which is a typical length of time for a college president. He or she should also have the authority to reduce the mobility of his or her support staff without any repercussions to their careers. Likewise, the commandant of cadets (one of two one-star generals ranking directly under the superintendent) should serve a minimum of five years.

The issue of greater tenure for faculty must also be addressed as a remedy for structural instability. The US Naval Academy has tenured civilian faculty, as well as senior military professors. The US Military Academy at West Point has academy professors to likewise ensure continuity and experience. Individuals, once appointed to these positions, can be expected to remain at their respective academy for the bulk of their professional careers and can develop the expertise necessary to provide continuity and leadership through difficult challenges. USAFA, by contrast, has neither. Two relatively modest proposals to provide four-year rolling appointments for USAFA civilian faculty and increase assignments for military doctoral faculty are steps in the right direction.

In addition to moving these proposals forward, civilian faculty members who have been at USAFA for over 10 years (fortunately, that number is growing) should be given a greater role in Academy governance. They represent an untapped wealth of institutional memory and professional experience that, if properly utilized, can go a long way towards effective leadership on the difficult issue of religious expression at a military academy. Similarly, the existing professional development path for Air Force officers who wish to become long-term academics at the Academy should be expedited, approved, and put in place.

Most of the issues concerning religious intolerance and possible unconstitutional actions in the military can be laid at the feet of demographics. Evangelical Protestant Christianity is disproportionately represented at various levels of the military and the chaplain corps; other faiths, along with individuals who profess no affiliation or no religion at all, are underrepresented. (The United States, for example, is approximately 80 percent Christian, while 92 percent of USAFA cadets are. Jews make up 0.4 percent of the Air Force but 1 percent of the United States, and while 10 percent of the US population professes no religion, only 0.6 percent of the Air Force does.) Some have speculated this is an artifact of the post-Vietnam era, when mainline religious denominations that opposed the war dropped out of the chaplain corps, while evangelicals saw the military as a "mission field" and an opportunity to expand their influence. Regardless of the reasons, it seems clear that a greater balance among religious perspectives can only benefit the armed forces. There is no reason, as far as we know, why the military can't more aggressively recruit those from underrepresented religious traditions, including Jews, Catholics, Muslims, and atheists. Such diversity would dissuade religious assertions and improve teamwork, cohesiveness, and the military mission overall.

In an environment like the military, ritual and symbolism are just as important as structure, perhaps even more so. Mission statements and guidance from the senior leadership, even if they seemingly state the obvious, matter a great deal. In this regard, much of the sense of isolation felt by junior military members who don't share the views of the religious majority would be eased if they could be reassured of a few seemingly obvious but critical points.

The biggest issue for nonmajority military members is the perception, whether well founded or not, that they are seen as second-class citizens, soldiers, and human beings. Statements from commanders and senior leadership throughout the past few years have not effectively addressed this concern. Beyond the mere platitudes about respect, dignity, and teamwork, a direct and forceful affirmation of an essential aspect of military service is needed: All men and women in uniform operate under the same presumption of high ethical standards, loyalty, patriotism, and integrity, regardless of professed religious belief or lack thereof.

The Oath of Equal Character

We would therefore like to see all officers in positions of command publicly attest to the truth of the following statement. We call it the "Oath of Equal Character." (Note: The oath is written from a Christian's perspective, but we would expect *Muslim, Jew, atheist, Buddhist, Hindu, Wiccan, nontheist,* or any other chosen identification to be inserted as applicable.)

> *I am a <Christian>. I will not use my position to influence individuals or the chain of command to adopt <Christianity>, because I believe that soldiers who are not <Christians> are just as trustworthy, honorable, and good as those who are. The standards of those who are not <Christians> are as high as mine. Their integrity is beyond reproach. They will not lie, cheat, or steal, and they will not fail when called upon to serve. I trust them completely and without reservation. They can trust me in exactly the same way.*

It does no good to say, as some clearly will, that the above states the obvious. Our interaction with cadets and officers from nonevangelical, nonmajority faith traditions tells us that they believe their character is impugned on a regular basis because of their differing belief system. If something like the statement above had been articulated clearly and forcefully from the senior leaders at the Air Force Academy, from all Air Force chaplains, and indeed from all Air Force commanders, the religious climate of the Air Force would be very different—and better—today.

Consider, for example, how the following actual situations might have been different had the Oath of Equal Character been involved:

- In 2004 flyers promoting Mel Gibson's *The Passion of the Christ* were placed on tables at the Academy's dining facility during the mandatory lunch formation. What if they had been accompanied by copies of the Oath of Equal Character?

- PowerPoint slides at a succeeding lunch formation intended to address religious issues displayed New Testament verses. What if instead they had displayed the Oath of Equal Character?

- Some USAFA instructors are alleged to have begun classes with a statement of faith and/or started examinations with prayer. What if classes had spent time discussing the Oath of Equal Character instead?
- What if, instead of asserting the Air Force chaplaincy's "right to evangelize the unchurched" in a 12 July 2005 New York Times article, the two-star general and head chaplain of the Air Force had recited the Oath of Equal Character?

Beliefs remain a right and a privilege, and freedom of conscience is among the oldest and most precious freedoms enshrined in the history of America's founding. But all members of the armed forces have taken an oath of allegiance to the Constitution of the United States. If they believe that their comrades who don't share their religious beliefs aren't as good as those who do, then they should leave the military and seek another career. Equating the morality of all to the religion of some is incompatible with ensuring effective armed forces for the United States of America.

Timeline

April 2003: An e-mail message goes out to all Air Force Academy (USAFA) cadets, faculty, and staff from senior leadership promoting the National Day of Prayer. It includes the directive: "Ask the Lord to give us the wisdom to discover the right, the courage to choose it, and the strength to make it endure. The Lord is in control. He has a plan for each and every one of us. If we seek His will in our lives, we will find the 'peace that passes all understanding.' May God bless the Air Force Academy, our great Air Force, this great nation, and you."

December 2003: The Christian Leadership Ministries (a division of the Campus Crusade for Christ) publishes an annual advertisement in *The Academy Spirit*, the USAFA base newspaper, as they've done for the previous 12 years. The full-page advertisement includes the message: "We believe that Jesus Christ is the only real hope for the world. If you would like to discuss Jesus, feel free to contact one of us!" The ad then lists the names of over 200 faculty and staff, including many senior leaders.

February 2004: Based on write-in comments in the annual faculty and staff climate survey citing concerns of religious insensitivity, the superintendent directs his staff to start looking into potential problems in this area. Around the same time, thousands of flyers promoting the movie *The Passion of the Christ* appear in the cadet academic and dining facilities. This garners major attention and catalyzes the need for senior leadership to address the appropriate role of religion in official duty environments.

February 2004: Multiple internal inquiries and investigations are made to learn the extent of religious bias, proselytizing, and discrimination within the organization. During this period, experts from the Yale Divinity School are brought in to observe and comment on the pastoral care provided during basic cadet training, applicants' initial introduction to the USAFA curriculum.

November 2004: The USAFA chaplaincy unveils a new training program called Respecting the Spiritual Values of Persons (RSVP). Shortly thereafter, the head football coach displays a banner in the locker room that reads: "I am a member of Team Jesus."

November 2004: The acting secretary of the Air Force directs a task force from the Pentagon to visit USAFA and prepare a report regarding the religious climate.

January–May 2005: All cadets, faculty, and staff complete the 50-minute RSVP training. RSVP II, the second in a proposed series of training sessions on religious respect, is announced.

May 2005: A Protestant chaplain resigns her commission and speaks out in the major media against the established practices of proselytizing at USAFA.

June 2005: The Air Force issues its *Report of the Headquarters Review Group Concerning the Religious Climate at the U.S. Air Force Academy.*

June–August 2005: A committee of academics is assembled to create the RSVP II training.

September 2005: The Air Force releases "Interim Guidelines Concerning Free Exercise of Religion in the Air Force."

October 2005: Former cadets (including Michael Weinstein) file a lawsuit against the Air Force for religious discrimination. The Air Force then withdraws a document previously circulated at the Chaplain School that included the statement: "I will not proselytize from other religious bodies, but I retain the right to evangelize those who are not affiliated."

November 2005: Senior leadership at USAFA changes over.

October 2006: Congress repeals Air Force and Navy guidelines on religion. Three days later, the Air Force releases new guidelines. A federal court throws out Weinstein's suit on grounds that graduates couldn't claim their First Amendment rights were violated since they no longer attended the Academy. Weinstein vows to re-file a more expansive suit in federal court.

April 2007: USAFA hosts a debate between Weinstein and Jay Sekulow (American Center for Law and Justice) on finding the balance between religious freedom and official neutrality in the military.

July 2007: The Office of the Inspector General publicly releases a report finding high-ranking Army and Air Force personnel violated regulations when they participated in a promotional video for the Christian Embassy while in uniform and on active duty.

About the Authors

At the time of original publication, **Lt Col Jim Parco** was an associate professor at the United States Air Force Academy. His research in the areas of game theory, management, strategic interaction, and complex systems has been published in the top journals in economics, management, and psychology. He was the 2007 recipient of the Thomas Jefferson Award for the Preservation of Religious Freedom.

Dr. Barry Fagin graduated magna cum laude from Brown University in 1982 and received a PhD in computer science from the University of California at Berkeley in 1987. He is currently professor of computer science at the Air Force Academy in Colorado Springs. Throughout his career, Dr. Fagin has maintained a lifelong interest in connecting the world of ideas to the world of politics. He is a Fulbright Scholar and the founder of Families Against Internet Censorship, a successful plaintiff in the Supreme Court case of *Reno v. ACLU*, for which he later received the National Civil Liberties Award. He is a featured columnist in the *Colorado Springs Gazette*, and since the original publication of this article, a recipient of the 2009 Thomas Jefferson Award for the Preservation of Religious Freedom. Dr. Fagin is the author of over 30 scholarly papers covering areas of computer science and public policy. He is the co-inventor of the Crandall-Fagin multiplication algorithm, used to discover the world's largest prime numbers.

CHAPTER 5

AGAINST ALL ENEMIES, FOREIGN AND DOMESTIC

Chris Rodda

Top 10 Ways to Convince the Muslims We're on a Crusade

10. Have Top US Military Officers, Defense Department Officials, and Politicians Say We're in a Religious War.

We couldn't have gotten off to a better start on winning hearts and minds back in 2003, when US Army Lt Gen William "Jerry" Boykin decided to go on a speaking tour of churches, publicly proclaiming in uniform that the global war on terrorism (GWOT) was really a battle between Satan and Christians, and making comments like, "We in the Army of God, in the House of God, the Kingdom of God have been raised for such a time as this." Of course, Boykin knew what he was talking about. After all, a decade earlier he had captured the dangerous Somali warlord Osman Atto and was very clear about the reason that happened—"I knew that my God was a real God, and his was an idol."

President Bush, in spite of the fact that Boykin believed he was "in the White House because God put him there," wasn't too pleased with these remarks, but still, the general's friends stood by him—friends like then-Cong. Robin Hayes (R-NC), who, speaking at a Rotary Club meeting in his hometown a few years later, pronounced that stability in Iraq ultimately depended on "spreading the message of Jesus Christ, the message of peace on earth, good will towards men," and "everything depends on everyone learning about the birth of the Savior."[1]

While few such statements have been as overt or widely publicized as those of Boykin and Hayes, plenty of other military leaders and policy makers are on record espousing similar views. When asked what effect such statements have on the US military operations in Iraq and Afghanistan, a retired Air Force officer appearing on MSNBC in a segment about the remarks of Congressman Hayes answered:

> Well, it's not helpful if this stuff gets back to the Iraqis, and of course in the days of the internet and the blogosphere out there it's likely that it could. And you

---

Portions of this article were originally published in the *Daily Kos* on 18 September 2009.

know our troops have enough problems over there just doing their jobs. Having to defend what a U.S. congressman might say, because you know, when you bring up the idea of proselytizing Christianity, to a lot of Muslims, that's very offensive, and if we can keep religion out of what we're trying to do over there, which is very difficult, it would be a lot easier for our troops. . . . If you're trying to be a unit trainer to, say, an Iraqi battalion and the battalion religious advisor, the imam, would come in and say look what a congressman said, it just takes away from what we're trying to do.[2]

Nevertheless, some representatives of our government continue to present the war on terror as a spiritual battle, promoting the specious notion that victory in Iraq and Afghanistan is somehow necessary to preserve our own religious freedom here in America. "Thomas Jefferson would understand the threat we face today—tyranny in the name of religion," asserted a top Army official at a West Point graduation ceremony. "Your sons and daughters are fighting to protect our citizens . . . from zealots who would restrain, molest, burden, and cause to suffer those who do not share their religious beliefs, deny us, whom they call infidels, our unalienable rights."[3] And, finding it vitally important for Congress to recognize "the importance of Christmas and the Christian faith," another congressman made his case: "American men and women in uniform are fighting a battle across the world so that all Americans might continue to freely exercise their faith."[4] As of yet, nobody making such statements has offered any explanation of *how* the outcome of this war could possibly affect the free exercise of religion by Americans.

9. Have Top US Military Officers Appear in a Video Showing Just How Christian the Pentagon Is.

In addition to providing propaganda material to our enemies, public endorsements of Christianity by US military leaders can also cause concern among our Muslim allies. It might have seemed like a good idea at the time, but the situation became very awkward for Air Force Maj Gen Pete Sutton shortly after he appeared in a promotional video for the Christian Embassy.[5] Dressed in uniform and using their official titles, several high-ranking military officers and DOD civilians gave testimonials and made statements such as "we're the aroma of Jesus Christ," which were publicly available on the Christian Embassy's Web site. What Sutton didn't know when he appeared in this video was that he would soon be assigned as the US European Command's chief of defense cooperation to Turkey, a country in which religion and government are strictly separated. According to the DOD Inspector General's report on the investigation of allegations relating to the video:

> Maj Gen Sutton testified that while in Turkey in his current duty position, his Turkish driver approached him with an article in the Turkish newspaper 'Sabah.' That article featured a photograph of Maj Gen Sutton in uniform and described him as a member of a radical fundamentalist sect. The article in the online edition of Sabah also included still photographs taken from the Christian Embassy video. Maj Gen Sutton's duties in Ankara included establishing good relations with his counterparts on the Turkish General Staff. Maj Gen Sutton testified

that Turkey is a predominantly Muslim nation, with religious matters being kept strictly separate from matters of state. He said that when the article was published in Sabah, it caused his Turkish counterparts concern, and a number of Turkish general officers asked him to explain his participation in the video.[6]

Unfortunately, there is no shortage of uniformed military personnel endorsing fundamentalist Christian organizations and military ministries, some of which have clearly publicized missions that include proselytizing Muslims. These videos are easily found on the Internet, providing plenty of potential propaganda material for recruiting by extremists.

8. Plant Crosses in Muslim Lands and Make Sure They're Big Enough to Be Visible from Really Far Away.

As Gen Norman Schwarzkopf recounted in his autobiography, *It Doesn't Take a Hero*, back in 1990, when US troops were deployed to Saudi Arabia for Operation Desert Shield, an attempt by a Christian missionary organization to use the military to proselytize Saudi Muslims led the Pentagon to issue strict guidelines on religious activities and displays of religion in the region. It was left to the discretion of individual company commanders to determine how visible religious services should be, depending on their particular location's proximity to Saudi populations. In some cases, decisions were made not to display crucifixes or other religious symbols, even at worship services. There were a few complaints about these decisions, but the majority of the troops willingly complied, understanding that these decisions were being made for their own security. According to General Schwarzkopf, even his request that chaplains refrain from wearing crosses on their uniforms received an unexpectedly positive reaction, with the chaplains not only agreeing with the policy, but also going a step further by calling themselves "morale officers" rather than chaplains.

But now, in Iraq and Afghanistan, General Schwarzkopf's commonsense policies and priority of keeping the troops safe have been replaced by a flaunting of Christianity by Christian troops and chaplains who feel that nothing comes before their right to exercise their religion, even if it means putting the safety of their fellow troops at risk. Numerous photos, some posted on official military Web sites, show conspicuously displayed Christian symbols, such as large crosses, being erected on and around our military bases in Iraq and Afghanistan.[7]

Large Christian murals have been painted on the outside of the T-barriers surrounding a chapel on Forward Operating Base (FOB) Warhorse in Iraq. In addition to being a highly visible display of Christianity to Iraqis on the base, photos of these murals were posted on an official military Web site.[8] It is even more important that the regulation prohibiting displays of any particular religion on the grounds of an Army chapel—a regulation that protects the religious freedom of our Soldiers by keeping chapels neutral and welcoming Soldiers of all faiths—be strictly enforced on our bases in Iraq and Afghanistan. Yet there is clear and credible evidence that those in charge routinely overlook such regulations.

7. Paint Crosses and Christian Messages on Military Vehicles and Drive Them through Iraq.

For those Iraqis who may not see the overt stationary displays of Christianity on and near US military bases in their country, there have been plenty of mobile Christian messages painted on our tanks and other vehicles that patrol their streets.

The title of Jeff Sharlet's May 2009 *Harper's Magazine* cover story, "Jesus Killed Mohammed: The Crusade for a Christian Military," actually comes from one such vehicular message—the words "Jesus killed Mohammed" were painted in *large red Arabic lettering* on a Bradley fighting vehicle, drawing fire from nearly every doorway as it was driven through Samarra. Other vehicles have sported everything from the Islamic crescent overlaid with the internationally recognized red circle and slash "no" sign to large crucifixes hanging from gun barrels. A military public relations office even officially released a photo of the tank named "New Testament."[9]

6. Make Sure That Our Christian Soldiers and Chaplains See the War As a Way to Fulfill the Great Commission.

To many fundamentalist Christians, the "Great Commission" from Matthew 28:19—"Go and make disciples of all nations"—trumps all man-made laws, including military regulations. It's hard to find a military ministry whose mission statement doesn't, in one way or another, include fulfilling the Great Commission. Thus, it is not surprising that many service members who've been influenced by these military ministries are conflicted about their mission, a conflict often leading some of these service members to disregard the military's prohibition on proselytizing.

Campus Crusade for Christ's (CCC) Military Ministry,[10] a parachurch ministry active at all of the largest US military training installations, the service academies, and on ROTC campuses, frequently states its goal of turning the US military into a force of "government-paid missionaries for Christ." The vision statement of another organization, Military Missions Network,[11] is "an expanding global network of kingdom-minded movements of evangelism and discipleship reaching the world through the military of the world."

Describing the duties of a CCC Military Ministry position at Lackland Air Force Base and Fort Sam Houston in Texas, for example, the organization's Web site stated, "Responsibilities include working with Chaplains and Military personnel to bring lost soldiers closer to Christ, build them in their faith and send them out into the world as government paid missionaries."[12]

CCC's Valor ministry,[13] which primarily targets future officers on ROTC campuses, states, "The Valor ROTC cadet and midshipman ministry reaches our future military leaders at their initial entry points on college campuses, helps them grow in their faith, then sends them to their first duty assignments throughout the world as 'government-paid missionaries for Christ.'"[14]

In a promotional video filmed at the US Air Force Academy, a USAFA CCC program director pronounced that CCC's purpose is to "make Jesus Christ the issue at the Academy," and for the cadets to be "government paid missionaries" by the time they leave.[15]

According to a CCC Military Ministry instructional publication uncovered in 2007, CCC's mission is not simply to provide Bible studies to allow Christians in the military to exercise their religion, as its defenders claim. The instructions state, "We should never be satisfied with just having Bible studies of like-minded believers. We need to take seriously the Great Commission."[16]

Whatever one's position on the issue of evangelism, the undeniable fact is that all of the above quotes, as well as the video filmed at the Air Force Academy, were found on the Internet, which, of course, means that any extremist looking for recruiting tools could also find this easily accessible "evidence" that the US military is being groomed to be a force of crusaders.

5. Post Photos on the Internet of US Soldiers with Their Rifles and Bibles.

CCC's indoctrination of basic trainees at Fort Jackson, South Carolina, the Army's largest basic training installation, is a program called "God's Basic Training," in which the recruits are taught that "The Military = 'God's Ministers'" and that one of their responsibilities is "to punish those who do evil" as "God's servant, an angel of wrath."[17]

Until being exposed (and taken down), the Fort Jackson CCC Military Ministry had a Web site containing not only its Bible study materials, but also numerous photos of smiling trainees posed with their rifles and Bibles.[18] Obviously, no explanation is necessary to see the propaganda value of photos like these.

4. Invite Virulently Anti-Muslim Speakers to Lecture at Our Military Colleges and Service Academies.

In June 2007, anti-Muslim activist Brigitte Gabriel, author of *Because They Hate*, was allowed to deliver a lecture at the Joint Forces Staff College (JFSC).[19] In February 2008, the 3 Ex-Terrorists,[20] a trio of self-proclaimed former Muslim terrorists turned fundamentalist Christians, appeared at the US Air Force Academy's 50th Annual Academy Assembly, in spite of the fact that their claims about their terrorist pasts have long been questioned by both academics and terrorism experts.[21]

Gabriel's JFSC lecture, which was broadcast to the world on C-SPAN, eventually ended up on YouTube,[22] and articles about the ex-terrorists' Air Force Academy presentation, which included details such as Walid Shoebat's pronouncement that converting Muslims to Christianity was a good way to defeat terrorism, also ended up online,[23] providing yet more "evidence" to extremists that the US military's training includes teaching cadets, officers, and senior noncommissioned officers (NCO) that Islam is evil and must be stopped.

3. Have a Christian TV Network Broadcast to the World That the Military Is Helping Missionaries Convert Muslims.

Travel the Road, a popular Christian reality TV series that airs on the Trinity Broadcasting Network (TBN), follows the exploits of two "extreme" missionaries who travel to remote, and often dangerous, parts of the world to fulfill their two-part mission to "(1) Vigorously spread the gospel to people who are either cut off from active mission work, or have never heard the gospel," and "(2) Produce dynamic media content to display the life of missions, and thus, through these episodic series electrify a new generation to accomplish the Great Commission."

The second season of the series ended with three episodes filmed in Afghanistan. To film these episodes, the missionaries were embedded with US troops as "journalists," staying on US military bases and accompanying and filming troops on patrols—all for the purposes of evangelizing Afghan Muslims and producing a television show promoting the Christian religion. As the first of the program's three Afghanistan episodes clearly showed, these missionaries were able to waltz into Afghanistan without any of the advance approval and planning required for embedded journalists and, within two days, be embedded with an Army unit.

A question that many will ask is whether or not the Army knew what these missionaries were up to. According to ABC News *Nightline*, which did a segment on the embedded missionaries, the answer from one of the missionaries was yes: "They knew what we were doing. We told them that we were born again Christians, we're here doing ministry, we shoot for this TV station and we want to embed and see what it was like."[24]

USCENTCOM's General Order 1A (now GO-1B) prohibits any and all proselytizing in its area of responsibility (AOR) and applies to civilians accompanying US troops as well as military personnel. Yet despite this directive, the US Army facilitated the evangelizing of Afghans by these Christian missionaries, which included the distribution of New Testaments in the Dari language. Numerous Soldiers and NCOs, as well as several officers, including one general, appeared in the program.[25]

While the Army's participation in the *Travel the Road* program is certainly one of the most prominent examples of broadcasting to the world that the US military was aiding missionaries who were trying to convert Muslims, it is regrettably not the only example.

In September 2008, the Discovery Channel's Military Channel aired a two-hour program titled *God's Soldier*. Filmed at FOB McHenry in Hawijah, Iraq, the program's credits identified that it had been "produced with the full co-operation of the 2-27 Infantry Battalion 'Wolfhounds.'" The co-producer of the program was Jerusalem Productions, a British production company whose "primary aim is to increase understanding and knowledge of the Christian religion and to promote Christian values, via the broadcast media, to as wide an audience as possible."

Bible verse text captions appearing between segments of the program included "I did not come to bring peace, but the sword" and "put on the full armor of God so that when the day of evil comes, you may stand your ground."

This was one of the prayers uttered by the program's star, CPT Charles Popov, an evangelical Christian Army chaplain, during a scene in which he was blessing a group of Soldiers about to go out on a patrol: "I pray that you would give them the ability to exterminate the enemy and to accomplish the task that they've been sent forth by God and country to do. In Christ's name I pray. Amen." That prayer was followed by a scene in which the chaplain, sounding an awful lot like the Campus Crusade Bible study described earlier, said to the Soldiers: "Every soldier should know Romans 13, that the government is set up by God, and the magistrate, or the one who wields the sword—you have not swords but 50 cals and [unintelligible] like that—does not yield it in vain because the magistrate has been called, as you, to execute wrath upon those who do evil."

The scene that tops them all, however, is one in which Popov is setting up a nativity pageant for Christmas—using the unit's Iraqi interpreters to play some of the roles. The chaplain described this as some sort of cultural exchange, with US troops recognizing Ramadan, and Muslim interpreters, in turn, celebrating Christmas. The notion of this merely being a harmless cultural exchange is absurd. US Soldiers participating in a Muslim religious observance are not risking death by doing so, while Muslims, in a country where many consider converting to Christianity a death penalty offense, are. Broadcasting to the world via the Discovery Channel that US Army personnel were putting Muslims in a Christmas pageant not only provides more fodder for radical Islam extremists, but also exposes the Iraqis who are helping the US military to grave danger.[26]

2. Make Sure Bibles and Evangelizing Materials Sent to Muslim Lands Have Official US Military Emblems on Them.

It's not hard to imagine what message is being communicated to the Iraqis and Afghans when hundreds of thousands of Bibles with official US military emblems show up in their countries. Some of these military Bibles are produced by private organizations, and others are officially authorized by the military. One of the officially distributed editions has both the Multi-National Corps-Iraq and I Corps seals imprinted on a camouflage background cover. And it doesn't stop with Bibles.[27]

A chief warrant officer from the 101st Airborne Division, for example, referring to a special military edition of a Bible study daily devotional published and donated by Bible Pathways Ministries, told Mission Network News that "the soldiers who are patrolling and walking the streets are taking along this copy, and they're using it to minister to the local residents," and that his "division is also getting ready to head toward Afghanistan, so there will be copies heading out with the soldiers." Just like the many civilian missionaries who see the wars in Iraq and Afghanistan as a window of opportunity to evangelize Muslims, the warrant officer continued, "The soldiers are being placed in strategic

places with a purpose. They're continuing to spread the Word." This daily devotional, admittedly being used by the 101st Airborne Division "to minister to
the local residents," has the official military branch seals on its cover, giving the
impression that it is an official US military publication. And while these logos
are sometimes used without permission and may have been on this particular
book, the Iraqis and Afghans don't know that.[28]

The chiefs of chaplains even designed one of the Bibles sporting the official
military logos. An organization called Revival Fires Ministries has, "at the request of the Chief Chaplains of the Pentagon," been promoting, collecting
money for, and shipping these Bibles to Iraq since 2003. A formal arrangement
between the Pentagon and Revival Fires has allowed these Bibles to be shipped
via military airlift.

To promote these Bibles, a Navy chaplain, whose own anti-Muslim book
was taken off the market when it was revealed that much of its content had
been plagiarized and some of the endorsements on its cover fabricated, has
improperly appeared in uniform at three of Revival Fires' rancorously anti-
Muslim camp meetings[29] and also endorses the ministry on the Web sites of
both its founder, Cecil Todd, and his son, evangelist Tim Todd. At one point,
the chaplain's photo and endorsement appeared right next to the following
statement on the younger Todd's Web site: "We must let the Muslims, the
Hare Krishnas, the Hindus, the Buddhists and all other cults and false religions
know, 'You are welcome to live in America . . . but this is a Christian nation . . .
this is God's country! If you don't like our emphasis on Christ, prayer and the
Holy Bible, you are free to leave anytime!'"[30]

1. Send Lots of Arabic, Dari, and Pashtu Language Bibles to Convert the Muslims.

Arguably worse than any English language Bibles stamped with official US
military emblems are the countless thousands of Arabic, Dari, and Pashtu Bibles making their way into Iraq and Afghanistan, often with the help of US
military personnel.

In his autobiography, General Schwarzkopf recounted his 1990 run-in with
one fundamentalist Christian organization—an incident that made it clear
that the Saudis' fears and complaints of Christian proselytizing were not unfounded. While some of the Saudis' fears, as the general explained, had resulted
from Iraqi propaganda about American troops disrespecting Islamic shrines,
the attempt by this religious organization to get US troops to distribute tens of
thousands of Arabic language New Testaments to Muslims was real.

> The Saudi concern about religious pollution seemed overblown to me but under-
> standable, and on a few occasions I agreed they really did have a gripe. There was a
> fundamentalist Christian group in North Carolina called Samaritan's Purse that had
> the bright idea of sending unsolicited copies of the New Testament in Arabic to our
> troops. A little note with each book read: "Enclosed is a copy of the New Testament
> in the Arab language. You may want to get a Saudi friend to help you to read it." One

day Khalid[31] handed me a copy. "What is this all about?" he asked mildly. This time he didn't need to protest—he knew how dismayed I'd be.

This was the incident that, as mentioned earlier, led to the implementation of strict guidelines on religious activities of military personnel in Muslim countries.

A recent al-Jazeera English news report showed US troops at Bagram Airfield in Afghanistan discussing the distribution of Dari and Pashtu language Bibles to the local Afghans.[32] While the US military claimed that these Bibles were destroyed and that this was an isolated incident, countless other examples seem to indicate that these incidents are anything but isolated.

In the newsletter of the International Ministerial Fellowship (IMF), an Army chaplain described the evangelizing he was doing while passing out food in the predominantly Sunni village of Ad Dawr: "I am able to give them tracts on how to be saved, printed in Arabic. I wish I had enough Arabic Bibles to give them as well. The issue of mailing Arabic Bibles into Iraq from the U.S. is difficult (given the current postal regulations prohibiting all religious materials contrary to Islam except for personal use of the soldiers). But the hunger for the Word of God in Iraq is very great, as I have witnessed first-hand."[33]

Another Army chaplain, in an article titled "Kingdom Building in Combat Boots," wrote: "But the most amazing thing is that I was constantly led to stop and talk with Iraqis working at the Coalition Provisional Authority. I learned their names, became a part of their lives, and shared Jesus Christ by distributing DVDs and Arabic Bibles."[34]

And here's one from a private organization, boasting of the help it gets from military personnel to distribute its Bibles: "OnlyOneCross.com recently sent a case of Arabic Bibles to a Brother who is working in a detention center in Iraq."[35]

Another organization, the Salvation Evangelistic Association, now has the Soldiers they converted at Fort Leonard Wood, Missouri, distributing the Arabic Bibles for them: "Many young men in training at Fort Leonard Wood were converted to Christ. The Lord led us on to preaching in Army camps in the US, Korea, and the Philippines. We are now supplying Arabic Bibles for distribution by our troops in Iraq."[36]

Then there was a lieutenant colonel, whose religious zeal was so extreme that a missionary had to explain to him that he was putting his troops at risk. The missionary's organization had already shipped 20,000 Arabic-language "Soul-Winning Booklets" into theater with more on the way. The lieutenant colonel, who knew the missionary from the states, had gone to his hotel with 15–20 armed troops and literally blocked off an entire city block with tanks and Humvees to secure the area. He offered to use his troops to protect the missionaries who were there on an evangelical mission to convert the Muslims. The missionary later remarked, "I had to tell [the lieutenant colonel] that it would probably be best if he and his unit left as soon as possible. . . . The Iraqi people in the hotel and those on the street were to say the least, very concerned. I did not want to bring that much attention to the hotel for fear that the terrorists would target the area as well."[37]

In a video from Soldiers Bible Ministry, an Army chaplain boasts about managing to get Swahili Bibles into Iraq to evangelize Muslim workers from Uganda employed by the US military, in spite of the regulations prohibiting this. Referring to this shipment of Bibles, the chaplain said, "Actually, they're in Baghdad right now. Somehow the enemy tried to get 'em hung up there. There was a threat they were gonna get shipped back to the States and all that. We prayed, and they're gonna be picked up in a couple of days. God raised someone up right there in Baghdad that's gonna go—a Christian colonel that's stationed there in Baghdad, and he's gonna go and get the Bibles."[38] Despite its disregard of military regulations, Soldiers Bible Ministry is officially endorsed by the Army's chief of chaplains, with the following statement on his Web site: "Thanks so much for your invaluable ministry of the Word to our Soldiers."[39]

In addition to Bibles, other Arabic language Christian books are being shipped into Iraq for distribution by our troops. The January 2009 newsletter of World-wide Military Baptist Missions, for example, included photos of its English-Arabic proselytizing materials, an English-Arabic New Testament, and an English-Arabic Gospel of John. This is from the caption for these photos: "In 2008, we shipped over 226,000 gospel tracts, 21,000 Bibles, New Testaments and gospels of John (to include English-Arabic ones!) and 404 'discipleship kits' to service members & churches for use in war zones, on ships and near military bases around the world."[40]

Clearly, converting the Iraqis and Afghans is a pet project of numerous private organizations, some with the help of the military, as well as military personnel and military ministries. In one case, a DOD-authorized chaplain endorsing agency actually set up a well-organized network of 40 of its chaplains in Iraq to receive and distribute Arabic Bibles and an Arabic gospel tract titled "Who Is Jesus" for a private missionary organization.[41] All of these groups and individuals have found ways to circumvent the prohibition on sending religious materials contrary to Islam into the region. There are literally thousands of people involved, and hundreds of thousands of Arabic and other native language Bibles, tracts, videos, and audio cassettes have made their way into Iraq and Afghanistan, along with Christian comic books, coloring books, and other materials to evangelize Muslim children. The line between joining the military and joining the ministry has seemingly become increasingly blurred for many.

Joining the Military = Joining the Ministry

To Campus Crusade for Christ, basic training installations and the military service academies are "gateways"—the places that young and vulnerable military personnel pass through early in their careers. This was the explanation of its gateway strategy that appeared on CCC's Military Ministry Web site: "Young recruits are under great pressure as they enter the military at their initial training gateways. The demands of drill instructors push recruits and new cadets to the edge. This is why they are most open to the 'good news.' We target specific locations, like Lackland AFB and Fort Jackson, where large numbers of military members transition early in their career. These sites are excellent locations to pursue our strategic goals."[42]

According to CCC's executive director, "We must pursue our particular means for transforming the nation—through the military. And the military may well be the most influential way to affect that spiritual superstructure. Militaries exercise, generally speaking, the most intensive and purposeful indoctrination program of citizens."[43]

At Fort Jackson, the largest Army basic training installation, trainees attending CCC's "God's Basic Training" Bible studies are taught that by joining the military, they've become ministers of God. This is also taught by CCC's Valor ministry, which targets future officers on ROTC campuses.

A Valor ministry video titled "God and the Military" is a presentation given at Texas A&M by a Texas pastor to an audience of cadets and an assortment of officers from the various branches of the military. The pastor's presentation opens:

> I, a number of years ago, was speaking at the University of North Texas—it happens to be my alma mater, up in Denton, Texas—and I was speaking to an ROTC group up there and when I stepped in I said, "It's good to be speaking to all you men and women who are in the ministry," and they all kind of looked at me, and I think they wondered if maybe I had found the wrong room, or if they were in the wrong room, and I assured them that I was speaking to men and women in the ministry, these that were going to be future officers.[44]

The stated mission of CCC's ministry for enlisted personnel is "Evangelize and Disciple All Enlisted Members of the US Military. Utilize Ministry at each basic training center and beyond. Transform our culture through the US Military."[45]

Cadence International[46] is another large military ministry that targets young service members, seeing those who are likely to be deployed to war zones as low-hanging fruit. One of the reasons given by Cadence for the success of its "strategic ministry" "Deployment and possibly deadly combat are ever-present possibilities. They are shaken. Shaken people are usually more ready to hear about God than those who are at ease, making them more responsive to the gospel."[47]

Organizations like CCC's Military Ministry and Cadence could not succeed in their goals without the sanction and aid of the military commanders who allow them to conduct their missionary recruiting activities on their installations. And there is no shortage of military officers who not only condone but also participate in and promote these activities. The Officers' Christian Fellowship, an organization consisting of over 15,000 officers and operating on virtually every US military installation worldwide, which has frequently stated its goal to "create a spiritually transformed US military with Ambassadors for Christ in uniform, empowered by the Holy Spirit,"[48] has actually partnered with CCC's Military Ministry.

In addition to the military-wide organizations like Campus Crusade, there are also a number of coercive religious programs on individual bases. A basic training schedule from Fort Leonard Wood described "Free Day Away," a program attended by all trainees during their fifth week of training, as follows: "Soldiers spend the day away from Fort Leonard Wood and training in the town of Lebanon. Free Day Away is designed as a stress relief that helps soldiers return to training re-motivated and rejuvenated."

Omitted from this event description was that this day was actually spent at the Tabernacle Baptist Church and included a fundamentalist religious service. All facilities that the trainees were permitted to go to during this free time (a bowling alley, a convenience store, etc.) are owned by the church. Numerous Soldiers have reported that they were unaware that this part of their "training" was run by a church until they were being loaded onto the church's buses that came to pick them up, and those who wanted to opt out of the church service once they were there were not permitted to do so.

While claims are made that Free Day Away and other religious programs and events conducted at basic training installations are not mandatory, these words make little or no difference to the trainees. As anyone who has gone through basic training is well aware, no trainee wants to stand out, and almost none would risk being singled out as different or difficult by speaking up and telling their drill sergeant that they don't want to attend a program or event because it goes against their religious beliefs.

Spiritual Fitness

"Spiritual fitness" is the military's new code phrase for promoting religion, and the religion being promoted is Christianity. There are spiritual fitness centers, spiritual fitness programs, spiritual fitness concerts, spiritual fitness runs and walks, and so forth.

This year, for example, Fort Eustis, Virginia, and Fort Lee, Virginia, have been holding a spiritual fitness concert series. At Fort Eustis, it's actually called the "Commanding General's Spiritual Fitness Concert Series." This is a Christian concert series. All of the performers are Christian recording artists. Photos from one of the Fort Lee concerts show crosses everywhere, and one photo's caption even says that the performer "took a moment to read a Bible passage" during her set.[49] In some cases, attendance at Christian concerts held at basic training installations has been mandatory for the Soldiers in training.[50]

In March 2008, a program was presented at a commander's call at RAF Lakenheath, England. This commander's call was mandatory for an estimated 1,000 service members, and the PowerPoint version of the presentation was e-mailed to an additional 4,000–5,000 members. The "spiritual fitness" segment of this presentation was titled "A New Approach to Suicide Prevention: Developing Purpose-Driven Airmen," a takeoff on Rick Warren's *The Purpose Driven Life*. The presentation also incorporated creationism into suicide prevention. One slide, titled "Contrasting Theories of Hope, 2 Ultimate Theories Explaining Our Existence," has two columns, the first titled "Chance," and the second "Design," comparing Charles Darwin and "Random/Chaos" to God and "Purpose/Design." Darwin, creationism, and religion are also part of a chart comparing the former Soviet Union to the United States, which concludes that "Naturalism/Evolution/Atheism" lead to people being "in bondage" and having "no hope," while theism leads to "People of Freedom" and "People of Hope/Destiny."[51]

Strong Bonds

Strong Bonds is an Army-wide evangelistic Christian program operating under the guise of a predeployment and postdeployment family wellness and marriage-training program. Strong Bonds events are typically held at ski lodges, beach resorts, and other attractive vacation spots, luring Soldiers who would never attend a religious retreat to sign up for the free vacation.

The materials officially authorized by the Army for Strong Bonds are not religious, but there's a loophole. These authorized materials are only required to be used for a minimal number of the mandatory training hours, leaving the remaining mandatory training hours open for other materials selected by the chaplain running the retreat. In some cases, the chaplains do stick to the authorized materials and keep the program nonreligious, but this is not the norm.

At one Strong Bonds weekend, the attendees, upon arrival, were handed a camouflage box called "Every Soldier's Battle Kit." This kit was imprinted with the name New Life Ministries and the ministry's phone number and Web site, and contained *The Life Recovery Bible* and four volumes by a Christian author. They were also given several Christian devotional books and *The Five Love Languages* by pastor Gary Chapman, who is described on his Web site as "the leading author in biblical marriage counseling." Pastor Chapman's book was used as the core of the Saturday portion of the training, at which a video of Chapman, full of Bible verses and a call to "love your partner like Jesus loved the church," was also shown.[52]

DOD contracts also show the frequent hiring of Christian entertainers and speakers for Strong Bonds events. One base, for example, contracted, at a cost of $38,269, an organization called Unlimited Potential, Inc.[53] to provide "social services" for a Strong Bonds event. Unlimited Potential, Inc. is an evangelical baseball ministry that has a military ministry whose mission is "to assist commanders and chaplains in providing religious support to military service members and their families by sharing the life-changing Gospel of Jesus Christ through the medium of baseball" and "to use our God-given abilities in baseball to reach those who do not have a personal relationship with Jesus Christ." This same ministry has been "serving Christ through baseball" at a number of other Army bases in the United States, as well as many bases overseas.

Godspam

The use of official military e-mail to send religious messages is another ongoing problem. These e-mails range in content from Bible verses and evangelistic Christian messages to "invitations" from superiors to worship services and Bible studies.

One recent e-mail, widely distributed to an Air Force installation's e-mail list, contained an essay by the executive director of the Officers' Christian Fellowship. The essay began by posing the question, "Why do you serve in our military?" The answer was:

> We serve our Lord by serving our nation, our family or prospective future family, and so that we have something that we can share with God's people in need. But what is the greatest need? Why do we serve our God as Joshua exhorted? We serve our God because of what Jesus did for us on the Cross. We are blessed to be able, through our lives in the military, to demonstrate the message of salvation to those who have not heard or received it. It was by God's grace through faith that we were brought fully into His family and presence. Our love for Him motivates us to serve Him in our military, to serve and work for our families, and to serve and work to enable the message of salvation to reach those who have yet to accept Him as Lord and Savior.

In another recent case, an Air Force colonel sent out an e-mail to a large number of subordinates containing a link to an "inspirational" video. Not only was the video an overt promotion of Christianity, but the Web site linked to was a far right Catholic Web site containing material attacking the president and vice president of the United States, including an image of the president depicted as Adolf Hitler.[54]

Often, command staff and NCOs forward religious e-mails to a base or a unit on behalf of a chaplain. A recent example of this was a flyer for a Bible study titled "Moses the Leader: How Would You Like to Lead 1,000,000 Whiners?" Numerous recipients of this e-mail complained about its negative stereotype of Jews, as well as the fact that it was e-mailed to the base e-mail list by command staff.

Occasionally, officers and NCOs send out e-mails inviting their subordinates to religious events that they themselves are hosting, putting the recipients in the position of wondering if not attending their superior's religious event will negatively affect their career, and if those who do attend will be shown favoritism.

For example, the Soldiers of a platoon in Iraq recently received an e-mail that had a flyer[55] attached to it for a Christian men's conference being hosted by their platoon sergeant. The flyer had the unit and division emblems on it, and the sender of the e-mail, an E-7, listed himself as a minister and the host of the event.

This platoon sergeant had been sending out religious e-mails almost daily, including one with an attachment titled "Psalm 23 (For the Work Place)," which began, "The Lord is my real boss, and I shall not want," and ended with, "When it's all said and done, I'll be working for Him a whole lot longer and for that, I BLESS HIS NAME!!!!!!"[56] Another contained several Bible verses, preceded by the following statement: "There are many things that work to keep us from completing our life-missions. Over the years, I've debated whether the worst enemy is procrastination or discouragement. If Satan can't get us to put off our life missions, then he'll try to get us to quit altogether."

Overt Promotions of
Christianity in Military Publications

Numerous chaplains, as well as a few commanders and other officers and NCOs, are taking advantage of their military base newspapers and unit newsletters as another forum for promoting Christianity. While some would argue that protection of free speech applies and that anyone can publish virtually anything anywhere, when the publication is an officially sponsored base newspaper and the authors are members of the military, the perception is an official endorsement of these religious messages.

In an article titled "Living in Victory," a publication of the Louisiana National Guard, one chaplain explained how having Jesus as "your reference point to victory is crucial," how "victory is not something that is ahead of us, but has already been accomplished by Jesus' completed victory on the cross," and why "when you experience defeat, it just shows you that you need to quickly get your branch reconnected to the Vine, who is the Victorious Life of Christ in you." He summed up his piece by telling the troops that they "are Champions 'in Jesus Christ.'"[57]

In a column about Independence Day in a Marine unit newsletter, the chaplain got off to a good start, explaining in his opening paragraph how our independence from England led to "people having the right to worship in accordance with their own faith tradition," and that the First Amendment is "the reason the military has chaplains to uphold every service member's . . . right to worship in accordance to their particular faith group tradition." The rest of his article, however, was all about promoting one "particular faith group tradition"—his.

> I always remind people that we live in a fallen world, darkened by sin and evil because mankind wanted their independence from God. I also remind people of the incredible cost our Heavenly Father paid with the sacrifice of his one and only Son who died in our place in order that whomever [sic] would believe in Him would not perish but have everlasting life (John 3:16). In other words, our Heavenly Father through his Son paid the ultimate price, even death on a cross in order that whomever [sic] would believe could live a life independent from sin. Therefore, because of this great sacrifice paid by the Son of God any and every person can walk in victory beyond the struggles, skeletons in one's closet, and temptations that can keep us from being men and women of honor, courage and commitment.[58]

Writing about the upcoming move of the headquarters of an Air National Guard fighter wing, a chaplain assistant compared the move to Moses, the tabernacle, and the Christian Holy Spirit. She wrote:

> I have been studying about the life of Moses and recently studied how the Israelites set up the tabernacle. I won't go into all of the details about the tabernacle, but I do want to tell you about the "cloud" since I found the cloud to be very interesting and perfect for our upcoming Wing HQ move. . . .

> The cloud was a gift to the Israelites that the Lord had given to them for protection from the hot and cold. This cloud is like the Christian Holy Spirit that we have available to us today. The cloud was a gift and the Holy Spirit is a gift that all human beings can receive. The Holy Spirit helps us to make decisions and enables us to know when we need to move just like the cloud did for the Israelites.[59]

Sometimes, in addition to promoting Christianity, the articles get political, as in this example from one Army base newspaper. In an article titled "Virtue of Truth," the chaplain condemns all the "sins" of our "progressive" culture—freedom of choice, gay marriage, and so forth. He then injects the word "progressive" into a quote from the apostle John, a word that appears nowhere in the Bible verse he quotes, and adds the word "progressive" again before a quote from Pope John Paul II, although that word was not used by the late pontiff.

> At the heart of all sin is pride. This is the kind of pride that makes itself the arbiter of right and wrong. This is good to remember in an age when euthanasia is called mercy, suicide termed "creative medicine" and abortion described as "freedom of choice." All three are really murder.

> Today, marriage is too often considered outdated as an institution and divorce is considered the better option. Even more disturbing, opposition to same-sex marriage is thought to be bigoted and intolerant. This makes adultery and sodomy very uncomfortable terms in some people's lexicon.

> In contrast with today's attitudes, the apostle John reminds us: "Anyone who is so 'progressive' as not to remain in the teaching of the Christ does not have God; whoever remains in the teaching has the Father and the Son" (2 John 9).[60]

The last example comes from an article titled "The Opportunity to Follow Is Afforded to Us All," written by an Air Force master sergeant:

> There's a tremendous biblical illustration of the ever-present duplicitous nature of followership between leading and accepting and executing orders.

> This passage tells of a military leader in command of 100 followers. One day this leader, who is not a religious man, compassionately sends messengers to ask Jesus to pray for a dying subordinate. Jesus, so motivated by this compassionate appeal, deviates from his intended course to visit this kindhearted leader. However, just prior to Jesus' arrival to the installation, the leader sends his followers to stop Jesus from coming to his installation, deeming himself not worthy of hosting such an esteemed visitor. This is where the leader communicates through his followers the most convicting principle of true followership. His principled statement is, "I know authority because I am under the authority of my superior officers, and I have authority over my soldiers. I only need to say, 'Go,' and they go, or 'Come,' and they come." This very powerful confession prompts Jesus to clearly identify the next principle of responsible followership. The scripture reads, "when Jesus heard this, he was amazed and said to the crowd following him, 'I tell you, I have not seen faith, or confidence, like this in all the land . . .' The leader's statement truly reflects the heart of followership. Followership is firmly rooted in confident obedience. And followership and leadership are transitional meaning to pass back and forth between positions. This compassionate military leader knew that even though he was not a religious man, demonstrating his willingness to follow Jesus' command without question would save his follower's life.[61]

The master sergeant who wrote the above is from the wing's Equal Opportunity Office—the very office where an Airman would go for help if he or she had a complaint about an inappropriate promotion of religion, like this article written by this master sergeant.

Religious Programs for Military Children

Nobody would disagree that military personnel and their families should have the opportunity to worship as they choose. This is the justification for the military providing chaplains and chapels, and it is a reasonable one. But just how much support of religion is necessary to ensure this access to worship opportunities?

Countless DOD contracts show that what the government is providing for religion on military bases goes far beyond chaplains and chapels and, in many cases, far beyond what would be available to most civilians in their communities or towns. If a civilian church doesn't happen to have any talented musicians in its congregation, for example, the congregation might have to deal with having less than professional quality music at their services. Not so in military chapels. If chapels want better music, they hire professional musicians and music directors, contracted by the DOD. If a civilian church wants to start a youth program or provide religious education classes, it might have to find volunteers to run them. Military chapels hire base religious education directors, also paid for with DOD contracts.

And, while the contracting of these religious "service providers" is in itself highly questionable, the larger problem is that these contracts are almost exclusively open only to Christians. Contract descriptions, in complete disregard of the Constitution's "no religious test" clause, make this abundantly clear by including requirements such as "contractor shall ensure all programs and activities are inclusive of all Christian traditions," and the contractor will "use a variety of communications medium that shall appeal to a diverse group of youth, such as music, skits, games, humor, and a clear, concise, relevant presentation of the Gospel."[62]

The most egregious practices are found in the programs for the children of military personnel. These youth programs, many funded by DOD contracts, are designed to target and evangelize the "unchurched" among our military youth. The tactics employed by these government-contracted Christian ministries to achieve this goal range from luring teenagers with irresistible events and activities to infiltrating the off-post public middle and high schools attended by military children. One of these organizations, Youth for Christ Military Youth Ministry, actually goes as far as stalking military children, following their school buses to find out where they live and what schools they go to.

Incredibly, even the job descriptions in some DOD contracts make it clear that stalking kids is expected. One recently posted Army base position required that the contractor target "locations and activities where youth live and spend time, such as neighborhood community centers, school and sports and recreational activities, etc." to draw in "youth that are not regularly affiliated with established chapel congregational youth programs."[63]

According to a video interview[64] of Fort Riley's religious education director about one of the base's exclusively Christian youth programs, the mission of the program, called Spiritual Rangers, is "to train young men to be Godly leaders by instilling in them biblical character, values and principles and thus giving them a sense of what it truly means to be a man." This video, which was aired on the base's local cable access channel, described a program where teenage boys get to do things like using the base's close combat tactical trainer, engagement skills trainer, and helicopter flight simulator—in other words, the coolest video games *ever*! And all a kid on Fort Riley has to do to play them is hang out with the "godly" men and memorize some scripture.

Military Community Youth Ministries (MCYM),[65] whose Club Beyond program "seeks to celebrate life with military kids and introduce them to the Life-giver, Jesus Christ," has received millions of dollars in DOD contracts and operates on dozens of US military bases, both overseas and in the United States.

MCYM's Contracting Officer's Performance Evaluation, a form to be filled out each year by a "person duly appointed with the authority to enter into and to administer contracts on behalf of the government" at the installations where the organization is contracted, not only shows that MCYM's mission is to target non-Christian children, but also that the contracting officer actually rates MCYM on its success in this constitutional violation. These are two of the questions on the evaluation form:

1. MCYM staff are expected to conduct outreach ministry to teens who have no relationship with the chapel or established churches. What is your assessment of this ministry objective?

2. MCYM staff are expected to present the Gospel to teens with due respect to their spiritual traditions, i.e. to engage in evangelism but not proselytization. This means that they are not to endorse a particular theology or denomination or creed excepting that which is generally accepted as representing the principle tenents [*sic*] of the Christian faith with a focus on introducing teens to Jesus Christ and to help teens develop in their faith in God. What is your assessment of this ministry objective?[66]

Saying that they "engage in evangelism but not proselytization" is questionable at best. MCYM narrowly defines refraining from proselytization as not trying to convert someone from one Christian denomination to another and places no restrictions on evangelizing those teenagers who need some "introducing" to Jesus Christ.

One of MCYM's "partner" organizations is Youth for Christ's Military Youth Ministry. Actually, Youth for Christ (YFC) and MCYM are one and the same. Both have the same address and phone number, and the YFC Military Youth Ministry mission statement states only one mission—to partner with MCYM: "The Mission of Youth For Christ Military Youth Ministry is to partner with Military Community Youth Ministries (MCYM) in assisting and equipping Commanders, Chaplains, Parents, Volunteers and local Youth for Christ (YFC) chapters on behalf of reaching military teens with the Good News of Jesus Christ."[67]

YFC Military Youth Ministry is just the arm of MCYM that goes after military children who attend off-post public schools, and its first step in obtaining a contract from the military is to convince a chaplain that his or her base needs its services. To do this convincing, YFC provides a fill-in-the-blank template for a YFC "steering committee" to write up an assessment to present to the installation chaplain. The first part of completing this assessment is for the YFC steering committee to attempt to get a meeting with the local high school principal. This is done with a cold call to the principal in which committee members say, according to the script provided by YFC, that they are assisting the base chaplains, even though this phone call appears to be made prior to approaching the chaplains:

> Example when you call the principle [*sic*] of the local high school: Hello my name is and I am assisting the chaplains of Fort _____ by putting together several facts concerning adolescent culture and youth serving organizations in our community. Could I drop by and ask a few questions?

Here are a few more sections of YFC's assessment template, including the instruction to essentially stalk the children by following their public school buses:

> 3. a. _____ High School. The principle [*sic*] is _____. I spoke with _____ and he indicated that he would be willing/unwilling to allow me campus access. He did indicate that he would be glad to allow me to support students by attending practices, games, rehearsals and school activities on an "as invited" basis. My general impression is that _____ and will continue to develop my relationships at the High School.
>
> b. _____ Middle School. The principle [*sic*] is _____.

ACCESSMENT [*sic*]:

6. Demographics

a. High School: This is a completely unscientific measurement but I followed the buses around for three days. Each morning four buses leave the installation in [*sic*] route to the high school. There are approximately _____ students on these buses. Students are primarily picked up in the _____, _____ and _____ neighborhoods. Students appeared to be equally spread over the four different grade levels with slightly more/less 9th and 10th graders.

b. Middle School: See a above.[68]

Like MCYM, Malachi Youth Ministries,[69] the youth division of Cadence International, is funded by DOD contracts. In addition to teenagers, Cadence International also targets the younger children of military personnel, partnering with Child Evangelism Fellowship (CEF) "to anchor children in the hope of Jesus and lead them to living fully devoted to Him" by getting the elementary school children into Good News Clubs on their bases and in their schools.[70]

Cadence and CEF have the "mutual goal of reaching every child of the US military around the world," and clearly they will have the support and aid of the military itself to achieve this goal, based on statements like this one from the deputy installation chaplain at one large Army base, who, in a video promoting CEF, proclaimed, "The harvest is ready, and I mean it's out there in more abundance than we have ability to harvest."[71]

Religious Tests

In addition to the unconstitutional "religious tests" found in job requirements for DOD contracts, there are a number of service members who have
expressed concerns about the requirement to disclose their religion on forms
whose purposes would include no legitimate reason to contain any information
about their religion. Two examples are the Army Officer Record Brief (ORB)
and the Air Force Single Unit Retrieval Format (SURF). The ORB and the
SURF are forms whose purpose is to provide information on the career history,
education, and special skills of officers. The information contained in these
forms is used for job placement, award nominations, applications to military
training programs and colleges, and so forth. The religion of an officer should
never be a factor in career decisions or recommendations, yet the Army's ORB
now contains a block for the officer's religion, and the Air Force's SURF, a recently implemented electronic form, also lists the officer's religion.

Fear of Making Complaints
through Military Channels

The almost universal problem faced by military personnel who encounter any
of the problems listed above is the fear of what might happen if they report a
violation of regulations or bring a complaint to their superiors or the Equal Opportunity Office. Service members who fear harassment from both peers and
superiors, negative effects on their careers, and occasionally even physical harm
often refrain from reporting violations of regulations regarding religion, even
when those violations are personally impacting their or their family's lives. Few
ever decide to file official complaints, allowing military spokespersons, when an
issue is reported or uncovered, to say that it was an isolated incident and to
quickly point out how few official complaints have been filed. Clearly, the number of official complaints filed, usually said to be less than 100, is unrealistically
small given that over 15,000 service members have contacted the Military Religious Freedom Foundation for assistance from 2005 to 2009. The disparity in
these numbers is something that cannot be ignored.

Recommendations

After dealing with thousands of service members and carefully examining
virtually every military regulation that would apply to their concerns and complaints, the Military Religious Freedom Foundation has concluded that there
are very few situations in which the existing regulations are the problem. The
problem is that these existing regulations are not being followed or enforced.

One important exception, however, relating to the proselytizing of Muslims
in Iraq and Afghanistan, must be noted here. Because CENTCOM's General
Order 1B, in its list of prohibited activities in the CENTCOM AOR, lists only
"proselytizing of any religion" as being prohibited, Christian military personnel

intent on converting Muslims are getting around this crucial prohibition. How? By saying that the order only prohibits proselytizing, but not evangelizing, and claiming that activities such as distributing Arabic and other native-language Bibles are merely evangelizing and thus do not violate the order. Simply changing the wording of GO-1B to "*evangelizing or proselytizing* of any religion" would leave no loophole for those who rely on semantics to continue their attempts to convert the Iraqis and Afghans to Christianity.

Setting the Record Straight Regarding the Military Chaplaincy

Ever since chaplains praying in Jesus' name at nonreligious military functions and ceremonies became a hot-button issue, a distorted version of the history of the chaplaincy has emerged. This altered history of the chaplaincy has one purpose—to make it appear that the military chaplaincy has existed continuously since the Revolutionary War, with no problems or objections until recent years. This is accomplished by simply leaving a few minor gaps in the history, such as most of the nineteenth century.

MYTH: The chaplaincy has been an essential part of the military since the Revolutionary War.

FACT: The military chaplaincy was almost nonexistent between the end of the Revolutionary War and the Civil War.

There really wasn't much of a military chaplaincy at all during the War of 1812 or up through and including the Mexican-American War. Naval commanders were authorized to appoint chaplains, but many of these were not ordained ministers, and their purpose was as much to be instructors in everything from reading and writing to navigational skills as it was to be preachers. Some officers even saw their authority to appoint chaplains as a way to get a personal secretary and chose them for their ability to perform that job, with little regard for their religious qualifications.

During the War of 1812, there was only one Army chaplain for as many as 8,000 men, and, with the exception of the 1818 appointment of a chaplain at West Point who doubled as a professor of history, geography, and ethics, there were no new Army chaplains until 1838, when a small number of post chaplains were authorized. But these post chaplains were not members of the military. They were civilian employees hired by the post's administrators, and like their counterparts in the Navy, they were hired mainly as teachers and also served as everything from librarians to mess officers to defense counsel during courts-martial. Post chaplains, since they were not in the military, were not assigned to a military unit, but to their post, so when the Mexican-American War began, they did not accompany the troops.

In 1847, Congress passed a law transferring control over post chaplains from the post administrators to the secretary of war, giving the secretary of war the authority to require a chaplain to accompany his post's troops into the field whenever a majority of the troops were deployed. Those chaplains who refused to go were fired. This 1847 law caused a bit of a problem, however, because it neglected to actually give anyone the authority to appoint chaplains. In fact, when President Polk appointed two Catholic priests as "chaplains" in an effort to stop the propaganda that the war was an attack upon the Mexicans' religion, he made them as political appointments rather than chaplain appointments, saying that there was no law authorizing Army chaplains.

The total number of Army chaplains during the Mexican-American War was 15, including the two Catholic priests who weren't actually chaplains. The chaplaincy grew much larger during the Civil War, of course, with the appointment of a chaplain for each regiment. But when the war ended, the chaplaincy was reduced to the 30 post chaplains authorized in 1838, even though the regular Army was twice the size it had been in 1838. Six additional chaplains were authorized for the six black regiments of the regular Army, but this was reduced to four in 1869. The number of chaplains authorized for the Army would remain 34 until 1898.

MYTH: *There were no problems with or objections to chaplains until recent years.*

FACT: *There was a widespread campaign to completely abolish the chaplaincy in the mid-1800s.*

By the late 1840s, opposition to government-paid chaplains was growing, and a vigorous campaign to abolish both the military and congressional chaplaincies would go on for well over a decade, supported by both members of the military and civilians, including churches and religious leaders. Hundreds of petitions, signed by thousands of Americans, were sent to Congress during the 1840s and 1850s calling for an end to all government-paid chaplains. A large part of the American public of the mid-1800s objected to chaplaincy establishments on constitutional grounds; religious organizations objected to them on both religious and constitutional grounds; and military personnel, including chaplains, had complaints of religious coercion and discrimination uncannily similar to those heard today.

Take, for example, the following statement, which was written in 1858: "Mr. Hamlin presented the memorial of Joseph Stockbridge, a chaplain in the navy, praying the enactment of a law to protect chaplains in the performance of divine service on shipboard, according to the practices and customs of the churches of which they may be members."[72] Given the current disputes over chaplains' prayers, this statement could just as easily be from 2010.

A common complaint in the military during the nineteenth century was the takeover of the chaplaincy by Episcopalians. Once the Episcopalians gained control, all members of the military, regardless of their religion or denomination, began to be forced or coerced to attend Episcopalian worship services, and non-Episcopalian chaplains were being forced to perform these services.

While the particular "bully" denomination may have changed since the petition of the naval officers in 1858, the issue has not. In the mid-1800s it was the

Episcopalians; in 2010 it's fundamentalist Protestants. And, as in the mid-
1880s, this is also not an issue of Christians versus non-Christians. The overwhelming majority of the petitions received by the Congresses of the 1840s
and 1850s were written and signed by Christians and Christian religious organizations, just as the majority of complaints received by the Military Religious
Freedom Foundation—96 percent of them—are from self-identified Christians, both Protestant and Catholic.

Beginning in 1848, hundreds of petitions poured into both houses of Congress. The first of these petitions to be presented in the Senate was from a
Baptist association in North Carolina:

> Mr. Badger presented the memorial, petition, and remonstrance of the ministers
> and delegates representing the churches which compose the Kehukee Primitive
> Baptist Association, assembled in Conference with the Baptist Church at Great
> Swamp, Pitt County, North Carolina praying that Congress will abolish all laws
> or resolutions now in force respecting the establishment of religion, whereby
> Chaplains to Congress, the army, and navy, are employed and paid to exercise
> their religious functions.

> Mr. Badger said he wished it to be understood that he did not concur in the ob-
> ject of this memorial. He thought the petitioners were entirely wrong. But as the
> petition was couched in respectful language, he would ask for its reading and
> would then move that it be laid on the table and printed.[73]

Five years later, as a member of the Senate Judiciary Committee, Senator
Badger, a devout Episcopalian, would write a very pro-Christian report dismissing the countless petitions received by that time to abolish the chaplaincy—a report that is frequently quoted by today's Christian nationalists to
show just how very religious and pro-Christian Congress was in the nineteenth
century. These historical revisionists simply neglect to mention that Badger's
report, and a similar report written a year later by an equally religious member
of a House committee,[74] had anything to do with a campaign to abolish the
chaplaincy. Acknowledging the historical context of these reports would, of
course, contradict their claims that there were no complaints or questions about
the constitutionality of government religious establishments until modern-day
secularists decided to wage a war on Christianity.

Obviously, Senator Badger, who had already stated in 1848 that he "did not
concur in the object" of the Baptists' petition to abolish the chaplaincy, was not
someone who was going to be objective in considering the many similar petitions he was asked to report on in 1853. So it was no big surprise that Badger's
report dismissed the petitions, stating that "the whole view of the petitioners
seems founded upon mistaken conceptions of the meaning of the Constitution," and that the Founding Fathers "did not intend to spread over all the
public authorities and the whole public action of the nation the dead and re-
volting spectacle of atheistical apathy."[75]

In 1860, Congress addressed the issue of commanders forcing chaplains to
conduct worship services of a faith tradition other than their own with a provision
stating, "Every chaplain shall be permitted to conduct public worship according to

the manner and forms of the church of which he may be a member."[76] They did not, however, address the issue of the hijacking of the chaplaincy of one denomination, even though an investigation had shown the complaints to be valid.

Instead of moving forward, Congress soon took a giant step backwards, mandating in August 1861, in the act that authorized the appointment of regimental chaplains for the Union Army, that all chaplains be Christians.[77] A similar provision was in the act for the regular Army—the act passed in July 1861 authorizing the president to raise a volunteer force stated that a chaplain "must be a regular ordained minister of a Christian denomination."[78] No prior legislation authorizing chaplains had ever mandated that chaplains had to be of a particular religion or even that they had to be ordained ministers. Apparently, the earlier Congresses were familiar with that pesky "no religious test" clause in the Constitution, applying it even to the office of chaplain. The criteria for a chaplain in the 1838 law authorizing post chaplains, for example, was simply that "such person as they may think proper to officiate as chaplain."[79]

But the 1861 law requiring chaplains to be Christians was quickly and successfully challenged. The usual practice at the time for appointing Army chaplains was for each regiment to elect its own chaplain, and a regiment from Pennsylvania had elected a Jewish cantor. When the Young Men's Christian Association exposed this grievous violation of the 1861 chaplain law, the Jewish chaplain resigned rather than face the humiliation of losing his commission. But the regiment decided to test the constitutionality of the law. This time they chose a rabbi, knowing full well that his application for a commission would be denied. After a public outcry over the denial of the rabbi's commission, which included numerous petitions from Jewish organizations, groups of citizens, and even the members of one state legislature, the provision requiring chaplains to be Christians was repealed.[80] A few months later, in September 1862, President Lincoln legally commissioned the first Jewish chaplain.

Another issue during the mid-nineteenth-century chaplain battle was over a naval regulation from 1800 giving commanders the authority to force their subordinates to attend religious services.[81] It had been enacted during the very religious Adams administration and remained in force in 1858. This example is often used by historical revisionists to show that "it is simply inconceivable that the members of the First Congress, who drafted the Establishment Clause, thought it to prohibit chaplain-led prayer at military ceremonies, having passed legislation not only approving that practice, but indeed requiring service members to attend divine services." However, what these revisionists fail to mention is that, in 1858, this act was protested by a group of naval officers[82] who successfully petitioned Congress to amend it to make religious services optional.

As already mentioned, most of the protests against government-paid chaplains came from Christians, and it's absolutely remarkable how similar the opinions of these nineteenth century Christians were to those of the modern-day "secularists" who are currently trying to destroy Christianity. The following was written by Rev. William Anderson Scott, one of the most prominent Presbyterian ministers of his day, in his 1859 book *The Bible and Politics*. Reverend

Scott's book was written in large part to refute the arguments being used by those who wanted the Bible in public schools, another issue that is far from new, but it also addressed the issue of government-paid chaplains, including the following from a section on military chaplains:

> Is it constitutional to take the public money to pay a chaplain for religious services that are not acceptable to a majority of the rank and file of the army? I do not think so. If the majority of a regiment, or of the men on board a man-of-war, should elect a chaplain, then, possibly, the Government might make an appropriation to pay him, though I doubt whether this is constitutional, and I do not believe it the best way. I believe that the supplying of religious consolations to the members of our Legislature, and to the officers and men of our army and navy, according to our organic laws, should be left to themselves, just as it is to our merchant ships and to our frontier settlements—that is, to their own voluntary support. Our blacksmiths, police officers, Front-street merchants, lawyers and physicians all need the blessings of religion; but they must provide for their own individual wants. And, in the same way, I would leave the army and the navy and the legislatures, and I would do so the more readily, because the different churches and voluntary religious societies would then all stand truly on an equality, and hold themselves ready to help in furnishing such supplies. Suppose a regiment is ordered to the wilderness, let the men elect a chaplain and pay him themselves. Then they will be more likely to profit by his services. Or let a missionary society, by the vote of the citizen soldiers, be asked to send them a minister of religion. If the government appoints a Protestant chaplain, is it a disobedience of orders for a Catholic to refuse to accept of his services? I see nothing but difficulty and the engendering of constant sectarian feuds and bad feeling, if the Federal Government touches anything that is religious.[83]

Clearly, this nineteenth century Presbyterian minister must have been trying to destroy Christianity and turn the military into a bunch of atheists.

What Would the Founding Father of the US Military Think?

The version of history in which the inconvenient events of the 1800s are simply ignored typically begins with the many instances of George Washington issuing orders regarding chaplains and religious services and usually includes his 1776 directive for each regiment to procure a chaplain. What's omitted is that a year later, when Congress wanted to cut the number of chaplains from one per regiment to one per brigade, an act that would put many regiments under chaplains who were not of similar beliefs to the Soldiers, Washington and his generals strongly objected.

This is what Washington wrote to the Continental Congress in 1777 on behalf of his generals:

> It has been suggested, that **it has a tendency to introduce religious disputes into the Army, which above all things should be avoided, and in many instances would compel men to a mode of Worship which they do not profess.** The old Establishment gives every Regiment an Opportunity of having a Chaplain of their own religious Sentiments, it is founded on a plan of a more generous

toleration, and the choice of the Chaplains to officiate, has been generally in the Regiments. Supposing one Chaplain could do the duties of a Brigade, (which supposition However is inadmissible, when we view things in practice) that being composed of four or five, perhaps in some instances, Six Regiments, there might be so many different modes of Worship. I have mentioned the Opinion of the Officers and these hints to Congress upon this Subject; from a principle of duty and because I am well assured, *it is most foreign to their wishes or intention to excite by any act, the smallest uneasiness and jealousy among the Troops.*[84] (emphasis added)

Washington and his generals worried about the "smallest uneasiness" over religion and objected to anything that would "compel men to a mode of worship that they didn't profess." What would they have to say about what's going on in today's military? Regardless of the side one happens to be on, few would disagree that the current issues are causing far more than the "smallest uneasiness."

Notes

1. "Hayes: Most Troops Will Be Home by 2008," *Concord Standard and Mount Pleasant Times*, 21 December 2006.

2. Lt Col Rick Francona, retired US Air Force intelligence officer, appearing on MSNBC, 28 December 2006.

3. Pete Geren, then secretary of the Army (commencement remarks, United States Military Academy, West Point, NY, 31 May 2008), http://www.army.mil/-news/2008/06/02/9573-west -point-commencement-remarks-by-secretary-of-the-army-pete-geren/. Secretary Geren was also among the civilian DOD officials who appeared in the Christian Embassy video.

4. Rep. Trent Franks (R-AZ), remarks on "Religious Freedom," 11 December 2007, Congressional Record, H15291.

5. Christian Embassy is the arm of Campus Crusade for Christ operating at the Pentagon. The Christian Embassy promotional video can be viewed at http://www.militaryreligiousfreedom.org/ Media_video/christian-embassy/index.html.

6. DOD Inspector General, "Alleged Misconduct by DoD Officials Concerning Christian Embassy," Report No. H06L102270308, 20 July 2007, http://www.militaryreligiousfreedom.org/ press-releases/christian_embassy_report.pdf.

7. Photos archived at http://www.militaryreligiousfreedom.org/dodspp.

8. "Mural Painter," http://www.riley.army.mil/NewsViewer.aspx?id=579. Photos also archived at http://www.militaryreligiousfreedom.org/dodspp.

9. Photos archived at http://www.militaryreligiousfreedom.org/dodspp.

10. http://www.militaryministry.org/.

11. http://www.militarymissionsnetwork.com.

12. Web page archived at http://www.militaryreligiousfreedom.org/dodspp.

13. http://www.valormovement.com.

14. Web page archived at http://www.militaryreligiousfreedom.org/dodspp.

15. Video at http://www.militaryreligiousfreedom.org/video/USAF.mov.

16. Military Ministry of Campus Crusade for Christ, *Movement Model of Ministry Volume 2*.

17. "God's Basic Training" Bible study. Page images archived at http://www.militaryreligious freedom.org/dodspp.

18. Photos archived at http://www.militaryreligiousfreedom.org/dodspp.

19. From Brigitte Gabriel's lecture at the Joint Forces Staff College on 13 June 2007: Questioning a statement in Gabriel's book, a student asked, "Should we resist Muslims who want to seek political office in this nation?" Gabriel replied:

Absolutely. If a Muslim who has—who is—a practicing Muslim who believes the word of the Koran to be the word of Allah, who abides by Islam, who goes to mosque and prays every Friday, who prays five times a day—this practicing Muslim, who believes in the teachings of the Koran, cannot be a loyal citizen to the United States of America. . . . A Muslim is allowed to lie under any situation to make Islam, or for the benefit of Islam in the long run. A Muslim sworn to office can lay his hand on the Koran and say "I swear that I'm telling the truth and nothing but the truth," fully knowing that he is lying because the same Koran that he is swearing on justifies his lying in order to advance the cause of Islam. What is worrisome about that is when we are faced with war and a Muslim political official in office has to make a decision either in the interest of the United States, which is considered infidel according to the teachings of Islam, and our Constitution is incompatible [*sic*] with Islam—not compatible—that Muslim in office will always have his loyalty to Islam.

Among her many other derogatory statements, Gabriel referred to Dearborn, Michigan, as "Dearbornistan" because of its large Muslim community, and, in a comment about racial profiling, said that American Muslims "are good at nothing but complaining about every single thing."

20. http://www.3xterrorists.com.

21. Bethany Duemler, "Alleged Ex-PLO Raises Eyebrows," *Chimes* (newspaper of Calvin College, where the 3 Ex-Terrorists appeared), 9 November 2007, http://www-stu.calvin.edu/chimes/article.php?id=3125; "Doubt Cast on Anani's Terrorist Claims," *The Windsor Star*, 20 January 2007, http://www.canada.com/windsorstar/news/story.html?id=4a479502-4490-408e-bdb5-f2638619a62c; and Neil MacFarquhar, "Speakers at Academy Said to Make False Claims," *New York Times*, 7 February 2008, http://www.nytimes.com/2008/02/07/us/07muslim.html?scp=1&sq=Shoeb at&st=nyt.

22. http://www.youtube.com/watch?v=2WN5rqKkhUU; http://www.youtube.com/watch?v=ipqO_ke-NH4; http://www.youtube.com/watch?v=3l_Mc-0MaZM; http://www.youtube.com/watch?v=thfYB-VejSQ; and http://www.youtube.com/watch?v=XIokAQa1Xs4.

23. Maria Luisa Tucker, "'Reformed Muslim Terrorists' Preach Christ to College Kids," *Village Voice*, 19 February 2008, http://www.villagevoice.com/2008-02-19/news/reformed-muslim-terrorists-preach-christ-to-college-kids/1.

24. Video at http://www.youtube.com/watch?v=2MibbDnH8BM.

25. Video at http://www.youtube.com/watch?v=CFqIPjj3ciU.

26. Video at http://www.youtube.com/watch?v=UVPcjVvvMQU.

27. Photos of military Bibles archived at http://www.militaryreligiousfreedom.org/dodspp.

28. "Ministry Provides Hope in Second Run of Bible Devotional," *Mission Network News*, 26 November 2007, http://www.mnnonline.org/article/10592.

29. In November 2008, the Military Religious Freedom Foundation wrote to the secretary of defense, calling for the DOD inspector general to promptly initiate an investigation into the background and activities of Navy chaplain LCDR Brian K. Waite and requesting that any existing association between the US military and Revival Fires Ministries be immediately terminated. That letter can be found at http://www.militaryreligiousfreedom.org/Gates_Letter.pdf. Video of Lieutenant Commander Waite at a Revival Fires camp meeting and links to additional information regarding this situation can be found at http://www.militaryreligiousfreedom.org/newsletters/2008-11/video.html.

30. Web page archived at http://www.militaryreligiousfreedom.org/dodspp.

31. Lt Gen Khalid Bin Sultan al-Saud, commander of Saudi Arabia's air defense forces, appointed by King Fahd as General Schwarzkopf's counterpart.

32. Video at http://www.militaryreligiousfreedom.org/Media_video/al_jazeera/index.html.

33. "IMF Chaplains Serving in Iraq," *Gathering*, Spring 2004, http://www.i-m-f.org/pdfs/Gatherings/Spring2004.pdf.

34. LTC Lyn Brown, "Kingdom Building in Combat Boots," *Heart & Mind*, Bethel Seminary, Summer 2005, http://www.bethel.edu/publications/heartmind/2005-summer/bethel-army-boots/.

35. http://www.onlyonecross.com.

36. http://www.larryclayton.org/index.php?option=com_content&view=article&id=27&Itemid=2.

37. http://www.lightsofliberty.us/iraq.html.

38. Video at http://www.youtube.com/watch?v=0B7pBbkZpq0.

39. http://soldiersbibleministry.org/Default.aspx?tabid=1783.

40. Worldwide Military Baptist Missions "Prayer Letter," archived at http://www.military religiousfreedom.org/dodspp.

41. For numerous reasons in addition to the distribution of Arabic Bibles, the Military Religious Freedom Foundation has demanded that the DOD revoke the ecclesiastical endorsing authority of this endorsing agency. The letter to the secretary of defense and enclosures detailing the reason for this demand can be found at http://www.militaryreligiousfreedom.org/press-releases/gates_letter.html.

42. Web page archived at http://www.militaryreligiousfreedom.org/dodspp.

43. Campus Crusade for Christ Military Ministry Life and Leadership newsletter, October 2005.

44. *God and the Military,* video, filmed in 1997, re-released on DVD in 2005 for distribution by Campus Crusade for Christ Military Ministry.

45. http://www.militaryministry.org/about/strategic-objectives/.

46. http://www.cadence.org.

47. http://www.cadence.org/home/who-we-are/a-strategic-ministry.

48. Until January 2009, the Officers' Christian Fellowship's official vision statement was "a spiritually transformed military with ambassadors for Christ in uniform, empowered by the Holy Spirit, living with a passion for God and a compassion for the entire military society." Its mission statement was "Christian officers exercising biblical leadership to raise up a godly military." Examples of the use of these statements are archived at http://www.militaryreligiousfreedom.org/dodspp.

49. Photos archived at http://www.militaryreligiousfreedom.org/dodspp.

50. Several sources confirm that, in some cases, concerts by Eric Horner, a Christian artist who regularly performs at military bases, have been mandatory for basic trainees.

51. http://www.militaryreligiousfreedom.org/powerpoint/Lakenheath.ppt.htm.

52. Report of a US Army major in the National Guard who attended this Strong Bonds event.

53. http://www.fedspending.org.

54. http://www.nytimes.com/2009/03/15/washington/15video.html.

55. Archived at http://www.militaryreligiousfreedom.org/dodspp.

56. Archived at http://www.militaryreligiousfreedom.org/dodspp.

57. Chaplain Maj. Jeff Mitchell, "Living in Victory," *The Engineer Express,* 225th Engineer Brigade, Louisiana National Guard, 15 July 2009.

58. Chaplain Bailey, 31st Marine Expeditionary Unit newsletter, July 2009.

59. MSgt Diane Watters, *In Formation,* newsletter of the 187th Fighter Wing of the Alabama Air National Guard, February/March 2009.

60. Chaplain Capt Paul-Anthony Halladay, "Virtue of Truth," *The Guidon,* base newspaper of Fort Leonard Wood, 15 April 2009.

61. MSgt Stephen Love, "The Opportunity to Follow Is Afforded to Us All," 460th Space Wing at Buckley Air Force Base, 18 March 2009, http://www.buckley.af.mil/news/story.asp?id=123138478.

62. Community-wide Outreach Youth Ministry Program for High School Students, Fort Bragg, North Carolina, Solicitation Number: W9124709T0004, 17 October 2008.

63. Ibid.

64. http://www.militaryreligiousfreedom.org/media_video/spiritual_rangers/index.html.

65. http://www.mcym.org.

66. Archived at http://www.militaryreligiousfreedom.org/dodspp.

67. http://www.yfcmym.org.

68. Archived at http://www.militaryreligiousfreedom.org/dodspp.

69. http://www.malachi.org.

70. http://www.cefonline.com/content/category/4/102/343/.

71. Video archived at http://www.militaryreligiousfreedom.org/dodspp.

72. *Journal of the Senate of the United States of America,* vol. 50, 35th Cong., 2nd Sess. (Washington, DC: William A. Harris, 1858–59), 53.

73. *The Congressional Globe*, 30th Cong., 2nd Sess., 13 December 1848, 21.

74. *Reports of Committees of the House of Representatives Made During the First Session of the Thirty-Third Congress*, vol. 2, H. Rep. 124 (Washington, DC: A. O. P. Nicholson, 1854).

75. *Reports of Committees of the Senate of the United States for the Second Session of the Thirty-Second Congress, 1852-53,* S. Rep. 376 (Washington, DC: Robert Armstrong, 1853), 4.

76. George P. Sanger, ed., *The Statutes at Large, Treaties, and Proclamations of the United States of America*, vol. 12, 36th Cong., 2nd Sess. (Boston: Little, Brown and Co., 1863), 24.

77. Ibid., 37th Cong., 1st Sess., 288.

78. Ibid., 270.

79. Richard Peters, ed., *The Public Statutes at Large of the United States of America*, vol. 5, 25th Cong., 2nd Sess. (Boston: Little, Brown and Co., 1856), 259.

80. Sanger, ed., *Statutes at Large*, vol. 12, 37th Cong., 2nd Sess., 595.

81. Peters, ed., *Public Statutes at Large*, vol. 2, 45.

82. *Journal of the House of Representatives of the United States*, vol. 54, 35th Cong., 1st Sess. (Washington, DC: James B. Steedman, 1857 [sic]), 792.

83. Rev. W. A. Scott, DD, *The Bible and Politics: Or, An Humble Plea for Equal, Perfect, Absolute Religious Freedom, and Against All Sectarianism in Our Public Schools* (San Francisco: H.H. Bancroft & Co., 1859), 78.

84. George Washington to the president of Congress, 8 June 1777, in John C. Fitzpatrick, ed., *The Writings of George Washington from the Original Manuscript Sources 1745-1799*, vol. 8 (Washington, DC: Government Printing Office, 1933), 203.

About the Author

Chris Rodda is the senior research director for the Military Religious Freedom Foundation and a writer on issues related to religion and politics. Focusing for many years on the issue of the politically motivated revisionism and distortion of American history by the Religious Right, she authored the book *Liars For Jesus: The Religious Right's Alternate Version of American History*, vol. 1, the first of a projected three-volume series debunking the historical myths and lies found everywhere from homeschooling textbooks to congressional debates and legislation to Supreme Court opinions. She is a regular contributor at Talk2Action.org and a blogger on the *Huffington Post*.

The Military Religious Freedom Foundation is a 501(c)(3) founded by Mikey Weinstein in 2005. Weinstein is a 1977 graduate of the US Air Force Academy. MRFF does not seek to rid the military of all religion, as its critics would have people believe. In fact, 96 percent of the service members who seek the assistance of MRFF are Christians, and the work of the foundation is endorsed by a number of religious organizations representing a variety of faiths. For more information, visit http://www.militaryreligious freedom.org.

CHAPTER 6

RELIGIOUS RIGHTS AND MILITARY SERVICE

Jay Alan Sekulow
Robert W. Ash

Congress shall make no law respecting an establishment of religion, or prohibiting the free exercise thereof.

—US Constitution, Amendment 1

Introduction

We live in a very litigious society, where almost anyone can sue another for virtually any offense, real or imagined. DOD policy makers are not immune from such litigation. In fact, there are growing numbers of persons and advocacy groups in the United States actively seeking to remove from public life—including in the armed services—virtually all symbols and expression of religion and America's religious heritage by advocating strict separation of church and state.[1] Many of these groups are already actively engaged in filing lawsuits against DOD and its leaders over various concerns about religious expression in the armed services.[2] Still others have threatened lawsuits.[3] Persons and groups have every right to hold and zealously advocate such views, but many of their views on church-state separation go well beyond what the Constitution and US law require. In fact, they endanger the very freedoms the First Amendment was intended to protect. Indeed, protecting free exercise of religion is particularly important in the armed services because it is a key component in developing and strengthening the warrior ethos, an indispensible factor in fighting and winning our nation's wars. This chapter will examine a number of issues of concern regarding free exercise of religion and religious expression in the armed services. It also will suggest ways of protecting service members' free exercise and expressive rights while maintaining good order and discipline.

General Legal Principles

Separation of Church and State

When discussing free exercise of religion and its limits in the US armed forces, one quickly encounters arguments citing the phrase "separation of church and state." Those making such arguments often use that phrase when what they are really referring to is the establishment clause in the First Amendment.[4] In truth, the phrase "separation of church and state" is found nowhere in the US Constitution. Instead, that phrase comes from a letter written in 1802 by Pres. Thomas Jefferson to members of a Baptist association in Danbury, Connecticut.[5] Hence, rather than wasting time trying to determine the meaning of a phrase that does not exist in the Constitution, time would be better spent determining what the drafters of the First Amendment meant by "establishment of religion," a phrase that does exist in the Constitution.

One of the methods used by the Supreme Court of the United States for interpreting the meaning and legal reach of the First Amendment is to examine how early Congresses acted in light of the amendment's express terms. One can begin to understand what the establishment clause allows (and disallows) by examining what transpired in the earliest years of our nation during the period when Congress drafted the First Amendment and after the states ratified it.[6] For example, "the First Congress, as one of its early items of business, adopted the policy of selecting a chaplain to open each session with prayer,"[7] and a "statute providing for the payment of these chaplains was enacted into law on September 22, 1789."[8] Moreover, within days of legislating to pay congressional chaplains from the federal treasury, "final agreement was reached on the language of the Bill of Rights."[9] From these facts, the Supreme Court concluded that, whatever its ultimate meaning and reach, the establishment clause was not intended to forbid paid, legislative chaplains and their daily, public prayers.[10] The *Marsh* Court concluded that chaplain-led prayer opening each day's session in both houses of Congress "is not . . . an 'establishment' of religion," but rather "a tolerable acknowledgment of beliefs widely held among the people of this country."[11] Additionally, the First Congress—the same Congress that drafted the First Amendment—established the tradition of clergy-led prayer at presidential inaugurations (which, in truth, constitute military change-of-command ceremonies, where the nation's new commander in chief assumes office from his predecessor).[12] These practices have continued to this very day.

Early national leaders also acted in ways that some today argue expressly violate the establishment clause. For example, Pres. George Washington issued proclamations of thanksgiving to Almighty God during his presidency,[13] and Pres. John Adams called for a national day of fasting and prayer.[14] Pres. Thomas Jefferson—a man often described as a strong defender of strict church-state separation—signed multiple congressional acts to support Christian missionary activity among the Indians.[15] Further, during his presidency, Jefferson also developed a curriculum for schools in the District of Columbia which used the Bible and a Christian hymnal as the primary texts to teach reading,[16] and he signed the Articles of War, which "earnestly recommended to all officers and soldiers, diligently to attend divine services."[17] Once the US Navy was formed, Congress also enacted legislation directing the holding of, and attendance at, divine services aboard US Navy ships.[18] As one honestly examines governmental acts contemporaneous with the adoption of the First Amendment, it is difficult to deny that, in the early days of our republic, church and state existed relatively comfortably (and closely) together, with contemporaries of the drafters of the First Amendment showing little concern that such acts violated the establishment clause. As the *Marsh* Court aptly recognized, actions of the First Congress are "contemporaneous and weighty evidence" of the Constitution's "true meaning."[19]

More recent court decisions have confirmed that strict separation between church and state is not required by the Constitution. In fact, the government must often yield what it might otherwise be able to do to ensure that free exercise rights are protected. In *Corporation of Presiding Bishop v. Amos*,[20] the Supreme Court noted that "this Court has long recognized that the government may (and sometimes must) accommodate religious practices and that it may do so without violating the Establishment Clause."[21] Furthermore, permissible religious accommodation need not "come packaged with benefits to secular entities."[22] The Supreme Court has also noted that strict separation could lead to absurd results. In *Zorach v. Clauson*,[23] the Court stated that the First Amendment

> does not say that in every and all respects there shall be a separation of Church and State. . . . Otherwise the state and religion would be aliens to each other—hostile, suspicious, and even unfriendly. . . . Municipalities would not be permitted to render police or fire protection to religious groups. Policemen who helped parishioners into their places of worship would violate the Constitution. . . . A fastidious atheist or agnostic could even object to the supplication with which the Court opens each session: "God save the United States and this Honorable Court."[24]

Rather than a bright-line rule, the so-called "wall" separating church and state "is a blurred, indistinct, and variable barrier depending on all the circumstances of a particular relationship,"[25] and the location of the line separating church and state must be determined on a case-by-case basis.[26] Hence, *strict* church-state separation has never been required in the United States and is not required now.

The United States as a Nation of Laws

The United States is a nation governed by the rule of law. We are also a nation with a robust, yet diverse, religious heritage. That religious heritage is reflected throughout our society—including within the armed forces of the United States. In *Zorach v. Clauson*, the Supreme Court noted that "we are a religious people whose institutions presuppose a Supreme Being."[27] The *Zorach* Court continued with that theme: "[The government] sponsor[s] an attitude . . . that shows no partiality to any one group and that lets each flourish according to the zeal of its adherents and the appeal of its dogma."[28] Elsewhere, the Supreme Court has held that "the First Amendment's Religion Clauses mean that religious beliefs and religious expression are too precious to be either proscribed or prescribed by the [government]."[29] As noted in *Locke v. Davey*,[30] the establishment clause and the free exercise clause are frequently in tension.[31] Yet, the Court has long said that " 'there is room for play in the joints' " between them.[32] In other words, there are some state actions permitted by the establishment clause but not required by the free exercise clause. Moreover, neutrality in religious matters requires that the state neither favor nor disfavor religion. The First Amendment clearly proscribes favoring religion over non-religion or one religion over others, but it likewise proscribes favoring non-religion over religion.[33] In *Rosenberger v. Rector and Visitors of University of Virginia*,[34] the Court noted that government neutrality is respected, not offended, when even-handed policies are applied to diverse viewpoints, including religious viewpoints.[35]

In the area of religious expression, the Supreme Court has held that "private religious expression receives *preferential* treatment under the Free Exercise Clause" (emphasis in original).[36] In fact, "discrimination against speech because of its message is presumed to be unconstitutional."[37] Of special note, the Supreme Court has "not excluded from free-speech protections religious proselytizing . . . or even acts of worship . . . ."[38] Further, "the [government's] power to restrict speech . . . is not without limits. The restriction must not discriminate against speech on the basis of viewpoint . . . and the restriction must be 'reasonable in light of the purpose served by the forum.' "[39] These views are fully in line with the well-established principle that "there is a crucial difference between government speech endorsing religion, which the Establishment Clause forbids, and private speech endorsing religion, which the Free Speech and Free Exercise Clauses protect."[40] The *Mergens* Court aptly noted that it is not a difficult concept to understand that the government "does not endorse or support . . . speech that it merely permits on a nondiscriminatory basis."[41]

The Military in American Society

Another key legal principle to keep in mind concerns the uniqueness of the military in American society. "'It is the primary business of armies and navies to fight or be ready to fight wars should the occasion arise' . . . and this Court has recognized the limits of its own competence in advancing this core national interest."[42] "Both Congress and this Court have found that the special character of the military requires civilian authorities to accord military commanders some flexibility in dealing with matters that affect internal discipline and morale."[43] In 10 US Code, § 654, Congress expressly noted in its findings that the military is a "specialized society" that "is characterized by its own laws, rules, customs, and traditions, including numerous restrictions on personal behavior, that would not be acceptable in civilian society."[44]

Within that specialized military society, the Department of Defense has chosen to strongly support free exercise of religion by the men and women in uniform, and that DOD position deserves due deference from the courts.[45] In DOD Instruction 1300.17, *Accommodation of Religious Practices within the Military Services*, DOD lays out its policy on free exercise:

> The U.S. Constitution proscribes Congress from enacting any law prohibiting the free exercise of religion. The Department of Defense places a high value on the rights of members of the Military Services to observe the tenets of their respective religions. It is DoD policy that requests for accommodation of religious

practices should be approved by commanders when accommodation will not have an adverse impact on mission accomplishment, military readiness, unit cohesion, standards, or discipline.[46]

The military services concur in the DOD policy. In Air Force Policy Directive 52-1, *Chaplain Service*, the Air Force acknowledges free exercise of religion as "a basic principle of our nation" and then declares that "the Air Force places a high value on the rights of its members to observe the tenets of their respective religions. In addition, spiritual health is fundamental to the well being of Air Force personnel . . . and *essential for operational success*" (emphasis added).[47] The Air Force defines "religious accommodation" as follows:

> Allowing for an individual or group religious practice. It is Air Force policy that we will accommodate free exercise of religion and other personal beliefs, as well as freedom of expression, except as must be limited by compelling military necessity (with such limitations being imposed in the least restrictive manner feasible). Commanders should ensure that requests for religious accommodation are welcomed and dealt with as fairly and as consistently as practicable throughout their commands. They should be approved unless approval would have a real, *not hypothetical*, adverse impact on military readiness, unit cohesion, standards, or discipline.[48] (emphasis added)

Similarly, the Department of the Navy (DON) is fully committed to accommodating the religious practices of Sailors and Marines:

> The DON recognizes that religion can be as integral to a person's identity as one's race or sex. The DON promotes a culture of diversity, tolerance, and excellence by making every effort to accommodate religious practices absent a compelling operational reason to the contrary. . . .
>
> DON policy is to accommodate the doctrinal or traditional observances of the religious faith practiced by individual members when these doctrines or observances will not have an adverse impact on military readiness, individual or unit readiness, unit cohesion, health, safety, discipline, or mission accomplishment.[49]

In Army Regulation 600-20, *Army Command Policy and Procedures*, the Army recognizes the importance of an individual's spiritual state for "providing powerful support for values, morals, strength of character, and endurance in difficult and dangerous circumstances."[50] Like its sister services, the Army "places a high value on the rights of its Soldiers to observe tenets of their respective religious faiths. The Army will approve requests for accommodation of religious practices unless accommodation will have an adverse impact on unit readiness, individual readiness, unit cohesion, morale, discipline, safety, and/or health."[51]

Though not part of DOD, as a uniformed service, the US Coast Guard also supports the free exercise rights of its personnel: "It is Coast Guard policy that commanding officers shall provide for the free exercise of religion by all personnel of their commands."[52]

The remainder of this chapter will focus on the following areas: (1) the importance of the free exercise of religion to developing and strengthening the warrior ethos; (2) the role and responsibility of military commanders and other leaders in maintaining and protecting the moral and spiritual health of their units, including protecting the free exercise rights of the men and women they lead; (3) the general role of chaplains in assisting commanders in executing the commanders' programs to protect and assist free exercise of religion and the role of the individual chaplain in meeting the unique needs of service members from the individual chaplain's own faith group while assisting adherents of other faith groups, and of no faith, to obtain the specific help they may be seeking; (4) the rights enjoyed by all members of the armed forces to exercise their faith; (5) specific examples of permissible religious exercise in the military; (6) specific examples of impermissible religious conduct in uniform; and (7) recommendations to policy makers on how to protect the religious rights of men and women in uniform while maintaining good order and discipline.

Military Roles, Responsibilities, and Rights

Free Exercise of Religion Is Essential for Developing and Strengthening the Warrior Ethos

Gen George S. Patton aptly noted the following: "Wars may be fought with weapons, but they are won by men. It is the *spirit* of the men who follow and the man who leads that gains the victory" (emphasis added).[53] Every professional organization has a purpose, its *raison d'être*. To fulfill that purpose, an organization must establish a specific culture to which its individual members subscribe and in which they flourish.[54] The military is the only institution in civilized society whose ultimate purpose is "to kill people and break things."[55] This organizational purpose is unique among professions; not surprisingly, the military has therefore developed a culture that is also unique. This culture, the very "spirit" embodied by military service members, referred to in General Patton's quotation above, has been dubbed the "warrior ethos."

The warrior ethos comprises beliefs and attitudes that have been passed down through generations of professional war fighters from time immemorial.[56] These beliefs and attitudes can generally be broken into three disciplines: physical, mental, and moral.[57] Physical prowess has long been a necessary trait of a successful warrior. Whether for a Spartan warrior 2,400 years ago[58] or a current member of the US armed services, the rigors of warfare demand that the military professional subscribe to an intense physical regimen.[59] Similarly, professional warriors have cultivated and mastered a specific mental discipline required by the profession of arms. This discipline includes proficiency in one's military specialty[60] as well as a mental toughness that is characterized by "[the ability] to sustain the will to win when the situation looks hopeless and shows no indication of getting better."[61] Lastly, professional war fighters exhibit a certain moral discipline, an "unrelenting and consistent determination to do what is right."[62] War brings difficult choices. Warriors must stand firm, despite temptation to the contrary, in their moral conviction to "*win with honor*" (emphasis added).[63]

There are innumerable examples that define the physical, mental, and moral disciplines of the warrior ethos; yet they may be accurately summarized by the following excerpt from the Soldier's Creed: "I will always place the mission first. I will never accept defeat. I will never quit. I will never leave a fallen comrade."[64] Moral discipline is of utmost importance for the professional warrior—and to the nation. It is critical that one understand the importance of this discipline. Only then can one discern how the conviction to win with honor is developed and, finally, how it is maintained.[65]

What differentiates a murderer from a professional warrior? Both take the life of another human being. *Why they kill* differentiates the one from the other. The murderer may kill on a whim or after detailed planning but usually for his own purposes, while the warrior's killings are constrained by purposes of state and are limited to certain defined instances on the battlefield. What defines the warrior's constraints is *moral discipline*.[66] Without such discipline, that which distinguishes the warrior from the murderer becomes negligible. Moral discipline (1) protects the general population from the warrior's killing and (2) guards the warrior from the psychological damage inherent in being a murderer.[67] Moral discipline is, in essence, the "glue" that holds the warrior ethos together and allows the individual warrior to commit otherwise objectionable acts with honor and integrity.

How then is moral discipline developed and maintained? While some may despise or belittle the thought, for many, there is an important underlying spiritual aspect to the moral discipline of the warrior ethos. This is not to say that a prerequisite for becoming a great warrior is to be religious; there have been, and undoubtedly still are, great professional military men and women who are nonreligious. Nevertheless, it is incontrovertible that many—indeed, most[68]—military service members derive their moral beliefs of right and wrong from personal religious beliefs and values.[69] Hence, to successfully develop and maintain the moral discipline of the warrior ethos within its organizational structure, the military must provide religious care and encourage religious free exercise amongst its members. To do otherwise places at risk the development of those qualities that define and motivate the warrior ethos in the US armed forces.

Leaders of military units must understand that, for the vast majority of those serving within their various commands, the moral discipline of the warrior ethos is inexorably linked with their religious faith.[70] Thus, to create and maintain an effective fighting force, leaders must make provision for the spiritual well-being of their subordinates.[71] The US military has recently taken great care to rekindle a warrior ethos that was, at one time, thought to be endangered.[72] To neglect (or, worse yet, to suppress) the religious aspect of moral discipline would eviscerate the warrior ethos and would significantly degrade the military culture necessary for winning on the battlefield.[73]

Role and Responsibility of Military Commanders and Leaders at All Levels in Ensuring Free Exercise Rights

As noted above, life in the military is markedly different from life as a civilian. Good order and discipline are required in the military to ensure that our armed forces will be able to carry out their vital duty to defend the United States whenever called upon to do so. Critical to ensuring the readiness of our armed forces are the various leaders assigned at all levels of command within each of the armed services. The US military has produced countless military commanders and other leaders who lead by example and model servant leadership for their subordinates. Such leaders take an active interest in their subordinates and their welfare. They demand high standards in training—both of themselves and of the men and women they lead. Further, such leaders give freely of themselves and of their time to mentor their subordinates so that they are properly prepared for the rigors of military life, including, when required, the rigors of combat when life and death decisions demand utmost courage and integrity. Given its level of responsibility, a commander's life is not an easy life. In effect, commanders at every level are responsible for all that their commands do and fail to do; they are responsible for developing and honing the warrior ethos in their commands.

Among the many responsibilities that fall on commanders' shoulders is the *responsibility for the moral and spiritual welfare of their subordinates and their family members.*[74] Irrespective of the individual commander's personal religious faith (or lack thereof), he[75] is nonetheless responsible for ensuring that his subordinates' moral and spiritual needs (as well as those of the subordinates' families) are identified and met. Hence, *it is the commander's responsibility to develop the moral/religious program for his command.* It is not (as is often thought) the military chaplain's responsibility, although the chaplain, as a special staff officer, exists in part to advise and assist the commander in developing and carrying out the commander's program. Moreover, as with every other command responsibility and command program, the commander is responsible to periodically—and personally—check to ensure that his religious program is being properly executed and is achieving the results intended. Failure to do so constitutes dereliction of duty and is a betrayal of the high trust we place in commanders.

Good commanders are team builders. They lead by example.[76] They model caring servant leadership. They spend time and share hardships with their subordinates.[77] They are present where the weather is foulest and the training is toughest. They are there at the toughest times to see that the needs of the men and women in their charge are being adequately met. They are there to ensure that ongoing training meets required standards.[78] They are there to make on-the-spot corrections, where needed, and to give individual and collective praise, where appropriate. They speak to—and with—their subordinates. They listen to what their subordinates have to say, treat them with respect, and answer their questions.[79] Good commanders share the good times—and the bad times—with the men and women they command. By spending time and sharing hardships with their subordinates, good commanders establish mutual trust and confidence.[80] Moreover, American commanders—beginning with Gen George Washington—have recognized that proper moral and spiritual health is a force multiplier on the battlefield, that it enables and emboldens men and women to perform beyond their perceived individual limitations to achieve superior, collective results.[81] And successes in wartime begin with training in peacetime. Thus, effectively caring about moral and spiritual health in peacetime contributes to victory and success in wartime—when it really counts.[82]

Role of Military Chaplains in Furthering Free Exercise

Military chaplains are unique members of the US armed forces. By law, they are commissioned officers without command.[83] As such, the chaplain has no command authority, meaning that the chaplain lacks lawful authority "to order a subordinate unit to execute directives or orders."[84] Each chaplain is a member of the clergy of a specific faith group and serves in uniform to represent and propagate the specific teachings of that faith group.[85] Because Christianity, as represented in its myriad forms, is the most widely practiced religion in the United States,[86] it is also the religion with the most adherents within the US armed forces. Hence, to meet the spiritual needs of the US armed forces, the majority of US military chaplains represent some denominational variant of the Christian faith. Yet because beliefs and practices even among Christian groups and denominations differ widely,[87] it is not fully accurate to speak of "Christianity" per se as the largest faith group represented within the US armed forces. Instead, one should note the relative sizes of the various Christian denominational groups for purposes of comparison—especially when charging that the military is favoring one faith group over another.[88]

Military chaplains wear multiple hats. They serve, first and foremost, to meet the free exercise needs of the men and women in the US armed services.[89] This has been true from the earliest days of our national history and predates the founding of the republic. Consequently, *military chaplains are selected precisely because they represent specific faith groups and specific theological beliefs*. Each military chaplain is commissioned to meet the free exercise needs of adherents of his specific faith group. As members of the *clergy*, military chaplains are not "fungible" assets. Jewish chaplains are not capable of ministering the rites of the Catholic faith to Catholic service members; Methodist chaplains are not capable of ministering the rites of the Islamic faith to Muslim service members; Buddhist chaplains are not capable of ministering the rites of the Baptist faith to Baptist service members; and so on. Nor may they be compelled to do so.[90]

In their free exercise role, military chaplains also wear a second hat. In addition to assisting adherents of their own faith group, military chaplains exist to support service members of other faiths, or no faith, in obtaining the spiritual and/or other assistance that they seek. In that context, military chaplains must be familiar with the beliefs and needs of other faith groups and must do whatever they can to assist the service member in contacting a chaplain or civilian clergyman of that service member's faith when faith-specific needs require it.[91]

Military chaplains, as commissioned officers in their respective service, wear a third hat as well. They fulfill a non-faith-specific role. In addition to their faith group responsibilities, military chaplains are special staff officers who assist their respective commanders in developing and carrying out the commanders' moral/religious programs.[92] They are also trained in the areas of counseling and are often relied upon by their commanders to be a nonthreatening resource to whom service members can turn when they need advice, are in trouble, have emergencies, and so forth.[93]

Because the government commissions military chaplains due to their membership in specific faith groups (i.e., to meet the free exercise needs of the men and women in uniform), and because it is constitutionally inappropriate for the government to delve into the details of religious belief and clergy qualification within a specific faith group (i.e., to avoid violating the establishment clause by entangling the government in religious matters), DOD relies on civilian ecclesiastical endorsing agencies to ensure that chaplains seeking to serve in the armed forces meet the religious standards required by their respective faith groups.[94] Were a chaplain to lose his denominational endorsement, he would be separated from the military.[95] Hence, *denominational affiliation is the irreducible essence of membership in the chaplaincy of the US armed forces, and as such, military chaplains are intentionally hired, and hence expected, to represent a specific denominational view within the military.* Military chaplains are, in the final analysis, members of the clergy of their specific faith groups who conduct their ministries in uniform.

Finally, neither being paid a salary by the military nor wearing a uniform while performing chaplain duties converts a chaplain's religious message into government speech which must be squelched to avoid violating the establishment clause. As the court in *Rigdon v. Perry*[96] aptly noted, "while military chaplains may be employed by the military to perform religious duties, it does not follow that every word they utter bears the imprimatur of official military authority; if anything, the content of their services and counseling bears the imprimatur of the religious ministries to which they belong."[97] From that, the *Rigdon* court concluded that there was "no need for heavy-handed censorship, and any attempt to impinge on the [chaplain's] constitutional and legal rights [wa]s not acceptable."[98]

Rights of Individual Service Members to Exercise Their Faith

When discussing an individual service member's right to free exercise of religion, it must be clearly understood that "free exercise of religion" means what it says—free exercise. Free exercise may not legitimately be limited to what some government official or civilian advocacy group or attorney may think it *should* mean—or is willing to tolerate.[99] Further, the right to free exercise of religion applies to all members of the armed services—including general or flag officers, commanders, and chaplains—because the First Amendment guarantees the right to free exercise to every American, irrespective of that person's station in life.

Subject to the demands of military service[100] and the need to maintain good order and discipline,[101] free exercise of religion for service members includes, but is not necessarily limited to, the following: the right to believe or not believe; the right to engage in corporate or individual worship; the right to study religious texts, both individually and with others; the right to fellowship with members of the same faith; the right to discuss and share basic truths of one's faith, both with fellow adherents of that faith and with nonadherents as well; the right to teach one's faith as truth; the right to observe religious holidays, feasts, ceremonies, and so forth; the right to attend religious retreats and conferences; the right to invite others to participate in a religious activity associated with one's faith, such as a Bible study, a bar mitzvah, or a holiday celebration (like a Seder meal or a Christmas party or an Iftar celebration); the right to pass on one's faith to one's own children and other children placed for that purpose in one's care (such as in Sabbath school, Sunday school, catechism classes, or youth groups like Young Life or Club Beyond); and the right to participate in activities sponsored by local religious groups or parachurch groups (like the Knights of Columbus, the B'nai B'rith, the Navigators, or the Officers' Christian Fellowship).

For certain groups and individuals, sharing their faith with others is a religious command. To officially proscribe the sharing of a chaplain's (or other service member's) faith may itself run afoul of the establishment clause in that government officials sit in judgment of what constitutes acceptable religious belief and activities and what does not. This is not to say that a religious activity might not, under some circumstances, upset good order and discipline, just as a secular activity may do so. When that occurs *in either case*, of course, commanders may intervene, but commanders must be careful not to limit free exercise merely because some individual or group does not appreciate or want to be bothered by the message shared.[102] Persons can be offended by both religious and secular sentiments.[103] Tolerance must be a two-way street. Just as adherents of the majority religious faith must understand and respect the rights of those of minority faiths, or no faith, so too must those of minority faiths and of no faith understand and respect the rights of those professing the majority faith.

Examples of Permissible Religious Exercise

Praying by Chaplains at Military Ceremonies and Other Events

Many of the complaints about religious exercise in the military center around prayers proffered by military chaplains at ceremonies or other events where adherents of many different faiths, or persons of no faith, are present.[104] Yet such prayers have been permitted since the founding of our nation. Further, the fact that the first Congress established the tradition of clergy-led prayer at presidential inaugurations—in themselves, change of command[105] ceremonies between outgoing and incoming commanders in chief—indicates that contemporaries of the First Amendment did not regard such prayers as violating the establishment clause. Moreover, in light of the fact that the first Congress commissioned the first chaplain of the Army,[106] and subsequent Congresses appointed the first Navy chaplain and directed that divine worship take place aboard Navy ships,[107] it is inconceivable that those who drafted the First Amendment intended it to prohibit chaplain-led prayers at military ceremonies. The *Marsh* Court has aptly recognized that actions of the first Congress are "contemporaneous and weighty evidence" of the Constitution's "true meaning."[108]

Given our long and unbroken history of permitting prayers to solemnize military ceremonies and other events, calling on chaplains to continue such historical practice today merely reflects long-held traditions and constitutes "tolerable acknowledgment[s] of beliefs widely held among the people of this country."[109] Hearing such prayers is also the price one pays for living in a pluralistic society that honors free exercise of religion and free expression of religious sentiments. It is, in fact, a testimony to the religious tolerance that we have been able to achieve in the United States and is something that should be recognized and applauded, not rejected and forbidden.

Some worry that prayers said at military ceremonies will cause discomfort to, or offend, attendees of different faiths, or of no faith. Yet potential discomfort about things one does not like to hear is, once again, the price one pays for the rights of free speech and free exercise in a pluralistic society. The First Amendment protects speech, including religious speech; it does not—and was never intended to—protect potential hearers against discomfort at what is spoken. Generally, if everyone agrees with what is said, such sentiments need no constitutional protection. Only speech and sentiments which are disfavored or disliked require such protection. In *Lee*, the Supreme Court explicitly declared that it did "not hold that every state action implicating religion is invalid if one or a few citizens find it offensive. People may take offense at all manner of religious as well as nonreligious messages, but offense alone does not in every case show a violation."[110] Hence, one must proceed cautiously when one tries to proscribe speech based on highly suspect and subjective standards, such as the potential "discomfort" of the hearers.[111]

The US Navy, for example, has an unbroken tradition of saying a prayer aboard each Navy ship each day.[112] That tradition is consistent with the sanctions of Congress concerning religious activity on board naval ships that were enacted shortly after the adoption of the First Amendment.[113] That in itself is strong evidence that such prayers were not considered as violating the establishment clause. Similarly, the US Naval Academy has a 164-year tradition of having a Navy chaplain recite a short prayer before noon meals at the Naval Academy.[114] These activities are long-standing traditions in the US Navy and serve to remind Sailors and Marines of their proud heritage as well as accommodate "beliefs widely held" by the American people.[115]

Praying by Chaplains as Their Faith Tradition Requires or Permits

Some argue that to avoid giving offense chaplains must—at a minimum—offer only "nonsectarian" prayers when praying at events where adherents of other faiths, and persons of no faith, are present. There are numerous problems with such an argument. One problem is that it is not clear how or when an otherwise "sectarian" prayer becomes "nonsectarian"—or who is to judge. As the Tenth Circuit has aptly noted, "All prayers 'advance' a particular faith or belief in one way or another" if for no other reason than "the act of praying to a supreme power assumes the existence of that supreme power."[116] A second problem is that offense at what is being said has never been a valid reason to proscribe such speech. The same is true today. Were our government or the US armed forces ever to adopt the nonsectarian prayer standard, they would then be in violation of the establishment clause by preferring one form of prayer (the nonsectarian form) over alternative forms of prayer (the sectarian forms). Such a policy would not only violate the establishment clause but also the free exercise and free speech rights of every chaplain.

The Supreme Court has held that "the First Amendment's Religion Clauses mean that *religious beliefs and religious expression are too precious to be either proscribed or prescribed by the [Government]*" (emphasis added).[117] *Lee* involved the giving of a "nonsectarian" prayer at a high school graduation ceremony. Much of the criticism about the prayer in *Lee* centered not only on the fact that school officials selected which clergyman would deliver the prayer but also on the inappropriateness of the school principal's telling the rabbi that he should render a "nonsectarian" prayer.[118] The *Lee* Court concluded, "The question is not the good faith of the school in attempting to make the prayer acceptable to most persons, *but the legitimacy of its undertaking that enterprise at all*" (emphasis added).[119] This comment applies with equal force to the oft-expressed desire that military chaplains deliver "nonsectarian" prayers in settings where adherents of other faith groups are present. No one questions the military's good intentions, but as the *Lee* Court concluded, adopting such a policy is simply unconstitutional.

Further, any attempt to restrict religious speech (such as a prayer) to avoid causing offense to the hearer is sure to fail. First, the free speech clause of the First Amendment protects free expression from government interference. And there is no language in the First Amendment that protects a hearer from being offended. In truth, inoffensive speech needs no protection. If everyone were to agree with the sentiment expressed, no one would challenge it, and no protection would be needed. It is *offensive* speech that needs protection. Praying in Jesus' name is offensive to some but not to others. Invoking the name of Allah also offends some people but not others. Still others—atheists and agnostics—may be offended by any and all prayer, no matter to what deity it may be directed. Hence, try as one might, one cannot avoid offending someone. Advocating a "cause no offense" strategy will surely fail. More importantly, it is unconstitutional.

As Supreme Court Justice O'Connor aptly noted in *Elk Grove Unified School District v. Newdow*,[120] "given the dizzying religious heterogeneity of our Nation, adopting a subjective approach would reduce the [reasonable observer] test to an absurdity. Nearly any government action could be overturned as a violation of the Establishment Clause if a 'heckler's veto' sufficed to show that its message was one of endorsement."[121] Further,

> there is always someone who, with a particular quantum of knowledge, reasonably might perceive a particular action as an endorsement of religion. A State has not made religion relevant to standing . . . simply because a particular viewer of a display [or hearer of a religious sentiment] might feel uncomfortable.
>
> *It is for this reason that the reasonable observer in the endorsement inquiry must be deemed aware of the history and context of the community and forum in which the religious [activity] appears.*[122] (emphasis added)

Likewise, service members are deemed to be "reasonable observers." Consequently, they are deemed to know that chaplains represent different faith groups and traditions and that prayers offered at certain military ceremonies are part of military tradition meant to solemnize the event, *not to endorse the faith or religious sentiments of the chaplain delivering the prayer*. Thus, the establishment clause is not violated by an individual chaplain's private choice of words for a prayer to solemnize a military ceremony.

Prayers at presidential inaugurations (which constitute, in fact, change of command ceremonies at the highest level of the armed forces) have been delivered by clergymen of many different faiths and have frequently included references to Jesus or the Trinity.[123] *Marsh* refutes the contention that clergy-led, ceremonial prayer violates the establishment clause merely because a particular prayer might reference monotheistic terminology or beliefs. In *Marsh*, the Court rejected the argument that selection by the Nebraska legislature of a Presbyterian clergyman who chose to pray in the "Judeo-Christian" tradition violated the establishment clause. The Court declared: "We cannot, any more than Members of the Congresses of this century, perceive any suggestion that choosing a clergy man of one denomination advances the beliefs of a particular church."[124] The Court noted that "the content of the prayer is not of concern to judges where, *as here*, there is *no indication that the prayer opportunity has been exploited to proselytize* or advance any one, *or disparage* any other, faith or belief" (emphasis added).[125] The same holds true in the military. Moreover, were the government to outlaw prayer altogether at military ceremonies and other events, it would demonstrate hostility, not neutrality, towards religion in light of the long history of such prayers in the military and in light of the Supreme Court's recognition that solemnizing, nonproselytizing prayers do not violate the establishment clause.

Many of the complaints about prayers in the military revolve around the issue of praying "in Jesus' name."[126] Not every Christian chaplain feels compelled to pray explicitly in Jesus' name, but some do. Such differences reflect the religious pluralism not only within American society but also within Western Christianity. Ending a prayer in Jesus' name (or a similar phrase)—without more—is not proselytizing. To proselytize is defined as "to make or try to make converts."[127] To assert that merely adding the words "in Jesus' name" to a prayer said in the presence of adherents of different faiths, or persons of no faith, constitutes proselytizing is absurd. Orthodox Christian theology teaches that Jesus *is* God[128]—hence, praying in Jesus' name is another form of praying in God's name. There is no principled reason why invoking Jesus by name is any different than invoking the name of Adonai or Allah or Vishnu, something few are suggesting should be forbidden.

Saying a prayer that ends in Jesus' name clearly identifies the religious faith of the person praying, just as beginning a prayer with the words "in the name of Allah the compassionate, the merciful" identifies the person praying as a Muslim, or invoking the "God of Abraham" before reciting the Shema identifies the person praying as Jewish. None of these prayers—without more—can be remotely construed as constituting proselytizing. Yet were any of these chaplains to pray in such a manner that the prayer was meant to convince the hearer to adopt the chaplain's specific faith, such a prayer would constitute proselytizing, whether Jesus, the God of Abraham, or Allah were specifically mentioned or not. Hence, fixating on praying explicitly in Jesus' name, without more, is without merit.

Because chaplains are intentionally brought into the armed forces as members of different religious faith groups, the military knows and indeed expects that those chaplains will proclaim and practice the tenets of their respective religious faiths in the military.[129] Hence, in such circumstances, as an accommodation to the chaplain's religious obligations, the chaplain must be allowed leeway to pray as his conscience and faith tradition require.[130]

The Constitution prohibits any federal official—including senior civilian leaders, military commanders, and senior chaplains—from directing that a chaplain either pray or refrain from praying in a certain manner, except when required to maintain good order and discipline in the respective service. *This position comports fully with the Constitution—it avoids government entanglement with religion, religious beliefs, and religious practices, while upholding the free speech and free exercise rights of military chaplains.*

Chaplains May Prefer Their Own Faith Group in Appropriate Circumstances

Although chaplains exist in part to assist commanders in executing their command religious programs for all service members in their respective commands, there are nevertheless times when a chaplain may legitimately focus exclusively on his own faith group. The most obvious example is when the chaplain is conducting worship services for adherents of his respective faith and others who are interested in attending such services. Yet chaplains, as staff officers charged with implementing the commander's religious program, should also be free to advertise religious activities of a specific denominational character via e-mail (and other communications channels) to the same extent that nonreligious activities are permitted to be advertised. For example, a Southern Baptist chaplain should be able to advertise a retreat aimed at Southern Baptist service members and their families; a Jewish chaplain should be able to advertise High Holy Day service opportunities to Jewish service members; a Muslim chaplain should be able to advertise events surrounding the observance of Ramadan; and so forth. In each instance, the advertisement need not be inclusive of other faith groups, or sensitive to those of no faith, and the chaplain should be able to freely share religious sentiments about the events advertised. Moreover, such advertising does not run afoul of the establishment clause.[131]

The same is true when a chaplain is teaching the truths of the chaplain's specific faith group to interested service members or their family members. Chaplains are selected by faith group to meet the religious needs of adherents of that faith group. Hence, the chaplain need not be inclusive of nonadherents during such times and may be exclusive, without violating the Constitution.

Commanders and Other Leaders May Speak of Religious Matters with Subordinates

Given the hierarchical nature of the military, some argue for the complete prohibition of superiors' discussing their faith with their subordinates or otherwise engaging in religious endorsements in the company of subordinates. Although senior officers and noncommissioned officers must be careful not to impose their religious views on subordinates, an absolute prohibition on all sharing of faith by a superior to a subordinate is patently unconstitutional and an egregious violation of the free exercise and free speech clauses.[132] Aside from the difficulty in defining exactly when discussion of religious matters would cross the line from protected religious expression to prohibited "proselytizing" and "religious endorsements," however such terms are defined, the First Amendment clearly protects such activity.[133]

Opponents of such interaction simply ignore the fact that it is the *commander* who bears full responsibility for the *moral and spiritual* welfare of his subordinates and their family members.[134] Such persons also fail to take into account that frequent, intimate interaction with one's subordinates is what helps to solidify one's command and create a healthy, effective unit.[135] Hence, speaking on topics of morality and spirituality with subordinates is a necessary part of the commander's job,[136] irrespective of the commander's personal belief system. Further, some of those who complain about such interaction are hypersensitive or hostile to religious matters and may see proselytizing or religious endorsement where there is none.[137] Individual hypersensitivity to religious discussions and sentiments must not be permitted to interfere with the commander's responsibility to develop and implement an effective program to meet the moral and spiritual needs of the men and women under his command.

An absolute ban on interaction between superiors and subordinates about religious matters, a ban that clearly violates the constitutional rights of free speech and free exercise, is worse than the putative disease. It denies the commander the access he needs to fulfill his responsibility to develop and implement an effective moral/spiritual program for his command. Surely, the military and civilian chains of command are fully capable of handling isolated incidents of abuse of a superior's position vis-à-vis a subordinate without resorting to a draconian sanction of prohibiting all such interaction between superiors and subordinates. When superiors overstep the bounds of their authority, for whatever reason, the means already exist in the US armed forces to appropriately sanction such behavior. Such means run the gamut from verbal or written reprimand to relief for cause, to administrative reduction in rank, to court-martial. Recent examples of investigating and/or disciplining senior military officers for misbehavior should suffice to demonstrate that the military services can take care of such problems as they arise, thereby avoiding the need for adopting an absolute policy of forbidding interaction between superiors and subordinates regarding issues of morality and spirituality.[138]

Moreover, there is no legitimate reason why commanders cannot mention their educational, professional, and religious backgrounds when introducing themselves to their subordinates. The *Army Leader Transitions Handbook*, a book for leaders based on the "best practices and proven techniques from military and civilian sources,"[139] declares, for example, that "talking to all your subordinates . . . about what is important to you and what you value as their leader will help establish trust."[140] The handbook recommends that military leaders discuss the following topics with their subordinates: (1) the leader's background;[141] (2) the leader's expectations and standards;[142] (3) the leader's values;[143] (4) the leader's view of ethics;[144] (5) the leader's objectives for the unit;[145] (6) the leader's thoughts on integrity;[146] (7) the leader's priorities;[147] (8) the leader's standards of discipline;[148] (9) the leader's thoughts on training, education, and safety;[149] (10) the leader's thoughts on leadership;[150] and (11) the leader's thoughts on caring for Soldiers and their families.[151] Sharing such thoughts is essential to informing one's subordinates of what is expected of them from the leader's perspective and what they can expect from the leader in return.[152]

Finally, an obvious example where commanders *must* speak to their subordinates about religious beliefs often occurs aboard ship. On board US Navy ships at sea, "divine services shall be conducted on Sunday[s] if possible."[153] Because so many Navy ships deploy without a "chaplain attached to the command[,] . . . [s]ervices led by laypersons are encouraged."[154] Regardless of whether a chaplain is embarked, the commanding officer is still responsible for ensuring "the religious preferences and the varying religious needs of individuals [are] recognized, respected, encouraged and ministered to."[155] Therefore, a commander must ensure that a religious lay leader is capable of adequately fulfilling a role like that of a chaplain so that the free exercise rights of his subordinates are protected. That commander must be free to communicate—in depth—with potential lay leaders to ensure the best quality spiritual care for those under his command.

All Service Members May Participate in Local Religious Groups and/or Parachurch Groups on Their Free Time

Despite the herculean efforts made by commanders and military chaplains to provide for the free exercise needs of all service members and their families, there are times when their efforts fall short of the service members' religious needs and desires. As such, when possible, service members often avail themselves of religious opportunities in nearby civilian communities and/or participate in parachurch groups to meet their spiritual needs. Many religious groups in communities located near military installations offer outreach programs to service members and their families, most of whom are far away from their families and friends. Such efforts are to be lauded and encouraged. There are a limited number of chaplains available at any military installation, and it is virtually impossible for them to meet the needs of each denomination or faith group represented by service members on that installation. Local and parachurch groups help to fill that gap. Such groups may also fill the gap by providing a greater array of religious opportunities throughout the week than can normally be provided by chaplains, thus accommodating the often chaotic schedules that define service members' lives. In some instances, without external help, chaplains would simply be unable to meet the spiritual needs of the men and women in uniform that constitute their respective flocks. For example, the Pentagon chaplain's office comprises three persons whose mission it is to serve the men and women assigned to and working in the Pentagon. Thus, three persons are expected to provide spiritual support to over 24,000 persons,[156] an impossible task. As such, the Pentagon chaplain must rely on volunteers—often from local religious and parachurch groups—to carry out his ministry. DOD and the armed services should applaud and encourage the efforts of such groups to minister to the spiritual needs of the men and women in uniform and their families. Working together, they help to ensure that the First Amendment's guarantee of free exercise of religion can be realized by those serving all of us in uniform.

Examples of Impermissible Religious Conduct

No Proselytizing Prayers or Disparaging Other Faiths

Prayers offered by chaplains at military ceremonies and other events are permissible as "a tolerable acknowledgment of beliefs widely held among the people of this country,"[157] even when they are clearly sectarian in nature. Hence, Christian chaplains who believe that they should pray "in Jesus' name" (or use a similar phrase like "through Jesus Christ our Lord") may do so without violating the establishment clause, just as Jewish chaplains may invoke the "God of Abraham, Isaac, and Jacob" and Muslim chaplains may invoke "Allah," without violating the Constitution. No chaplain, however, may proselytize while praying at such ceremonies or disparage other faiths.[158]

Teaching the strictures and beliefs of one's own faith, even when they contradict beliefs of another faith group, does not constitute disparaging the other faith, *provided that* such teaching occurs in a place where people freely gather on their own accord to receive such teaching. For example, a Christian chaplain's affirmative teaching to Christians and/or other interested persons that Jesus is the only way to heaven, a core Christian teaching, does not disparage Islam, despite Islamic teachings about Jesus to the contrary, just as a Muslim chaplain's affirmative teaching to Muslims and/or other interested persons that Mohammed is the last and greatest prophet of God, a core Islamic teaching not shared by Christians, does not disparage Christianity. Such faith-specific *teaching* is *inappropriate*, however, in settings where service members and their families are otherwise required to be present (i.e., where they are a captive audience).

No Compulsion in Belief or Practice

No official in the US government or armed forces—regardless of rank or station—has the right to compel or pressure any other person (1) to assent to any specific philosophy or religious belief or creed,[159] (2) to participate in a religious worship service (such as forcing someone to attend a chapel worship service—unless that person is on duty, for example, serving as a member of an honor guard or a color guard at a funeral or other ceremony), or (3) to engage in a religious act (even so simple an act as being asked to join hands with others when a short prayer of blessing is said over a Thanksgiving or Christmas meal in the military dining facility).

Merely being present at a military ceremony or event where a military chaplain says a solemnizing prayer, however, does not violate the First Amendment, since no person is being compelled or pressured to *assent* to any belief, no person is being asked to *participate* in religious worship, and no person is being asked to *engage* in a religious act.[160]

Likewise, no official in the US Government or armed forces—regardless of rank or station—has the right to compel or pressure a chaplain (or any other person, such as a lay religious leader on a naval vessel or someone else asked to pray) to pray in any particular manner. Instead, the chaplain or other person should be free to follow his conscience and the traditions of his specific faith group and to pray as he deems appropriate in the circumstances. Allowing a person to pray as he desires does not violate the establishment clause, whereas directing how he prays or pressuring him to pray in a certain way does violate the establishment clause.[161]

No Forcing of Subordinates to Hear Unwanted Religious/Philosophic Message as Part of Captive Audience

No commander or leader may require a subordinate to attend or remain in a meeting or other gathering (i.e., create a captive audience) when the commander or leader intends to use the opportunity to convince those in attendance to adopt or assent to his religious faith or secular philosophy.

This should not be understood to preclude a commander or leader from being able to mention his religious faith or upbringing when introducing himself to subordinates for the first time.[162] Such information informs the commander's/leader's subordinates about himself and his standards and is permissible, *provided that* the commander or leader makes clear that he will not judge his subordinates on anything other than that person's duty performance, character, and integrity.

Recommendations

Teach and Foster Tolerance of Differences, Including Religious Differences, during All Phases of a Service Member's Military Career

All of the armed services are in the team-building business. Each service must take men and women from all walks of life and all types of backgrounds and meld them into an effective team. Part and parcel of such a process is educating service members about their differences and building understanding, tolerance, and respect for each other despite those differences. Such differences manifest themselves, *inter alia*, through race, ethnicity, creed, gender, and culture. They mirror the American motto: *E pluribus unum*. Each service member must learn to tolerate and respect the differences exhibited by his fellows in uniform.

The same is true with respect to religion and chaplains. Religiously, we are a heterogeneous nation, and the military and its chaplains reflect that heterogeneity. Adherents of different faiths approach God differently. That is reflected in many ways, including how they pray. Rather than try to restrict how an individual chaplain prays at certain public events, the chaplain should pray consistent with his conscience and religious tradition. This presents a great opportunity to demonstrate, recognize, and celebrate diversity within the military.

All of the armed services have both entry-level schooling for enlisted service members and for officers as well as follow-on schooling as officers and enlisted service members increase in rank and assume greater responsibilities. Part of the team-building process is noting our differences and encouraging service members of all ranks to respect and tolerate those differences. Each member of the military takes an oath to defend the Constitution of the United States against all enemies, foreign and domestic. It should be a relatively simple task to teach enlisted service members and officers about the First Amendment's religion clauses and how they play themselves out in the individual service member's daily life. Service members can be taught that commanders are responsible to develop and implement moral and religious programs to meet their free exercise needs; that military chaplains traditionally offer prayers at various military ceremonies (such as at change of command ceremonies) to solemnize such events; that, due to the heterogeneous nature of religious beliefs in the United States, they are apt to hear prayers said from various religious perspectives; and that such prayers are evidence of the religious tolerance that our country has been able to achieve over time, *not an indication that our govern-ment, DOD, or the armed services favor a certain faith group or belief.*

Reminding the men and women in uniform that chaplains come from dif-fering religious traditions and that their prayers reflect those traditions should be embraced and celebrated, since what we have achieved in the United States differs markedly from many cultures where certain religious groups are often denigrated and marginalized, if not outright persecuted. Because commanders set the tone within their commands, they too should receive training at com-mand and staff schools concerning the roles of the chaplains within their com-mands as well as their responsibilities to ensure that their subordinates and their families may freely exercise their religious faiths. Commanders play the key role in ensuring that a chaplain's free speech and free exercise rights are not violated as well as ensuring that those under their commands understand that allowing a chaplain to pray as he deems appropriate does not constitute governmental sanction of any particular faith group or religious belief. If this is done even-handedly by commanders, there should be no reason—real or perceived—to direct how a chaplain should pray. Likewise, there should be no reason for any service member to misinterpret or misunderstand why a prayer is being offered or how the respective armed service views such prayer. After all, it is not a dif-ficult concept to understand that the government "does not endorse or support . . . speech that it merely permits on a nondiscriminatory basis."[163] Similarly, re-minding the men and women in uniform that their colleagues in uniform also reflect differing religious faiths, including no faith, and that such differences reflect our tolerant society should also be embraced and appreciated.

Tolerance is a two-way street, and military commanders must act as vigorously to protect the majority's free exercise rights as they do to pro-tect the rights of those in the minority. It is a given that the majority reli-gious faith in the United States (and, hence, in the armed forces) is the Chris-tian faith, in all its myriad forms. As such, it is the Christian message that will—simply by virtue of the sheer numbers of its adherents—be foremost among the religious sentiments publicly expressed in the military. That does not mean that the military is "favoring" the Christian faith merely because it is so visible, and commanders must always remember that their support of a ser-vice member's free exercise rights does not mean that the military is establish-ing religion. Facilitating the free exercise rights of Christians (and of adherents of other faith groups) is a command responsibility and, without more, does not implicate the establishment clause.

Because the largest religious faith in the US armed forces is some variant of the Christian faith, most complaints are lodged against Christian chaplains and their prayers. Yet despite opponents' attempts to lump all Christians to-gether in one basket, if one listens closely, one will note that there are a wide variety of messages being shared and proclaimed because not all professing Christians share the same theology, practices, or biblical interpretation.[164] Hence, to determine whether improper religious favoritism really exists, one must identify the specific Christian denomination that is allegedly being improperly advanced; it is not enough to assert that "Christianity" per se is being favored, as is the habit of some.[165]

In sum, a well-planned and executed program for educating service members—at all phases of their careers—about our religious heritage, chaplains and their roles, commanders' responsibilities for the moral and spiritual welfare of those they command, and the First Amendment will reduce confusion about religious expression in the military and increase appreciation for what we as a nation, unlike too many others, have been able to achieve in the area of religious tolerance. This relatively easy fix should resolve problems of perceived religious discrimination. Regarding those isolated times when actual religious discrimination occurs, DOD and the uniformed services have ample tools to remedy such violations, and those tools should be used as required.

Trust Military Leaders to Know What Works in Training Effective Teams to Fight Our Nation's Wars

One final topic needs to be addressed: that of training and preparing service members to assume the warrior ethos described earlier and to carry out their vital mission of national defense. Each military service is organized, equipped, and staffed to meet recognized military needs. Through long experience, military professionals learn how to train the men and women in uniform to accomplish the missions assigned to them. Because of the uniqueness of military life, what military leaders require for success has no civilian analog. It is, therefore, imperative that military leaders have the freedom to operate and train in ways that meld disparate individuals and units into combat-ready fighting formations, capable of achieving victory, whenever required. To do this, military commanders need sufficient leeway to apply principles proven over time and lessons learned from previous combat to conduct intense, realistic training in peacetime to ensure that our forces are ready to defeat the enemy in wartime. To that end, both the Congress and the courts have recognized that military commanders need flexibility to hone their forces to fighting trim.[166]

The defense of the nation is the highest priority of government,[167] and the Supreme Court has correctly recognized "the limits of its own competence in advancing this core national interest."[168] Many of the complaints raised against DOD in US courts involve service members dissatisfied with, and complaining about, something they experienced as part of their training.[169] In such circumstances, the trainee is, in effect, criticizing the training being conducted. This in itself constitutes a challenge to the military chain of command, suggests a potential breakdown in good order and discipline within the affected unit, and counsels caution before jumping in to remedy the alleged "violation" of the complaining service member's rights. It is wrong (as a matter of policy and common sense) for civilian advocacy groups and civilian attorneys to sue in court seeking to apply civilian standards to military units. Life in the military and life in the civilian world are different, and they need to remain different.

The armed forces of the United States have a proven record of success honed over time. Training methods are entrusted to persons in each service who have proven themselves capable of assuming such heavy responsibilities. Courts and civilian society should defer to their experience and training and should not second-guess their judgment merely because it does not mirror what might be acceptable in civilian society.

In sum, military commanders are entrusted with training our sons and daughters to defend the nation as required. Senior military commanders are masters of the profession of arms. They are competent, smart, and dedicated. They are committed to defending the nation and the Constitution, to the point of laying down their lives on behalf of us all. They deserve our trust in developing and implementing the training regimens that they—in their professional opinions—believe will protect us. When commanders determine that a solemnizing prayer at certain ceremonies is appropriate as a team-building tool, for example, they are acting in accordance with military traditions that predate the founding of the republic, traditions that have been considered important to team-building throughout our history and are consistent with long-held values of the majority of our population—both in civilian society and in uniform. Given the unique nature of the military, such reasoned judgments should be supported, not challenged in court. Nothing in the Constitution requires that Americans shed their religious beliefs and heritage once they don a military uniform, and military commanders have recognized the positive role of religious faith on morale and service consistently over the course of our history.[170] Commanders and leaders at all levels of our armed forces are responsible for the moral and spiritual health of their commands, and they deserve our support and our deferring to their professional judgment when it comes to planning and implementing those training regimens that they believe are necessary to defend the nation.

* * * * *

In conclusion, the foregoing examples and recommendations are consistent with our history and fully in accord with the Constitution and laws of the United States. An aggressive education program performed at every level of the service member's career should remove any misunderstanding about religious observance and expression in the military and should help each service member to understand and appreciate the degree of religious liberty and tolerance that our nation, unlike many others, has been able to achieve.

Notes

1. Representative among individuals advocating strict church-state separation are the following: Rev. Barry W. Lynn—see his *Piety & Politics: The Right-Wing Assault on Religious Freedom* (New York: Random House, 2007), advocating the importance of the strict-separationist viewpoint and decrying challenges to that philosophy by the "religious right"; Michael "Mikey" Weinstein, see Michael L. Weinstein and Davin Seay, *With God on Our Side: One Man's War Against an*

Evangelical Coup in America's Military (New York: Thomas Dunne Books, 2006), detailing Weinstein's legal fight against a perceived Evangelical Christian takeover of the military, generally, and the US Air Force Academy, specifically; Christopher Hitchens, "GI Jesus: The Real Problem with Military Chaplains," *Slate*, 2 October 2006, http://www.slate.com/id/2150801/?nav=ais, criticizing the National Defense Authorization Act of 2007 due to "lawmakers arguing seriously over how much religious instruction and rhetoric should be permitted in the [military] ranks and how explicitly monotheistic that instruction and rhetoric ought to be." Representative among groups advocating strict separation of church and state are the following: Americans United for Separation of Church and State, Freedom from Religion Foundation, American Civil Liberties Union, and Military Religious Freedom Foundation.

2. See, for example, *Chalker v. Gates*, No. 08-CV-2467-KHV-JPO (D. Kan. filed 25 September 2008).

3. See, for example, letter demanding the cessation of the Naval Academy's traditional noon-meal prayer from Deborah A. Jeon, legal director, ACLU of Maryland, to Vice Adm Jeffrey Fowler, superintendent, US Naval Academy, 2 May 2008, on file with author.

4. US Constitution, Amendment I ("Congress shall make no law respecting an establishment of religion").

5. Thomas Jefferson, president of the United States, to Danbury Baptist Association of Connecticut, letter, 1 January 1802, in *The American Republic: Primary Sources*, ed. Bruce Frohnen, 2002, 72, 75.

6. Most agree that, at a minimum, the establishment clause was intended to prohibit the creation of a national church for the United States, such as existed in England. Nevertheless, one must keep in mind that the First Amendment did not preclude individual states from adopting a state church or a state religion. See Carl Zollman, *American Church Law* (St. Paul, MN: West Publishing Co., 1933) (first edition in 1917), 2–4. In fact, Massachusetts was the last state to disestablish its state church, and it did so of its own accord in 1833, more than 40 years after the ratification of the First Amendment. Kelly Olds, "Privatizing the Church: Disestablishment in Connecticut and Massachusetts," *Journal of Political Economy* 102, no. 2 (1994): 277, 281–82.

7. *Marsh v. Chambers*, 463 U.S. 783, 787-88 (1983).

8. Ibid., 788.

9. Ibid. (citation omitted). The First Amendment is part of the Bill of Rights.

10. Ibid. See also ibid., 790 ("It can hardly be thought that in the same week Members of the First Congress voted to appoint and to pay a chaplain for each House and also voted to approve the draft of the First Amendment for submission to the States, they intended the Establishment Clause to forbid what they had just declared acceptable.").

11. Ibid., 792.

12. *See Newdow v. Bush*, 355 F. Supp. 2d 265, 270 n.5, 286–87 (D.D.C. 2005).

13. For example, Catherine Millard, *The Rewriting of America's History* (Camp Hill, PA: Horizon Books, 1991): 61–62.

14. Proclamation of President John Adams (6 March 1799), in *A Compilation of the Messages and Papers of the Presidents 1789–1897*, vol. 1, James D. Richardson, ed., 1899, 284–86.

15. See Daniel L. Dreisbach, *Real Threat and Mere Shadow: Religious Liberty and the First Amendment* (Westchester, IL: Crossway Books, 1987): 127, noting that the 1803 treaty with the Kaskaskia Indians included federal funds to pay a Catholic missionary priest; noting further treaties made with the Wyandotte and Cherokee tribes involving state-supported missionary activity.

16. John W. Whitehead, *The Second American Revolution* (Charlottesville, VA: The Rutherford Institute, 1982), 100, citing J. O. Wilson, *Public School of Washington*, vol. 1 (Washington, DC: Columbia Historical Society, 1897): 5.

17. Charles E. Rice, *The Supreme Court and Public Prayer: The Need for Restraint* (New York: Fordham University Press, 1964): 63–64.

18. *Act of March 2, 1799*, ch. XXIV, 1 Stat. 709, requiring commanders of ships with chaplains on board "to take care that divine service be performed twice a day, and the sermon preached on Sundays"; and *Act of March 23, 1800*, ch. XXXIII, 2 Stat. 45, directing commanders of ships to require the ship's crew "to attend at every performance of the worship of Almighty God."

19. *Marsh v. Chambers*, 463 U.S. at 790 (citation omitted); see also *United States v. Curtiss-Wright Export Corporation*, 299 U.S. 304, 328 (1936), noting that understanding "placed upon the Constitution . . . by the men who were contemporary with its formation" is "almost conclusive" (citation omitted).

20. *Corporation of Presiding Bishop v. Amos*, 483 U.S. 327 (1987).

21. Ibid., 335, quoting *Hobbie v. Unemployment Appeals Commission of Florida*, 480 U.S. 136, 144–45 (1987).

22. Ibid., 338.

23. *Zorach v. Clauson*, 343 U.S. 306 (1952).

24. Ibid., 312–13; See also ibid., 314, noting "no constitutional requirement which makes it necessary for government to be hostile to religion and to throw its weight against efforts to widen the effective scope of religious influence."

25. *Lemon v. Kurtzman*, 403 U.S. 602, 614 (1971).

26. Ibid.

27. *Zorach v. Clauson*, 343 U.S. at 313.

28. Ibid.

29. *Lee v. Weisman*, 505 U.S. 577, 589 (1992).

30. *Locke v. Davey*, 540 U.S. 712 (2004).

31. Ibid., 718.

32. Ibid., quoting *Walz v. Tax Commission of New York*, 397 U.S. 664, 669 (1970).

33. See, for example, *Abington School District. v. Schempp*, 374 U.S. 203, 299 (1963) (Justice Brennan, concurring), noting that the state may "neither favor nor inhibit religion."

34. *Rosenberger v. Rector and Visitors of University of Virginia*, 515 U.S. 819 (1995).

35. Ibid., 839.

36. *Capitol Square Review & Advisory Board v. Pinette*, 515 U.S. 753, 767 (1995).

37. *Rosenberger v. Rector and Visitors of University of Virginia* 515 U.S. at 828, citing *Turner Broadcasting System, Inc. v. FCC*, 512 U.S. 622, 641–43 (1994).

38. *Capitol Square Review & Advisory Board v. Pinette*, 515 U.S. at 760 (citations omitted).

39. *Good News Club v. Milford Central School*, 533 U.S. 98, 106–07 (2001) (internal citations omitted).

40. *Board of Education v. Mergens*, 496 U.S. 226, 250 (1990).

41. Ibid.

42. *Loving v. United States*, 517 U.S. 748, 778 (1996) (Justice Thomas, concurring), quoting *United States ex rel. Toth v. Quarles* 350 U.S. 11, 17 (1955).

43. *Brown v. Glines*, 444 U.S. 348, 360 (1980).

44. 10 U.S.C. § 654 (a)(8)(A) & (B) (2006).

45. Ibid.

46. DOD Instruction 1300.17, *Accommodation of Religious Practices Within the Military Services*, 2009, para. 4.

47. Air Force Policy Directive (AFPD) 52-1, *Chaplain Service,* 2006, introduction.

48. Ibid., attachment 1.

49. Secretary of the Navy Instruction (SONI) 1730.8B, *Accommodation of Religious Practices*, 2008, paras. 1 & 5.

50. Army Regulation (AR) 600-20, *Army Command Policy*, 2009, para. 3-3.b.(4).

51. Ibid., para. 5-6.a.

52. Commandant of the Coast Guard Instruction M1730.4B, *Religious Ministries within the Coast Guard*, 1994, para. 5.a.

53. Carlo D'Este, *Patton: A Genius for War* (New York: HarperCollins, 1996); 221 (citation omitted); see also John Paul Jones, personal journal entry (1787), in Augustus C. Buell, *Paul Jones: Founder of the American Navy*, vol. 1 (London: K. Paul, Trench, Trubner & Co., 1900): 286–87 ("Men mean more than guns in the rating of a ship.").

54. See Mats Alvesson, *Understanding Organizational Culture* (London: Sage Publications, 2002): 1–2; see also Army Field Manual (FM) 6-22, *Army Leadership*, 2006, § 4-46.

55. Don Snider, "U.S. Civil-Military Relations and Operations Other Than War," in *Civil-Military Relations and the Not-Quite Wars of the Present and Future*, ed. Vincent Davis (Carlisle Barracks, PA: Strategic Studies Institute, Army War College, 1996): 1, 3.

56. Christopher Coker, *The Warrior Ethos: Military Culture and the War on Terror* (New York: Routledge, 2007): 141, comparing the warrior cultures of the ancient Chinese, Greek, Roman, and Japanese societies; and Army FM 6-22, *Army Leadership*, § 4-47, § 4-51.

57. Jamison Yi, "MCMAP and the Marine Warrior Ethos," *Military Review*, November–December 2004, 17, illustrating a "synergy of disciplines" via Venn diagram; see also Air Force Recruiting Service, *Air Force Warrior Facts: Expand Your Training* 2 (n.d.) ("It takes a strong *mind*, *body*, and *spirit* to become an Air Force warrior" [emphasis added]); Army FM 6-22, *Army Leadership*, § 4-47 to § 4-52; and H. Michael Gelfand, *Sea Change at Annapolis: The United States Naval Academy, 1949–2000* (Chapel Hill: The University of North Carolina Press, 2006): 9, listing part of the US Naval Academy's mission as "develop[ing] [midshipmen] morally, mentally, and physically" (citation omitted).

58. Humfrey Michell, *Sparta* (Cambridge: Cambridge University Press, 1964): 165.

59. Yi, "MCMAP and the Marine Warrior Ethos," 21 ("Physical discipline consists of armed and unarmed combat techniques combined as part of the USMC Physical Fitness Program . . . [which] develops a Marine's ability . . . [to] overcom[e] physical hardship and obstacles under any climatic condition.").

60. Ibid., 23.

61. Army FM 6-22, *Army Leadership*, § 4-49.

62. Ibid., § 4-52.

63. Ibid.

64. Ibid., § 4-48.

65. For example, ibid., § 4-53 ("The Warrior Ethos is crucial but also perishable. Consequently, the Army must continually affirm, develop, and sustain it.").

66. Shannon E. French, *The Code of the Warrior: Exploring Warrior Values Past and Present* (Lanham, MD: Rowman & Littlefield, 2005): 1–3.

67. Ibid., 3–4, 9–10.

68. David R. Segal and Mady Wechsler Segal, "America's Military Population," *Population Bulletin*, December 2004, 25, table 5, reporting the combined percentage of Protestants, Catholics, and "Other Christians" alone at 68 percent as of 2001; Hindus, Muslims, Buddhists, and Jews were also reported but comprised less than 0.5 percent each of the total number; see also Barry S. Fagin and James E. Parco, "A Question of Faith: Religious Bias and Coercion Undermine Military Leadership and Trust," *Armed Forces Journal*, January 2008, 40, 42, recognizing that "for many, if not most, in the military, religion is part and parcel of their original decision to serve, their loyalty to country and family, and their source of strength in times of great stress."

69. Army FM 6-22, *Army Leadership*, § 4-57 ("Beliefs matter because they help people understand their experiences. Those experiences provide a start point for what to do in everyday situations. Beliefs are convictions people hold as true. Values are deep-seated personal beliefs that shape a person's behavior. Values and beliefs are central to character."); see also ibid., § 4-59 ("Beliefs derive from upbringing, culture, *religious* backgrounds, and traditions. As a result, different moral beliefs have, and will, continue to be shaped by diverse *religious* and philosophical traditions" [emphasis added].); and French, *The Code of the Warrior*, 3.

70. See ibid., § 4-59; see also note 68, noting that over two-thirds of US service members claim religious affiliation.

71. Army FM 6-22, *Army Leadership*, § 4-58 ("Army leaders should recognize the role beliefs play in preparing Soldiers for battle.").

72. Coker, *The Warrior Ethos*, 132–33; and Yi, "MCMAP and the Marine Warrior Ethos," 17.

73. Even those otherwise opposed to overt religious expression in the military recognize the importance of religious faith and values to members of the armed forces:

> Members of the military live with the fact that they could be asked to surrender their lives at any moment. Those who see combat face life-and-death issues on a regular basis and are

forced to grapple with fundamental questions of existence in a way those they protect likely will never face.

This means that for many, if not most, in the military, religion is part and parcel of their original decision to serve, their loyalty to country and family, and their source of strength in times of great stress. . . . [I]t's unrealistic to expect the spiritual beliefs of soldiers to vanish once they put on a uniform.

In Fagin and Parco, "A Question of Faith," 42.

74. For example, Operational Naval Instruction 1730.1, *Chaplains' Manual*, 1973, § 1301(1); US Air Force, "Revised Interim Guidelines Concerning Free Exercise of Religion in the Air Force," 2006, § 3.D.1; Army FM 1-05, *Religious Support*, 2003, § 1-16.

75. The use of "he" and "his" throughout this chapter is simply for convenience and is not intended to demean or denigrate women in uniform or their military service. Women serve with distinction throughout the US armed services in virtually every job category, including as commanders and chaplains.

76. For example, Combined Arms Center, Center for Army Leadership, *Army Leader Transitions Handbook*, 2008, 20. ("You are the role model. . . . Your example speaks for what is acceptable and what is not.") The handbook "contains best practices and proven techniques *from military and civilian sources*" (emphasis added), 1.

77. Ibid., 14 ("Leave plenty of time for visits to see Soldiers at their duty stations or in training"), 15, 18, 20 ("Meet your troops at ranges, on guard duty and during squad and crew training. Do physical training with different groups regularly.").

78. Ibid.

79. Ibid., 14, 20 ("Never pass up an opportunity to talk with your Soldiers."), 25.

80. Ibid., 19, 26.

81. See Order No. 50 of George Washington to the Continental Army at Valley Forge, 2 May 1778, in *Revolutionary Orders of General Washington*, ed. Henry Whiting (New York: Wiley and Putnam, 1844), 74–75 ("While we are duly performing the duty of good soldiers, we certainly ought not to be inattentive to the higher duties of religion. To the distinguished character of a Patriot, it should be our highest glory to add the more distinguished character of a Christian."); see also "The Christmas Message and Prayer Sent the Third Army, 1944," in Brenton Greene Wallace, *Patton & His Third Army*, (1946; repr., Mechanicsburg, PA: Stackpole Books, 2000): app. 7, 231, detailing the prayer sent by Gen George Patton to the Third Army.

82. Don M. Snider, "Intrepidity. . . . and Character Development within the Army Profession," *Strategic Studies Institute*, January 2008, 2, http://www.strategicstudiesinstitute.army.mil/pdffiles/PUB847.pdf ("The soldier's heart, the soldier's spirit, the soldier's soul are everything. Unless the soldier's soul sustains him, he cannot be relied on and he will fail himself, his commander, and his country in the end. It is not enough to fight. It is the spirit that wins the victory," quoting Gen George Marshall); see also *Army Leader Transitions Handbook*, 20, noting that the commander/leader is "the role model for the ethical and moral climate of the unit" and that the commander's/leader's "example speaks for what is acceptable and what is not" in the unit.

83. See 10 U.S.C. § 3581 (2006).

84. Army FM 1-05, *Religious Support*, § 3-106.

85. See, for example, Department of the Navy (DON), *United States Navy Regulations: 1990*, ch. 8, § 1, art. 0817(2) ("Chaplains shall be permitted to conduct public worship according to the manner and forms *of the church of which they are members*" [emphasis added].). Legislative chaplains are not so. Legislative chaplains exist, first and foremost, to seek divine blessings on, and to solemnize the proceedings of, legislators in enacting the statutes that govern us all, not to ensure free exercise of religion by legislators. In the legislative milieu, the chaplain is not hired to represent a specific denomination and, in fact, is not expected to do so. See Andy G. Olree, "James Madison and Legislative Chaplains," *Northwestern University Law Review* 102 (2008): 151.

86. For information concerning the prevalence of Christianity in the United States as a whole, see US Census Bureau, *Religious Composition of U.S. Population: 2007*, 2008, table 74, http://www.census.gov/compendia/statab/tables/09s0074.pdf, reporting the combined percentage of

Protestants and Catholics in the United States at 75.2 percent as of 2007. For information about the military, see Segal and Segal, "America's Military Population," note 68.

87. See *Religion Facts*, Comparison Chart of Christian Denominations' Beliefs, http://www .religionfacts.com/christianity/charts/denominations_beliefs.htm (accessed 5 May 2009).

88. Whatever else it was understood to mean when drafted and adopted, the establishment clause meant that none of the religious groups present at the founding of our nation would be elevated to become the established, national church of the United States. It is also important to recognize that military commanders have a responsibility to support the free exercise needs of the men and women in uniform. Merely because most of those serving in uniform happen to practice some variant of the Christian faith does not mean that DOD is favoring Christianity over other religious faith groups. As noted above, sheer numbers dictate most chaplains and resources are used to meet the needs of Christian service members and their families.

89. *Katcoff v. Marsh*, 755 F.2d 223, 234 (2d Cir. 1985).

90. See, for example, *Wooley v. Maynard*, 430 U.S. 705, 714 (1977), recognizing that freedom of expression includes the right to refrain from such expression; and Air Force Instruction (AFI) 52-101, *Chaplain Planning and Organizing*, 13 August 2005, § 2.1 ("Chaplains do not perform duties that are incompatible with their faith group tenets.").

91. For example, Secretary of the Navy Instruction 1730.7D, *Religious Ministry within the Department of the Navy*, 8 August 2008, para. 5(e)(3) ("Chaplains *care* for all Service members, including those who claim no religious faith, [and] *facilitate* the religious requirements of personnel of all faiths.").

92. DOD Directive 1304.19, *Appointment of Chaplains for the Military Departments*, 11 June 2004, para. 4.1.

93. Israel Drazin and Cecil B. Currey, *For God and Country: The History of a Constitutional Challenge to the Army Chaplaincy* (Hoboken, NJ: KTAV Publishing House, 1995): 35, 41.

94. Ibid., 32. DOD can, and does, set neutral criteria that all chaplains—irrespective of faith group—must meet, such as education, health, age, and experience requirements. DOD Instruction 1304.28, *Guidance for the Appointment of Chaplains for the Military Departments*, 11 June 2004, paras. 6.1–6.4. However, aside from such neutral criteria, DOD relies on the endorsement by the respective faith group that a chaplain nominee fully meets the religious requirements of his respective faith group.

95. Drazin & Currey, *For God and Country*, 32; and DOD Instruction 1304.28, *Guidance for the Appointment of Chaplains for the Military Departments*, para. 6.5.

96. *Rigdon v. Perry*, 962 F. Supp. 150 (D.D.C. 1997).

97. Ibid., 159.

98. Ibid., 165.

99. See, for example, *Thomas v. Review Board of the Indiana Employment Security Division*, 450 U.S. 707, 714 (1981) ("Religious beliefs need not be acceptable, logical, consistent, or comprehensible to others in order to merit First Amendment protection.")

100. The US armed forces operate 24 hours per day, every day of the year. As such, men and women will be assigned to duties at odd hours and times throughout the year. When those times conflict with regularly scheduled chapel worship times or other religious activities, those on duty will be required to forgo attending such religious activities in order to carry out their military duties. Affected service members may, of course, request an accommodation, but the granting of such an accommodation will ultimately depend on mission requirements. See, for example, AFPD 52-1, *Chaplain Service*, attachment 1; AR 600-20, *Army Command Policy*, para. 5-6.a; and SECNAV Instruction 1730.8B, *Accommodation of Religious Practices*, para. 5.

101. Good order and discipline are essential components of an effective military unit. William A. Cohen, *Secrets of Special Ops Leadership: Dare the Impossible, Achieve the Extraordinary* (New York: AMACOM, 2005): 98, quoting George Washington as saying, "Nothing is more harmful to the service than the neglect of discipline; for that discipline, more than numbers, gives one army superiority over another." Yet, admittedly, the phrase is somewhat vague. When attempting to maintain good order and discipline, commanders and leaders at all levels must ensure that religious service members are not singled out for special detriment, especially if those complaining about a

religious activity or expression of a religious sentiment are persons especially sensitive—or even hostile—to religion or a religious message. See, for example, *Americans United for Separation of Church & State v. City of Grand Rapids*, 980 F.2d 1538, 1553 (6th Cir. 1992), noting the existence of persons who see religious endorsement, "even though a reasonable person, and any minimally informed person, knows that no endorsement is intended."

102. See *Lee v. Weisman*, 505 U.S. at 597, noting that people "may take offense at all manner of religious as well as non-religious messages"; and *Americans United v. City of Grand Rapids*, 980 F.2d at 1553, noting the existence of those who see religious endorsement, "even though a reasonable person, and any minimally informed person, knows that no endorsement is intended."

103. Lee, 505 U.S. at 597, noting that people "may take offense at all manner of religious as well as non-religious messages."

104. See for example, *Chalker v. Gates*, Case No. 08-2467-KHV-JPO (D. Kan. filed 25 Sep 2008), where plaintiff complains, *inter alia*, about hearing "sectarian Christian prayers" being delivered at mandatory events. As an aside, one wonders what a "nonsectarian" *Christian* prayer would sound like and whether the plaintiff would have been satisfied had that kind of prayer been offered at the mandatory events. See also Ezra W. Reese, counsel to the Military Religious Freedom Foundation, to Thomas F. Kimble, acting inspector general, DOD, letter, 11 December 2006, http://www.militaryreligiousfreedom.org/MRFF%20Letters.pdf, complaining about the promulgation of a Christian video that featured several Pentagon officials extolling, *inter alia*, the virtues of prayer.

105. Technically, George Washington's inauguration as president under our current Constitution, being the first, was an assumption of command ceremony, not a change of command ceremony, but the principle is exactly the same.

106. *Military Establishment Act of 1791*, ch. XXVIII, § 5, 1 Stat. 222.

107. See *Act of March 2, 1799*, ch. XXIV, 1 Stat. 709, requiring commanders of ships with chaplains on board "to take care that divine service be performed twice a day, and the sermon preached on Sundays"; *Act of March 23, 1800*, ch. XXXIII, 2 Stat. 45, directing commanders of ships to require the ship's crew "to attend at every performance of the worship of Almighty God."

108. *Marsh v. Chambers*, 463 U.S. at 790 (citation omitted); see also *United States v. Curtiss-Wright Export Corporation*, 299 U.S. 304, 328 (1936), noting that understanding "placed upon the Constitution . . . by the men who were contemporary with its formation" is "almost conclusive" (citation omitted).

109. *Marsh v. Chambers*, 463 U.S. at 792.

110. *Lee v. Weisman*, 505 U.S. at 597.

111. See *Rosenberger v. Rector*, 515 U.S. at 828 ("Discrimination against speech because of its message is presumed to be unconstitutional." Citing *Turner Broadcasting System, Inc. v. FCC*, 512 U.S. at 641–43.).

112. See "Chaplain John Maurice Delivers Meaningful Shipboard Prayer on the Eve of the War in Iraq," *Military Christian*, Summer 2003, http://members.iquest.net/~c_m_f/cmfnew56.htm (accessed 6 May 2009); Navy Recruiting Command, Delayed Entry Program, "Daily Routine," http://www.cnrc.navy.mil/DEP/daily.htm (accessed 6 May 2009), including the traditional evening prayer in Navy recruits' daily schedules; and Robert S. Lanham, ENCM (SW/AW), USN, "I Love the Navy," *Goat Locker*, http://www.goatlocker.org/retire/lovenavy.htm (accessed 6 May 2009), noting, through poetry, a myriad of naval traditions, including the evening prayer.

113. See note 107 and accompanying text.

114. Jacqueline L. Salmon, "ACLU Might File Suit to End Lunch Prayer," *The Washington Post*, 26 June 2008, B04; see also Charles J. Gibowicz, *Mess Night Traditions*, 115 (2007).

115. *Marsh v. Chambers*, 463 U.S. at 792.

116. *Snyder v. Murray City Corporation*, 159 F.3d 1227, 1234 n.10 (10th Cir. 1998).

117. *Lee v. Weisman*, 505 U.S. at 589.

118. Ibid., 588.

119. Ibid., 588–89.

120. 542 U.S. 1 (2004).

121. Ibid., 34–35 (Justice O'Connor concurring).

122. *Capitol Square Review & Advisory Board v. Pinette*, 515 U.S. at 780; see also *Rosenberger v. Rector*, 515 U.S. at 828 ("It is axiomatic that the government may not regulate speech based on its substantive content or the message it conveys. . . . Discrimination against speech because of its message is presumed to be unconstitutional.").

123. See *Newdow v. Bush*, 355 F. Supp. 2d at 286–87.

124. *Marsh v. Chambers*, 463 U.S. at 793.

125. Ibid., 794–95.

126. For example, "Efforts Afoot to Protect Military Prayers," *WorldNetDaily*, 17 November 2005, http://www.worldnetdaily.com/news/article.asp?ARTICLE_ID=47432, describing the backlash following the US Air Force's decision to ban prayers in Jesus' name "in the wake of complaints from non-Christians at the Air Force Academy who believed Christians, both cadets and staff, were being too heavy-handed about their faith on campus."

127. *The New Lexicon Webster's Encyclopedic Dictionary of the English Language*, Deluxe ed. 1991, s.v. "proselytize."

128. See John 1:1, 14 ("In the beginning was the Word, and the Word was with God, and the Word was God. . . . The Word became flesh and made His dwelling among us."); and John 10:30 ("I and the Father are one.").

129. For example, AFPD 52-1, *Chaplain Service*, para. 3.4.

130. See *Hobbie v. Unemployment Appeals Commission of Florida*, 480 U.S. 136, 144–45 (1987), noting that "the government may (and sometimes must) accommodate religious practices and that it may do so without violating the Establishment Clause"; and *Marsh v. Chambers*, 463 U.S. at 791–92, approving legislative prayers from the "Judeo-Christian tradition."

131. *Rosenberger v. Rector*, 515 U.S. at 839, recognizing that government neutrality is respected, not offended, when evenhanded policies are applied to diverse viewpoints, including religious viewpoints.

132. See ibid., 828 ("Discrimination against speech because of its message is presumed to be unconstitutional." Citing *Turner Broadcasting System, Inc. v. FCC*, 512 U.S. at 641–43); *Capitol Square Review & Advisory Board v. Pinette*, 515 U.S. at 767, noting that "private religious expression receives preferential treatment under the Free Exercise Clause"; and *Zorach v. Clauson*, 343 U.S. at 313 ("We are a religious people whose institutions presuppose a Supreme Being. . . . [The Government] sponsor[s] an attitude . . . that shows no partiality to any one group and that lets each flourish according to the zeal of its adherents and the appeal of its dogma.").

133. See *Capitol Square Review & Advisory Board v. Pinette*, 515 U.S. at 760–61, noting that the free speech clause protects, *inter alia*, "religious proselytizing."

134. Operational Naval Instruction 1730.1, *Chaplains' Manual*, 1973, § 1301(1); US Air Force, "Revised Interim Guidelines Concerning Free Exercise of Religion in the Air Force," 2006, § 3. D.1; and Army FM 1-05, *Religious Support*, 2003, § 1-16.

135. See Center for Army Leadership, *Army Leader Transitions Handbook*, 14 ("Open communications early."), 18 ("Spend time . . . talking to Soldiers. . . . Never be too busy to stop and share thoughts and ideas with your subordinates."), 20 ("Never pass up an opportunity to talk with your Soldiers."), 25 ("Spend more time listening and talking to subordinates."), 26 ("As their leader, provide . . . an ear for listening. Listening to your subordinates gives individuals a share in the organization's future.").

136. See ibid., 11, identifying topics to be addressed with subordinates, including values, ethics, and integrity; and 20 ("You are the role model for the ethical and moral climate of the unit. Your example speaks of what is acceptable and what is not.").

137. See, for example, *Americans United v. City of Grand Rapids*, 980 F.2d at 1553, noting the existence of those who see religious endorsement, "even though a reasonable person, and any minimally informed person, knows that no endorsement is intended."

138. For example, Josh White, "4-Star General Relieved of Duty: Rare Move Follows Allegations of an Extramarital Affair," *The Washington Post*, 10 August 2005, A01; William Fisher, "Jesus Is Not Our Co-Pilot, Academy Insists," *AntiWar.com*, 20 June 2005, http://www.antiwar.com/ips/fisher.php?articleid=6484 (accessed 6 May 2009); and Dave Moniz and Blake Morrison, "General

Who Led Abu Ghraib Prison Guard Unit Has Been Suspended," *USA Today*, 25 May 2004, http://www.usatoday.com/news/world/iraq/2004-05-24-abuse-karpinski_x.htm.

139. Center for Army Leadership, *Army Leader Transitions Handbook*, 1.

140. Ibid., 19.

141. Ibid.

142. Ibid.

143. Ibid., 11.

144. Ibid.

145. Ibid.

146. Ibid.

147. Ibid.

148. Ibid.

149. Ibid.

150. Ibid.

151. Ibid.

152. Ibid., 15.

153. DON, *United States Navy Regulations: 1990*, ch 8, § 1, art 0817(2).

154. Ibid., art. 0817(3).

155. Ibid., art. 0817(2).

156. "The Pentagon," GlobalSecurity.org, http://www.globalsecurity.org/military/facility/pentagon.htm (accessed 6 May 2009).

157. *Marsh v. Chambers*, 463 U.S. at 792.

158. Ibid., 794–95. To proselytize is defined as "to make or try to make converts." *The New Lexicon Webster's Encyclopedic Dictionary of the English Language*, Deluxe ed. 1991, s.v. "proselytize." To disparage is defined as "to belittle, deprecate," Ibid., s.v. "disparage."

159. There are a number of suggested alternatives being proffered by well-meaning persons to resolve alleged violations of church-state separation. Yet some of the proposed cures are fraught with constitutional infirmities. Among suggested cures, for example, is a proposal to require all commanders to take an oath (called the "Oath of Equal Character"). Fagin and Parco, "A Question of Faith," 43. The Oath of Equal Character reads as follows:

> I am a [Fill in your belief system (e.g., Christian, Muslim, Jew, atheist, Buddhist, Hindu, Wiccan, nontheist, etc.)]. I will not use my position to influence individuals or the chain of command to adopt [Fill in your belief system (e.g., Christianity, Islam, Judaism, atheism, etc.)], because I believe that soldiers who are not [Fill in your belief system (e.g., Christians, Muslims, Jews, atheists, etc.)] are just as trustworthy, honorable and good as those who are. Their standards are as high as mine. Their integrity is beyond reproach. They will not lie, cheat or steal, and they will not fail when called upon to serve. I trust them completely and without reservation. They can trust me in the same way.

The underlying assumptions of the oath appear to suggest that all religious/philosophical belief systems are essentially equivalent and that the adherents of one religious/philosophical system essentially exhibit the same characteristics as adherents of every other religious/philosophical system. Aside from the fact that it is impossible to prove the truthfulness of the underlying assumptions contained in the oath—to wit, about the trustworthiness, dependability, integrity, and the like, of adherents of belief systems other than the oath taker's—and the fact that many could convincingly argue that readily available evidence indicates that such assertions are, in fact, demonstrably untrue, requiring the taking of such an oath would violate a whole host of constitutional provisions. *First*, it seeks to compel belief in the equivalence of different religions and between religion and non-religion. No government official may require that. Simply put, things are rarely equivalent, and some things are definitely not equivalent to others. For example, one could legitimately argue that a philosophy or religion that demeaned women would be inferior (and so not equivalent) to one that did not do so. Likewise, a philosophy or religion that preferred one race over another would be inferior (and so not equivalent) to a philosophy or religion that did not do so. *Second*, the undertaking seeks to compel speech with which one may disagree, and freedom of speech includes the right to refrain from expressing ideas with which one disagrees. See *Wooley v. Maynard*, 430 U.S. at 714 (recognizing that freedom of expression includes the right to refrain from such expression). *Third*, the undertaking seeks to replace the religious/philosophical views held by various commanders—as of right—with a view of religion and its adherents acceptable to the oath's proponents (and, they hope, ultimately the US government).

Yet, once government officials put their stamp of approval on a religious belief, they have violated the very establishment clause that they were sworn to uphold. The above oath, if required, would violate the free exercise, the free speech, and the establishment clauses of the First Amendment, irrespective of the good intentions of those proffering the suggestion. The Supreme Court stated in *Lee v. Weisman* that "the First Amendment's Religion Clauses mean that *religious beliefs and religious expression are too precious to be either proscribed or prescribed by the [Government]*" (emphasis added), 505 U.S. at 589.

Further, there seems to be a basic non sequitur in the argument. The authors correctly recognize that "beliefs remain a *right*," and "*freedom of conscience is among the oldest and most precious freedoms enshrined in the history of America's founding*" (emphasis added) (Fagin and Parco, 43). In the very next sentence, they acknowledge, correctly, that members of the armed forces take an oath to uphold the Constitution of the United States (including, one presumes, the First Amendment). But then they argue that military leaders who *believe* that adherents of other faiths are less likely to have good character than adherents of the leader's own faith/philosophy should leave the military and seek another career. What happened to the constitutional "right" of that leader to believe as he does? What happened to that leader's constitutionally protected "freedom of conscience"? On what legal basis do the authors conclude that those who do not share *their* views on how to resolve potential religious misunderstandings in the military have any less right to remain in the military than those who agree with them? The authors refer to the First Amendment, but that amendment protects the leader's right to believe as he wishes, not as the government or the authors may prefer. The First Amendment does not stand for what the authors contend. It protects the individual's right to believe against government coercion or government-supported orthodoxy, even when the individual's beliefs are strange or offensive.

160. Merely being present when a prayer is being said does not mean that one is assenting to the sentiments being expressed, that one is actively participating in religious worship, or that one is actively engaging in a religious act. Instead, the service member is an observer. People encounter and observe religious ceremonies all the time without their mere presence converting them into participants in the ceremonies. The same is true when present at military ceremonies or formations where a short, solemnizing prayer is said. Solemnizing prayers constitute only a minute part of such ceremonies and, thus, do not convert such gatherings into religious gatherings.

161. See *Lee v. Weisman*, 505 U.S. at 588–89, noting that it is inappropriate for a government official to tell a member of the clergy how to pray.

162. See Center for Army Leadership, *Army Leader Transitions Handbook*, 11, 15, and 19.

163. *Board of Education v. Mergens*, 496 U.S. at 250.

164. At its most obvious level in the West, one easily notes that Roman Catholics and Protestants share different theological views and practices. There continue to be theological differences separating Roman Catholics from Eastern Orthodox as well. Likewise, there are significant differences in theological beliefs and practices within Protestantism, such as between liturgical denominations (e.g., Episcopalians, Lutherans) and nonliturgical denominations (e.g., Baptists, Assemblies of God). Then, there are differences between denominations that believe that spiritual gifts (i.e., charismata) are still in use today (e.g., Church of God in Christ) and denominations that believe that such gifts are no longer in use (e.g., Independent Fundamental Churches of America). Further, there are religious groups that do not fall neatly into any category (e.g., Latter-Day Saints [Mormons], Christian Scientists). Even within groups with a common heritage, there can be significant theological differences (e.g., the Evangelical Lutheran Church in America versus the Lutheran Church-Missouri Synod or the Presbyterian Church [USA] versus the Presbyterian Church in America). To accommodate free exercise of religion as much as possible in the military, military chaplains represent many different Christian denominations, based in large part on the relative numbers of adherents of the respective denominations in uniform (i.e., denominations with greater

numbers of adherents in uniform are allotted more chaplains than denominations with fewer numbers). See also Religion Facts, Comparison Chart of Christian Denominations' Beliefs, http://www.religionfacts.com/christianity/charts/denominations_beliefs.htm (accessed 5 May 2009).

165. For example, complaint at 3-4, *Chalker v. Gates*, No. 08-CV-2467-KHV-JPO (D. Kan filed 25 Sep 2008), describing the "requirement for [P]laintiff... to attend military functions and formations where *sectarian Christian prayers* are delivered" (emphasis added).

166. See, for example, 10 U.S.C. § 164(c) (2006), delegating substantial authority to military combatant commanders in the performance of their duties; and *Goldman v. Weinberger*, 475 U.S. 503, 507, acknowledging that "the military need not encourage debate or tolerate protest to the extent that such tolerance is required of the civilian state by the First Amendment; to accomplish its mission, the military must foster instinctive obedience, unity, commitment, and esprit de corps."

167. See *Haig v. Agee*, 453 U.S. 280, 307 (1981), noting as "obvious and unarguable" that there is no governmental interest more compelling than security of the nation (citing *Aptheker v. Secretary of State*, 378 U.S. 500, 509 (1964).

168. *Loving v. United States*, 517 U.S. at 778.

169. See, for example, *Chalker v. Gates*, No. 08-CV-2467-KHV-JPO (D. Kan. filed 25 Sep 2008), complaining about "sectarian prayers" given at three required formations.

170. Order No. 50 of George Washington, in *Revolutionary Orders of General Washington*, 74–75 ("The Commander-in-Chief directs that Divine service be performed every Sunday at 11 o'clock, in each Brigade which has a Chaplain. Those Brigades that have none will attend the places of worship nearest them."); "The Prayer at Sumter," *Harper's Weekly: A Journal of Civilization*, 26 January 1861, http://www.sonofthesouth.net/leefoundation/major-anderson-ft-sumter_Dir/civil-war-prayer-fort-sumter.htm, describing the dramatic prayer offered by the command chaplain following Maj Robert Anderson's raising of the American flag over Fort Sumter just days before the post fell, signaling the start of the Civil War; "Proud to Pay Debt, says Gen. Pershing," *New York Times*, 1 December 1918 ("[General Pershing] paid tribute to the dead and wounded, urged the soldiers to thank God for the victory, and declared that a new vision of duty to God and country had come to all."); James H. O'Neill, "The True Story of the Patton Prayer: The Author of General Patton's Famous Third Army Prayer Reveals the Story of its Origin, Paying Tribute Both to the General's Trust in God and to the Power of Faith-filled Prayer," *The Review of the News*, 6 October 1971, reprinted in *The New American*, 12 January 2004, http://findarticles.com/p/articles/ mi_m0JZS/is_1_20/ai_n25081623?tag=untagged, describing General Patton as a self-proclaimed "firm believer in prayer" as he issued his famous Third Army Prayer to his subordinates; and Don M. Snider, "Intrepidity... and Character Development," 2 ("The soldier's heart, the soldier's spirit, the soldier's soul are everything. Unless the soldier's soul sustains him, he cannot be relied on and he will fail himself, his commander, and his country in the end. It is not enough to fight. It is the spirit that wins the victory," quoting Gen George Marshall).

About the Authors

Dr. Jay Alan Sekulow is chief counsel of the American Center for Law and Justice, a national public interest law firm specializing in constitutional litigation, including protecting religious freedom. Dr. Sekulow has argued a number of important First Amendment cases before the Supreme Court of the United States, including, most recently, *Pleasant Grove City v. Summum*, 129 S. Ct. 1125 (2009). He is also chief counsel of the European Centre for Law and Justice (ECLJ), a public interest law firm located in Strasbourg, France, which specializes in defending religious freedom. The ECLJ is also accredited to the United Nations as a non-governmental organization (NGO) and has been active in promoting freedom of religion worldwide. *The National Law Journal* has twice named Dr. Sekulow one of the "100 Most Influential Lawyers" in the United States. Dr. Sekulow earned his bachelor's and juris doctor degrees from Mercer University and his doctor of philosophy degree from the Regent School of Leadership Studies.

Robert Weston Ash is an assistant professor of law at the Regent University School of Law in Virginia Beach, Virginia, where he teaches courses in international law, national security law, comparative law, business associations, and First Amendment law. Mr. Ash also serves as senior litigation counsel for national security law at the American Center for Law and Justice. Mr. Ash received his bachelor of science degree from the United States Military Academy, his master of international public policy degree from the School of Advanced International Studies (SAIS) of the Johns Hopkins University, and his juris doctor degree from the Regent University School of Law. Mr. Ash served 22 years on active duty in the United States Army. His assignments included command of both armored cavalry and armor units, a tour on the history faculty at West Point, a tour as a Congressional Fellow in the office of Senator John McCain (R-AZ), a tour as the NATO desk officer in the War Plans Division of the Army staff in the Pentagon, and a tour as a strategist in the Office of the Secretary of Defense.

HOMOSEXUALITY

Evidence shows that allowing gays and lesbians to serve openly is unlikely to pose any significant risk to morale, good order, discipline, or cohesion.
—Report of the General/Flag Officer's Study Group

The president should not ask military leaders if they support lifting the ban.
—Aaron Belkin and colleagues, The Palm Center

As a matter of national security, we urge you to support the 1993 law regarding homosexuals in the military (10 USC 654) and to oppose any legislative, judicial, or administrative effort to repeal or invalidate the law.
—1,163 Flag and General Officers for the Military

President Clinton's convoluted "Don't Ask, Don't Tell" regulations were and still are inefficient and contrary to sound policy. In the civilian world it would be tantamount to a state law forbidding store and bar owners to check IDs before selling liquor to younger customers.
—Elaine Donnelly, Center for Military Readiness

HOMOSEXUALITY

Discrimination is often, but not always, a controversial and complex issue. Discriminating against color-blind pilot candidates, mandating that firemen have the physical strength to carry essential equipment, and requiring doctors and lawyers to have relevant professional degrees aren't contentious. It's both understood and accepted that people should be "qualified" to perform a job. When a person is deemed unqualified, he or she must be excluded. As a matter of military service qualification, discrimination isn't only permissible, it's essential.

Judgments based on criteria for the sole benefit of the employer render a situation less clear. Mere employer convenience has long been rejected as an appropriate justification to choose prospective employees in corporate America. However, applying the same generally accepted standard to the military isn't always prudent because of the exceptional responsibilities placed upon it for national defense. When it comes to military readiness, Congress has been considerably reticent to second-guess commanders' judgments regarding what appropriate service standards should be. Specifically, when considering military readiness, there has always been an understanding that commanders should have the latitude to discriminate against behaviors that threaten unit cohesion, morale, and general discipline. Logic informs us that military commanders have enough to worry about preparing their units for combat and should not be distracted by the desires of others outside the military domain. As an example, if commanders testify to the ineffectiveness of people beyond a certain age in handling the demands of military service, seldom will this judgment be questioned—particularly if it fits within an existing paradigm. The open question, however, is whether or not their logic is correct.

Relying on the intuition of military leadership can be a highly effective method of developing knowledge about issues that pose high degrees of ambiguity and uncertainty. Absent sufficient data to make a determination based on relevant facts, intuition is often all we have. However, once sufficient data emerges, it is incumbent on leaders to take a critical and unbiased look at it. Merely relying on one's intuition is no longer enough.

The issue of open homosexuality in the military emerged as a center-stage issue back in 1992 when President-elect Bill Clinton vowed to repeal the ban. Within months, the issue became sufficiently complicated and has remained particularly contentious for nearly two decades. Before the reader delves into the following chapters, which span the spectrum of perspectives, we offer several factors we believe have contributed to its complexity. First, it hits at the one basic human drive that has been a taboo subject throughout the ages—human sexuality. Second, a variety of sacred religious texts have taken a strong stand against homosexuality, equating it with immorality and sin. Third, the word "homosexual" tends to evoke strong visceral responses in many that bypass thought centers and strike an emotional chord. Thoughtful, well-meaning people have repeatedly arrived at vastly different conclusions as to what the best policy regarding open homosexuality in the military should be in the future.

Current Department of Defense (DOD) policy has taken on a variety of names over the years, but the moniker which emerged over time is "Don't Ask, Don't Tell." This policy has placed the military in quite a predicament, which is fully articulated in the chapters to follow.

We begin this section with a reprinted 2008 report from a blue-ribbon commission sponsored by the Palm Center in which four retired general and flag officers—**Hugh Aitken, Minter Alexander, Robert Gard, and Jack Shanahan** (one from each service)—explore the arguments of the current "Don't Ask, Don't Tell" policy and make recommendations on a way forward.

Matthew Cashdollar, a major in the US Army, bypasses the issue of whether or not gays should be allowed to serve openly and focuses on what would need to be done if the policy were rescinded. Major Cashdollar explores how an effective transition could be facilitated by noting lessons learned from other countries that successfully dealt with this issue, as well as how the previous integration of African-Americans into the military relates to the issue at hand.

Dr. Tammy Schultz, from the US Marine Corps War College, notes the missteps of the Clinton administration that led to "Don't Ask, Don't Tell" rather than President Clinton's desired result of allowing homosexuals to serve openly. Dr. Schultz addresses issues such as pay and benefits, living quarters, and how a change of policy could affect homosexuals already discharged from the US military because of sexual orientation.

Dr. Aaron Belkin, the director of the Palm Center at the University of California at Santa Barbara, and his colleagues offer a solution for the commander in chief to use the current "stop loss" legislation to repeal the ban though a presidential executive order. Following their proposed roadmap, Dr. Belkin and Dr. Frank provide a synopsis of the Palm Center's work and offer their perspective as to why the ban against open homosexuals serving in the military should be lifted.

In March 2009, an open letter to the president of the United States and Congress signed by more than **1,000 flag and general officers** suggested that the path to the open service of homosexuals is not quite as smooth as some might suggest. These senior military leaders directly state that homosexuality is incompatible with military service. They collectively contend a repeal would have a negative impact on morale, discipline, unit cohesion, and military readiness. They appealed to the president and members of Congress to recognize this issue as one of national security and to oppose any legislative, judicial, or administrative effort to repeal the law.

Elaine Donnelly, president and founder of the Center for Military Readiness, provides the final analysis of open homosexuality in the military by reminding us that the current ban on homosexuals is a matter of law, and as a nation of laws, we are bound to adhere to it. She provides a robust and complete perspective arguing that in the best interest of national defense, the only change that should be made to the existing military policy is to exclude homosexuals from service based on their congressionally mandated ineligibility to serve.

CHAPTER 7

REPORT OF THE GENERAL/FLAG OFFICERS' STUDY GROUP

Hugh Aitken   *Minter Alexander*
Robert Gard   *Jack Shanahan*

Executive Summary

A bipartisan study group of senior retired military officers, representing different branches of the service, has conducted an in-depth assessment of the "don't ask, don't tell" policy by examining the key academic and social science literature on the subject and interviewing a range of experts on leadership, unit cohesion, and military law, including those who are training our nation's future military leaders at the service academies. The Study Group emphasized that any changes to existing personnel policy must not create an unacceptable risk to the armed forces' high standards of morale, good order and discipline, and unit cohesion that are the essence of military capability.

The Study Group has made ten findings, including:

Finding one: *The law locks the military's position into stasis and does not accord any trust to the Pentagon to adapt policy to changing circumstances.*

Finding two: *Existing military laws and regulations provide commanders with sufficient means to discipline inappropriate conduct.*

Finding three: *"Don't ask, don't tell" has forced some commanders to choose between breaking the law and undermining the cohesion of their units.*

Finding four: *"Don't ask, don't tell" has prevented some gay, lesbian, and bisexual service members from obtaining psychological and medical care as well as religious counseling.*

Finding five: *"Don't ask, don't tell" has caused the military to lose some talented service members.*

Finding six: *"Don't ask, don't tell" has compelled some gay, lesbian, and bisexual service members to lie about their identity.*

Finding seven: *Many gays, lesbians, and bisexuals are serving openly.*

This essay is a 2008 report issued by four retired general and flag officers for the Palm Center at the University of California, Santa Barbara.

Finding eight: *"Don't ask, don't tell" has made it harder for some gays, lesbians, and bisexuals to perform their duties.*

Finding nine: *Military attitudes towards gays and lesbians are changing.*

Finding ten: *Evidence shows that allowing gays and lesbians to serve openly is unlikely to pose any significant risk to morale, good order, discipline, or cohesion.*

On the basis of these findings, the Study Group offers the following four recommendations:

Recommendation 1. Congress should repeal 10 USC § 654 and return authority for personnel policy under this law to the Department of Defense (DOD).

Recommendation 2. The Department of Defense should eliminate "don't tell" while maintaining current authority under the Uniform Code of Military Justice (UCMJ) and service regulations to preclude misconduct prejudicial to good order and discipline and unit cohesion. The prerogative to disclose sexual orientation should be considered a personal and private matter.

Recommendation 3. Remove from Department of Defense directives all references to "bisexual," "homosexual," "homosexual conduct," "homosexual acts," and "propensity." Establish in their place uniform standards that are neutral with respect to sexual orientation, such as prohibitions against any inappropriate public bodily contact for the purpose of satisfying sexual desires.

Recommendation 4. Immediately establish and reinforce safeguards for the confidentiality of all conversations between service members and chaplains, doctors, and mental health professionals.

Rationale

All policies that affect the military must be designed to promote readiness, and must be evaluated in terms of how well they measure up to that standard. The military, cultural, and political landscapes have shifted significantly in the years since the "don't ask, don't tell" (DADT) policy was adopted in 1993. As a result, Professor Charles Moskos, one of the principle authors of DADT, said in October 2007 that the time is ripe for "a bi-partisan Commission [to] look at the whole issue of homosexuals in the military. This should involve the consultation of prominent Americans who are known to be pro-military and have respected national reputations."[1]

The Study Group agrees that a reasoned conversation on this subject requires the counsel of former military officials who have the institutional experience and perspective to offer sound recommendations to Congress and to the public concerning whether and how the current policy should be reformed.

As senior retired military officers, representing different branches of the service, we came into the process with open minds. We were supportive of the policy and felt that it was important at this time, on the eve of its 15th anniversary, to give considered thought from a military perspective to the policy's current contribution to its stated goal: preserving military effectiveness. In our view, three conditions form the necessary foundation from which any re-examination of DADT should proceed: first, respect for military policy that maintains the armed forces' high standards of morale, good order and discipline; second, a willingness to examine the policy's present relationship to military effectiveness; and third, the ability to engage controversial issues through sustained, rational inquiry and fact-finding.

In 1993, when DADT was drafted, the policy was intended by DOD as an interim measure.[2] The policy was the result of political compromise in the aftermath of a presidential campaign promise. Military and political leaders viewed DADT as a stopgap measure.[3] While DADT was the right solution at the time it was enacted, the statute and the policy have remained in force for years with almost no significant change. This fact alone goes against the original intent of the statute and signals the importance of resuming an informed civil-military conversation. It stands to reason that after such a significant lapse in time, it is now appropriate and necessary to assess the effectiveness and goals of the statute and the policy.

On 28 February 2007, former Rep. Martin Meehan (D-MA) and a bipartisan group of 109 original cosponsors reintroduced the Military Readiness Enhancement Act in the House of Representatives to amend 10 USC § 654 to enhance the readiness of the armed forces by replacing the current policy concerning homosexuality in the armed forces with a policy of nondiscrimination on the basis of sexual orientation. The immediate prospects of this bill's passage are uncertain. But the perspective of senior military leaders ought to be consulted in this dialogue, and the Study Group offers this report as a small step in that direction.

The aims of the Study Group are (a) to review the DOD DADT policy and the law 10 USC § 654, Policy Concerning Homosexuality in the Armed Forces, to see if, over time, these two instruments are continuing to serve the best interests of the armed forces; (b) to provide objective, knowledgeable military judgment about the effects of the DOD policy and the law over time; and (c) to consider what steps, if any, should be taken by the military and Congress. It is not the intention of the Study Group to craft a new policy or to resolve questions raised by the possible continuation of the current policy. Rather, it has been the goal of the Study Group to review all available data and to hear and consider expert opinion in order to make recommendations on the overall current state of the DOD policy and law and their present impact on military personnel, leadership, and effectiveness.

This report is funded by the Palm Center at the University of California, Santa Barbara. The Palm Center's rigorous research has been published by distinguished military journals including *Parameters*, the official journal of the Army War College, and has been cited in major news venues around the world. As a think tank engaged in controversial social science research, Palm has also reached conclusions that are critical of military policy and that have, themselves, been critiqued by scholars with different opinions. In order to ensure the impartiality of this project, the Study Group insisted, and the Palm Center agreed, that as a condition of participation, the Study Group conclusions would be their own, and would be reported unmodified by Palm researchers or staff.

The Study Group has focused on two key areas concerning the policy on homosexuality in the military: (1) the "unacceptable risk" standard established in 10 USC § 654 and (2) DOD's policy of "don't ask, don't tell" implementation of the law through implementing regulations, in particular DOD Directive 1332.14. During meetings at the Army Navy Club in Washington, DC in August and September 2007, the Study Group heard testimony and comment from a wide array of experts and interested parties including architects of the 1993 policy; scholars of military personnel issues and military psychology; military commanders; service members discharged under the current policy; experts on foreign militaries and integration; foreign military commanders; and constitutional law and other legal experts. The Study Group carefully sought out expert opinion representing all viewpoints, including supporters and detractors, advocates and critics of the current policy.

The Study Group reviewed materials from the 1993 Congressional hearings and met with architects of the statute and the policy. The group examined in detail the language of the law with the help of lawyers and legal scholars. Finally, the Study Group reviewed the relevant policies in the Uniform Code of Military Justice (UCMJ) and discussed the relationship of the statute, the policy, and the UCMJ with military commanders who had experience implementing them in Iraq and elsewhere.

The Study Group examined the key academic and social science literature on the subject. This included the most recent quantitative information (polling data) available on military opinion and civilian attitudes; the most up-to-date research on unit cohesion and military psychology; and comparative work on foreign militaries. The group heard from academic experts on the history of sexual minorities in the military and on the history of DADT. The group spoke with and sought out the opinion of those who are training the nation's future military leaders at the service academies.

The study group was saddened that not a single expert who opposes gays in the military was willing to meet or talk with us in person. For each expert, the group offered to take written and/or in-person testimony, and offered to arrange and subsidize transportation to Washington, DC or to arrange videoconferencing or teleconferencing facilities. The group also asked experts who oppose gays in the military to provide additional names of experts who might participate. Because not a single one of these experts was willing to participate in person or to provide additional names of people who would, therefore the group devoted particular and extensive effort to the study of their published work and any written comments they were willing to submit for consideration.

History of "Don't Ask, Don't Tell"

The question of whether gays and lesbians should be allowed to serve in the US military has surfaced several times in the history of the United States. Up until World War II, homosexuals were not specifically named in military regulations. Those caught engaging in homosexual conduct were punished or separated—albeit inconsistently—under regulations proscribing certain kinds of sexual behaviors or under policies targeting socially disreputable conduct or social types. By the end of World War II, all services banned homosexuals and homosexual conduct, although enforcement continued to be unevenly applied.

A string of court cases in the 1970s and 1980s challenged inconsistencies in how the homosexual exclusion policy was being implemented. In response to some of these legal challenges, and in deference to political considerations, the Carter administration initiated the first Pentagon-wide ban on gays and lesbians in uniform. Implemented at the end of President Carter's term, DOD Directive 1332.14 effectively removed any discretion that different services or individual commanders previously enjoyed.[4] The new policy modified the language that had regarded gay people as "unsuitable for military service" stating instead that "homosexuality is incompatible with military service." The rationale given was that

> the presence of such members adversely affects the ability of the armed forces to maintain discipline, good order and morale; to foster mutual trust and confidence among service members; to insure the integrity of the system of rank and command; to facilitate assignment and worldwide deployment of service members who frequently must live and work under close conditions affording minimal privacy; to recruit and retain members of the armed forces; to maintain the public acceptability of military service; and to prevent breaches of security.[5]

In the late 1980s, the homosexual exclusion policy came under increasing public scrutiny. A purge of suspected lesbians at the Parris Island Marine training center in South Carolina added to the momentum of critics of the gay ban, and new gay, lesbian, and bisexual advocacy groups joined civil rights organizations, legal aid groups, and members of Congress to raise awareness of the consequences of the policy. After the first Gulf War, the press reported allegations that the military had sent known gays and lesbians to war, only to discharge them upon their return. The confluence of ongoing legal challenges to the policy and growing opposition in the court of public opinion, particularly on college campuses, where the presence of ROTC was routinely protested, caught the attention of lawmakers and candidates for office in the early 1990s.[6]

In October 1991, Gov. Bill Clinton, a Democratic contender for the White House, was asked during a speech at Harvard's John F. Kennedy School of Government about his position on the ban on gay service members. He answered that he opposed it and would lift it if he became president. Clinton framed his position in terms of "meritocracy," saying the nation could not afford to exclude capable citizens from helping their country even if some citizens did not like them. In contrast, those opposed to lifting the gay ban, including many members of the military and of religious and other socially conservative organizations, cast the issue as one of "national security" and "military readiness," arguing that such a change would put lives needlessly at risk by compromising the high standards of discipline, morale, and unit cohesion on which a strong military relies.[7]

After Clinton won the election in November 1992, his campaign promise on gays in the military dominated the news cycle for months. Opposition from the military was fierce, as was resistance from other sectors of American society. Senator Sam Nunn, chairman of the Senate Armed Services Committee, and General Colin Powell, chairman of the Joints Chiefs of Staff, insisted that homosexual conduct must not be permitted in the military, and they pointed out that the Uniform Code of Military Justice, which bans certain sexual acts such as sodomy which are commonly associated with homosexuals, could only be changed by an act of Congress. President-elect Clinton argued that a person's status—as opposed to his or her conduct—should not be a bar to service. He continued to assert his intention to lift the ban outright and to allow gay, lesbian, and bisexual Americans to serve their country without concealing their identity.[8]

In January 1993, just days after Clinton's inauguration, the new president came to a compromise with the Joint Chiefs of Staff and members of Congress to suspend certain aspects of the homosexual exclusion policy while studying the issue for a six-month period. The most notable change for the interim period was that recruits would no longer be asked if they were homosexual as a pre-condition for enlistment. But investigations of homosexuality would continue, and, if found out, gays and lesbians would be transferred into the "standby reserves," where they would receive no pay or benefits.[9]

President Clinton then ordered his secretary of defense, Les Aspin, to study how best to reform the policy in a way that would end discrimination on the basis of sexual orientation while remaining consistent with the standards of discipline and order necessary to maintain military readiness. Policy options were supposed to take the Uniform Code of Military Justice into consideration.

Secretary Aspin ordered two major studies that spring. One study was by a panel of general/flag officers called the Military Working Group (MWG), which Aspin appointed and instructed to deliver a report by July 1993. The RAND Corporation's National Defense Research Institute, a private think tank created by members of the military following World War II, commissioned the other study. The two organizations delivered competing proposals, with the MWG suggesting a policy that retained the finding that "homosexuality is incompatible with military service," and RAND concluding that sexual orientation should be considered "not germane" in determining who should be allowed to serve.[10]

While military experts were preparing their reports, Congress separately held hearings on the matter, led by Senator Nunn. The hearings, both in the House and Senate, took place over several months and invited testimony of numerous parties, including national security experts, legal scholars, sociologists, members of Congress, and current and former members of the armed forces. The Senate also conducted field hearings to discuss the matter with enlisted personnel on ships and submarines.[11]

On 19 July 1993, the Clinton White House announced its policy: "don't ask, don't tell, don't pursue." In a Ft. McNair speech, Clinton made permanent the temporary suspension of asking potential recruits if they were gay or lesbian. In a memo signed by Secretary Aspin, the Department of Defense directed that applicants for military service "not be asked or required to reveal their sexual orientation."

The policy called for the separation of service members "for homosexual conduct," which was defined to include "a statement by a service member that demonstrates a propensity or intent to engage" in homosexual acts. Acts are defined as "any bodily contact" between members of the same sex undertaken "for the purpose of satisfying sexual desires" or which a "reasonable person would understand to demonstrate a propensity or intent to engage in homosexual acts." The policy explained that "an open statement by a service member that he or she is a homosexual" would be taken to demonstrate a "presumption that he or she intends to engage in prohibited conduct." Therefore, both statements to that effect and the prohibited conduct itself would result in separation.[12]

Congress debated and then voted on a variety of versions of Clinton's policy, finally passing a version in September that hardened the language by making a number of changes. In particular, the new Senate language did not mention "don't pursue" and did not echo the Clinton policy's assertion that "sexual orientation is considered a personal and private matter and homosexual orientation is not a bar to entry or continued service unless manifested by homosexual conduct," while it did call gays an "unacceptable risk" to the military, and allowed the secretary of defense to re-instate "asking" if deemed appropriate. The Senate version required the separation of service members found to have engaged in or attempted to engage in homosexual acts, defined to include statements that they are gay or bisexual.[13] The House passed an identical measure by a vote of 301 to 134, and, in November 1993, President Clinton signed the legislation (the National Defense Authorization Act of fiscal year 1994) into law. Over the next several months, the Pentagon wrote implementing regulations that updated prior Department of Defense directives, and the statute and regulations were implemented in February 1994.

Study Group Findings

Finding one: *The law locks the military's position into stasis and does not accord any trust to the Pentagon to adapt policy to changing circumstances.*

As a result of the way in which the DADT law is written, the Defense Department is restricted from adjusting its policy to suit military needs or readiness. The Study Group finds that it is the practical and everyday flexibility of military commanders that leads some to mistakenly assume that DADT is working. However, the policy is not working; rather, it is the flexibility of military leaders, often ignoring or violating the policy, who are making the system work. The Defense Department needs the latitude to develop and adapt a policy that meets its needs. The framing of the current law does not recognize military flexibility or accord the Pentagon the authority to adjust its policies. The justification for the restrictions on homosexuals found in 10 USC § 654 is contained in the 15 Congressional findings provided in the beginning narrative of the statute. The last finding sets the rationale for the law: "The presence in the armed forces of persons who demonstrate a propensity or intent to engage in homosexual acts would create an unacceptable risk to the high standards of morale, good order and discipline, and unit cohesion that are the essence of military capability." The "unacceptable risk" standard was carefully established by Congress in 1993 based on expert testimony from trustworthy military leaders. The basis for their advice to Congress lies in the attitudes toward homosexuality by members of the armed forces serving at that time or earlier. Witnesses confirmed to us that attitudes of the members of the armed forces concerning homosexuality have changed since 1993. The Study Group was informed that only about 20 percent of those serving in 1993 when the law was passed remain in the service today. If DOD needed to adjust the policy because of the changing attitudes, it would be unable because of the specificity of the law. The Study Group believes that Congress should return the authority to the Defense Department to establish personnel policies that meet the needs of the military.

Finding two: *Military laws and regulations provide commanders with sufficient means to discipline inappropriate conduct.*

Many types of conduct are not appropriate for military settings. The Uniform Code of Military Justice, as well as Pentagon regulations, provide commanders with numerous and sufficient means for disciplining inappropriate public displays of affection, fraternization, adultery, or any other conduct which is prejudicial to the maintenance of good order, discipline, morale and unit cohesion. In addition, the Defense of Marriage Act prevents the federal government from recognizing same-sex marriages for any purpose, even if recognized by any particular state.

Finding three: *DADT has forced some commanders to choose between breaking the law and preserving the cohesion of their units.*

The Study Group was concerned to learn that DADT puts some commanders in a double bind in their everyday workplace, as they weigh the need to follow the law against the importance of keeping their teams together.

The Study Group heard from a heterosexual officer who returned recently from a tour of duty in Iraq. He told the group that one of his best noncommissioned officers was probably a lesbian, and that if he had been presented with credible evidence of her homosexuality, he would have been forced to choose between following the law and keeping his unit intact. For this officer, unit cohesion was marked by the need to retain a qualified, meritorious lesbian service member. When asked which choice he would have made, he said that he would have opted to break the law. Experts in military law attested, "The statute makes it mandatory to follow up if told." Yet, a former noncommissioned officer confirmed, "There were times I should have said something. I didn't. I helped people manage their career." He acknowledged, "I was breaking the law myself."

Related to this issue, legal and military experts confirmed that even though DADT requires commanders to take action upon learning of a subordinate's homosexuality, "no commander has been admonished for not following up." Therefore, in practice, because many reported cases are based only on rumors or unobserved behavior, commanders can have a great deal of discretion about whether to launch an investigation into someone's sexual orientation. This is one factor that can lead to uneven and sometimes arbitrary enforcement. One noncommissioned officer told us, "You get accustomed to being 'open' at one duty station, then you're transferred to another, stricter, more conservative environment, and there you have problems." For gay, lesbian, and bisexual service members, the unpredictability of enforcement can add a burden to their ability to perform their duties.

Finding four: *DADT has prevented some gay, lesbian, and bisexual service members from obtaining psychological and medical care as well as religious counseling.*

The Study Group was surprised to learn about the lack of confidentiality accorded gay, lesbian, and bisexual service members in conversations with doctors, chaplains, counselors, and other professionals in whom heterosexual military members can freely confide.[14] The policy also creates ethical dilemmas for professionals attempting to balance their obligation to obey federal law and their obligation to professional ethics.

Despite the general supposition that conversations with clergy are supposed to be confidential in the military, gay, lesbian, and bisexual service members have been investigated and discharged when chaplains reported the contents of their private conversations to commanders. Professor Tobias Barrington Wolff published a study on this topic that found that in 2000 "the Pentagon actually instructed gay soldiers to speak with clergy if they had questions about the policy, implicitly suggesting that confidentiality would be respected. But this instruction provided little security, as the military has continued to initiate discharge proceedings against gay soldiers when chaplains report the statements that the soldiers make during counseling sessions." He adds, "some military commanders instruct doctors and therapists that they are required to report any soldier who speaks about being gay during treatment."[15] Vincent Patton, former Master Chief Petty Officer of the Coast Guard, confirmed that confidentiality in the Chaplains Corps is a serious issue. He indicated that gay, lesbian, and bisexual service members might have more need of clergy support, since they may have less family support, but that confiding in the chaplaincy can prompt a discharge.

In the case of doctors and mental health practitioners, there is no formal pretense of general confidentiality, and gay, lesbian, and bisexual service members incur risk when they speak about their sexual identities to healthcare professionals. One service member, Rhonda Davis, a former noncommissioned officer who was discharged from the Navy for being gay, told us, "As an E-6, I had become a leader, and as a leader, troops came to me for advice and guidance. I had many gay troops working for me, and some of them I saw suffer a great deal because of this policy. One gay troop had a sexually transmitted disease and he asked what he should do about it. I advised him, of course, to see a doctor, but he called it to my attention that if he did, he could be kicked out of the Navy.

Another troop was having a relationship problem with her girlfriend—she threatened committing suicide—and I told her to see a counselor or chaplain, but then I realized that wasn't a good idea because talking about her girlfriend would violate the 'don't ask, don't tell' policy. No matter what I told these troops, nothing was the right answer and I felt like a hypocrite."

False accusations can also threaten the career or well being of heterosexual service members and can produce a generalized atmosphere of fear and suspicion. "In one remarkable incident in 2001, an Air Force airman sought the assistance of a military psychiatrist after a civilian raped him. The psychiatrist announced that the airman must be gay if he allowed himself to be raped, and he threatened to out the soldier to his command if he spoke about being gay during their therapy session."[16]

A February 2007 report of the American Psychological Association (APA) called attention to the increasing mental health needs of all military personnel and their families. The report found that many service personnel and their family members are going without mental health care because of the limited availability of such care and because of barriers to accessing that care. According to the APA, more than 30 percent of all service members who have been deployed to the Iraq and Afghanistan theaters meet the criteria for a mental disorder but less than half of those with mental health concerns seek help. According to Col Thomas Kolditz, a psychologist who chairs the Behavioral Sciences and Leadership Department at the US Military Academy at West Point, "Insofar as DADT makes it less likely for gay, lesbian, and bisexual service members to seek treatment, it exacerbates this [existing] problem." Not only are service members prevented from seeking healthcare, but also health professionals are prohibited from doing their job.

By inhibiting access to religious, medical and psychological services, DADT poses a risk to the well being of some service members. In addition, this denial of confidentiality raises serious questions of professional ethics and constitutional protections. Therefore, confidentiality for such professional consultations should be returned to gay and lesbian service members.

Finding five: *DADT has caused the military to lose some talented service members.*

To meet President Bush's goal of adding 95,000 new service members over the next five years, the military needs to add more than 18,000 new troops each year. According to Dr. Jan Laurence, who retired recently from her position as director of research in the Office of the Undersecretary of Defense for Personnel and Readiness, personnel shortages are so serious that "we're looking at converting positions to civilian because we need people." She emphasized that "we are in dire straits." Given Dr. Laurence's professional background, the Study Group places special emphasis on her conclusions.

In response to such shortages, the number of convicted felons who enlisted in the US military almost doubled in the past three years, rising from 824 felons in fiscal year 2004 to 1,605 in fiscal year 2006. The data indicate that from 2003 through 2006, the military recruited 4,230 convicted felons to enlist under the "moral waivers" program, which enables otherwise unqualified candidates to serve. In addition, 43,977 individuals convicted of serious misdemeanors such as assault were recruited to enlist under the moral waivers program during that period, as were 58,561 illegal drug abusers.

At the same time, according to a report prepared by the Government Accountability Office, nearly 800 people with skills deemed "mission-critical" by the Pentagon have been dismissed under DADT. This figure includes 268 in intelligence, 57 in combat engineering, 331 in medical service delivery, and more than 322 language experts, at least 58 of whom specialized in Arabic.[17] It is counterproductive to military readiness to discharge qualified gay, lesbian, and bisexual service members at the same time that we are filling ranks with service members brought in under the moral waivers program.

A recent UCLA Law School study found that had DADT not been instituted in 1994, approximately 4,000 lesbian, gay and bisexual military personnel would have been retained each year. Of that group, an average of 1,000 men and women were discharged each year as a direct result of the policy, and 3,000 would likely have stayed in the military if they could have served openly and without fear of discharge.

By contrast, 2 percent of presumably heterosexual service members who responded to a recent Zogby poll said that they would not have joined the military if gays and lesbians were allowed to serve openly, a total that would amount to about 4,000 lost recruits per year across the 14 years the policy has been in effect.

If all of these statistics are to be taken at face value, then the repeal of DADT would be a wash in terms of recruiting and retention, with 4,000 heterosexuals refraining from joining the military each year, and 4,000 gays, lesbians, and bisexuals joining and remaining in the force.

These statistics, however, must be read critically. Approximately two-thirds of service members in the Canadian and British forces said that they would not work with gays, but when gay bans were lifted in both of those countries, recruiters reported no mass resignations and no increased difficulties, and even reported slightly enhanced recruiting and retention performance. According to several studies including official Ministry of Defence analyses, less than a handful of service members resigned from the British armed forces after the repeal of the British ban, despite the fact that two-thirds of British service members had previously told survey researchers that they would not work alongside gays.[18] The vast literature on retention and enlistment propensity in the United States does not even include the lifting of the gay ban as a potential determinant in its research. However, the Zogby poll, which did include this factor in a list of motives for joining and staying in the military, found that out of 10 possible motives, the repeal of DADT was ranked 10th in importance.

In the worst case scenario, if it turns out to be true that the numbers cancel out and 4,000 heterosexuals refrain from enlisting, while 4,000 gays, lesbians, and bisexuals do enlist, the group nevertheless points to the many official military pronouncements about the importance of building and maintaining a diverse force to represent the values of a free, pluralist democracy. Building and maintaining a diverse force is a central component to winning the war on terror because the diversity of the armed forces can serve as a living example to peoples living under authoritarian rule, and demonstrate that pluralism and tolerance offer a better way of life.[19]

Finding six: *DADT has compelled some gay, lesbian, and bisexual service members to lie about their identity.*

The Study Group was concerned to discover that DADT encourages dishonesty for some gay, lesbian, and bisexual service members. While some are able to serve in silence and refrain from saying anything about their sexual orientation, many are forced to assert a false identity. According to Professor Tobias Wolff, an expert in constitutional law who has done extensive research on DADT, "It is impossible to be 'agnostic' about one's sexual identity in the course of normal interaction. Rather, a presumption of heterosexuality pervades most settings."[20] Imagine, for example, whether it would be realistic for a married, heterosexual service member to never admit that she has a husband. While theoretically possible, in practice such concealment could be a difficult pretense to maintain given the constant banter and genuine concern for loved ones that takes place among service members.

Several noncommissioned officers who met with us confirmed that while they were never officially asked about their sexual orientation, dating was a topic that came up frequently in informal settings. To escape suspicion in such circumstances, they often felt that they had to lie. One noncommissioned officer who served for 20 years in the Air Force, including a tour in Afghanistan, said that whenever he was asked, "I lied." He added forcefully: "I did not like lying."

The policy puts some gay, lesbian, and bisexual service members in a quandary and undermines the personal integrity essential to honor and trust.

Finding seven: *Many gays, lesbians, and bisexuals are serving openly.*

Despite the fact that DADT causes many gay, lesbian, and bisexual service members to lie about who they are, many others do serve openly. An estimated 65,000 gay, lesbian, and bisexual persons are currently serving on either active or reserve duty, and it is estimated that there are another one million gay, lesbian, and bisexual veterans.[21] A 2006 Zogby poll of troops who served in Iraq and Afghanistan found that nearly one in four US troops (23 percent) say that they know for sure that someone in their unit is gay or lesbian. Of those who say they know for certain that they serve with a gay or lesbian service member, 59 percent said they learned about the person's sexual orientation directly from the individual. More than half (55 percent) of the troops who know a gay peer said that the presence of gays or lesbians in their unit is well known by others. One of the most distinguished academic experts on the military in the country told us, "One thing I have been disabused of is that gays survive by being in the closet. If there are large numbers of gays completely in the closet, I haven't seen it."

As one noncommissioned officer told us, "Of course, I never walked into a room and announced 'I'm a lesbian,' but people aren't stupid, and they always picked up on the fact I didn't have a boyfriend or husband or kids—and eventually, when we were all hanging out at the Enlisted club bonding as friends and shipmates—my secret would come out. As my friends spoke casually of their husbands and wives, I often spoke of some girl I was dating at the time." She concluded, "The reason I could be honest with my Navy friends is because I generally found that people respected me for my work ethic, my integrity, and for my character. I am a good person, and a workaholic—and they could see that."

Finding eight: *DADT has made it harder for some gays, lesbians, and bisexuals to perform their duties.*

Those who do choose to adhere to the policy and lie about their identity sometimes become the target of suspicion or scorn from their peers and this can impact individual and unit performance. As a noncommissioned officer told us:

> I had two gay friends while I was stationed in Spain. One man, E., was very open [about being gay], like me. The other one, T., followed the "don't ask, don't tell" policy nearly to the letter of the law. T. told me that he was gay, but to his co-workers he lied about having girlfriends. But everyone hated him. I asked the guys at work why they harassed T. when none of them harassed E. or me. They said the problem wasn't the fact T. was gay, the problem was he was a liar. And to them, that meant he was a coward. They were personally insulted that he lied to them. In this case, DADT is a dual-edged sword: if you follow it, you're mistrusted; if you don't, you play Russian roulette every day with your career.

Stories such as this suggest to us that service members may be more disturbed about serving with dishonest peers than about serving alongside gays and lesbians. It places young professionals, homosexual and heterosexual, in an unworkable situation.

For those who are open about their sexual orientation, however, other risks present themselves. One former service member told us, "'Don't ask, don't tell' had only been around a little more than a year by the time I enlisted ... but that didn't stop me from being honest with most of my fellow shipmates about my sexual orientation." However, she explained, this meant

> the guy in the office down the hall who had asked me out on a date, only to find out later that I'm a lesbian, could have ended my career. My troop whom I yelled at constantly for being late—who knew I'm a lesbian—could have ended my career. Any number of people, at any time, could have had the power to end my career. Even when I felt comfortable with people, it was always in the back of my mind that anyone at anytime could turn on me and turn me in. My Navy career was always somewhat at their mercy, and that was an incredible burden to bear. Many people know you're gay, but look the other way because they know you're a good sailor.

The Study Group finds that the policy can produce an atmosphere of uncertainty and suspicion for all concerned.

Finding nine: *Military attitudes towards gays and lesbians are changing.*

The existing law and DOD DADT policy on homosexuals serving in the armed forces are based on the attitudes of service members. In 1993, 40 percent of the public supported allowing "openly gay men and lesbian women" to serve in the military.[22] Civilian and military opinion has shifted in the intervening years, indicating much more acceptance for gays and lesbians serving. Recently, national polls have been administered by at least five different polling organizations that have asked members of the public whether gays and lesbians should be allowed to serve openly. All survey results show that between 58 and 79 percent of the public believe that gays and lesbians should be allowed to serve openly.

One conservative polling organization hired by Fox News found that 64 percent of the public, including 55 percent of Republicans, believe that gays and lesbians should be allowed to serve openly, and other pollsters have confirmed that a majority of Republicans now believe that gays and lesbians should be allowed to serve openly.[23] A majority of regular churchgoers say that gays and lesbians should serve openly.

Gallup found that 91 percent of young adults say that gays and lesbians should be allowed to serve openly. Of course, as Moskos pointed out in a message to the Study Group, "Public opinion is not what counts. Attitudes of soldiers does." While it is impossible to know with certainty when a change of attitude of "soldiers" will occur, it seems implausible, given where military and public opinion stands, to imagine that DADT will continue in perpetuity.

Finding ten: *Evidence shows that allowing gays and lesbians to serve openly is unlikely to pose any significant risk to morale, good order, discipline, or cohesion.*

The justification for DADT is contained in 15 Congressional findings which establish the rationale for the law, and which conclude that "the presence in the armed forces of persons who demonstrate a propensity or intent to engage in homosexual acts would create an unacceptable risk to the high standards of morale, good order and discipline, and unit cohesion that are the essence of military capability." While this may have been true in 1993, there are indications that this may no longer be the case. In 1993, the finding of "unacceptable risk" was based on the views of currently serving service members and military leaders, and on the experiences of foreign militaries. However, the group was not able to find any evidence to suggest that the finding of unacceptable risk remains valid.[24]

Colonel Tom Kolditz, chairman of the Department of Behavioral Sciences and Leadership at the US Military Academy at West Point, is one of the Army's top experts on leadership and cohesion. He served 18 years as a Field Artillery officer including two years of battalion command and seven years as a professor at West Point. From 1995 through 1997, he worked in the Human Resources Directorate of the Army G-1, where Army policy related to DADT is managed. He completed service as one of four doctoral level researchers supporting the Secretary of the Army's Senior Review Panel on Sexual Harassment following the Aberdeen Proving Ground scandal. And he is the author of a number of well-received studies on leadership and cohesion, including a Strategic Studies Institute monograph titled "Why They Fight," and a book released this past June titled *In Extremis Leadership: Leading as if Your Life Depended on It.*

Colonel Kolditz told us, "Cohesion is important to Army leaders, especially in combat. Current Army leadership doctrine, FM 6-22 requires Army leaders to 'build high-performing and cohesive organizations.' Among the principal issues in cohesion research is the relative contribution of task cohesion (the ability for individuals to work in teams to accomplish tasks) versus social cohesion (personal relationships among team members)." Kolditz emphasized that there is a current emphasis on training cross-cultural skills in concert with a tolerance for diversity among soldiers and leaders to enable US success in current missions around the world. He elaborated that

> I've taught a course in cross cultural leadership and diversity for the Eisenhower Leader Development Program, a graduate program taught to Army Captains at West Point in concert with Columbia Teachers' College, and most recently adapted the course as an extended lecture in the Yale School of Management. A core instructional element of that course is that people can develop cross cultural leadership skills not only by being in foreign cultures, but by practicing their skills and abilities at home, across diversity areas, such as race, gender, class, age, religious affiliation, physical ability, and sexual orientation. I introduce the concept by saying that people who have a hard time communicating and working with, say, Amish people in Lancaster County, Pennsylvania, will certainly have a difficult time working with Sunni Muslim police administrators in Baghdad.

He added, "I could just as easily substitute an example based on sexual orientation." Kolditz emphasized that he is unaware of any evidence suggesting that heterosexuals cannot form bonds of trust with gays, lesbians, and bisexuals.

A scholar at the RAND Corporation who is a leading academic expert on unit cohesion confirmed that "I do not know of any evidence" that suggests that gays undermine cohesion. A heterosexual officer who returned recently from Iraq explained that the friction resulting from the prosecution of service members found to be gay is far greater than the friction that results from simply knowing a gay person. And retired Master Chief Petty Officer Vincent Patton confirmed that service members have told him that "we had unit cohesion till this [gay] person was kicked out."

Given the differences between foreign armed forces and the US military, the Study Group does not place too much stock in lessons learned from overseas. That having been said, it is worth noting that the British Ministry of Defence has completed two official studies of the repeal of the British gay ban and that while some units did experience minor friction, overall the policy transition posed no serious challenges whatsoever. The Study Group heard testimony from top uniformed and academic experts on gays, lesbians and bisexuals in the Israeli and British militaries who confirmed that they were not aware of any detriment to morale, good order, discipline or cohesion that followed from allowing gays, lesbians, and bisexuals to serve openly. In fact, Britain has recently begun actively recruiting gay, lesbian, and bisexual men and women for service in the Royal Navy.[25]

While polls show that a majority of American service members say that they would prefer that DADT remain in place, only a small minority of those polled say that they are personally uncomfortable interacting and working with gays and lesbians. This represents a major shift from 1993. General Wesley Clark confirms that the "temperature of the issue has changed over the decade. People were much more irate about this issue in the early '90s than I found in the late '90s, for whatever reason, younger people coming in [to the military]. It just didn't seem to be the same emotional hot button issue by '98, '99, that it had been in '92, '93."[26] In 2005, a West Point Cadet received an award for writing the best senior honors thesis in his department for a study arguing that DADT is inconsistent with the military's emphasis on fairness and equal treatment.[27] And former chairman of the Joint Chiefs of Staff John Shalikashvili publicly announced that despite his original support for DADT, he no longer believes that the policy serves the military's interest. In a January 2007 *New York Times* op-ed, he noted, "I now believe that if gay men and lesbians served openly in the United States military, they would not undermine the efficacy of the armed forces."[28]

Finally, a new statistical analysis of 545 service members returning from Iraq and Afghanistan finds that there is no correlation between knowing a gay unit member and the level of readiness or cohesion in the unit.[29]

Recommendations

Recommendation 1. Congress should repeal 10 USC § 654 and return authority for personnel policy under this law to the Department of Defense.

Recommendation 2. The Department of Defense should eliminate "don't tell" while maintaining current authority under the Uniform Code of Military Justice and service regulations to preclude misconduct prejudicial to good order and discipline and unit cohesion. The prerogative to disclose sexual orientation should be considered a personal and private matter.

Recommendation 3. Remove from Department of Defense directives all references to "bisexual," "homosexual," "homosexual conduct," "homosexual acts," and "propensity." Establish in their place uniform standards that are neutral with respect to sexual orientation, such as prohibitions against any inappropriate public bodily contact for the purpose of satisfying sexual desires.

Recommendation 4. Immediately establish and reinforce safeguards for the confidentiality of all conversations between service members and chaplains, doctors, and mental health professionals.

Return authority to DOD.

Letter from the Palm Center

The General/Flag Officers' Study Group project emerged out of conversations with a number of offices in Congress, Democratic and Republican, who wanted to be sure, as the Military Readiness Enhancement Act moves forward, that senior military voices were consulted throughout the process. The project also emerged out of a recognition that this may be the time, on the eve of the 15th anniversary of the "don't ask, don't tell" policy, to resume an informed civil-military conversation on the issue. The military perspective on the policy's current contribution to its stated goal, preserving military effectiveness, is of utmost importance.

Therefore, a nonpartisan national study group, comprised of retired General/Flag Officers from different branches of the service, who have the institutional experience and perspective to offer sound recommendations, was assembled to study the effectiveness of "don't ask, don't tell." The Study Group was to review available evidence, consider arguments from all sides, and issue a public report. The goal of the Study Group was to explore two key areas concerning the "don't ask, don't tell" policy: 1) the "unacceptable risk" standard established in the law and 2) DOD's implementation of the law through implementing regulations. The Study Group would then make recommendations based on their findings about the current state of the policy and its present impact on military personnel, leadership, and effectiveness.

We are grateful to all those who have assisted the Study Group, especially the national and international experts who agreed to share their expertise in person and directly address its questions. We are also grateful to those scholars who could not meet in person, but nevertheless provided the Study Group with taped and written comments. Many of the experts who agreed to speak with the Study Group were centrally involved in the policy conversations that culminated in the passage of "don't ask, don't tell" in 1993, and we appreciate their generosity and willingness to return to these issues with the benefit of hindsight and to offer their analysis.

Our further thanks go to Col Richard Klass and Brant Shalikashvili, both of whom played central roles in the project. Finally, we thank the research scholars at the Palm Center who compiled current information and data for the Study Group's review, especially Dr. Nathaniel Frank, senior research fellow. Funding for the General/Flag Officers' Study Group Project has been provided by the University of California, Santa Barbara.

Aaron Belkin, *Director*
Jeanne Scheper, *Research Director*
Indra Lusero, *Assistant Director*

Notes

1. Personal correspondence with Prof. Charles Moskos, 6 October 2007.

2. "I would not say that the policy that we are implementing here today is the policy that will be forever." "I would not say that this is going to be forever." News conference, Secretary of Defense Les Aspin and Jamie Gorelick, general counsel, Department of Defense, Regarding the Regulations on Homosexual Conduct in the Military, 22 December 1993.

3. John Holum confirmed this in his comment to the Study Group, stating that "don't ask, don't tell" was "not meant to be long term" and "doesn't make sense as a long term policy."

4. US Department of Defense Directive (DODD) 1332.14, *Enlisted Administrative Separations*, 28 January 1982, 1:H: "Homosexuality is incompatible with military service. The presence in the military environment of persons who engage in homosexual conduct or who, by their statements, demonstrate a propensity to engage in homosexual conduct, seriously impairs the accomplishment of the military mission. The presence of such members adversely affects the ability of the military services to maintain discipline, good order, and morale; to foster mutual trust and confidence among service members; to ensure the integrity of the system of rank and command; to facilitate assignment and world-wide deployment of service members who frequently must live and work under close conditions affording minimal privacy; to recruit and retain members of the military services; to maintain public acceptability of military service; and to prevent breaches of security."

5. National Defense Research Institute, *Sexual Orientation and U.S. Military Personnel Policy: Options and Assessments* (Santa Monica, CA: RAND Corporation, 1993).

6. Randy Shilts, *Conduct Unbecoming: Gays and Lesbians in the U.S. Military* (New York: St. Martin's Press, 1993).

7. Curtis Wilkie, "Harvard Tosses Warmup Queries to Clinton on Eve of New Hampshire Debate," *Boston Globe*, 31 October 1991.

8. *Policy Concerning Homosexuality in the Armed Forces, Hearings before the Committee on Armed Services*, US Senate, 103rd Cong., 2nd sess., US Government Printing Office, 1994.

9. Thomas Friedman, "Compromise Near on Military's Ban on Homosexuals," *New York Times*, 29 January 1993.

10. Office of the Secretary of Defense, "Summary Report of the Military Working Group," July 1993; and National Defense Research Institute, *Sexual Orientation and U.S. Military Personnel Policy: Options and Assessments* (Santa Monica, CA: RAND Corporation, 1993).

11. *Policy Concerning Homosexuality in the Armed Forces, Hearings before the Committee on Armed Services*, US Senate, 103rd Cong., 2nd sess., US Government Printing Office, 1994.

12. Office of the Secretary of Defense, "Policy on Homosexual Conduct in the Armed Forces, Memorandum for the Secretary of the Army, Secretary of the Navy, Secretary of the Air Force, Chairman, Joint Chiefs of Staff," *Weekly Compilation of Presidential Documents*, vol. 29, 19 July 1993.

13. 10 USC 654, Policy Concerning Homosexuality in the Armed Forces.

14. In October 2003, the APA Board of Directors established the Task Force on Sexual Orientation and Military Service to address issues including a) confidentiality for military service members within military mental health systems, b) the education of service members regarding federal law and military mental health services, c) the training of military psychologists, d) consultation with military mental health providers, and e) the DOD implementation of the Policy Concerning Homosexuality in the Armed Forces (1993) and specifically "don't ask, don't tell."

15. Tobias Barrington Wolff, "Political Representation and Accountability," *Iowa Law Review* 89 (2004): 7.

16. Ibid.

17. Bryan Bender, "Gays' Ouster Seen Leaving Gap In Military," *Boston Globe*, 24 February 2005: 1; and Government Accountability Office (GAO), *Financial Costs and Loss of Critical Skills Due to DOD's Homosexual Conduct Policy Cannot Be Completely Estimated*, GAO 05-299 (Washington, DC: GAO, February 2005).

18. Ministry of Defence, *Tri-Service Review of the Armed Forces Policy on Homosexuality and Code of Social Conduct*, December 2002; and Ministry of Defence, *A Review on the Armed Forces Policy on Homosexuality*, 20 October 2000.

19. Remi Hajjar, "A New Angle on the US Military's Cultural Awareness (CA) Campaign: Connecting In-Ranks' Diversity to CA" (paper prepared for the Meetings of the Inter University Seminar on Armed Forces and Society, Chicago, IL, 26–28 October 2008).

20. Amici Brief, 9.

21. Gary Gates, "Gay Men and Lesbians in the U.S. Military: Estimates from Census 2000," Urban Institute, 28 September 2004.

22. See David Burrelli and Charles Dale, *Homosexuals and U.S. Military Policy: Current Issues* (Washington, DC: Congressional Research Service, March 2006): 6, citing a July 1993 NBC/*Wall Street Journal* poll.

23. Dana Blanton, "Majority Opposes Same-Sex Marriage," *Fox News*, 26 August 2003.

24. Aaron Belkin, "Don't Ask, Don't Tell: Is the Gay Ban Based on Military Necessity," *Parameters*, Summer 2003, 108–19; GAO, *Homosexuals in the Military: Policies and Practices of Foreign Countries*, GAO/NSIAD-93-215 (Washington, DC: GAO, June 1993); and "What Is Known about Unit Cohesion and Military Performance?" in National Defense Research Institute, *Sexual Orientation and US Military Personnel Policy: Options and Assessment* (Santa Monica, CA: RAND, 1993). RAND's report was begun at the request of US Secretary of Defense Les Aspin sometime after 29 January 1993 and completed before 19 July 1993.

25. Sarah Lyall, "Gay Britons Serve in Military with Little Fuss, As Predicted Discord Does Not Occur," *New York Times*, 21 May 2007, A: 1, Foreign Desk.

26. NBC *Meet the Press*, transcript, 15 June 2003.

27. Cadet Alexander H. Raggio, "Don't Ask, Don't Tell, Don't Be: A Philosophical Analysis of The Gay Ban in the US Military" (honors thesis, Department of English, US Military Academy at West Point, April 2005).

28. John M. Shalikashvili, "Second Thoughts on Gays in the Military," op-ed, *New York Times*, 2 January 2007.

29. Zogby International, 2006.

Study Group Members

Brig Gen Hugh Aitken, USMC, retired—General Aitken's distinguished career spanned five decades, beginning with his enlistment in 1946. He served in Korea, where he was company commander with the 1st Marine Division, and Vietnam, where he joined the 1st Marine Division as the assistant G-3. He attended the Army War College and served as deputy director, Plans Division. In 1975, he became the executive assistant to the DC/S for Manpower. Promoted to brigadier general in March 1978, he became the director, Manpower Plans and Policy Division. In August 1978, he was assigned as the assistant division commander, 2nd Marine Division. He was assigned duty as the director, Manpower Plans and Policy Division at Headquarters Marine Corps in September 1979, serving in this capacity until his retirement in 1980.

Lt Gen Minter Alexander, USAF, retired—Lt Gen Minter Alexander retired in 1994 as the deputy assistant secretary of defense for military personnel policy after more than 30 years of service. The general is a command pilot with more than 4,000 flying hours, including 800 combat hours. His personnel policy background covered critical issues that included planning and implementing the post–Cold War drawdown of 500,000 military personnel. His awards and decorations include the Defense Distinguished Service Medal, the Air Force Distinguished Service Medal, Silver Star with oak leaf cluster, Defense Superior Service Medal, Legion of Merit with oak leaf cluster, Distinguished Flying Cross with oak leaf cluster, Bronze Star Medal, Meritorious Service Medal, Air Medal with 18 oak leaf clusters, and Air Force Commendation Medal.

LTG Robert Gard, USA, retired—Lt Gen Robert Gard retired in 1981 after serving as president of the National Defense University in Washington, DC. He started his military education at West Point, graduating in 1950, and earned his PhD in political economy and government from Harvard in 1962. He served in Vietnam, Germany, and Korea, and was the military assistant to two secretaries of defense. Since retirement from the Army he has been a professor and director of Johns Hopkins University's School of Advanced International Studies in Bologna, Italy, and president of the Monterey Institute of International Studies. He currently serves as senior military fellow at the Center for Arms Control and Non-Proliferation in Washington, DC.

VADM Jack Shanahan, USN, retired—VADM Jack Shanahan retired in 1977 after 35 years of service during which time he served in the Pacific in World War II, in Korea off the coast, and in Vietnam, including tours in the Tonkin Gulf and as commander of the Coastal Surveillance and Interdiction Force. In addition to the standard campaign awards, he holds the Joint Chiefs Commendation Medal (two awards), the Legion of Merit (three awards, one with the Combat V), the Distinguished Service Medal (two awards), the Navy Commendation Medal, and the Navy Combat Action Medal.

Invited Experts

Mark Agrast, senior fellow, Center for American Progress

Graham Beard, commander, Head of Diversity and Equality in the Royal Navy, UK

Phillip Carter, served as an officer in the US Army, including nine years of active and reserve service with military police and civil affairs units. In 2005 and 2006, he deployed to Iraq with the Army's 101st Airborne Division where he served as an adviser to the Iraqi police.

Rhonda Davis, former Petty Officer 1st Class, US Navy, was discharged under "don't ask, don't tell."

Elaine Donnelly, president, Center for Military Readiness*

---

*Declined invitation. Referred Study Group to previously published work.

Chai Feldblum, PhD, professor at Georgetown University Law School

Jamie Gorelick, former deputy attorney general of the United States, served as the general counsel of the Department of Defense in 1993 and 1994.

John Holum, former under secretary of state for international security and arms control and director of the Arms Control and Disarmament Agency, was in charge of the gays in the military issue for the incoming Clinton administration between the 1992 election and the 1993 inauguration.

Lt Cmdr Craig Jones, Royal Navy

Danny Kaplan, PhD, is an officer in the Israel Defense Forces and a leading academic expert on gays in the Israeli military.

Col Thomas Kolditz is professor and head of the Department of Behavioral Sciences and Leadership at the US Military Academy at West Point.

Lawrence Korb, PhD, senior fellow at the Center for American Progress, was assistant secretary of defense (manpower, reserve affairs, installations and logistics) during the Reagan administration, from 1981 through 1985.

Dave Lebsack retired recently as master sergeant in the US Air Force. His 20-year career included a tour of duty in Afghanistan.

Jan Laurence, PhD, served as director of research and analysis in the Office of the Under Secretary of Defense for Personnel and Readiness from 2005 through 2007.

Lt Col Robert Maginnis served as vice president for policy and director of the Military Readiness Project at the Family Research Council.[+]

Eugene R. Milhizer, associate dean and associate professor of Law, Ave Maria School of Law[+]

Laura Miller, PhD, RAND Corporation and a member of the Army Science Board as well as the Board of Directors, Inter-University Seminar on Armed Forces and Society

Charles Moskos, PhD, is an emeritus professor of sociology at Northwestern University and was a principal architect of the "don't ask, don't tell" policy.[+]

Dr. Vincent Patton III, MCPOCG US Coast Guard, served as the Eighth Master Chief Petty Officer of the US Coast Guard.

Ronald Ray was deputy assistant secretary of defense during the Reagan administration.[**]

David Segal, PhD, director, Center for Research on Military Organization and professor of sociology at the University of Maryland

Peter Sprigg is vice president for policy at the Family Research Council.[+]

Maj Melissa Wells-Petry served as counsel to the Readiness Project at the Family Research Council.[+]

John Allen Williams, PhD, is professor of political science at Loyola University Chicago and chair and president of the Inter-University Seminar on Armed Forces and Society. He retired as a captain in the US Naval Reserve with 30 years of commissioned service.[*]

Tobias Barrington Wolff is a professor of law at the University of Pennsylvania and a leading scholar on the constitutionality of "don't ask, don't tell."

William Woodruff is professor of law at Campbell University, a Christian university in North Carolina.

---

[*]Declined invitation to appear or to submit written comment.
[+]Declined invitation to appear from the Study Group. Submitted written comment.
[**]Declined invitation. Referred Study Group to previously published work.

CHAPTER 8

NOT YES OR NO, BUT WHAT IF
IMPLICATIONS OF OPEN
HOMOSEXUALITY IN THE US MILITARY

Matthew P. Cashdollar

Introduction

Since the introduction of Department of Defense Directive (DODD) 1332.14, *Enlisted Administrative Separations*, commonly known as "Don't Ask, Don't Tell" (DADT), a tremendous amount has been written arguing both sides of the issue. With a change of the presidential administration, it is quite possible that a review of the current DOD Instruction 1332.14 (August 2008) may lead to its repeal, allowing homosexuals to serve openly in the US military. Although President Obama's staff has intimated he wants gays and lesbians to serve openly in uniform, he is undoubtedly aware of the quagmire in which President Clinton found himself in 1993 when he attempted to implement unilateral change. Although the current administration is feeling pressure from the gay community, its stated intent is to work with both the Pentagon and Congress to gain the favor of the American people.[1]

Based on recent actions and statements of the current administration, President Obama seems to recognize that this transformation will require an extensive review of current policies and regulations to establish a framework that assists in the facilitation of open homosexuality into the US armed forces. Until now, very little has been written on how the US military would facilitate lifting the ban, demonstrating a perceived attitude of "if we plan for it, we will have to do it."[2] This chapter is not a case *for* or *against* DADT, but rather an examination of the policies and logistical changes or adjustments that would facilitate a transition for both straight and gay service members.

The following analysis will address three main areas. The first is an examination of international examples used to explore feasibility. British, Canadian, and Australian military examples are evaluated as a benchmark of a common cultural and social heritage as well as institutional similarities (e.g., all-volunteer, regular armed forces).[3] A comparison to the US military's integration of blacks following World War II illustrates how the executive order issued by President Truman in 1948 initiated a complicated and sometimes unwelcome process that eventually led to a fully integrated armed service. Next, and arguably the most critical area, is how best to prepare the current force for the lifting of the ban. This can be accomplished by incorporating lessons learned from the review of foreign militaries, focusing on leadership attitudes, extensive and parallel training across all services, elucidating appropriate and inappropriate behavior, and equal and fair enforcement of the new policy. The last section is a consideration of possible areas of concern that could cause difficulty with the integration. It addresses how the military might deal with activism (from both religious and pro-gay groups), the issue of transsexuality, as well as other implications which might emerge with adoption of a new policy.

Background

Prior to 1992, the overarching policy for the US military was to ban homosexuals from service if identified at reception or administratively separate them upon discovery after enlistment or commissioning. The policy and treatment of homosexuals varied greatly throughout the twentieth century. During the interwar years and prior to the end of World War II, enlisted personnel suspected of or charged with homosexual acts were discharged under Section VIII without honor.[4] Following the war, in 1945 the War Department policy changed to punishment by courts-martial or hospitalization. If the service member was hospitalized, his or her fate after "treatment" was either to be returned to duty, separated, or court-martialed.[5]

Even though separation was still the most common practice, the policies continued to change over the next two decades. The most liberal policy of the immediate post-war era was allowing enlisted members identified as having homosexual tendencies, but not yet committing a sexual offense, to be discharged honorably. Officers were allowed to resign under honorable conditions.[6]

During the 1950s, the Army adopted more stringent regulations dividing homosexuals into three classes. Classes I and II were homosexuals charged with offenses of assault or coercion or who engaged or attempted to engage in homosexual acts. These service members were either court-martialed or separated from service. The third class included personnel who exhibited, professed, or admitted homosexual tendencies. Class III could receive an honorable or general discharge.[7]

Regulations continued to be adjusted during the 1960s and 1970s, eventually developing loopholes allowing commanders to disapprove the decisions of separations boards. Even though Army policy was that homosexuality is incompatible with military service, the officer elimination regulation implied that separation was discretionary.[8] This provided means for service members to challenge discharges in the courts. Cases such as *Matlovich v. Secretary of the Air Force* (1978) and *Ben-Shalom v. Secretary of Army* (1980) caused the Army to revise the enlisted regulations in 1981 to create a separate chapter specifically for homosexuality. The Department of Defense followed in 1982 by issuing a directive that rendered homosexuality an exclusionary policy uniform throughout the services, thus eliminating any loopholes and banning the enlistment, commissioning, or service of anyone identified as homosexual.[9]

It was not until Pres. William Jefferson Clinton cast the US military into turmoil by announcing during a speech at Fort McNair on 19 July 1993 that he intended to extend civil rights to homosexuals. This included rescinding the ban on gays and lesbians serving in the US armed services, which was met with a firestorm of criticism from the Joint Chiefs of Staff and several prominent members of Congress led by Sam Nunn. The chairman of the Joint Chiefs of Staff, Gen Colin Powell, stated that "active and open homosexuality by members of the armed forces would have a negative effect on military moral and discipline."[10] Retired Gen Norman Schwarzkopf testified at the Senate Armed Services Committee hearings, suggesting that if the ban was lifted, the troops "will be just like many of the Iraqi troops who sat in the deserts of Kuwait, forced to execute orders that they didn't believe in." A great deal of pressure was placed on President Clinton from both sides of the issue. Vice Pres. Al Gore insisted that Clinton shouldn't compromise and should lift the ban as a matter of principle. The president worried that even private statements about homosexuality would be largely prohibited and that a compromise would be too restrictive on something he considered a private matter. But he also realized that in the end, there would have to be some restrictions on conduct.[11]

Despite being aligned under the Democratic Party, Congress moved quickly to override the newly elected president. To find a solution in the form of compromise, Secretary of Defense Les Aspin formed an internal military working group and charged the panel to come up with a suitable plan for accommodating homosexuals in the military.[12] In August 1993, the president, with the concurrence of the secretary of defense and the Joint Chiefs of Staff, announced a compromise policy called "Don't Ask, Don't Tell, Don't Pursue."[13] Congress held further hearings that led to legislation being passed on 22 December 1993, providing regulations for the military to enforce the new law. The new policy contained two main differences from the previous one. First, it included the phrase "a person's sexual orientation is considered a personal and private matter and is not a bar to service unless manifested by homosexual conduct."[14] The second difference was that military recruiters and commanders were no longer allowed to ask potential recruits or service members questions related to their sexual orientation, and commanders were restricted from actively seeking to identify homosexuals in their units unless the service member was involved in homosexual conduct.[15]

The current military policy has changed very little over the past 15 years and is seen by pro-gay organizations as failing to provide equal rights to homosexuals. Despite numerous legal challenges to the standing policy, most senior military leadership support DADT. Groups on both sides of the argument cite polls that support their argument, but even the combination of these polls show the overwhelming majority of US service members are against overturning the policy and allowing open homosexuality in the military.[16] However, among the general civilian population, the trend seems to be more tolerant of homosexuals

serving openly. Public opinion polls have shown that the once-contentious issue has become more accepted in American society, even though the majority
of military service members remain opposed.[17] No matter what the polls indicate, Pres. Barack Obama has made it clear that his administration is in support
of amending DADT to allow gays and lesbians to serve openly in the US
military. "The key test for military service should be patriotism, a sense of duty,
and a willingness to serve. Discrimination should be prohibited," read an entry
on his preinaugural transition Web site.[18] With a Democratic majority in the
Senate and House, it is likely that President Obama may have the votes to
overturn DADT. The fundamental question of interest is not what the new
policy should be, but rather how it should be implemented and administered
for the good of all American service members.

Studies of Foreign Countries

Numerous studies have focused on how other countries have changed their
policies to allow gays and lesbians to serve openly in their armed forces. Some
of the more notable include the RAND study,[19] Government Accountability
Office (GAO) report,[20] and several studies from the Palm Center.[21] Examination of international examples provides valuable confirmation of feasibility, and
the British, Canadian, and Australian armed services are excellent examples of
totally integrated militaries. Although these are only a few of the countries
studied by the aforementioned reports, about half of the international countries
do not have laws banning the military service of homosexuals. Britain, Canada,
and Australia are used for comparison because of their institutional similarities
and their common cultural heritage with the United States. Britain, Canada,
and Australia each have all-volunteer forces, have an active combined troop
level above 60,000, and have recently changed their own respective military
social policies to allow homosexuals to serve openly.

Canada

Canada was the first of the three nations to change its policy toward open
homosexuality in its armed forces. The Canadian military is an all-volunteer
force, consisting of approximately 77,800 active forces and 33,700 reserves. Men
constitute 86 percent of the force and women 14 percent. Women are permitted
to serve in combat and noncombat positions.[22] The Canadian forces are routinely committed to peacekeeping operations and have been involved with operations in Cambodia, Cyprus, Lebanon, Somali, and the former Yugoslavia.

The close relation between the Canadian Defense Forces and the social construct of the nation has directly resulted in the overturning of the ban on gays in
the Department of National Defense (DND). The development of both civilian and military policies runs parallel, with the major change in the late 1980s,
when the Canadian courts determined that sexual orientation was covered by
the Charter of Right and Freedoms. This was followed in 1993 with the DND

revoking its policy and removing all restrictions on homosexuals. The DND began a review of conduct associated with the change in regulations which included inappropriate sexual conduct, personal relationships, and harassment. Government officials hoped that resulting data would help them create identical standards of conduct for homosexual and heterosexual service members.[23]

At the inception of the new policy, the DND did not recognize partner benefits, but it later reversed the decision. In 1998, the DND received 17 claims for medical, dental, and relocation benefits for homosexual partners.[24] Current information on the effects of lifting the ban in the Canadian military is incomplete. The DND has not conducted any follow-up studies; however, a briefing for the director of personnel policy titled "Effects of Cancellation of Canadian Forces Policy Restricting Service of Homosexuals" revealed two significant findings:

1. Results from a poll that asked service members the following question on human rights issues and policies: "How satisfied/dissatisfied are you with the Canadian Forces policy on sexual orientation?" Of the 3,202 respondents:

   a. 3.8 percent had no opinion;

   b. 8.5 percent were either dissatisfied or very dissatisfied;

   c. 24.4 percent were neutral; and,

   d. 43.3 percent were either satisfied or very satisfied.

2. Attitudinal reaction to the policy change on sexual orientation appears to be mixed, but not more so than to other social policy change. Behavioral and conduct data compiled by several agencies in National Defense Headquarters yield little or no evidence to suggest that allowing homosexuals to serve in the Canadian forces has been problematic, either in terms of their behavior or their treatment by other members. This finding must be qualified, however, by the observation that it is not known to what extent homosexual members generally refrain from making their sexual orientation known, in which case the behavioral and conduct indicator might not be reliable and the effect of the policy change on such variables as unit cohesion and morale would be extremely difficult, if not impossible, to measure.[25]

Prof. Aaron Belkin of the Palm Center, a gay-rights organization at the University of California in Santa Barbara, published a study in *Parameters* during the summer of 2003 that showed lifting the ban in the Canadian DND had little to no impact in military performance, readiness, or cohesion among the Canadian armed forces. In the report, a DND civilian official, Steve Leveque, commented that including gays and lesbians in the Canadian forces is "not that big a deal for us . . . on a day-to-day basis, there hasn't been much of a change." The study also found that after the ban was lifted, there was no "mass coming out of the closet" of service members. This did change with time, but the majority of gay and lesbian soldiers refrained from acknowledging their homosexuality.[26]

Australia

The Australian Defense Force (ADF) is a small, all-volunteer force of approximately 70,000 active-duty personnel. Although Australia participated heavily in World War II, its military presence in the Korean and Vietnam conflicts was much more restricted and has been reduced in modern times to peacekeeping operations and providing a limited number of troops to the campaigns in Iraq and Afghanistan. The ADF is quite similar to the CDF in that both are primarily for national defense and limited peacekeeping or humanitarian assistance operations.

The ADF did not have an official ban on service members openly admitting that they were homosexuals, and recruits were not asked their sexual orientation upon entry. However, if a service member self-identified as homosexual, he or she was asked to resign, and in most cases the service member complied. If the service member refused to resign, the ADF would commence actions to remove the person from military service. The Australian government ended the prohibition of homosexuals from their armed forces in 1992 by implementing a new policy stating that unacceptable sexual behavior applies to all service members regardless of their sexual orientation.[27] Credit is given to the passage of the Sex Discrimination Act and the Human Rights and Equal Opportunity Commission Act for forcing the change to the ADF policy. Australia does not have any laws prohibiting sodomy that have interfered with the new law. The ADF charged individual commanders with implementation of the new policy, which is monitored routinely through the chain of command.

Great Britain

The British military and its corresponding social structure are probably the most similar to those of the United States. It has an all-volunteer force consisting of approximately half a million personnel and has been the closest ally to the United States in most military operations since World War II. British attitudes toward open homosexuality in their military also closely align with those of Americans. Opinion polls for both the military and civilian population show the majority against open service for gays and lesbians.[28]

The development of British policy on homosexuality followed a similar track as it did in the United States. Several committees were assembled to consider the legal position of homosexual practice; they concluded in 1957 that homosexuality, in some circumstances, should no longer be a crime.[29] This conclusion was based on a key argument that homosexuality is a matter of private morality and the law should not intrude. The government rejected the proposal that the military was not ready for this change. Ten years later, the Sexual Offences Act of 1967 allowed that in most cases, "homosexual practice conducted between two consenting adults, over the age of 21, was a private matter and no longer a crime."[30] However, the policy did not translate to the military, which viewed it as a civilian law and continued to consider homosexuality a criminal act subject to military law.

As the years progressed, so did the push for decriminalization of homosexuality in the British armed services. A select committee on the Armed Forces Bill recommended a change of policy that was approved by the British government,[31] leading to an acceptance of the Criminal Justice and Public Order Act of 1994. The Sexual Offences Act of 1967 was officially repealed and the existing social policy changed, rendering homosexual acts under civil law no longer criminal under military law. The minister of defence emphasized, however, that military personnel would no longer be prosecuted under military law but could still be discharged from the service.

During the 1990s, numerous court challenges were brought against the British government by Stonewall, a gay activist group who represented numerous gay and lesbian service members previously discharged from the British military. These lawsuits eventually found their way to the European Union Human Rights Court, which, in November 2000, adopted a directive prohibiting discrimination on the basis of sexual orientation in public and private employment. This decision forced the United Kingdom to drop its ban on open homosexual service in the British armed forces, and on 12 January 2000, the ban was replaced with a new, sexual orientation–free general code of conduct, fully titled the "Armed Forces Code of Social Conduct" (AFCSC). The Ministry of Defence (MOD) summarized the new code in the following way: "The policy to bar homosexuals from the Armed Forces was not legally sustainable and has now been replaced with a new policy which recognizes sexual orientation as a private matter. It was formulated with the full consultation and support of the three Service Chiefs and is firmly underpinned by a code of social conduct that applies to all regardless of their sexual orientation."[32]

The new British AFCSC is a three-page document that outlines appropriate behavior, regardless of sexual orientation. The following is an excerpt from the AFCSC on sexual conduct:

> Examples of behavior which can undermine such trust and cohesion, and therefore damage the morale or discipline of a unit (and hence its operational effectiveness) include: unwelcome sexual attention in the form of physical or verbal conduct; over-familiarity with the spouses or partners of other service personnel; displays of affection which might cause offence to others; behavior which damages or hazards the marriage or personal relationships of service personnel or civilian colleagues within the wider defense community; and taking sexual advantage of subordinates.[33]

The code also provides guidance through the "service test," which assists commanders in determining how to handle inappropriate behavior. AFCSC provides the following direction in the fifth paragraph of the document:

> When considering possible cases of social misconduct, and in determining whether the Service has a duty to intervene in the personal lives of its personnel, Commanding Officers at every level must consider each case against the following Service Test:

"Have the actions or behavior of an individual adversely impacted or are they likely to impact on the efficiency or operational effectiveness of the Service?" In assessing whether to take action, Commanding Officers will consider a series of key criteria. This will establish the seriousness of the misconduct and its impact on operational effectiveness and thus the appropriate and proportionate level of sanction.[34]

The AFCSC provides commanders with simple and commonsense guidance that describes appropriate social conduct and allows unit commanders discretion to address inappropriate behavior as the situation dictates, keeping in mind that the MOD views sexual orientation as a private matter.

The ability of the British MOD to establish a common and understandable regulation that provided gays and lesbians the ability to serve openly in the military became a key to the success of the change in policy. The new regulation allowed homosexuals to serve openly but did not provide partner benefits or family housing to gay couples. Stephen Deakin, a senior lecturer at the Royal Military Academy Sandhurst, credits the success of the program to the ability of the MOD to keep it "low key," maintaining the attitude of "don't ask, don't tell" in most cases.

Nevertheless, over the past decade, the British government has been unable to keep the issue out of play. The push for civil partnerships is a new topic on the British social agenda and may have implications for partner benefits for service members. Several cases have also been made public of British military personnel having sex change operations and continuing service in the military and service members marching in uniform in gay rights parades. In addition, the Stonewall organization is working with the British military to create a plan specifically for recruitment and retention of gays and lesbians into the military ranks. It is unclear if these recent events will have an effect on the current attitudes of British military personnel toward homosexual service mates, but the evolution of such issues is to be considered by the US military when drafting a plan for the repeal of DADT.

As demonstrated in all three studies, the success of the new policy can be attributed to the emphasis on equal standards and conduct and the development of new regulations that are easy to understand and implement. The Australian and British policies focus on individual rights, with neither the ADF nor the British MOD offering partner benefits, making a clear distinction between individual rights and group rights. This is in contrast to the Canadian model, which seems to have more emphasis on group rights and does provide partner benefits.

History of Integration in US Military

While the integration of open homosexuality in the US military is not directly comparable to the desegregation of blacks in the late 1940s, the experiences of racial integration provide insights into the military's ability to adapt to change.[35] Numerous pro-gay organizations cite the integration of blacks as case in point for the integration of open homosexuality into the US military. The argument is that if such a controversial social issue as racial discrimination can be overcome, then it is time for the elimination of sexual discrimination and repeal of the DADT policy.[36] Although they can be viewed as two different issues, the examination of the trials and processes that ended racial segregation may provide examples of how to achieve a better and more accommodating policy for allowing homosexuals to serve openly in the US military.

When Pres. Harry Truman gave the executive order to desegregate the US military in 1948, few were accepting of the idea. Polls taken both within the military and in civilian society reflected a strong resistance to the idea. In 1943, 90 percent of white civilians and 18 percent of black Soldiers believed that whites and blacks should be assigned to separate units.[37] Truman believed the time was at hand and, by Executive Order 9981, gave the mandate to provide equal treatment and opportunity to all persons in the armed services. The first paragraph of the order reads:

> It is hereby declared to be the policy of the President that there shall be equality of treatment and opportunity for all persons in the armed services without regard to race, color, religion or national origin. This policy shall be put into effect as rapidly as possible, having due regard to the time required to effectuate any necessary changes without impairing efficiency or morale.[38]

It is likely that few in America would disagree today that the president's decision in 1948 was the right and necessary thing to do. However, as Jim Garamone points out in an *American Forces Press Service News* article, "Looking back on the order after 60 years, one might think it was a slam-dunk decision, but it was not. In fact, Truman and Defense Secretary James Forrestal were about the only two U.S. leaders who favored the proposal."[39] There was an extreme amount of resistance not only from American society, but more importantly from within the military and its leadership. Army Chief of Staff Gen Omar N. Bradley stated that "desegregation will come to the Army only when it becomes a fact in the rest of American society," while Secretary of the Army Kenneth Royall argued in favor of maintaining segregation, saying that the Army "was not an instrument for social evolution."[40]

In fact it wasn't until October 1953 that the Army announced that 95 percent of African-American Soldiers were serving in integrated units.[41] The key to the eventual success of the action was the presidential order to establish the creation of a committee in the National Military Establishment. The committee's role was "on behalf of the President to examine the rules, procedures and practices of the armed services in order to determine in what respect such rules, procedures and practices may be altered or improved with a view to carrying out the policy of this order."[42] The first committee established was the Fahy Committee, named after the committee chairman, US Solicitor General Charles H. Fahy.

Even with the backing of President Truman and the establishment of numerous committees and programs to facilitate the integration of minorities, the path from 1948 to the present day was a long and rocky road. It can be argued that if left to the services, desegregation would have taken a much longer time to achieve.

Numerous articles shed light on the integration of blacks into the armed services from the 1940s through the 1970s. Alan Osur's *Black-White Relations in the U.S. Military 1940–1972* and *Social Research and the Desegregation of the U.S. Army Project Clear*, edited by Leo Bogart, provide examples of the successes and problems encountered and may serve as a template for the integration of homosexuals.

- Implementation of the program was greatly hampered within the Army because of a lack of a single strategic action plan. Different commands proceeded at separate paces and were even allowed to create their own programs. US forces were slower to desegregate the Eighth Army, with other units showing complacency toward the executive order and doing nothing toward racial integration. It wasn't until 30 years after Executive Order 9981 was issued that a servicewide training program on equal opportunity and race relations was integrated.[43]

- Integration takes time, and the more it was "forced" on the services, the more they resisted. The Air Force and Navy, who had begun limited integration during and immediately following World War II, saw better reception of blacks after the executive order was given in 1948. The US Army, whose senior leadership fought desegregation, took a great deal longer to accomplish the directive's end state. Truman's executive order identified the need for immediate integration, but also identified time to adjust to the transformation. "This policy shall be put into effect as rapidly as possible, having due regard to the time required to effectuate any necessary changes without impairing efficiency or morale."[44] It can be said that the DADT policy has been a bridge to open service of homosexuals. The past 15 years of the current policy, although viewed by both sides of the argument as a reluctant concession, have served to provide time for attitudes both in the military and in civilian society to become more receptive to homosexuals serving openly in the armed forces.

- Integration of blacks worked relatively well in Korea during wartime operations, where blacks were more readily accepted, especially when they showed great merit. Noncombat units such as those stationed in Europe saw a slower process and experienced more problems.[45] The dependence of individuals on one another in combat situations nullified the prejudices among most service members, allowing unit cohesion to grow. This dependence upon another individual was not as prevalent in units in peacetime Europe, where cohesion was much slower to form. If integration of homosexuals is similar to the desegregation of the 1940s, then perhaps unit cohesion would also nullify homosexual prejudice in combat situations where straight and gay service members would be in close quarters, given the similarity to men and women currently serving together in de facto combat operations in Iraq and Afghanistan.

- The work environment within garrison saw fewer issues with regards to race, as opposed to non–work related and off-post events, where problems with race were more prevalent. This included areas like nonmilitary social functions, billeting, and housing. To a lesser scale, this is similar to the differences seen in combat units and garrison units in the aforementioned paragraph. The work environment may not have produced as strong a need to bond as was seen on the battlefield, but it did foster an atmosphere in which blacks were seen as a Soldier, Sailor, Marine, or Airman and not just someone of another race. Unfortunately, this did not always hold true outside the gates of a military base or confines of a ship. The attitude of tolerance or acceptance was much less prevalent off the military installation, especially when negatively influenced from the surrounding community, particularly in the southern United States.

The Integration Process:
The Dos and Don'ts

Despite the overused maxim that if we fail to learn from history, we will be doomed to relive it, it is critical that the armed services develop the right criteria to evaluate the open sexuality issue. Failing to do so could lead to an ineffective, prodigious, or incomplete policy that alienates more than it integrates. The following examples of lessons learned are not exclusive, but do provide a good basis with which to begin.

Lesson One

The process will take time, and immediate acceptance of the change may not be welcomed by all service members, especially those with strong religious convictions. Positively protecting one group very easily discriminates against another group that belongs to a different category.[46] It is imperative that the US military services provide a safe and secure atmosphere for gay and lesbian service members, as well as demanding tolerance and restraint to foster the good of the group. The DOD must walk a fine line between supporting equal treatment without going as far as to endorse homosexuality. Findings in the RAND report show a program that endorses a "homosexual lifestyle" may lead to possible alienation and resentment of the policy by straight service members.[47]

Lesson Two

A "low-key" approach was the universal explanation touted in the success of the three foreign militaries studied. In all three countries, the opposition's fear of "homosexuals swinging from the rafters" did not materialize, which assisted in calming the anxiety about the change. It is believed that the vast majority of homosexual service members decided not to reveal their identity, and thus it became a nonissue. The few gay or lesbian service members that did make known their sexual identity usually had a good relationship with their coworkers, and the knowledge made little or no difference in the unit's cohesion.

Lesson Three

Having support from the leadership of all branches is critical to implementation of the plan. Failure to do so will only cause confusion and hamper the process of integration. The RAND study finds that implementation is most successful where the message is unambiguous, consistently delivered, and uniformly enforced. The military must send a message of reassurance to the force that the new policy is not a challenge to traditional military values and will not create undue disruption. Leadership is critical to this regard. Once the policy is in place, any open dissention, like that displayed by some Army leadership during the integration of blacks, cannot be tolerated. Senior leadership that have irreconcilable differences with the new policy may have to make the decision to foster support or tender their resignation.

Lesson Four

A comprehensive and strategic action plan, universal to all five armed services, is required. Currently, the five services conduct different programs pertaining to the treatment of homosexuals or considerations of others classes. The disparity in the service training plans would require the DOD to establish a universal plan that is administered by a disinterested, independent group, preventing the perception of bias for or against homosexuality as a lifestyle. Monitoring of the process should be established to identify any problems and to address those problems immediately. RAND's research suggested several key actions that are essential to a smooth and successful policy change.

- Training efforts that provide leaders with the information and skills needed to implement the policy are essential. Emphasis should be placed on conduct, not on teaching tolerance or sensitivity. For those who believe that homosexuality is primarily a moral issue, efforts to teach tolerance would simply breed more resentment. Attitudes may change over time, but behavior must be consistent with the new policy from the first day. Belkin's article points out that none of the four countries his report included attempted to force its service members to accept homosexuality, but rather insisted that service members refrain from abuse and harassment and focused on equal treatment.

- The military must establish a standard of professional conduct that requires all personnel to conduct themselves in ways that enhance good order and discipline. The British Armed Forces Code of Social Conduct is a good example that establishes a clear and decisive message and provides guidance to commanders in dealing with inappropriate conduct. It is important that the DOD create a clear and concise policy, not shrouded in lawyer jargon and less complicated than the current DADT policy.

- The policy selected should be implemented immediately. Any sense of experimentation or uncertainty gives those opposed to the change an opportunity for continued resistance.

Lesson Five

The DOD must coordinate a planning committee that includes not only gay rights groups like the Palm Center and Servicemembers Legal Defense Network (SLDN), but also organizations that represent the views of the majority of service members. Religious and conservative groups need to be included to ensure that differing beliefs and moral issues are considered. The Obama administration has prided itself on promoting the inclusion of all parties in resolution of conflict, and this issue should be no different. Once the decision is made, all parties need to come together to find a solution that benefits the whole group and not just a single party.

Lesson Six

Of all lessons learned, perhaps the most important message is that of judging individuals on their own merit. The more we hold people accountable for their valued attributes, the less we pay attention to skin color, religion, or sexual preference. This was exactly the case with the integration of blacks during the Korean War. Long after the transition, President Truman issued Executive Order 9981. There was a significant difference between forces fighting on the Korean Peninsula and the garrison units in the United States and Europe. Troops fighting at the Pusan Perimeter or at the Chosin Reservoir learned to become discriminatory about the important traits that matter in battle. These Soldiers and Marines learned not to care about another person's skin color, as much as if they could shoot straight, show courage under fire, or carry their share of the load. Continental United States (CONUS) and European Command (EUCOM) service members were not as exposed to this opportunity to break free from stereotypes and prejudice, and therefore it took longer to accomplish the true nature of what integration was meant to achieve. This goes directly against what most opponents of open homosexuality in the military claim to be a valid argument, the disruption of good order and discipline in combat situations. The example of the integration of blacks during the Korean conflict is not proof that the opposing argument is irrelevant, but it does provide a reasonable example of what can be achieved when we begin to be less tolerant of superficial judgments in lieu of a person's ability to accomplish an assigned mission.

Conclusion

This chapter has not addressed all issues that are pertinent to ensuring a smooth transition for our military. There are several subjects that still have to be addressed prior to a repeal of DADT or in the years following the inception of a new policy. The subject of transgender soldiers in the British military has grabbed headlines in England, with the MOD adopting a policy that allows service members to remain in their current position after sex modification surgery. This is not only a moral question, but it would also be a health care benefits (TRICARE) question.

If the US military abolishes DADT, it may need to deal with existing sodomy laws in numerous states, which could have Tenth Amendment implications. Furthermore, partner benefits and recognition of same-sex marriage would likely emerge as key issues. State's rights, coupled with the *Defense of Marriage Act* 5/96 H. R. 3396 (May 1996), would present a major roadblock for a military policy that attempted to recognize same-sex marriage for the purpose of partner benefits.

Even in its most modest form, the right to sexual privacy may have implications for other sexual behaviors addressed in military regulations, such as adultery. If sexual behavior is to be viewed as a purely private matter, the US military may need to address current adultery laws and whether prohibiting sexual acts outside of one's own marriage is still incompatible with military standards.[48] A relationship between two male soldiers in accordance with established sexual behavioral regulations may not have a drastic effect on unit cohesion or morale; however, the permitted relationship of a soldier's spouse and a coworker would most likely have serious implications on unit integrity.

Reports and studies have addressed the easy questions for the past 15 years. Not allowing homosexuals to serve openly in the military and forcing all service members to accept homosexuality as a way of life are both flawed perspectives, and any policy requiring such is likely to create more problems than it solves. With a new administration willing to push for the repeal of DADT, the US military must recognize the fact that open homosexuality could become reality sooner than later and begin the proper planning and preparation that is required for such a monumental transition. Despite all the reports and studies, it is impossible to say that open homosexuality will have a smooth transition with no issues or that it will ultimately not succeed. But if we fail to conduct a true examination of all factors that can affect such a complicated issue, we are choosing to ignore potential problems, and in doing so we fall short of the ultimate objective. In the end, it is not an issue of promoting homosexual rights or defending religious and moral beliefs; it's about creating a policy that benefits the US military and the dedicated Soldiers, Sailors, Airmen, and Marines that serve therein.

Notes

(All notes appear in shortened form. For full details, see the appropriate entry in the bibliography.)

1. Flaherty, "Obama Consults Advisers on Military Gay Ban," 14.
2. Hopkins, "Out of the Closet," 36.
3. Wolfenden, *Introduction of a Code of Social Conduct in the Armed Forces*, 4.
4. Davis, "Military Policy Toward Homosexuals," 27.
5. Army Regulation (AR) 615-368, para. 2.b.
6. Ibid., para. 3.b.
7. AR 635-89, para. 3.
8. Army Regulation 635-100, para. 5, note 141.

9. DOD Directive 1332.14, note 1.

10. *Assessment of the Plan to Lift the Ban on Homosexuality in the Military: Hearings.*

11. Drew, *On the Edge*, 248–50.

12. Military Working Group, Memorandum.

13. Cleveland and Ohl, 11.

14. US Code 654, Public Law 103-160, 30 November 1993, 107 Stat. 1671.

15. Peterson, "Homosexuality, Morality and Military Policy," 6.

16. *Military Times*, "Politics, Civilians and Policy."

17. Pew Research Center, "Less Opposition to Gay Marriage."

18. Benoit, "Obama Era Expected to End Taboo on Gays in US Military."

19. The RAND (Research and Development) Corporation is a nonprofit global policy think tank first formed to offer research and analysis to the US armed forces. The organization has since expanded to working with other governments, private foundations, international organizations, and commercial organizations.

20. General Accounting Office (GAO), *Homosexuals in the Military.*

21. The Palm Center, a think tank at the University of California, Santa Barbara, produces scholarship designed to enhance the quality of public dialogue about critical and controversial public policy issues. Since 1998, the Center has been a leader in commissioning and disseminating research in the areas of gender, sexuality, and the military.

22. GAO, "Homosexuals in the Military," 29.

23. Ibid., 30.

24. Belkin, "Don't Ask, Don't Tell," 4.

25. Wenek, "Effects of Cancellation of Canadian Forces Policy," briefing note, 2.

26. Belkin, "Don't Ask, Don't Tell," 110–12.

27. GAO, *Homosexuals in the Military*, 19.

28. Strachan, *The British Army, Manpower and Society*, 123.

29. Deakin, "The British Army and Homosexuality," 2.

30. United Kingdom Revised Statute, Sexual Offences Act of 1967, chapter 60.

31. *Select Committee on the Armed Forces Bill*, London, HMSO, 1991, Para. 38.

32. British MOD, *Policy on Homosexuality.*

33. British MOD, *Armed Forces Code of Social Conduct.*

34. Ibid.

35. RAND, "Changing the Policy Toward Homosexuals," 3.

36. Rimmerman, *Gay Rights, Military Wrongs*, 24.

37. Lee, "Employment of Negro Troops," 305.

38. Executive Order 9981, Desegregation of the Armed Forces.

39. Garamone, "Historian Charts Six Decades of Racial Integration."

40. Harry S. Truman Library and Museum, "Desegregation of the Armed Forces."

41. Ibid.

42. Executive Order 9981, Desegregation of the Armed Forces.

43. Hopkins, "Out of the Closet," 16.

44. Executive Order 9981, Desegregation of the Armed Forces.

45. Osur, "Black-White Relations," 3.

46. Deakin, "The British Army and Homosexuality," 17.

47. RAND, "Changing the Policy Toward Homosexuals," 4.

48. Strachan, *The British Army, Manpower and Society*, 207.

Bibliography

Army Regulation (AR) 615-368. *Enlisted Men, Discharge, Undesirable Habits or Traits of Character*, 7 March 1945.

AR 615-368. *Enlisted Men, Discharge, Undesirable Habits or Traits of Character*, 14 May 1947.

AR 635-89. *Personnel Separations, Homosexuals*, 21 January 1955.

AR 635-100, *Personnel Separations, Officer Personnel*, 19 February 1969.

Assessment of the Plan to Lift the Ban on Homosexuality in the Military: Hearings before the Military Forces & Personnel Subcommittee of the Subcommittee on Armed Services. 103d Cong., 1st Sess., 1993.

BBC News. "Army Acts to Promote Gay Rights," 13 July 2008. http://news.bbc.co.uk/1/hi/uk/7504238.stm (accessed 21 January 2009).

Belkin, Aaron. "Don't Ask, Don't Tell: Is the Gay Ban Based on Military Necessity?" *Parameters*, US Army War College Quarterly, Summer 2003.

Benoit, Daphne. "Obama Era Expected to End Taboo on Gays in US Military." Agence France Press, 7 January 2009. http://www.sldn.org/news/archives/obama-era-expected-to-end-taboo-of-gays-in-military/ (accessed 8 January 2010).

Bogart, Leo, ed. *Social Research and the Desegregation of the U.S. Army Project Clear*. Chicago: Markham Publishing Co., 1969, 49–50, 319–21.

British Ministry of Defence. *Armed Forces Code of Social Conduct*, 2000.

———. *Policy on Homosexuality*, 2 November 2001.

Cleveland, Fred E., and Mark A. Ohl. "Don't Ask, Don't Tell: Policy Analysis and Interpretation" (master's thesis, Naval Postgraduate School, 1994). http://handle.dtic.mil/100.2/ADA283306 (accessed 8 January 2010).

Davis, Jeffrey S. "Military Policy Toward Homosexuals: Scientific, Historic and Legal Perspectives." The Judge Advocate General's School, US Army, April 1990.

Deakin, Stephen. "The British Army and Homosexuality." In *The British Army: Manpower and Society in the Twenty-First Century*, edited by Hew Strachan. New York: Frank Cass Publishers, 1999.

Defense Of Marriage Act. 5/96 H.R. 3396, Chapter 115 of Title 28, US Code section 1738B. Library of Congress, http://thomas.loc.gov/cgi-bin/query/z?c104:H.R.3396.ENR (accessed 24 February 2009).

Department of Defense Directive 1332.14. *Enlisted Administrative Separations*, 1982.

Drew, Elizabeth. *On the Edge: The Clinton Presidency*. New York: Simon and Schuster, 1994.

Eberhart, Dave. "Military Opposes Obama's Pro-Gay Stance," 8 January 2009. http://www.newsmax.com/headlines/obama_military_gays/2009/01/08/169333.html.

Executive Order 9981. Desegregation of the Armed Forces, 26 July 1948.

Flaherty, Anne. "Obama Consults Advisers on Military Gay Ban." Associated Press, 3 March 2009.

Garamone, Jim. "Historian Charts Six Decades of Racial Integration in U.S. Military." American Forces Press Service. http://www.defense.gov/news/newsarticle.aspx?id=50560 (accessed 31 December 2008).

General Accounting Office (GAO). *Homosexuals in the Military: Policies and Practices of Foreign Countries*. GAO/NSIAD-93-215. Washington, DC: GAO, 1993.

Harry S. Truman Library and Museum. "Desegregation of the Armed Forces: Chronology," 28 December 2008. http://www.trumanlibrary.org/whistle stop/study_collections/desegregation/large/index.php?action=chronology.

Hopkins, Yvette C. "Out of the Closet: Addressing Policy Options." School of Advanced Military Studies, US Command and General Staff College, Fort Leavenworth, Kansas, 2001.

Lee, Ulysses. *The Employment of Negro Troops*. Washington, DC: Center of Military History, US Army, 1966.

Military Times. "Politics, Civilians and Policy." Poll, 31 December 2007. http://www.militarycity.com/polls/2007activepoll_politics.php (accessed 31 December 2007).

Military Working Group. To the secretary of defense. Memorandum, "Recommended DOD Homosexual Policy Outline," 8 June 1993.

Osur, Alan M. "Black-White Relations in the U.S. Military 1940–1972." *Air University Review* 33, no. 1 (November–December 1981): 69–78.

Peterson, Michael A. "Homosexuality, Morality and Military Policy." Master's thesis, Naval Postgraduate School, Monterey, CA, March 1997.

Pew Research Center for the People & the Press. "Less Opposition to Gay Marriage, Adoption and Military Service." Washington, DC, 22 March 2006.

RAND. "Changing the Policy Toward Homosexuals in the U.S. Military." Research brief, 2000. http://www.rand.org/pubs/research_briefs/RB7537/index1.html.

Rimmerman, Craig A., ed. *Gay Rights, Military Wrongs: Political Perspectives on Lesbians and Gays in the Military*. London: Taylor & Francis, 1996.

Strachan, Hew, ed. *The British Army, Manpower and Society into the Twenty-first Century*. London: Frank Cass, 2000.

Wenek, Karol. "Effects of Cancellation of Canadian Forces Policy Restricting Service of Homosexuals." Briefing for the director of personnel policy, Canadian Armed Forces, 1995.

Wolfenden, Robin. *Introduction of a Code of Social Conduct in the Armed Forces*. London, UK: Government Centre for Management Center and Policy Studies, 2002.

About the Author

MAJ Matthew Cashdollar is currently the deputy G4 for the 4th Infantry Division at Fort Carson in Colorado. His 24-year military career includes service as a noncommissioned officer and warrant officer. He was born in York, Pennsylvania, graduated from Troy State University with a BA, and holds a master's degree in military operational art and science from Air University.

CHAPTER 9

The Sky Won't Fall

Policy Recommendations for Allowing Homosexuals to Serve Openly in the US Military

Tammy S. Schultz

As a candidate, Barack Obama promised to reverse the ban on gays and lesbians openly serving in the US military, comparing overturning the law commonly referred to as "Don't Ask, Don't Tell" (DADT)[1] to "the integration of blacks in the armed forces as both a moral issue and an achievable goal."[2] Once he was elected, President-elect Obama's press secretary, Robert Gibbs, unwaveringly stated, "You don't hear politicians give a one-word answer much," but "yes," the president would reverse the ban.[3] On 27 January 2010, President Obama reiterated his campaign pledge during his first State of the Union address regarding gays and lesbians in the US military. The following week at an historic Senate Armed Services Committee hearing, Chairman of the Joint Chiefs of Staff, Admiral Mike Mullen, said, "No matter how I look at the issue, I cannot escape being troubled by the fact that we have in place a policy which forces young men and women to lie about who they are in order to defend their fellow citizens." Never before had the highest ranking member of the US military spoken out in favor of allowing gays and lesbians to openly serve. At the same hearing, Secretary of Defense Robert Gates said, "We have received our orders from the commander in chief, and we are moving out accordingly."[4] Gates also announced that he had asked Pentagon legal counsel Jeh Johnson and Army General Carter Ham to lead a yearlong study on how the military would lift its ban on openly gay service members.[5]

These perspectives are sure to ignite contradictory voices on the issue of homosexual service, as does this book. This chapter seeks to aid policy makers by providing policy lessons and recommendations to ease the transition from DADT to a US policy that allows citizens to serve in the military regardless of sexual orientation.[6] This chapter presents practical policy recommendations for easing the transition to a US military where homosexuals serve openly. These lessons are drawn from historical examples, such as desegregation, the fuller inclusion of women in the US military, DADT, and other militaries' lifting of the ban.[7]

The Red Herring of Military Readiness

Before addressing policy recommendations, it is important to analyze one issue that will *not* be addressed in the policy recommendations, largely because of its red herring status in this debate: military readiness. The military measures readiness based on three major areas: equipment, training, and personnel. It is the final two areas, training and personnel, where those who want the ban to stay in place focus, using the following arguments: Allowing homosexuals to serve openly would hurt morale, which, in turn, would erode unit cohesion and undermine military readiness. Furthermore, straight service members not comfortable with serving alongside homosexual counterparts will leave the military in droves, causing the number of military personnel to precipitously drop. Given that the United States is currently engaged in two hot wars and a global counterinsurgency, this argument must be taken seriously. In doing so, however, the argument's bankruptcy becomes clear.

Opponents of allowing homosexuals to serve openly usually point to opinion polls that indicate military personnel do not accept homosexuals and would not serve with them. Yet using such polls as an indicator of military readiness commits a logical fallacy—to measure readiness (or cohesion as an integral part of readiness), one should assess readiness or cohesion, not merely opinion. A perfect example of this logical fallacy is the use of a 2008 *Military Times* poll by Elaine Donnelly, who noted that 10 percent of respondents said that they would not reenlist or extend their service if DADT were overturned. (Incidentally, 71 percent said that they would continue to serve, and 6 percent had no opinion; 14 percent indicated that they would consider leaving.) Donnelly continued that "if the poll's findings approximate the number of military people who would leave" if DADT is overturned (and her argumentation makes it clear that she believes this to be the case), "the voluntary exodus would translate into a loss of almost 527,000 personnel—a figure approaching the size of today's active-duty Army."[8] This is a leap of heroic proportions. There is a big difference between clicking on the bubble of an online survey that one would leave the service and actually doing so. Even negating the difficulty of leaving one's brothers and sisters in the profession of arms, there is the more mundane issue of retirement pay that requires 20 years of service and transitioning into a civilian sector during a time of economic downturn. And, as noted, one cannot make assumptions regarding readiness merely using opinion polls. Indeed, when the US military itself measured homosexuals' impact on unit cohesion, sexual orientation had no effect on military effectiveness.[9] A study by the RAND Corporation also found homosexuals have no adverse effect on military readiness.[10]

Beyond US studies, however, is empirical evidence from other countries facing similar policy decisions. In Britain, for instance, resistance to the inclusion of openly serving homosexuals was similar to the poll numbers in the United States before Britain lifted its ban. Yet only a handful of soldiers actually left Britain's military as a result of the change in policy.[11] In fact, the predictions of a heterosexual mass exodus of military personnel in the countries that allow homosexuals to serve openly never happened.[12] In a study of foreign militaries with open homosexuals serving who were known to their combat units, no evidence was found "of deterioration in cohesion, performance, readiness, or morale. Generals, ministry officials, scholars, and [nongovernmental organization] observers all have said that their presence has not eroded military effectiveness."[13] Since the Dutch military became the first to allow open service in 1974, not a single study has indicated a decline in performance that can be empirically linked to homosexuals in *any* of the approximately 70 countries who allow open service.[14]

Those who make such inferences arguably do not understand military culture. Even while some may simultaneously disagree with a civilian order, plans for the policy's implementation are being made. For instance, although many military leaders disagreed with the talk of a Bosnian intervention, including a chairman of the Joint Chiefs of Staff,[15] the military was simultaneously planning for the intervention.[16] Given healthy civilian-military relations, such planning to implement policies even when the military leadership disagrees makes sense. If it were not so, the United States would clearly have a far greater problem than homosexuals openly serving in the military.

The United States enjoys the most professional all-volunteer force the world has ever seen. Yet as Rear Adm John Hutson, who was involved in the DADT process in 1993, said, unit cohesion arguments assume the United States suffers from an unprofessional, bigoted force. Leaders, said Rear Admiral Hutson, "welcomed their homophobia and used it as an excuse for inaction."[17] Opponents of integrating minority groups have used these arguments before: "Whites feared that 'mixing of the races' would result in an epidemic of sexually transmitted disease; and increase in antiracial violence and criminal activity by African Americans; the breakdown of morale, order, and discipline, resulting in weakened national defenses; mass exit from the military by whites; and greater difficulty recruiting whites for service."[18] The parallel to arguments made today against homosexuals openly serving is striking.

US Army Lt Col John H. Sherman delivered a speech entitled "Command of Negro Troops" in November 1944, a speech that later became required reading for every officer who commanded African-Americans. In the speech, Lieutenant Colonel Sherman acknowledged the propensity to assume that the force could not include anything other than white males and remain a professional fighting force:

> At the start we must recognize that in any large group there are likely to be officers who have long considered that their attitudes on the Negro question are their own business: A matter personal to them, settled and unchangeable—settled for them and by them long before they entered the Army. But the Army has a definite policy and requirement on this matter, just as it has on other matters. . . . An Officer of the Army has no more freedom to speak or act by old habit on this matter, than a buck private has to stand or walk by old habit when on Review. Also: It is fundamental that a good Officer takes any duty which Higher Authority sees fit to assign to him, masters the job and his preferences relative to it, and does it well, without complaint or question. . . . no officer who allows his prejudices to render

him ineffective on his assigned duty can ever properly be assigned to any other duty which he might find more pleasing to him, for it is not the Army's policy thus to reward insubordination or weakness.[19]

Scholarly works back the lieutenant colonel's approach. In a study of the US military, University of Washington political scientist Dr. Elizabeth Kier found that whether group members like one another need not have any bearing on organizational performance. What does matter is if the individuals are committed to the same goals or mission.[20] Another study of US police and fire departments found that attitudes did not equal behavior.[21] In any organization the size of the US military, it is not wise to assume homogeneity of group belief. For instance, not all service members agree on any given deployment, but most go serve the nation's interest at risk of life. From basic training's first day, drill instructors drive self-interest as much as possible out of new recruits, or at least try to make self-interest subservient to the unit's goals, making it possible for service members to put mission accomplishment over self-preservation. Indeed, it goes to the very heart of mission above self, which mandates that personal beliefs do not necessarily translate into personal behavior.

Given the current operations in which the United States is engaged, dispassionately examining the readiness issue is important, and those who have done so found no adverse effect. The fact that the United States is at war should actually make it *more* likely that readiness will not be adversely affected. As anyone who has served in a combat zone can recount, when bullets fly, the proficiency of soldiers matters most—not the private life of the individual—for it is that expertise that might mean the difference between life and death.

When examined based on its empirical versus emotive merits, therefore, the readiness issue falls to the side. Other issues remain. Various minority groups have been successfully incorporated into US and foreign militaries before, and the US experience from the 1990s provides clues for how best to implement a new policy. Drawing on lessons from these experiences can inform DADT's reversal.

Will and Leadership

One of the most important lessons involves the will of the various actors involved and the criticality of leadership. Given the constitutional allocation of responsibilities and the military's unique place in society, a wide variety of actors must be involved in the process of allowing homosexuals to serve openly, and the directives, laws, and other guidance cannot be contradictory. For the executive branch, the main players on this issue are the president, the secretary of defense, and the Joint Chiefs of Staff, to include the chairman.

If President Obama truly seeks to allow homosexuals to serve openly, his support and attention cannot waiver. When President Clinton began backpedalling on his campaign promise regarding gays in the military, the opposition smelled blood in the water. The president "conveyed to many that even *he* did not stand strongly behind the policy goal of allowing homosexuals to serve openly."[22] Presidential guidance and attention at every step, from study to implementation, were also notably absent during DADT.[23] Presidential involvement proves necessary not only because of the controversy surrounding the issue, but also because of the wide range of directives, laws, and other guidance that must be synchronized, as they were in the case of desegregation.

Larry Korb, a former Reagan and Clinton defense official closely involved in the DADT process, holds that Clinton's taking office during a time devoid of major crisis (the first president to do so in 60 years) and a weak electoral mandate (in 1992, Clinton received only 43 percent of the popular vote) negatively correlated to his power as the executive.[24] President Obama stepped into office during a time of multiple crises and with a strong electoral mandate after Pres. George W. Bush vastly expanded the executive's power. Even if President Obama's honeymoon was short-lived, the ingredients for successful implementation of allowing homosexuals to serve openly still exist, so long as the right policy lessons from previous experience are heeded.

As the commander in chief, the president must put the full force of his office behind the change, just as Pres. Harry Truman did with desegregation. An executive order should be issued that allows homosexuals to serve openly. Additionally, the president must coordinate the various actors on the executive's side and consult with, but not defer to, the legislative branch throughout the process.

The secretary of defense must also be involved. Although this will probably not occur due to his expected short tenure in the Obama administration, it would be wise to enact these changes under Secretary of Defense Robert Gates for a number of reasons. As a moderate Republican, he would add to the policy's bipartisan tone. He knows the building and has built a history of healthy civil-military relations. And no matter who replaces him, there will be a learning curve that would be made immensely steeper by this issue being on the new secretary's plate at the outset. That is not to say that a new secretary of defense cannot (or should not) take on this issue, rather that he or she must understand that it will take personal involvement to ease formulation and implementation.

The Joint Chiefs of Staff should also be involved, and they have the right to consult with Congress on policy issues related to national security when they disagree with the president. Accordingly, these top military commanders must be brought into the process at the start. Adm Mike Mullen made his position for DADT's reversal clear in the 2 February 2010 Senate hearing, but he did not disclose the other Joint Chiefs of Staff members' opinions. One member, however, the US Marine Corps commandant Gen James T. Conway, was reported to be "the most outspoken opponent of permitting gay men and women to serve openly in the US military."[25] The concerns of general and flag officers should, of course, be heard and addressed, not to stymie the president's decision, but to better implement it.

A Pentagon review of DADT is being conducted, according to an Admiral Mullen aid, to "make sure we move forward in a deliberate, measured fashion, that [Admiral Mullen] has the opportunity to provide his best military advice in that process and that the advice is based on facts and not emotion."[26] Former chairman and secretary of state Gen Colin Powell stated in an interview that "we should definitely reevaluate [DADT] . . . it's been fifteen years and attitudes have changed."[27] Additionally, previous chairman John M. Shalikashvili switched his stance from when he was serving and recommended that homosexuals be allowed to serve their country without fear of discharge.[28] Involving such respected retired officers in the consultations will be critical, as will the study's inclusion of military voices who desire the change (unlike the way past Pentagon studies of DADT have "loaded the deck" with opponents to the reversal).

The unique constitutional powers given to the legislative branch mandate lawmakers' involvement as well. An executive order alone is not enough to allow homosexuals to serve, as such an order could be overturned by Congress and DADT itself was a legislative act.[29] Both the executive and legislative branches, therefore, should act together. At least for now, the Democratic Party enjoys a majority in both the Senate and the House, as well as occupies the White House, which should make this coordination easier. That said, it is critical to easing implementation that the effort is bipartisan. Signs that such bipartisanship will occur already exist. Rep. Ellen Tauscher (D-California) introduced the Military Readiness Enhancement Act, which has over 180 cosponsors, not all of whom are Democrats. The late Senator Ted Kennedy hoped to introduce a similar bipartisan bill but died before being able to do so, leaving a gaping hole in Senate leadership on this issue. Freshman Senator Kirsten Gillibrand, a Democrat from New York who holds the seat once held by Robert F. Kennedy, has pushed the issue forward.[30] Pressure groups from both sides will undoubtedly play a role, but it is up to Congress to keep the hearings and process factual.

In 2009, Senate and House leaders seemed to be waiting for the military to tell the legislature what to do regarding DADT. Congressional leadership, while certainly taking into account the military's opinion, cannot defer to the military on this issue, a point which an aid to Admiral Mullen makes: "It's important to remember that 'don't ask, don't tell' is a law, and the military will obey the law."[31] As noted, if that were not the case, the implications for civil-military relations would be dire. Leadership across the US government, both military and civilian, must be involved in crafting the change, but at the end of the day, the civilians need to set the course.

It is important to note that the judicial branch is not mentioned in the set of actors included in the policy recommendations for two reasons: as a general rule, the judicial branch should not make policy but ensure that policies have the force of law, and having the courts mandate inclusion of openly serving homosexuals would have a detrimental effect on the successful integration of homosexuals. Practically speaking, a court decision would, almost by definition, circumvent or at least abbreviate the policy study, formulation, and implementation steps necessary for success. Furthermore, conferring a "special class" on homosexuals serving adds "a host of more troubling problems on the part of the majority group."[32] One of the key lessons from previous attempts to integrate minority groups is that such a change in policy is easier to implement "if it is perceived as benefiting *all* members of the force."[33] A more appropriate role for the courts is ensuring that the law be faithfully implemented.

Leadership from both branches, and both parties, must actively push and organize for the change. It is not inevitable that DADT will be reversed, and approaching it that way sets up the process for failure. When Clinton took office, proponents of allowing homosexual service did not organize as well as the right and were outhustled.[34] With the reversal of political fortunes, it is less likely that such disarray on the left will be seen with this round. It is critical, however, that this fight not be approached in a partisan fashion, both for the good of the policy and, frankly, the Republican Party.

A recent *Washington Post*-ABC News poll documented that 75 percent of Americans believe that homosexuals should be allowed to serve openly.[35] Compare that groundswell of public support to those a*gainst* desegregation (63 percent), and it becomes easier to make this an American cause vice a Democratic one.[36] If the Republicans decide to make this issue one to galvanize their hardcore base, they very well risk losing moderates within their party as well as independents. Recent Pew Research Center data shows social conservatism falling across party lines as well as among independents since around 1987, with only 22 percent of those polled identifying as Republicans, with independents largely favoring the Democrats on social values and religion.[37] Turning DADT's reversal into a partisan issue would be political risky for the Republicans, who increasingly need a more moderate message to attract more voters.

On the other side of the political aisle, the Democratic leadership should be wary of abandoning promises made to reverse DADT. Supporters of the president are increasingly leery of what they consider to be backpedaling on the issue of gay rights.[38] From a self-interested standpoint, Democratic politicians do not want to anger a base that makes substantial campaign contributions and, as importantly, goes to the polls in much higher percentages than their straight counterparts.[39] Additionally, as noted above, independents favor the Democrats' stance on these types of social issues. Making up approximately one-third of the electorate, independents (also called "undeclared" voters) hold significant political power.[40] Should the Democrats decide not to honor promises that most of the population support, such as the reversal of DADT, they risk losing these moderate voters.

A bipartisan approach would be best in terms of reversing DADT, as well as for both parties in terms of not alienating the 75 percent of Americans who think that homosexuals should be able to serve openly. If the issue does become partisan, though, the Republicans have more to lose than the Democrats. Putting country above party, however, should be the approach on all sides.

Study and Planning

As is clear from the transparent bias of this chapter in favor of allowing homosexuals to serve openly, this author does not believe that much new research needs to be conducted on *if* homosexuals should be allowed to serve openly.[41]

For the past 15 years, various hearings, commissions, studies, and research have shown that such a policy reversal will not bring all of the negative consequences that critics claim will occur. Impartial studies should be conducted *before* the policy is formulated, however, and these studies should focus on *why* the opposition exists, so as to better execute policy implementation, as well as how to best implement the policy.[42] During DADT, opposition groups used studies commissioned and ceasefire periods to circle wagons, rearm, and build opposition.[43] That must not happen this time.

In the DADT hearings, Senator Sam Nunn (D-Georgia) raised several "thorny issues" that he used to derail the DADT process.[44] These issues should be studied before the policy formulation and generally involve two large categories: pay and benefits, and service. Each will be examined in turn.

The first category of these thorny issues, pay and benefits, should be studied to ensure that homosexuals are placed on equal, rather than special, footing. Israel became the first country to offer survivor benefits to same-sex partners in 1997.[45] Given that the United States is at war, same-sex partners should receive the same benefits as their heterosexual colleagues. One question, however, is how to implement such a policy when marriage is not a federal right, and only around half a dozen states allow same sex unions. This issue should be studied and a recommendation made for what the standard of a same-sex union is for those service members whose state of residence does not allow them to wed. Many states allow "common law" marriages for straight couples who decide not to marry, with a minimum cohabitation requirement. The same requirement could be made of homosexual couples.

Other benefits include health and life insurance, as well as base housing.[46] All should be provided to same-sex partners using the baseline requirements for a partnership identified for survivor benefits. The military currently screens all new members for the HIV virus, so this issue, raised by some,[47] is not the reason to provide health and life insurance—fairness and equality mandate these rights. Along the lines that equal treatment by no means equates to special treatment, homosexuals should be afforded the same heath and life insurance benefits, as well as the base housing accorded to their rank based on housing availability. The budgetary impacts and proper implementation should be studied to prepare for this change.

Another issue raised is that of homosexuals living in separate quarters. There are many reasons that this is untenable and will not solve the perceived issue of heterosexual privacy. One, if the experiences of other militaries or US police and fire departments are indicative, few homosexuals will actually come out of the closet once DADT is reversed.[48] Unlike being African-American or female, one can hide sexual orientation. One may counter why, then, should the policy be changed, and the answer comes down to the simple matter of fear of being discharged if discovered. Given that not all homosexuals will self-identify as such, providing separate housing is not even a realistic option.

Two, on the issue of gays preying on straights in shared housing, this seems particularly far-fetched given their vastly smaller numbers and the empirical evidence. The idea of a homosexual raping a heterosexual in an environment where approximately 90 percent of the population is straight (never mind, according to those who support the ban, potentially homophobic) makes such shower horror stories appear devoid of logic.[49] The overwhelming amount of documented evidence for violence involving homosexuals is against homosexuals, not homosexuals preying on unsuspecting heterosexual colleagues (that said, the same rules regarding heterosexual sexual harassment should apply to homosexuals as well).

The second major category of thorny issues demanding study involves service. Around 13,000 service members have been discharged under DADT since 1994.[50] The question of reinstatement of these individuals needs to be determined before the policy is enacted, in part because some of these discharged service members will undoubtedly ask to be reenlisted. They should be welcomed into a force struggling under high operation tempos, but the issue of pay and grade for those who return should be examined. Another consideration is the length of time passed since being discharged under DADT. Given that the ousters under this policy began approximately 15 years ago, some may desire to rejoin who have not been on the military's rolls for over a decade. The study should identify how long is too long for reenlistment and/or what type of "reblueing" (i.e., retraining) these individuals require.

In terms of those homosexuals already serving in the US military, the question of affirmative action should be decided before the policy is implemented. Given that some homosexuals undoubtedly serve closeted at high ranks, affirmative action is not believed to be necessary or even desirable given the belief that the change will be more welcome if homosexuals share the same rights and responsibilities of all other service members.

Costs must be estimated when exploring all of these issues regarding pay, benefits, and service. In times of economic recession, shrinking defense budgets, and growing entitlement spending, cost is not a matter to be taken lightly, nor is it simply a straightforward calculation. Some believe that defense budgets will remain stagnant, while others hold that defense expenditures will drop for some time to come.[51] Given that approximately 65,000 homosexuals currently serve in the US military,[52] the cost of benefits for these individuals should be included in cost projections. The experiences of other countries' militaries indicate that few homosexuals apply for medical, dental, or relocation benefits for their partners, which suggests that these costs estimates need not include all 65,000 homosexual service members.[53]

Within this bigger expenditure picture, however, are the costs of implementing DADT. A blue ribbon commission reported in February 2006 that it costs the United States at least $363.6 million to discharge homosexuals from the US military, costs that include "lost benefit" (losing the service of the trained individual) and "implementation" (investigations and review boards).[54] True costs are difficult to estimate, as the blue ribbon commission acknowledged, although there are many compelling reasons to believe that its estimate is low.[55] In sum, the cost savings produced by overturning DADT may well help offset the costs of providing equal benefits.

Policy Formulation

Strong leadership with the will to change DADT, as well as solid empirical studies that point the way, are critical to the policy formulation stage. Having identified who is involved and at least a part of their deliverables, it is now necessary to offer the speed of this process, as well as what the general "look" of the policies should be.

The timing and speed of the policy formulation are critical. Just how fast the timeline should be from study to formulation to implementation is disputed.[56] Formulating the policy too hastily means that not enough of the thorny issues have been resolved, but not going fast enough may signal a lack of commitment. Additionally, the longer President Obama takes into his first administration to accomplish the task, the riskier the proposition. In addition to some not wanting to spend political capital very close to an election, some may decide to await the results of the 2012 election before truly throwing weight behind the policy's formulation and implementation. Worse, those who do not like DADT's reversal may decide to wait out the administration. Nothing can completely alleviate these concerns, but speed may minimize their impact. With Pentagon studies currently underway, the policy formulation phase must begin in earnest, but no later than early 2010. Forward momentum must be maintained, and the president cannot waiver anywhere during this process.

As to what the assorted directives, laws, and other guidance should look like, the KISS (keep it simple, stupid) principle should apply. The lessons from police and fire departments show that "nondiscrimination policies were most readily implemented when they were simple, clear, consistent, and forcefully stated."[57] As suggested when discussing the judicial branch above, homosexuals should not receive special status, but should be treated like their heterosexual colleagues: "Military experience with African Americans and women ... argues for a simple policy under which homosexuals are treated no differently in terms of work assignments, living situations, or promotability."[58] Instead of laws attempting to anticipate every single situation an officer might face, "codes built on general principles of fairness, respect, honor, decorum, and the need to avoid the creation of hostile environments were far more practical and effective."[59] Rather than devise all new standards for homosexuals, the same rules should apply to all service men and women, regardless of sexuality.

During the DADT debate, the Clinton administration attempted to frame the issue as one of status (sexual orientation) versus conduct (sexual acts). This paradigm, however, was very soon muddled by exactly what DADT meant, and even being homosexual counted as misconduct.[60] The new law must create "equal standards" for all service members regardless of sexual orientation, with "an emphasis on conduct."[61] Indeed, the military already has a code of conduct regarding sexuality for heterosexuals: No dating within one's chain of command is allowed, and officers cannot have relations with enlisted forces. These same rules should apply to homosexuals. The new directives, laws, and other guidance must make the enforcement systems explicit, and incentives should promote the following of the new policy.[62] The multiplicity of actors, laws, orders, and guidance requires a strong executive team to pull together these different strands during the policy formulation process—a team that will continue to meet once policy implementation begins.

Committed Implementation

Although implementation seems to be at the end of the policy process, in fact, it must be considered from the start. A strong implementation plan is the *sine qua non* of a policy reversing DADT, and "if the will, skill, and capacity to mount a meaningful implementation plan are lacking, then policy development is at best a sham and at worst may be harmful to those that the policy seeks to help."[63] Indeed, a weak implementation plan will increase opposition to the policy.[64] Thinking about an implementation plan even as the policy is being formulated is key to increase the speed of implementation. And the faster the policy is implemented, the greater chance opposition will crumble in the military once service members see how their daily lives are, and critically are *not*, impacted.[65]

For the secretary of defense's part, Department of Defense (DOD) directives must be on the shelf ready to go *before* the implementation stage begins. During DADT, critical directives were being written as implementation was occurring and were not ready to go beforehand, which implied a lack of commitment and undermined the ultimate policy.[66] Article 125 of the Uniform Code of Military Justice (10 U.S.C.A. ß925), which forbids sodomy, will also have to be rewritten.[67] Having a coherent package of guidance and directives ready to go will take significant effort and resources, especially during a time of war, and the Executive should propose (and Congress fund) the resources necessary to ensure the process is a success. Moreover, DOD must give the job to high fliers within the administration who hold high rank and enjoy direct access to top leaders, who also must be involved in this process, in part to ease implementation, and in part to show President Obama's seriousness of purpose.[68]

Implementing the new policy will also require that military personnel be trained on the new policy. To ease desegregation of the services, Secretary of Defense Melvin Laird established the Defense Race Relations Institute (DRRI) in 1971, which was later renamed the Defense Equal Opportunity Management Institute (DEOMI). Still in existence today, DEOMI should be utilized in training efforts. For DADT, an antidiscrimination policy was not written until 2000 under Secretary of Defense William Cohen, but no implementation plan occurred. The Pentagon had the directives written, but they were never issued because Under Secretary of Defense for Personnel and Readiness David Chu said that they were not necessary.[69] Both the guidance and directives should come down simultaneously to overturn DADT. This training should occur vertically (up and down the chain of command) as well as horizontally (across the services).[70] Additionally, training should be targeted to the level of command, with flag and general officers receiving special training on policy implementation and lower levels of command focusing on interaction with troops. Service leadership received no special training on integrating African-Americans until Secretary of Defense William Perry ordered it in 1994.[71] This mistake should not reoccur.

Another major question is what the training should include. The focus should be on behavior, not beliefs, which was essential to desegregation for those who objected to serving with African-Americans on religious grounds.[72] For police and fire departments, training that worked best included "accurate information on who homosexuals are, how they come to be that way, and how they lead their lives," training that was particularly helpful if actually led by a homosexual and even better if he or she was a respected member of the force.[73] The training should also be directly tied to the organization's mission lest the service members become resentful—how does the new policy serve a "legitimate need of the military"?[74] One answer: more Soldiers, Sailors, Airmen, Marines, and Coast Guardsman equal longer dwell times, which will increase mission effectiveness and sustainability of the all volunteer force.

As with the integration of other minority groups, military leadership will be critical. Leaders create a command climate, and previous successes (desegregation) and failures (Tailhook) can be traced back to leadership. When military commanders get behind a new policy, the change seems less of a threat since the implementation is coming from within the organization rather than appearing to be forced from the outside.[75] For those officers who currently serve closeted, should they choose to come out, being "treated with respect from above" makes it all the more likely that they will "be treated with respect from below."[76] Members of the US military have deference for the chain of command fostered from day one of basic training, as well as an understanding that democracy dictates civilian control of the military. When these values have been tested in the past, including the integration of African-Americans and women, the US military has risen to the challenge.

One way leaders can create driving forces for change is "by drawing on those aspects of the existing culture that *are* compatible" with open homosexual service.[77] For police and fire departments, "fairness, respect, honor, decorum, and the need to avoid the creation of hostile environments" proved useful compatible values, all of which are applicable to the military.[78] The "dominance of mission over individual preferences and characteristics"[79] is an essential ingredient to civilian control of the military—service members do not pick which deployments to support, nor should they judge the person to their left or right based on anything other than merit and skill. In speaking about African-Americans, Lt Col John H. Sherman said, "Show them the Army as a great Fraternity in which men of all races, creeds and colors come together to serve in the Brotherhood of the uniform of the Army of the United States: the salute its pass sign; mutual service and shared hardship its ritual; and final rest beneath the Flag of our Country its end and reward."[80] Homosexuals already serve their country, and undoubtedly, some have already died in service. They should be allowed to do so without fear of being discharged for who they are, and a committed implementation process can bring this change about.

Let Them Serve (Openly)

As the first African-American commander in chief, Barack Obama is uniquely positioned to amend the policy. He has already evoked analogies of open homosexual service to desegregation and can emphasize this point by using 26 July to roll out the policy. (President Truman signed Executive Order 9981 desegregating the services on 26 July 1948). Allowing homosexuals to serve openly will not immediately result in vast changes. For Truman's 1948 desegregation order, an order strongly opposed by the Joint Chiefs of Staff, "it took twenty-five years before all the services adopted the spirit of the directive."[81] Although following these policy lessons learned over time should speed up the successful integration of homosexuals, the time lag will undoubtedly still exist.

Since President Obama entered office and the 111th Congress assumed power, over 700 more service members have been discharged under DADT.[82] As John Fitzgerald Kennedy said, "In giving rights to others which belong to them, we give rights to ourselves and to our country." It is long past time to recognize that homosexuals currently do honorably serve their country, such as the first Marine seriously wounded in Iraq, SSgt Eric Alva. Alva, who happened to be gay, was medically discharged after losing his leg. In speaking on Capitol Hill urging the reversal of DADT, he said, "I'm an American who fought for his country and for the protection and the rights and freedoms of all American citizens—not just some of them, but all of them."[83] The United States should not only accept such sacrifices, but honor those who make them.

Notes

1. Pub.L. 103-160 (10 U.S.C. § 654) is the law enacted by Congress in 1993, commonly referred to as "Don't Ask, Don't Tell." Elaine Donnelly claims that the phrase "Don't Ask, Don't Tell" misrepresents what the law actually says and that instead this law codified an outright ban on homosexuals in the military that always existed. She thus retitles the law as the "Military Personnel Eligibility Act." In fact, the "don't ask" part of the law was nonbinding but given in a "Sense of Congress" included in 10 U.S.C. § 654, which also said that the secretary of defense could "reinstate that questioning." For purposes of following common usage, this chapter will use the "Don't Ask, Don't Tell" (DADT) shorthand. See Jeanne Scheper, Nathaniel Frank, Aaron Belkin, and Gary Gates, "'The Importance of Objective Analysis' on Gays in the Military: A Response to Elaine Donnelly's 'Constructing the Co-Ed Military,'" *Duke Journal of Gender Law & Policy* 15 (2008): 442–43; and Elaine Donnelly, "Constructing the Co-Ed Military," *Duke Journal of Gender Law & Policy* 14 (May 2007): 815–952.

2. Thom Shanker and Patrick Healy, "A New Push to Roll Back 'Don't Ask, Don't Tell,'" *New York Times*, 30 November 2007.

3. Bryan Bender, "Obama Seeks Assessment on Gays in the Military," *Boston Globe*, 1 February 2009.

4. Elisabeth Bumiller, "Top Defense Officials Seek to End 'Don't Ask, Don't Tell'," *New York Times*, 2 February 2010.

5. William Branigin, "Supreme Court Turns Down 'Don't Ask' Challenge," *Washington Post*, 8 June 2009.

6. Plenty of works exist that elucidate the normative debate regarding allowing gays to serve openly, many of which are cited below. For purposes of transparency, this author believes that the ban should be lifted, primarily for the following six reasons: First, *every* empirical example in the real world of the ban being lifted has not resulted in the dire predictions foretold regarding readiness and morale; none of these policies have been reversed, and there is no documented evidence of a drop in military readiness. As for the argument that those militaries do not resemble the combat-tested US military, and therefore are not applicable, that case is much harder to make after September 11th. Twenty-two countries that allow gays to serve openly have deployed alongside the United States in Iraq and/or Afghanistan, to include Great Britain, Canada, and Australia. Also, as anyone who has served in a combat zone can recount, combat makes it *more* likely that cohesion and increased camaraderie will occur regardless of background—not less. In other words, the combat experiences of today's US military increases, rather than decreases, the likelihood of success. Second, given that it is acknowledged that homosexuals currently do serve in the US military, it would better protect heterosexual privacy if the ban were lifted so people could know who was gay (never mind that one of the things new recruits give up is, in fact, their full privacy). Morale would also improve as all service members could choose to be more forthcoming and honest with their colleagues—a key aspect of trust, camaraderie, and cohesion. Third, most opinion polls show that societal views have changed since the infamous 1993 policy was enacted. Additionally, opinion polls are much more favorable to the inclusion of homosexuals than similar polls decades earlier regarding the inclusion of African-Americans, which obviously succeeded. Fourth, a ban assumes that homosexuality is a choice, and there is increasing scientific evidence that this is not the case. If homosexuality is not a choice, it makes regulating the behavior as immoral even less tenable. Fifth, the United States has a professional force, and the assumption that the military cannot successfully incorporate gays, as it did African-Americans and women, does a disservice to the profession. Sixth, as gays continue down the long path of receiving the same rights as other citizens, it is equally important that they share in the responsibilities of protecting the polity, to include open service of their country.

7. It is not claimed that the experience of African-Americans or women is the same as those of homosexuals. Reviewing the historical record, however, does reveal striking similarities in the *arguments used* against African-Americans and women and therefore may point to some lessons learned regarding policy formulation and implementation. Although clear differences exist between African-Americans, women, and gays, research shows that how a country integrates other races and genders in the military is indicative of how gays were effectively integrated. See Paul Gade, David Segal, and Edgar Johnson, "The Experience of Foreign Militaries," in *Out in Force: Sexual Orientation and the Military*, edited by Gregory M. Herek, Jared B. Jobe, and Ralph M. Carney (Chicago: University of Chicago Press, 1996), 115.

8. Elaine Donnelly, "*Military Times* Poll: Troops Oppose Gay Agenda for the Military," The Tank (blog), *National Review Online*, posted 2 January 2009.

9. Kate Dyer, ed., *Gays in Uniform: The Pentagon's Secret Reports* (Boston: Alyson Publications, 1990); and Randy Shilts, *Conduct Unbecoming: Gays & Lesbians in the U.S. Military* (New York: St Martin's Press, 1993).

10. RAND Corporation, *Sexual Orientation and U.S. Military Personnel Policy: Options and Assessment* (Santa Monica, CA: RAND, 1993).

11. The actual traceable number was between one and three Soldiers. See Nathaniel Frank, *Unfriendly Fire: How the Gay Ban Undermines the Military and Weakens America* (New York: Thomas Dunne Books, 2009), 147.

12. Nathaniel Frank, *Unfriendly Fire: How the Gay Ban Undermines the Military and Weakens America* (New York: Thomas Dunne Books, 2009), 147–48.

13. Aaron Belkin, "Don't Ask, Don't Tell: Is the Gay Ban Based on Military Necessity?" *Parameters*, Summer 2003, 116. For the approximately 70 countries who allow open service, see http://www.gaylawnet.com/laws/laws.htm. This does not mean that no problems occurred, but rather that these problems did not impact readiness. See Aaron Belkin and Melissa Levitt, "Homosexuality and the Israel Defense Forces: Did Lifting the Gay Ban Undermine Military Performance?" *Armed Forces and Society* 27, no. 1 (2001). Belkin's methodology has been attacked: see Joseph A. Craft, Letter to the Editor, "Legitimate Debate or Gay Propaganda?" *Parameters*, 22 June 2004, 132, http://www.carlisle.army.mil/usawc/Parameters/04summer/contents.htm; Elaine Donnelly, "Constructing the Co-Ed Military," *Duke Journal of Gender Law & Policy* 14 (May 2007), 927. Belkin forcefully responded to these attacks, however: Aaron Belkin, "Legitimate Debate, or Gay Propaganda? The Author Replies," *Parameters*, Summer 2004, http://www.carlisle.army.mil/usawc/Parameters/04summer/contents.htm.

14. Aaron Belkin, "Don't Ask, Don't Tell: Is the Gay Ban Based on Military Necessity?" *Parameters*, Summer 2003, 117.

15. Michael R. Gordon, "Powell Delivers a Resounding No on Using Limited Force in Bosnia," *New York Times*, 28 September 1992.

16. Gen Gordon R. Sullivan, interview with author, Washington, DC, 25 February 2005.

17. Cited by Nathaniel Frank, *Unfriendly Fire: How the Gay Ban Undermines the Military and Weakens America* (New York: Thomas Dunne Books, 2009), 122–23. See also Peter Singer, "The Damning Paradox of 'Don't Ask, Don't Tell,'" *Washington Examiner*, 2 June 2009.

18. Michael R. Kauth and Dan Landis, "Applying Lessons Learned from Minority Integration in the Military," in *Out in Force: Sexual Orientation and the Military*, edited by Gregory M. Herek, Jared B. Jobe, and Ralph M. Carney (Chicago: University of Chicago Press, 1996), 93.

19. John H. Sherman, "Command of Negro Troops," *Vital Speeches of the Day*, 15 January 1947, vol. 12, issue 7, 217.

20. Elizabeth Kier, "Homosexuals in the U.S. Military: Open Integration and Combat Effectiveness," *International Security* 23, no. 2 (Autumn 1998), 5–39; and Elizabeth Kier, "Rights and Fights: Sexual Orientation and Military Effectiveness," *International Security* 24 (Summer 1999), 194–201.

21. Paul Koegel, "Lessons Learned from the Experience of Domestic Police and Fire Departments," in *Out in Force: Sexual Orientation and the Military*, edited by Gregory M. Herek, Jared B. Jobe, and Ralph M. Carney (Chicago: University of Chicago Press, 1996), 150; and Robert J. MacCoun, "Sexual Orientation and Military Cohesion: A Critical Review of the Evidence," in *Out in Force: Sexual Orientation and the Military*, 157–76.

22. Gail L. Zellman, "Implementing Policy Changes in Large Organizations: The Case of Gays and Lesbians in the Military," in *Out in Force: Sexual Orientation and the Military*, edited by Gregory M. Herek, Jared B. Jobe, and Ralph M. Carney (Chicago: University of Chicago Press, 1996), 282.

23. Citing Lt Gen Minter Alexander, initially the head of the Military Working Group, supposedly set up to study the issue of homosexuals in the military during the DADT policy formulation period. Alexander said the working group "didn't have any empirical data," and that the military "knew the results of what was going to happen there. It was going to be very difficult to get an objective, rational review of this policy." See Nathaniel Frank, *Unfriendly Fire: How the Gay Ban Undermines the Military and Weakens America* (New York: Thomas Dunne Books, 2009), 115–17.

24. Lawrence J. Korb, "The President, the Congress, and the Pentagon: Obstacles to Implementing the 'Don't Ask, Don't Tell' Policy," in *Out in Force: Sexual Orientation and the Military*, edited by Gregory M. Herek, Jared B. Jobe, and Ralph M. Carney (Chicago: University of Chicago Press, 1996), 291 and 297.

25. Rowan Scarborough, "Marine Leads 'Don't Ask, Don't Tell' Fight: Commandant Resists Efforts to Lift Ban," *The Washington Times*, 2 November 2009, 1.

26. Manu Raju, "Gay Military Question Still Up in the Air," *Politico.com*, 2 September 2009.

27. Colin Powell, interview with Fareed Zakaria, CNN, 14 December 2008, http://www
.youtube.com/watch?v=HlThCfTJDgU.

28. John M. Shalikashvili, "Gays in the Military: Let the Evidence Speak," *Washington Post*, 19
June 2009; and John M. Shalikashvili, "Second Thoughts on Gays in the Military," *New York Times*,
2 January 2007, 17.

29. Nathaniel Frank, *Unfriendly Fire: How the Gay Ban Undermines the Military and Weakens
America* (New York: Thomas Dunne Books, 2009), 83.

30. Manu Raju, "Gay Military Question Still Up in the Air," *Politico.com*, 2 September 2009.

31. Ibid.

32. Paul Koegel, "Lessons Learned from the Experience of Domestic Police and Fire Departments," in *Out in Force: Sexual Orientation and the Military*, edited by Gregory M. Herek, Jared B.
Jobe, and Ralph M. Carney (Chicago: University of Chicago Press, 1996), 147.

33. Ibid.

34. Nathaniel Frank, *Unfriendly Fire: How the Gay Ban Undermines the Military and Weakens
America* (New York: Thomas Dunne Books, 2009), 77.

35. Kyle Dropp and Jon Cohen, "Acceptance of Gay People in Military Grows Dramatically,"
Washington Post, 19 July 2008, A03.

36. Aaron Belkin. "Don't Ask, Don't Tell: Is the Gay Ban Based on Military Necessity?" *Parameters*, Summer 2003, 115.

37. "Independents Take Center Stage in the Obama Era: Trends in Political Values and Core
Attitudes: 1987–2009" (Washington, DC: The Pew Research Center for the People & the Press, 21
May 2009).

38. Jonathan Capehart, "Okay, Obama. Now Let's Have a Speech on Gay Rights," *Washington
Post*, PostPartisan Blog, 4 June 2009, http://voices.washingtonpost.com/postpartisan/2009/06/
okay_obama_now_lets_have_a_spe.html.

39. In report by the Washington-based Committee for the Study of the American Electorate,
over 90 percent of gays and lesbians reported voting in the 2004 election. Gays were double the
turnout as compared to straights for the 2006 congressional midterm election. See Don Frederick
and Andrew Malcolm, "Top of the Ticket—Gay and Lesbian Power—Candidates Should Take
Note That This Is One Group That Knows How to Get Out and Vote," *Los Angeles Times*, 12 August 2007.

40. Adam Nagourney, "Independents Could Help Swing More Than One Primary Toward the
Unexpected," *New York Times*, 2 October 2007.

41. In fact, this research has made lawmakers from both sides of the aisle more savvy and educated. See the 23 July 2008 hearing in front of the House Armed Services Committee at http://
armedservices.house.gov/comdocs/schedules/2008.shtml (scroll to that date for transcripts and
audio/video recordings). Both Republicans and Democrats alike had questions regarding unem-
pirical arguments made against overturning the ban.

42. Nathaniel Frank, *Unfriendly Fire: How the Gay Ban Undermines the Military and Weakens
America* (New York: Thomas Dunne Books, 2009), 87; Gregory M. Herek, Jared B. Jobe, and Ralph
M. Carney, "Conclusion," in *Out in Force: Sexual Orientation and the Military*, edited by Gregory M.
Herek, Jared B. Jobe, and Ralph M. Carney (Chicago: University of Chicago Press, 1996), 306.

43. Nathaniel Frank, *Unfriendly Fire: How the Gay Ban Undermines the Military and Weakens
America* (New York: Thomas Dunne Books, 2009), 86.

44. Quoted in Nathaniel Frank, *Unfriendly Fire: How the Gay Ban Undermines the Military and
Weakens America* (New York: Thomas Dunne Books, 2009), 78.

45. Aaron Belkin, "Don't Ask, Don't Tell: Is the Gay Ban Based on Military Necessity?" *Parameters*, Summer 2003, 114.

46. Base housing would undoubtedly further decrease opposition to homosexual service since
one of the biggest nullifiers of homophobia is when someone knows a homosexual family, much
like knowing an African-American family in the 1950s.

47. Elaine Donnelly quoted by Roxana Tiron, "Lawmakers Grill Critic of Gays in Military," *The Hill.Com*, posted 23 July 2008.

48. Paul A. Gade, David R. Segal, and Edgar M. Johnson, "The Experience of Foreign Militaries," in *Out in Force: Sexual Orientation and the Military*, edited by Gregory M. Herek, Jared B. Jobe, and Ralph M. Carney (Chicago: University of Chicago Press, 1996), 108; Paul Koegel, "Lessons Learned from the Experience of Domestic Police and Fire Departments," in *Out in Force: Sexual Orientation and the Military*, 137; and Aaron Belkin, "Don't Ask, Don't Tell: Is the Gay Ban Based on Military Necessity?" *Parameters*, Summer 2003, 112.

49. The often cited figure that 10 percent of any population is homosexual comes from Alfred C. Kinsey, Wardell B. Pomeroy, and Clyde E. Martin, *Sexual Behavior in the Human Male* (Philadelphia, PA: W.B. Saunders, 1948), 610–66. It should be noted that the 10 percent figure came from the findings that 4 percent of men were more or less exclusively homosexual their whole lives, while another 6 percent had been more or less exclusively homosexual for at least three years between the ages of 16–55.

50. For the running tally, see Servicemembers Legal Defense Network's Web site at http://www.sldn.org/pages/about-dadt.

51. See Michael E. O'Hanlon, "Obama's Defense Budget Gap," *Washington Post*, 10 June 2009, A7; Edwin J. Feulner, PhD, "Spending Spree and Cutting Defense Don't Add Up," Special Report 62, Heritage Foundation, 20 July 2009; Mackenzie M. Eaglen, "The $64,000 Question: Is President Obama Actually Increasing the Defense Budget?," WebMemo No. 2411, Heritage Foundation, 27 April 2009; Greg Bruno, "The Fine Print on Defense Spending," Council on Foreign Relations, 8 April 2009; and Douglas W. Elmendorf, "The Long-Term Budget Outlook," Congressional Budget Office, Testimony before the Committee on the Budget, US Senate, 16 July 2009.

52. Gary Gates, *Gay Men and Lesbians in the US Military: Estimates from Census 2000* (Washington, DC: Urban Institute, 2004).

53. "Financial Analysis of 'Don't Ask, Don't Tell': How Much Does the Gay Ban Cost?" Blue Ribbon Commission Report, February 2006, 21, http://www.palmcenter.org/files/active/0/2006 -FebBlueRibbonFinalRpt.pdf.

54. Ibid., 23.

55. Ibid., 24.

56. The argument for being deliberate comes from Nathaniel Frank, *Unfriendly Fire: How the Gay Ban Undermines the Military and Weakens America* (New York: Thomas Dunne Books, 2009), 72. For the case of formulating and implementing the policy quickly, see Gail L. Zellman, "Implementing Policy Changes in Large Organizations: The Case of Gays and Lesbians in the Military," in *Out in Force: Sexual Orientation and the Military*, edited by Gregory M. Herek, Jared B. Jobe, and Ralph M. Carney (Chicago: University of Chicago Press, 1996), 275.

57. Paul Koegel, "Lessons Learned from the Experience of Domestic Police and Fire Departments," in *Out in Force: Sexual Orientation and the Military*, edited by Gregory M. Herek, Jared B. Jobe, and Ralph M. Carney (Chicago: University of Chicago Press, 1996), 146.

58. Gail L. Zellman, "Implementing Policy Changes in Large Organizations: The Case of Gays and Lesbians in the Military," in *Out in Force: Sexual Orientation and the Military*, edited by Gregory M. Herek, Jared B. Jobe, and Ralph M. Carney (Chicago: University of Chicago Press, 1996), 274.

59. Paul Koegel, "Lessons Learned from the Experience of Domestic Police and Fire Departments," in *Out in Force: Sexual Orientation and the Military*, edited by Gregory M. Herek, Jared B. Jobe, and Ralph M. Carney (Chicago: University of Chicago Press, 1996), 147.

60. Lawrence J. Korb, "The President, the Congress, and the Pentagon: Obstacles to Implementing the 'Don't Ask, Don't Tell' Policy," in *Out in Force: Sexual Orientation and the Military*, edited by Gregory M. Herek, Jared B. Jobe, and Ralph M. Carney (Chicago: University of Chicago Press, 1996), 299.

61. Aaron Belkin, "Don't Ask, Don't Tell: Is the Gay Ban Based on Military Necessity?" *Parameters*, Summer 2003, 111.

62. Gail L. Zellman, "Implementing Policy Changes in Large Organizations: The Case of Gays and Lesbians in the Military," in *Out in Force: Sexual Orientation and the Military*, edited by Gregory M. Herek, Jared B. Jobe, and Ralph M. Carney (Chicago: University of Chicago Press, 1996), 278. Disincentives should also apply, be explicit, and be enforced. See John H. Sherman, "Command of Negro Troops," *Vital Speeches of the Day*, 15 January 1947, vol. 12, no. 7, 217.

63. Gail L. Zellman, "Implementing Policy Changes in Large Organizations: The Case of Gays and Lesbians in the Military," in *Out in Force: Sexual Orientation and the Military*, edited by Gregory M. Herek, Jared B. Jobe, and Ralph M. Carney (Chicago: University of Chicago Press, 1996), 286.

64. Ibid.

65. Ibid., 275.

66. Ibid., 283.

67. As Senator Carl Levin noted in DADT hearings, sodomy, defined as oral or anal sex, is banned outright, meaning that the law applies equally to heterosexuals and homosexuals. See Levin and Senator John Kerry quoted in Nathaniel Frank, *Unfriendly Fire: How the Gay Ban Undermines the Military and Weakens America* (New York: Thomas Dunne Books, 2009), 91–92 and 98.

68. Gail L. Zellman, "Implementing Policy Changes in Large Organizations: The Case of Gays and Lesbians in the Military," in *Out in Force: Sexual Orientation and the Military*, edited by Gregory M. Herek, Jared B. Jobe, and Ralph M. Carney (Chicago: University of Chicago Press, 1996), 277.

69. Nathaniel Frank, *Unfriendly Fire: How the Gay Ban Undermines the Military and Weakens America* (New York: Thomas Dunne Books, 2009), 196–97.

70. Ibid., 196.

71. Michael R. Kauth and Dan Landis, "Applying Lessons Learned from Minority Integration in the Military," in *Out in Force: Sexual Orientation and the Military*, edited by Gregory M. Herek, Jared B. Jobe, and Ralph M. Carney (Chicago: University of Chicago Press, 1996), 100.

72. Ibid., 100–101. The same focus on behavior also worked for police and fire departments: Paul Koegel, "Lessons Learned from the Experience of Domestic Police and Fire Departments," in *Out in Force: Sexual Orientation and the Military*, edited by Gregory M. Herek, Jared B. Jobe, and Ralph M. Carney (Chicago: University of Chicago Press, 1996), 146–47.

73. Paul Koegel, "Lessons Learned from the Experience of Domestic Police and Fire Departments," in *Out in Force: Sexual Orientation and the Military*, edited by Gregory M. Herek, Jared B. Jobe, and Ralph M. Carney (Chicago: University of Chicago Press, 1996), 148.

74. Ibid., 151; and Gail L. Zellman, "Implementing Policy Changes in Large Organizations: The Case of Gays and Lesbians in the Military," in *Out in Force: Sexual Orientation and the Military*, edited by Gregory M. Herek, Jared B. Jobe, and Ralph M. Carney (Chicago: University of Chicago Press, 1996), 269.

75. Nathaniel Frank, *Unfriendly Fire: How the Gay Ban Undermines the Military and Weakens America* (New York: Thomas Dunne Books, 2009), 165.

76. Robert J. Maccoun, "Sexual Orientation and Military Cohesion," in *Out in Force: Sexual Orientation and the Military*, edited by Gregory M. Herek, Jared B. Jobe, and Ralph M. Carney (Chicago: University of Chicago Press, 1996), 172.

77. Gail L. Zellman, "Implementing Policy Changes in Large Organizations: The Case of Gays and Lesbians in the Military," in *Out in Force: Sexual Orientation and the Military*, edited by Gregory M. Herek, Jared B. Jobe, and Ralph M. Carney (Chicago: University of Chicago Press, 1996), 273.

78. Paul Koegel, "Lessons Learned from the Experience of Domestic Police and Fire Departments," in *Out in Force: Sexual Orientation and the Military*, edited by Gregory M. Herek, Jared B. Jobe, and Ralph M. Carney (Chicago: University of Chicago Press, 1996), 147.

79. Gail L. Zellman, "Implementing Policy Changes in Large Organizations: The Case of Gays and Lesbians in the Military," in *Out in Force: Sexual Orientation and the Military*, edited by Gregory M. Herek, Jared B. Jobe, and Ralph M. Carney (Chicago: University of Chicago Press, 1996), 275.

80. John H. Sherman, "Command of Negro Troops," *Vital Speeches of the Day*, 15 January 1947, vol. 12, no. 7, 220.

81. Lawrence J. Korb, "The President, the Congress, and the Pentagon: Obstacles to Implementing the 'Don't Ask, Don't Tell' Policy," in *Out in Force: Sexual Orientation and the Military*, edited by Gregory M. Herek, Jared B. Jobe, and Ralph M. Carney (Chicago: University of Chicago Press, 1996), 297.

82. Approximately 1.7 service members are discharged each day. For the current count, see http://www.sldn.org/content/discharged.

83. "Gay Veteran Calls for End of 'Don't Ask, Don't Tell,'" *CNN*, 1 March 2007, http://www.cnn.com/2007/POLITICS/02/28/gays.military/index.html.

About the Author

Dr. Tammy S. Schultz is the director of the National Security and Joint Warfare and an associate professor at the US Marine Corps War College. Dr. Schultz also conducts communication simulations at the State Department and is an adjunct professor at Georgetown University's Security Studies Program. She has served as a fellow at the Center for a New American Security, the US Army's Peacekeeping and Stability Operations Institute, and the Brookings Institution. She graduated summa cum laude from Regis University in 1995 and earned a master's from Victoria University in New Zealand in 1999. She received her PhD from Georgetown University in 2005. She is widely published and frequently quoted on defense and national security issues. She is on the Term Member Advisory Committee of the Council of Foreign Relations, a principal in the Truman National Security Project, and on the Executive Board of Women in International Security.

The author would like to thank her colleagues at the Marine Corps War College, especially Col Michael Belcher, for improving this work. Also, Dr. Becky Johnson of Marine Corps University's Command and Staff College provided invaluable comments. The librarians at the General Alfred M. Gray Marine Corps Research Center also helped with short-order research, in particular Pat Lane. Alexandra Singer proved once again more than capable of timely and intelligent research. All errors and omissions are, of course, solely the author's. The views expressed in this chapter are the author's, are not official, and do not necessarily reflect the positions of the US Department of Defense, the US Marine Corps, or the Marine Corps University.

CHAPTER 10

HOW TO END
"DON'T ASK, DON'T TELL"
A ROADMAP OF POLITICAL, LEGAL, REGULATORY, AND ORGANIZATIONAL STEPS TO EQUAL TREATMENT

Aaron Belkin
Gregory M. Herek
Diane H. Mazur

Nathaniel Frank
Elizabeth L. Hillman
Bridget J. Wilson

Executive Summary

Pres. Barack Obama has stated his intention to end the Pentagon policy known as "don't ask, don't tell," and allow gay men and lesbians to serve openly in the military. The federal statute governing this policy, Section 571 of the FY1994 National Defense Authorization Act, codified at 10 U.S.C. § 654, is titled "Policy Concerning Homosexuality in the Armed Forces" and has come to be known as "don't ask, don't tell."

While strong majorities of the public, and growing numbers within the military, support such a change, some political leaders and military members have expressed anxiety about what impact it will have on the armed forces. Scholarly evidence shows that the ban on service by openly gay personnel is unlikely to impair military effectiveness or to harm recruiting, retention, or unit cohesion. Yet questions remain as to how best to execute and manage the transition from exclusion to inclusion of openly gay personnel in a way that takes into consideration the concerns and sensitivities of the military community. In this report, we address political, legal, regulatory, and organizational steps that will ensure that the implementation process goes smoothly. We begin by suggesting six key points that should be kept in mind as policy makers consider the change.

1. The executive branch has the authority to suspend homosexual conduct discharges without legislative action.

The process of lifting the ban on gay service by openly gay personnel is both political and military in nature. While research shows that the planned policy

---

This article was originally published online at http://www.palmcenter.org/publications/all in May 2009. An appendix has been added to this edition by the first and second listed authors.

change does not pose an unmanageable risk to the military, how the transition is executed politically can affect how smoothly the change is implemented. The president has the authority to issue an executive order halting the operation of "don't ask, don't tell." Under 10 U.S.C. § 12305, "Authority of the President to Suspend Certain Laws Relating to Promotion, Retirement, and Separation," Congress grants the president authority to suspend the separation of military members during any period of national emergency in which members of a reserve component are serving involuntarily on active duty. We believe that issuing such an order would be beneficial to military readiness, as it would minimize the chances of replaying a debate that is already largely settled but could still inflame the passions of some in the military. Once gay people are officially serving openly in the military, it will become clear to those with concerns about the policy change that service by openly gay personnel does not compromise unit cohesion, recruiting, retention, or morale. This in turn will make it easier to secure the passage of the Military Readiness Enhancement Act (MREA) in Congress, which would repeal "don't ask, don't tell." While it would be optimal to see lawmakers embrace repeal by passing MREA, it may not be politically feasible to do so, despite overwhelming public support and Democratic control of Congress. Conservative Democrats in Congress may oppose MREA, and the White House may not wish to expend the political capital necessary to overcome their resistance. The executive option may end up costing the president less in political capital than the effort needed to push repeal through Congress. And it could help avoid the emergence of split military leadership, which could make the transition bumpier than it has to be.

2. Legislative action is still required to permanently remove "don't ask, don't tell."

Since MREA was first introduced in 2005, it has remained a stand-alone, unicameral bill. Passage of the bill would be the best way to permanently eliminate "don't ask, don't tell" for the following reasons: First, since the current policy is based on a statute passed by Congress, its permanent elimination will require legislative or judicial action. Second, the legislation as currently written would establish a uniform code of conduct across the military for all service members, gay and straight, without regard to sexual orientation. Evidence from foreign militaries indicates that this is one of the most important steps for the successful transition to a policy of inclusion. Finally, articulating the new policy in a federal statute will give the policy the imprimatur of broad public support and will create a clear set of standards and policies for service members and commanding officers. As stated in no. 1, above, pushing MREA through Congress may best be done after an executive order first halts discharges for homosexual conduct.

3. The president should not ask military leaders if they support lifting the ban.

The president has stated he wants to consult with the military leadership about lifting the ban on service by openly gay personnel. It is crucial that such consultation not take the form of yielding authority on this issue to the Defense Department, which could create a damaging wedge between the president and the military. A catch-22 is now paralyzing action on ending homosexual conduct discharges. Many members of Congress are fearful that supporting repeal could cost them political support, despite polls showing majority support for service by openly gay personnel even in conservative populations. Because of that fear, some lawmakers seek to shift responsibility for repeal to the Pentagon. But senior insiders in the Pentagon are unwilling to tackle "don't ask, don't tell" because they view the issue as a "hot potato" or "career killer," so they seek to shift responsibility back to Congress. A similar scenario is threatening to play out between the White House and the Pentagon, in which the current administration, despite having promised it will end the ban, wants the impetus for change to appear to come from the Pentagon, whose top leaders have indicated no such will for change. In other countries, militaries have acted to end discrimination only when so ordered, as was the case in the United States with respect to racial integration in the military. It is likely that reform in this case will happen only through action by civilian leadership. Since President Obama already has said that he plans to lift the ban, he will gain nothing from throwing this particular decision up to debate.

Already, interest groups have begun organizing to defeat the president's plan to lift the ban. Over 1,000 retired admirals and generals have signed a document opposing repeal, at the behest of a conservative group that is lobbying to retain the ban. While the document is not based on any research or new information, efforts such as this one will make the president's job more difficult and provide evidence for why decisive action is needed on this issue.

In 1993, members of President Clinton's transition team consulted extensively with all levels of the US military, ranging from the Joint Chiefs of Staff to enlisted personnel. Despite these efforts, the chiefs claimed that they had not been sufficiently consulted. This precedent suggests that, whether the Obama administration consults with the military or not, Pentagon leaders may feel or say they were inadequately consulted. Thus, despite the president's pledge to take military perspectives into account on this issue, he should realize that what the military needs most in this case is leadership. Any consultation with uniformed leaders should take the form of a clear mandate to give the president input about how, not whether, to make this transition.

4. The president should therefore take into consideration the following with regard to consulting the military:

- The president may be accused of not consulting with the Pentagon regardless of what steps he takes to reach out to the military.
- If he does consult, he may be told that most service members do not want the ban to be lifted, thus constricting his options when he decides to move forward with repeal.
- Significant support for repeal exists within the military, but there is enormous institutional pressure to avoid expressing that support, which hence does not get registered in consultation.

- A significant cadre of military leadership, although unwilling to acknowledge so in public, want lawmakers to mandate reform so as to give them cover.

- While many people in the military oppose policy change, the percentage that feels strongly that gay men and lesbians should not be allowed to serve openly is quite small, and research shows that there is a difference between what troops say they want in a poll and how they actually behave when taking orders.

- Even among opponents of repeal, most military members understand its inevitability.

- Extensive consultation of the armed forces could distract them from their efforts to secure our nation's security and expose them to the risk of being exploited by those who oppose change for moral or cultural reasons.

5. Studying the issue further would cause waste, delay, and a possible backlash.

Recent proposals to study *whether* to repeal the law are unwarranted. A significant body of scholarly research, which we summarize in section two of this report, already shows clearly that the ban is unnecessary, that it harms the military, and that repeal would improve the military. Even the question of how to repeal the law is not something that requires study. Research summarized in this report already explains how to implement change. And while some have suggested that the president could request a study on how, rather than whether, to end the ban, this was precisely what President Clinton ordered in 1993 with both the RAND study and the Military Working Group. Opening up these questions to study will allow time for mobilization of emotional constituencies who are more focused on a narrow moral agenda than on military readiness, as was the case in 1993.

6. Equal standards and leadership support are critical to a successful policy change.

Any legal or regulatory change should heed the two most important lessons from foreign militaries that have transitioned to open service. First, the military must adopt a single code of conduct for all service members, gay and straight, without regard to sexual orientation. Second, military leaders must signal clearly that they expect all members of the armed forces to adhere to the new policy, regardless of their personal beliefs.

Expected Impact of Service by Openly Gay Personnel

At the 23 July 2008 congressional hearings about the "don't ask, don't tell" law, former Rep. Nancy Boyda (D-KS) expressed frustration at the lack of evidence concerning the impact of service by openly gay personnel on the military. Referring to the testimony of service members and experts during the hearing, she said, "It's been people's stories, their feelings, opinions, and while it's been interesting, I'd like to see a little bit more . . . hard data."

There has never been a policy change that involved certain knowledge about outcomes, as the future is never perfectly predictable. That said, the data that former Congresswoman Boyda requested already exist. Evidence shows consistently that after gay men and lesbians are allowed to serve openly in the armed forces, military readiness will not be compromised. The data have been produced by a wide range of scholars at the Army Research Institute, the RAND Corporation, the Defense Personnel Security Research Center, and a large number of universities. No reputable or peer-reviewed study has ever shown that allowing service by openly gay personnel will compromise military effectiveness.

Three types of evidence can be used to assess the nature and likelihood of any impact to the military following the decision to allow service by openly gay personnel, and all three types of evidence suggest there will be no negative impact on the military. Those three areas of evidence are:

- Data about what happens in the US military when gay men and lesbians serve openly, notwithstanding the strictures of the current policy.
- Data from analogous institutions, including but not limited to foreign militaries, that allow gay men and lesbians to serve openly.
- Data about the unit cohesion rationale: the argument that unit cohesion will suffer if gay men and lesbians serve openly.

Data about What Happens in the US Military When Gay Men and Lesbians Serve Openly

The US military functionally suspended the gay ban during the first Gulf War by halting the gay discharge process. There have been no indications of any detriment to unit cohesion or readiness during that war. In fact, the cohesion and readiness of the troops during the first Gulf War have been widely praised. Researchers have followed units in which American troops worked with and even took orders from openly gay foreigners in integrated multinational units under the auspices of NATO, the United Nations, and other multinational organizations. They found no negative impact to cohesion and readiness. More recently, a survey was administered to 545 service members who fought in Afghanistan and Iraq. Respondents were asked about the presence of openly gay members of their units, and about their units' cohesion and readiness. A majority of respondents said they knew of, or suspected, gays in their units. Statistical analysis of results found that there was no relationship between the presence of openly gay troops and the cohesion or readiness of the unit.

Data from Analogous Institutions That Allow Gay Men and Lesbians to Serve Openly

Twenty-four foreign militaries allow gay men and lesbians to serve openly. None has reported any detriment to cohesion, readiness, recruiting, morale, retention, or any other measure of effectiveness or quality. Studies conducted by the militaries in Canada and Britain as well as scholarly studies published in peer-reviewed journals have confirmed the same finding: decisions to allow service by openly gay personnel had no negative impact on cohesion, readiness, recruiting, morale, retention, or any other measure of effectiveness or quality in foreign armed forces. In the more than three decades since an overseas force first allowed gay men and lesbians to serve openly, no study has ever documented any detriment to cohesion, readiness, recruiting, morale, retention, or any other measure of effectiveness or quality. No American police or fire department that allows gay men and lesbians to serve openly has reported any detriment to cohesion, readiness, recruiting, morale, retention, or any other measure of effectiveness, and scholarly research has confirmed the lack of any decline. No federal agency that allows gay men and lesbians to serve openly such as the CIA, FBI, or Secret Service has reported any detriment to cohesion, readiness, recruiting, morale, retention, or any other measure of effectiveness or quality.

Data about the Impact of Service by Openly Gay Personnel on Unit Cohesion

The "unit cohesion rationale" is the claim that heterosexuals will not form bonds of trust with gay people, and that if gay men and lesbians are allowed to serve openly, units will fail to develop a sufficient degree of cohesion; as a result, military effectiveness will suffer. Empirical data, however, show this assertion is not grounded in fact. A recent survey of 545 service members who served in Afghanistan and Iraq found that 72 percent reported that they are comfortable working with gay men and lesbians. Of the 20 percent who said they are uncomfortable, only 5 percent are "very uncomfortable," while 15 percent are "somewhat" uncomfortable. Senior members of the armed forces, both active duty and former, have concluded that no evidence has ever linked service by openly gay personnel to any impairment of military effectiveness. For example, Col Tom Kolditz, chairman of the Department of Behavioral Sciences and Leadership at the US Military Academy at West Point and one of the Army's top experts on leadership and cohesion, told a 2008 study commission of retired flag and general officers that he is unaware of any evidence suggesting that heterosexuals cannot form bonds of trust with gays, lesbians, and bisexuals.

Three additional observations deserve mention. First, while many service members indicate on surveys that they oppose lifting the ban, the relevant data point is not whether troops wish to serve with openly gay peers, but whether service by openly gay personnel will undermine military effectiveness. On one recent, nonrandomized survey, between 10 and 24 percent of service members indicated that they would leave or might leave the military if gay men and lesbians were allowed to serve openly. Social science research, however, shows that opinion polls do not predict the troops' behavior and that there is a significant gap between what is expressed in military surveys and the actual impact of policy change on behavior. In both Canada and Britain, two-thirds of male troops said that they would not work with gay men if gay bans were lifted in those countries. After the lifting of the bans, fewer than a half dozen people resigned in each case.

Second, while any policy change can generate certain disruptions, the very few "horror stories" that are sometimes used to oppose reform must not be confused with relevant empirical evidence. The question is not whether "bad apples" or isolated incidents cause problems in some units, but whether service by openly gay personnel presents problems that are any different or less surmountable than service by open heterosexuals. Conduct that is deemed inappropriate is deemed so regardless of the sexual orientation or gender of those involved. The military already has appropriate conduct laws and regulations which are neutral with respect to sexual orientation and gender to handle disruptions.

Finally, while the data show that allowing service by openly gay personnel will not undermine the military, research suggests that a number of positive benefits will accrue. Repeal of the law will: (1) make it easier for gay troops to do their jobs; (2) save hundreds of millions of dollars currently spent on training replacement troops; (3) prevent the loss of talented service members; (4) eliminate a source of negative media publicity for the military; and (5) promote unit cohesion both by minimizing unnecessary personnel loss and by enhancing a climate of honesty, respect, and obedience to a uniform code of conduct for all service members.

Scholarly Research on Military Readiness and Service of Openly Gay Personnel

Taken together, the evidence on the ability of countries to lift their gay bans without problems is overwhelming. Descriptions of relevant research are provided below, and full citations are included at the end of this report.

1. The US Navy's Crittenden Report from 1957 which found that gay troops did not present a security risk.

2. The Defense Department's Personnel Security Research and Education Center (PERSEREC) study from 1988, which found the same thing as the Crittenden Report and also concluded that the rationale for the ban was unfounded and not based on evidence.

3. A 1992 draft report by the General Accounting Office (GAO) (now the Government Accountability Office) suggesting that the military "reconsider the basis" of the gay exclusion rule.

4. A 1993 GAO study of four foreign militaries which found that "the presence of homosexuals in the military is not an issue and has not created problems in the functioning of military units."

5. A 1993 RAND study prepared by over 70 social scientists based on evidence from six countries and data analyses from hundreds of studies of cohesion that concluded that sexuality was "not germane" to military service and recommended lifting the ban.

6. A 1994 assessment of the Canadian Forces by the US Army Research Institute for the Behavioral and Social Sciences finding that predicted negative consequences of ending gay exclusion did not materialize following the lifting of the ban.

7. A 1999 article published in the journal International Security concluding that service by open gays and lesbians would not disrupt unit cohesion or combat effectiveness.

8. The assessments of the British Ministry of Defence in 2000 calling its new policy of equal treatment "a solid achievement" with "no discernible impact" on recruiting and no larger problems resulting from reform, and a 1995 assessment by a Canadian military office finding that there was no effect on readiness when the ban was lifted, despite enormous resistance and anxiety preceding the change.

9. Four independent academic studies conducted by the Palm Center at the University of California finding that lifting bans in Britain, Israel, Canada, and Australia had "no impact" on military readiness and that negative attitudes almost never translated into service member departures, recruiting problems, or other disruptions.

10. A 2008 report by a commission of retired general and flag officers who concluded that "allowing gays and lesbians to serve openly would pose no risk to morale, good order, discipline, or cohesion."

11. A 2009 statistical analysis by a RAND scholar and a University of Florida professor which shows that there is no correlation between whether or not a unit includes openly gay service members and the readiness or cohesion of the unit.

12. A report published in the flagship military journal, *Joint Force Quarterly*, by Col Om Prakash, an active duty Air Force officer researching at the National War College, deeming the policy a "costly failure," stating that "there is no scientific evidence to support the claim that unit cohesion will be negatively affected if homosexuals serve openly," and recommending that the government "examine how to implement the repeal of the ban" without further assessment of whether it should be lifted.

Presidential Authority to Suspend
Discharges for Homosexual Conduct

10 U.S.C. § 654, "Policy Concerning Homosexuality in the Armed Forces," states that a "member of the armed forces shall be separated from the armed forces under regulations prescribed by the Secretary of Defense if one or more of the following findings is made and approved in accordance with procedures set forth in such regulations": (1) "the member has engaged in, attempted to engage in, or solicited another to engage in a homosexual act or acts"; (2) "the member has stated that he or she is a homosexual or bisexual, or words to that

effect"; or (3) "the member has married or attempted to marry a person known to be of the same biological sex."

The president of the United States has authority under the laws of the United States and the Constitution to suspend all investigations, separation proceedings, or other personnel actions conducted under the authority of 10 U.S.C. § 654 or its implementing regulations. Below we explain the basis of such authority.

The Laws of the United States

Federal law recognizes that the president and Congress share authority to govern the military. In fact, by law currently in effect, Congress has already granted the president authority with respect to military promotions, retirements, and separations in a time of national emergency. This authority includes the power to suspend enforcement of laws such as 10 U.S.C. § 654. Under 10 U.S.C. § 12305, "Authority of the President to Suspend Certain Laws Relating to Promotion, Retirement, and Separation," Congress grants the president authority to suspend any provision of law relating to the separation of any member of the armed forces who the president determines is essential to the national security of the United States, during any period of national emergency in which members of a reserve component are serving involuntarily on active duty. The statute states:

> Notwithstanding any other provision of law, during any period members of a reserve component are serving on active duty pursuant to an order to active duty under authority of section 12301, 12302, or 12304 of this title, the President may suspend any provision of law relating to promotion, retirement, or separation applicable to any member of the armed forces who the President determines is essential to the national security of the United States.

This law is colloquially referred to as "stop-loss" authority, and it has been used to suspend the voluntary separation of members of the military who have reached the end of their enlistment obligation or have qualified for retirement. The law, however, gives the president authority to suspend "*any* provision of law" (emphasis added) relating to separation of members of the armed forces, including involuntary separations under 10 U.S.C. § 654. The Army has announced it will phase out the stop-loss program, which forcibly retains Soldiers who wish to leave after their tours. It is important to point out that this use of stop-loss has been particularly unpopular because it forces ongoing service by those who wish to leave the military, whereas the use of stop-loss to suspend homosexual conduct discharges would, by contrast, allow ongoing service by those who generally wish to remain in uniform.

10 U.S.C. § 12305 gives the president authority to suspend laws relating to separation of members of the military if two requirements are met. First, the suspension must occur during a period of national emergency in which members of the military reserve are involuntarily called to active duty under sections § 12301 (reserve components generally), § 12302 (ready reserve), and § 12304 (selected

reserve and certain individual ready reserve members). As of 7 April 2009, there were 93,993 members of reserve components or retired members serving on active duty after involuntary activation. Second, the president must make a determination that retention of members of the military—and suspension of any law requiring their separation—is essential to the national security of the United States. The conditions of 10 U.S.C. § 12305 are sensible because they give the president authority to suspend laws relating to separation when a national emergency has strained personnel requirements to the point that members of the reserve forces have been involuntarily called to active duty. The constitutionality of 10 U.S.C. § 12305 was upheld in *Santiago v. Rumsfeld*, 425 F.3d 549, 9th Cir., 2005.

Under 10 U.S.C. § 123, "Authority to Suspend Officer Personnel Laws during War or National Emergency," Congress grants the president similar authority to suspend laws relating to the separation of officer personnel.

The "don't ask, don't tell" policy itself, as codified by Congress, also grants authority to the Department of Defense to determine the procedures under which investigations, separation proceedings, and other personnel actions under the authority of 10 U.S.C. § 654 will be carried out. Section 654(b) states, "A member of the armed forces shall be separated from the armed forces under regulations prescribed by the Secretary of Defense if one or more of the following findings is made and approved in accordance with procedures set forth in such regulation." Under this section, the secretary of defense has discretion to determine the specific manner in which "don't ask, don't tell" will be implemented. Furthermore, the statute does not direct the military to make any particular findings of prohibited conduct or statements; it only states that members shall be separated under regulations prescribed by the secretary *if* such findings are made. The secretary has broad authority to devise and implement the procedures under which those findings may be made.

A recent decision of the Ninth Circuit Court of Appeals, *Witt v. Department of the Air Force*, 527 F.3d 806 (9th Cir. 2008), calls into question whether "don't ask, don't tell," as implemented by regulations prescribed by the secretary of defense, violates the due process rights of service members under the Fifth Amendment of the US Constitution. The court remanded the case for further findings on whether the separation of this specific service member would significantly further an interest in military effectiveness, and whether less intrusive means would be unlikely to further the same interest. The secretary has authority under 10 U.S.C. § 654 to determine whether regulations implementing the statute are consistent with the ruling in *Witt*, whether the regulations should be revised and, if necessary, whether amendments to the statute should be recommended for further consideration by Congress.

The Constitution of the United States

Federal law reflects that the president, the Congress, and the federal courts share constitutional power and responsibility for governance of the armed forces of the United States.

1. Under Article I, Section 8, Clauses 12–14, Congress has the power to raise and support armies, to provide and maintain a Navy, and to make rules for the government and regulation of the land and naval forces. Congress legislated under this authority in enacting 10 U.S.C. § 654.

2. Under Article II, Section 2, Clause 1, the president has the power to act as commander in chief of the armed forces of the United States.

3. Under Article III, federal courts have the power to decide all cases arising under the Constitution and the laws of the United States. Federal courts have the power to interpret law and ensure that the other branches of government act in accordance with the Constitution.

Although Congress has power to make rules to govern the military, it shares that power with the president, who, as commander in chief, has power to direct the operation of military forces. If Congress were understood to have sole power to remove members of the military from the chain of command operating under the direction of the president, particularly in a time of national emergency, the president's ability to carry out his constitutional obligations would be impaired. Therefore, the constitutional authority of the commander in chief includes at least shared authority to ensure that members of the military essential to national security are not removed from duty.

The Regulations of the Department of Defense

10 U.S.C. § 654 directs that the DADT policy be implemented under regulations prescribed by the secretary of defense. There are three principal Department of Defense implementing regulations in force: Department of Defense Instruction (DODI) 1304.26, *Qualification Standards for Enlistment, Appointment, and Induction* (11 July 2007); DODI 1332.14, *Enlisted Administrative Separations* (28 August 2008); and DODI 1332.30, *Separation of Regular and Reserve Commissioned Officers* (11 December 2008). Each of the military services has in turn issued regulations to implement Department of Defense guidance.

Department of Defense regulations governing the separation of members under 10 U.S.C. § 654 preserve discretion within the military chain of command to retain members under certain circumstances. "Enlisted Administrative Separations," for example, states at enclosure 3, paragraph 8.d (7)(c), page 21, "Nothing in these procedures . . . precludes retention of a Service member for a limited period of time in the interests of national security as authorized by the Secretary concerned." Military commanders have significant discretion to decide whether they should initiate investigations or separation proceedings, or whether no action should be taken at all: "They shall examine the information and decide whether an inquiry is warranted or whether no action should be taken" ("Enlisted Administrative Separations," enclosure 5, paragraph 3.b, page 39; "Separation of Regular and Reserve Commissioned Officers," enclosure 8, paragraph 3.b, page 23).

Regulatory Revisions That Should Accompany Policy Change

Service by openly gay personnel will require changes in administrative procedures that can be handled through the military's usual processes of revising, reissuing, and cancelling publications. The enforcement and administration of the homosexual conduct policy has spawned many rules and regulations, most of which can be changed easily to comply with an executive order suspending the policy. Below, we describe and propose revisions to the publications that currently enforce and administer the homosexual conduct policy—and control its collateral consequences—in the Defense Department, its components, the Department of Homeland Security, and the US Coast Guard. The process of publication review that is already in place can be used to make necessary changes. Pending a full review, interim guidance can be issued to suspend discharges, and other adverse personnel actions, under the policy.

Publications

Department of Defense and service publications referencing homosexual conduct include directives, instructions, manuals, secretarial memoranda, and local instructions. Most of these publications include but incidental references to the homosexual conduct policy and therefore would require only minor revisions. Others are specific to the policy and could be withdrawn or canceled. A few publications, primarily those related to separation procedures, will eventually require more substantial changes to implement permanent service by openly gay personnel. It is important to note the difference between discharges for homosexual conduct and action taken as the result of criminal conduct. Separations under "don't ask, don't tell" are not criminal but administrative and result, in the vast majority of cases, in an "honorable" discharge.

The major categories of relevant publications include:

1. **Criminal statutes, criminal procedure, and disciplinary codes as contained in the Uniform Code of Military Justice (UCMJ) and its implementing regulations.** No changes are required here. Congress should, however, consider adopting the recommendations of the Joint Services Committee on Military Justice to replace the consensual sodomy ban contained in Article 125 of the UCMJ with a ban in the *Manual for Courts-Martial* on all sexual acts that are prejudicial to good order and discipline. This would emphasize that a single standard of conduct applies to all military personnel.

2. **Personnel management directives and manuals that govern the policy and procedures for separation of officers and enlisted members under the homosexual conduct policies created by the Department of Defense, Army, Navy, Marine Corps, Air Force, and Coast Guard.**[1] The sections of these publications that govern discharge under the homosexual conduct policy can be reissued if Congress makes a statutory change. If an executive order suspends implementation of the policy, those sections

should be immediately suspended, subject to review for compliance with the order.

3. **Publications that govern documentation and record-keeping requirements as well as the collateral consequences of separation for homosexual conduct, including regulations regarding discharge documents, benefits, separation pay, and similar information.** These publications should be revised in the established process of administrative review pending permanent changes in the policy.

4. **Directives and orders that limit the use of information related to homosexual conduct in non-discharge-related areas such as law enforcement, security clearances, and medical care.**[2] These publications should be revised in the usual process of regularized review, pending permanent changes.

5. **Training materials and instructions intended to guide the implementation of the existing homosexual conduct policy, such as lesson plans, recruiting materials, and legal instructions.** These should be withdrawn and revised in accordance with new policy guidelines as those policies are articulated.

Existing Review Mechanisms

The armed forces are well practiced in adapting regulations and other administrative guidance to changed circumstances. Department of Defense and service department publications are subject to periodic review. Department of Defense directives (DODD) are reviewed prior to the four-year anniversary of their initial publication or last coordinated review to ensure they are necessary, current, and consistent with DOD policy, existing law, and statutory authority. Upon review, the DODD may be reissued, certified as current, or canceled. All DODDs certified as current shall be revised and reissued or canceled within six years of their initial publication or last coordinated revision. All Department of Defense instructions (DODI), Department of Defense manuals (DODM), and administrative instructions (AI) shall be reviewed every five years, and revised, reissued, or canceled (see DODI 5025.01, *DoD Directives Program*, 28 October 2007, para. 4). The *Manual for Courts-Martial* is reviewed annually, and updated and reissued as needed, by executive order. The Uniform Code of Military Justice is amended when necessary by Congress, most often in response to requests from the DOD but also as a result of external suggestions (as in the most recent major change to the code, the adoption of a revised sexual assault code in the new Article 120). Issuances that levy requirements or restrictions on the public, federal or government employees outside the DOD, and/or reserve components, or that have public or political interest should be considered for publication in the *Federal Register*. Publications addressing homosexual conduct have public and political interest that may mandate publication in the *Federal Register* for public comment. In general, a standard notice-and-comment period should be observed in revising these publications.

Recommendations

After the issuance of an executive order suspending all investigations, separations, and other personnel actions under 10 U.S.C. § 654 and its implementing regulations, the secretary of defense would issue appropriate guidance to implement the order. After that initial step, an orderly review of the relevant publications would ensue. No change is required to the military's criminal law or procedure, because no criminal statute or provision of the *Manual for Courts-Martial* (2008) makes specific references to homosexual conduct.[3] Publications that govern discharge under the homosexual conduct policy should be canceled or withdrawn.[4]

Publications related to the collateral consequences of the homosexual conduct policy should be reviewed to ensure compliance with a revised policy. The most extensive of those modifications will involve personnel/human resources management publications. Because of the hierarchy of tasking in the departments, however, most changes are generated as a matter of course once the initial guidance has been issued. The Department of Army (DA) publications range from administrative to technical and equipment publications and miscellaneous publication of such historical documents. Some of these will be unaffected by the executive order (EO), while others will require more extensive revision. Likewise, educational and training publications related to the homosexual conduct policy should be withdrawn and revised accordingly. There will be no need to train personnel on a policy that is no longer in effect.

Some observers have suggested that a change in the policy will require extensive retraining to prevent or limit harassment or abuse of openly gay or lesbian service members. Yet training materials already in use include specific instruction prohibiting harassment on the basis of sexual orientation.[5] This existing training is carried out during recruit training and officer candidate training, at intervals during the course of an individual's service, and upon reenlistment and is incorporated into the common task and common skills programs of the services.[6] As a result, the regulations directly speaking to training and EO issues are already institutionalized in regulations and functions. This means the functional elements of the policy and the regulations that set them can be modified from currently existing publications and tasking. Selected authorities include:

- DODI 1332.14, *Enlisted Administrative Separations* (28 August 2008)
- Air Force Instruction (AFI) 36-3208, *Administrative Separation of Airmen* (9 July 2004)
- AR 635-200, *Personnel Separations: Active Duty Enlisted Separations* (6 June 2005)
- Naval Military Personnel Manual (MILPERSMAN), Article 1920-040, *Involuntary Separation Pay (Non-Disability) Eligibility Criteria and Restrictions* (22 November 2005)

- Naval Military Personnel Manual, Article 1910-148, *Separation by Reason of Homosexual Conduct* (16 June 2008)
- Marine Corps Order P1900.16E, *Marine Corps Separation and Retirement Manual* (MARCORSEPMAN), para. 6207 (6 June 2007)
- Coast Guard Personnel Manual, *Homosexual Conduct*, COMDTINST M1000.6A (18 June 2007)
- DODI 5505.8, *Defense Criminal Investigative Organizations and Other DoD Law Enforcement Organizations Investigations of Sexual Misconduct* (24 January 2005)
- AR 25-30, *The Army Publishing Program* (27 March 2006)
- AR 600-20, *Army Command Policy* (18 March 2008)
- Marine Corps Administrative Message, R 220745Z, 2 August, MAR-ADMIN 451/02, Subject: Homosexual Conduct Policy Tasks and Responsibilities
- Chief of Naval Operations Instruction 5354.1F, *Navy Equal Opportunity* (25 July 2007)
- 42 U.S.C. § 217
- 33 U.S.C. § 3061

Organizational Changes That Should Accompany Policy Change

Social science research has proved invaluable to the US armed forces in confronting the challenges of racial and gender integration. The knowledge gained from these experiences, supplemented with insights from social science research that has focused specifically on sexual orientation and on the open service of gays and lesbians in militaries abroad, suggests a relatively small number of general guidelines for successfully implementing a new policy that permits openly lesbian, gay, and bisexual personnel to serve. These guidelines are listed below, with references to relevant bibliographic sources appended at the end of this report.

1. **The new policy should be stated in clear and simple terms that will be easily understood by all personnel.**

2. **The new policy should apply a single standard of conduct to all personnel, regardless of their sexual orientation.** The same standards for conduct should be applied to all personnel without regard to their sexual orientation or gender. The acceptability and appropriateness of specific conduct should be judged by a single standard, regardless of the sexual orientation or gender of the individual(s) involved. Implementing the policy will require that personnel receive guidance in setting such a standard, for example, explaining that mere disclosure of information that potentially reveals one's sexual orientation (such as

one's marital status, the gender of one's spouse or romantic partner, or
one's membership in a particular social or community group) does not
constitute misconduct. In addition, regulations for implementing a new
policy should emphasize that:

- each individual, regardless of sexual orientation, is to be judged on
  the basis of her or his performance relevant to military goals;
- all personnel must respect one another's privacy;
- interpersonal harassment—whether verbal, sexual, or physical—will
  not be tolerated, regardless of the gender or sexual orientation of the
  people involved;
- no service member will be permitted to engage in conduct that undermines military effectiveness.

3. **The benefits of the new policy for the armed forces and for individual
   personnel must be made clear.** Policies imposed from outside an
   organization can meet with resistance if they are perceived as incompatible
   with organizational culture. A new policy will work best if personnel are
   persuaded that it will not be harmful to the armed forces or to themselves,
   and may even result in gains. Toward this end, explanations of the new
   policy should be framed using themes reflecting military culture, such as
   the military's pride in professional conduct, its priority of mission over
   individual preferences, its culture of hierarchy and obedience, its norms
   of inclusion and equality, and its traditional "can do" attitude. In this
   regard, useful strategies can be drawn from past experiences with racial
   integration. In a 1973 training manual, for example, the goals of racial
   integration were framed in terms of accomplishing the Army's mission:

   > Equal and just treatment of all personnel exerts direct and favorable in-
   > fluence on morale, discipline, and command authority. Since these key
   > factors contribute to mission effectiveness, efforts to ensure equal treat-
   > ment are directly related to the primary mission.[7]

4. **Implementation plans for the new policy should include both pressure
   for compliance and support for effective implementation.** Compliance
   with the new policy will be facilitated to the extent that personnel
   understand that enforcement will be strict and that noncompliance
   will carry high costs, and thus perceive that their own self-interest lies
   in supporting the new policy. Consequently, the implementation plan
   should include clear enforcement mechanisms and strong sanctions
   for noncompliance, as well as support for effective implementation in
   the form of adequate resources, allowances for input from unit leaders
   for improving the implementation process, and rewards for effective
   implementation. Toward this end, the Defense Department should work
   to identify the most potent "carrots" and "sticks" for implementing the
   new policy. These include:

- the specific sanctions and enforcement mechanisms that will most effectively promote adherence to the policy;
- supporting mechanisms and resources that will be needed to assist personnel with enacting change; and
- the types of surveillance and monitoring of compliance with the new policy that will be most effective at different levels in the chain of command.

5. **Upper-level commanders must send strong, consistent signals of their support for the new policy and their commitment to ensuring compliance with it.** Commanders will play a critical role in supporting the junior ranking personnel who actually implement a policy, ensuring that the latter come to view it as consistent with their own self-interest and with their own self-image as members of a military culture. Thus, a new policy's effectiveness will depend on repeated strong statements of clear support from the highest levels of leadership.

6. **Junior ranking personnel must understand that their ongoing successful implementation of the policy will be noticed and rewarded and that breaches of policy by their subordinates will be considered instances of leadership failure.** Here again, strategies can be adapted from the military's efforts at racial integration. For example, the same training manual cited above clearly linked leadership abilities with successful implementation of policies for racial equality. After asserting that effective implementation of racial equality policies was integral to the accomplishment of the Army's mission and maintenance of the welfare of troops, the manual defined leadership success in terms of policy implementation:

> To a large extent, your success as a leader in the Army is going to depend on your ability to take men from a great variety of racial and cultural backgrounds, with all their racial suspicions and hostilities, and create in them the unity of spirit and action necessary for an effective fighting force. If you fail in this one task, you will have failed in creating high morale, *esprit*, unit efficiency, as well as failing to generate respect for your leadership by your troops. Your job, then, requires that you learn how to carry out your responsibilities for implementing basic Army policy regarding equal opportunity and treatment. If you do not know how, then your job is to learn.[8]

7. **Unit leaders should receive adequate training so they can address and solve challenges related to implementation.** Such training should stress that successful implementation of the policy is expected while imparting the knowledge and skills necessary to anticipate and identify implementation problems and to make adjustments that address implementation problems and improve the implementation process. Any discretion accorded to unit leaders in deciding how best to correct implementation problems must be bounded by behavioral monitoring and strict enforcement of a code of professional conduct.

8. **Unit leaders must be provided clear procedures for reporting problems, and they must believe that their superiors value accurate information about implementation problems.** It should be made clear that merely experiencing initial difficulties in implementing the new policy does not indicate a failure of leadership, provided that these problems are reported and appropriate steps are taken to resolve them.

9. **Plans should be developed for effectively monitoring and evaluating the new policy once it has been implemented.** It will be important to identify the key variables to be tracked for policy evaluation so that baseline data can be collected before a new policy is enacted. Examples of possible variables for monitoring include the number of openly gay or lesbian personnel serving, measures of unit performance (monitored in a way that will permit comparisons between units that do and do not have openly gay personnel and within-unit comparisons before and after having openly gay personnel), and incidents of anti-gay harassment and violence. In addition, conducting regular surveys of officers' and enlisted personnel's knowledge and understanding of the new policy, their attitudes toward it, and their experiences with it could be valuable for monitoring compliance, identifying problems, and formulating solutions.

Responses to 1993 Questions by Senator Sam Nunn

During a 27 January 1993 speech on the Senate floor, former Senator Sam Nunn posed a string of questions that he said would need to be answered before allowing military service by openly gay personnel. Some of his questions are answered elsewhere in this report or have been overtaken by changes in American society and abroad. For example, Nunn asked, "What has been the experience of our NATO allies and other nations from around the world? Not just in terms of the letter of their laws and rules, but the actual practice in their military services on recruiting, retention, promotion, and leadership of military members?" Elsewhere in this report we explain that none of the 24 foreign militaries that allow service by openly gay personnel has reported any overall detriment to recruiting, retention, cohesion, or any other aspect of readiness. A number of Nunn's "thorny questions," however, remain. We answer those questions here:

1. As society changes, should our military services reflect those changes in society? Even if civilians believe openly gay people should be allowed to serve, isn't that irrelevant? Military effectiveness will suffer if we make the military more like civilian society.

Our rules about military service have always reflected changes in society, and all of the national polls on the issue—more than a dozen—conducted over the past five years have shown that between 56 and 81 percent of the public favors allowing openly gay people to serve. Although that alone is an insufficient reason to change the law, military researchers have rightly worried about the widening of the "civil-military gap" and the impact of that gap on the mutual support of the civilian sector and the military. Furthermore, research shows that the current policy does not serve its intended purpose and creates burdens on individuals and the military. Changes in society merely punctuate the policy's ineffectiveness.

2. Should the military have a single code of conduct that applies to conduct between members of the same sex, as well as members of the opposite sex? Or are we going to have two separate codes of conduct for each of those groups?

The military already has a single code of conduct, which, after "don't ask, don't tell" is eliminated, will apply to all troops, straight and gay. This is a sufficient code to govern the behavior of all military members when applied equitably.

3. What if a gay service member makes a romantic overture to a straight colleague? What if a gay service member openly dates someone of the same sex on post or on base?

Asking for a date or conducting a romantic relationship should be governed by the same regulations that regulate heterosexual conduct. Standards should be the same for all service members and should not make distinctions based on sexual orientation.

4. What about displays of affection that are otherwise permissible while in uniform, such as dancing at a formal event?

In the British military, a servicewide code of conduct prohibits any behavior in the workplace that would compromise a unit's cohesion or readiness. Commanders are given discretion to apply that code on a situation-by-situation basis. As for non-workplace social events, the British have found that leadership, a norm of discretion among both gay and heterosexual service members, and the wish of military members to conform to their surrounding culture have taken care of almost every conceivable problem. In the British case, both gay and straight service members generally understand which conduct is appropriate and, based on traditions of honor, discipline, exemplary conduct, respect, and judgment, know how to avoid conduct that could be prejudicial to good order and discipline, whether on duty or at social events. When they fail to exercise proper conduct, existing disciplinary codes against conduct that is prejudicial to good order and discipline are enforced against them. In the United States, it is reasonable to expect that the military will face only minor adjustment problems that can be handled in the same way other personnel problems are handled.

5. What rules, if any, should be adopted to prohibit harassment on the basis of sexual orientation?

Standards governing sexual and other forms of harassment should be the same for all service members and should not mention sexual orientation. The military's equal opportunity system is capable of addressing this issue if given the appropriate authority to do so. The system does not involve lawsuits, and service members are barred from actions in tort incident to military service. Accordingly, the military equal opportunity system exists for remedies, not damages. Given as well that sexual harassment is sexual harassment regardless of the gender of the offending party, equal opportunity enforcement for gay and lesbian service members could easily be incorporated into extant equal opportunity systems and duties.

6. Should homosexual couples receive the same benefits as legally married couples? For example, nonmilitary spouses now are entitled to housing, medical care, exchange and commissary privileges, and similar benefits. Military spouses also benefit from policies that accommodate marriages, such as joint assignment programs. If homosexual couples are given such benefits, will they also have to be granted to unmarried heterosexual couples?

Not under current US law. The military, like all federal agencies, must comply with federal law with respect to marriage and partner benefits. The Defense Department currently relies on the Defense of Marriage Act as a controlling authority for its personnel decisions regarding civilian employees' same-sex partners. The same authority would govern decisions regarding service members' same-sex partners.

7. If discrimination is prohibited, will there be a related requirement for affirmative action recruiting, retention, and promotion to compensate for past discrimination?

No. Policy should be directed toward the future effectiveness of the armed forces, not historical questions, and new provisions should, in general, apply prospectively. With regard to those who have been separated and whose discharge did not involve misconduct and who still meet military standards for enlistment, the new statute should include a provision to waive reenlistment bars that exist in current law and permit correction of military service records if necessary.

Draft Executive Order Suspending
Discharges for Homosexual Conduct

By the authority vested in me as president by the Constitution and the laws of the United States of America, in order to retain members of the armed forces essential to national security, I hereby order as follows:

Sec. 1. Definitions. As used in this order:

1. "Implementing regulations" means Department of Defense Instruction 1304.26, *Qualification Standards for Enlistment, Appointment, and Induction* (11 July 2007); DOD Instruction 1332.14, *Enlisted Administrative Separations* (28 August 2008); DOD Instruction 1332.30, *Separation of Regular and Reserve Commissioned Officers* (11 December 2008); and all regulations of the armed forces issued under the authority of these instructions.

2. 10 U.S.C. § 654, "Policy Concerning Homosexuality in the Armed Forces," means the federal law commonly referred to as "don't ask, don't tell."

Sec. 2. Authority of the President. Under Article II, Section 2, Clause 1 of the Constitution of the United States, the president has authority as commander in chief to retain members of the armed forces serving under his command when essential to the national security of the United States. Under 10 U.S.C. § 123, "Authority to Suspend Officer Personnel Laws During War or National Emergency," and § 12305, "Authority of President to Suspend Certain Laws Relating to Promotion, Retirement, and Separation," Congress also has given the president authority to suspend any provision of law relating to the separation of any member of the armed forces who the president determines is essential to the national security of the United States, during any period of national emergency in which members of a reserve component are serving involuntarily on active duty.

Sec. 3. Findings.

1. Prior Proclamations and Executive Orders. On 14 September 2001, the president issued Proclamation 7463, Declaration of National Emergency by Reason of Certain Terrorist Attacks, and Executive Order 13223, Ordering the Ready Reserve of the Armed Forces to Active Duty.

2. Members of Reserve Components Serving on Active Duty. As of 7 April 2009, there were 93,993 members of reserve components or retired members serving on active duty after involuntary activation.

3. Military Readiness and National Security. Retention of members of the armed forces who may be subject to separation under the authority of 10 U.S.C. § 654, "Policy Concerning Homosexuality in the Armed Forces," is essential to the national security of the United States.

Sec. 4. Suspension of 10 U.S.C. § 654. Effective immediately, all investigations, separation proceedings, or other personnel actions conducted under the authority of 10 U.S.C. § 654 or its implementing regulations shall be suspended. No adverse action shall be taken under the authority of 10 U.S.C. § 654 or its implementing regulations after this period of suspension has ended if the adverse action is based on conduct engaged in or statements made during this period of suspension. This provision does not bar investigations, personnel actions, or disciplinary proceedings for misconduct.

Sec. 5. <u>Review of Implementing Regulations</u>. During this period of suspension, the secretary of defense shall review all implementing regulations prescribed under the authority of 10 U.S.C. § 654(b) in light of the Ninth Circuit Court of Appeals decision in *Witt v. Department of the Air Force*, 527 F.3d 806 (9th Cir. 2008). The secretary of defense shall determine whether the implementing regulations should be revised and, if necessary, whether amendments to 10 U.S.C. § 654 should be recommended for further consideration by Congress.

Sec. 6. <u>Entry Standards</u>. The secretary of defense shall ensure that the standards for enlistment and appointment of members of the armed forces reflect the policies set forth in this order.

Sec. 7. <u>General Provisions</u>. Nothing in this order shall prejudice the authority of the secretary of defense or military commanders to maintain good order and discipline as provided under other laws of the United States or other regulations of the armed services, provided such laws and regulations are enforced in a neutral manner, without regard to sexual orientation or the homosexual or heterosexual nature of conduct.

BARACK OBAMA

THE WHITE HOUSE,

[date]

Updated Appendix: A Note from the Director[9]

Aaron Belkin and Nathaniel Frank
The Palm Center

Some who claim that allowing gay troops to serve openly would compromise military readiness have recently sought to create the impression that there is serious scholarly evidence supporting this position. As part of their efforts, they aim to discredit the overwhelming evidence showing that openly gay service works. A primary criticism has focused on our center at the University of California in Santa Barbara, the Palm Center. Proponents of the gay ban frequently imply that our center is the only group whose research refutes claims that openly gay service would undermine the military. By casting the Palm Center as an activist organization, those who disagree with us hope to undercut inconvenient facts surrounding the debate. Yet even if the Palm Center had never been established, the research record would still reach the same conclusion: allowing gay men and lesbians to serve openly will not harm the military.

As we note in this chapter, a significant number of official military studies as well as research by reputable institutes in addition to Palm, including the RAND Corporation and the military's own Personnel Security Research and Education Center (PERSEREC), arrive at the same conclusion. Indeed, in October 2009, *Joint Force Quarterly* published an award-winning study by a National War College graduate and active-duty colonel who concludes that lifting the ban will not harm the military.[10] We believe the reason that so many scholars converge on the same finding is that the preponderance of evidence points in the same direction. Researchers at RAND, PERSEREC, and Palm, as well as other military and civilian studies listed here, and the well-regarded peer-reviewed journals like *International Security*, *Parameters*, and *Joint Force Quarterly* that have published those studies, were not working as part of a united front, yet all reached the same conclusions.

Despite the overwhelming evidence on one side of the ledger, however, there are still those who continue to argue that a repeal of the gay ban would cause harm. One of the most recent published articles advocating this position is by the president of the Center for Military Readiness, Elaine Donnelly, entitled "Constructing the Co-Ed Military."[11] Given that Ms. Donnelly is the leading voice of the opposition, we would like to focus on several of her claims, which we contend are not empirically sound. For those interested, a complete critique of her article can be found in the same journal in which her piece originally appeared.[12]

Wartime Service of Gay Soldiers

Fact: The military has routinely sent known gays and lesbians to war, despite rules forbidding known gays from serving under the assumption they are an "unacceptable risk" to the mission.

Opponents of gay service sometimes claim that the evidence of this phenomenon is thin or unpersuasive, but it is overwhelming. The evidence comes not just from anecdotes or from gay activists but from military experts. Retired Gen. John Shalikashvili, for instance, former chairman of the joint chiefs of staff, wrote in a 2009 *Washington Post* op-ed that "enforcement of the ban was suspended without problems during the Persian Gulf War, and there were no reports of angry departures."[13] Hoping to undermine the credibility of even the highest uniformed officer in the country, detractors have said that, since some Defense Department officials denied sending known gays to war, it must not have happened; or that because gays were officially exempted from the stop-loss order suspending separations during wartime, enforcement could not have been relaxed. But lawyers for gay troops cite at least 17 cases of service members in the first Iraq War who told their superiors they were gay but were informed they would still have to deploy. One lesbian reservist was even told she would have to provide documentation that she tried to marry another woman if she was to prove she was gay, even though, at the time, same-sex marriage was not legal anywhere in the world.

In the six months after the war, over 1,000 gays were discharged, many of whom were known to be gay at the time they were sent to fight. Citations of known gays being ordered by superiors to stay in the service come from a wide range of unconnected sources, including the Congressional Research Service, *Stars and Stripes*, the *Wall Street Journal, Minneapolis Star Tribune, Boston Globe, San Francisco Chronicle*, and heavily documented books such as *Conduct Unbecoming* and *Unfriendly Fire*.[14] According to a *Boston Globe* investigation, following 9/11 the military allowed an increasing number of service members identified as gay to remain in uniform—12 in 2003, 22 in 2004, 36 in 2005. And these were only the reported ones.[15] In 2006 and 2007, the Navy twice deployed a gay Hebrew linguist, Jason Knight, despite his public acknowledgment that he was gay. His dismissal form was marked "completion of service" rather than homosexual conduct, thus ensuring the Navy would be able to redeploy him in the future. Only after the Sailor became the subject of an article in *Stars and Stripes*, a military newspaper, did the Navy finally discharge him.[16]

Even David Burrelli, a congressional researcher who testified to the Senate about the "causes" of homosexuality, lumping it in with "asexuality, fetishes, and other paraphilias," admitted the military sent known gays to war. Although Burrelli said he could not confirm specific allegations cited in certain newspaper reports, he was persuaded by enough evidence of the phenomenon that he concluded, "The situation that arises during a time of deployment place[s] homosexuals in a no-win situation. They are allowed or ordered to serve at the risk of their own lives with the probability of forced discharge when hostilities end if their sexuality becomes an issue. By deploying suspected homosexuals with their units, the services bring into question their own argument that the presence of homosexuals seriously impairs the accomplishment of the military mission." He does not say "if" the military deploys known gays, but that, "by" doing so, the military undercuts its own argument against gay inclusion.[17]

According to the Army commander's handbook for reserve Soldiers obtained in 2005, there was no ambiguity. Under the section entitled "Personnel Actions during the Mobilization Process," it says that in cases of homosexuality, "if discharge isn't requested prior to the unit's receipt of alert notification, discharge isn't authorized. Member will enter AD [active duty] with the unit." When confronted with the document, the Defense Department admitted that it knowingly sent gays to war in the Middle East. Kim Waldron, a spokesperson at the US Army Forces Command at Fort McPherson, said the reason was to deny Soldiers an opportunity to leave the military on false pretenses. "The bottom line," she said, "is some people are using sexual orientation to avoid deployment. So in this case, with the Reserve and Guard forces, if a soldier 'tells,' they still have to go to war and the homosexual issue is postponed until they return to the U.S. and the unit is demobilized."[18] The rationale for sending known gays to war is to avoid giving troops—gay or straight—a "get out of jail free" card. Nevertheless, that doesn't change the fact that the military sends known gays to fight. If revealing that you are gay is a false pretense to leave the military, then one is left to wonder how it can also be a good reason to be kicked out.

Finally, the history of past wars and their discharge statistics make abundantly clear that known gays are sent to fight. In fact, during every war this country has fought, the gay ban has been relaxed and sometimes totally ignored or suspended. During World War II, the Army ordered commanders to "salvage" Soldiers who were facing separation for homosexual conduct with the aim of "conserving all available manpower," to cancel discharges, and to make convicted "sodomists" eligible for reassignment after prison. A psychiatric study during the war found that it was a common practice in the Army and Navy to permit virtually all gay troops to serve. In the peacetime years between World War II and 1950, the ousting of gays more than tripled. Yet during the Korean War, discharges in the Navy fell by half. In 1953, the year the truce was signed, they more than doubled again, and the same went for Vietnam, when discharges plummeted during the biggest buildups of troop strength in the late 1960s. In our own time, in addition to the numerous reports of known gay service during the first Iraq War, the discharge statistics since 1994 tell the same indisputable story of relaxing enforcement of the gay ban. Gay discharge rates increased nearly every year starting in 1994 and peaked in 2001, when the nation was attacked. When the nation went to war, those figures, as in the past, fell nearly every year since.[19]

Foreign Militaries' Experience with Gay and Lesbian Troops

A central part of the debate about openly gay service in the United States centers around whether those foreign militaries that are most similar to our own have experienced an overall detriment to readiness upon lifting their bans. The Palm Center has published four extensive studies on this question, focusing on Israel, Britain, Canada and Australia. We interviewed every available expert—more than 100 in total—including those who predicted prior to lifting their bans that disaster would ensue. These experts included defense ministry officials, senior officers, enlisted personnel, distinguished scholars, politicians, journalists, gay rights activists, and antigay activists. We examined hundreds of government documents, nongovernmental organization (NGO) reports, and newspaper articles to learn all the information we could about each case.[20]

We could not identify a single piece of evidence suggesting that any foreign military had experienced any overall detriment to readiness, cohesion, morale, or recruiting as a result of the lifting of a gay ban. Even experts such as Prof. Christopher Dandeker, one of the most well-regarded scholars of the British military, told us that they had been wrong in predicting negative consequences. Donnelly and colleagues, however, imply that it is only homosexual activists who report that policy transitions have been successful. In a recent study, Donnelly included a single footnote ostensibly confirming her point that the British integration of gays and lesbians had not been successful. The footnote referenced five media stories. But none of those stories linked gay troops to the problems cited.[21]

After the publication of that study, proponents of the ban identified a single newspaper article which, they maintain, finally proves their point. The article, published in 2007 by the *Daily Mail*, is entitled "Lifting Ban on Gays in Armed Forces Caused Resignations, Report Reveals." The article describes a previously unreported 2002 study by the Ministry of Defence and says that "Britain's armed forces faced a spate of resignations in protest when the government lifted the ban on homosexuals serving in the military." If true, the evidence would provide some support for their point. So we contacted the Directorate of Service Personnel Policy at the British Ministry of Defence to ask about the *Daily Mail* article. In response, we received an email which stated, "We were irritated by the article because it put a very negative slant on what was, in reality, a positive outcome."

Specifically, according to the 2002 report:

Navy: "When first announced the change in policy was not openly welcomed by many, but reaction was generally muted. Since then it has been widely agreed that the problems initially perceived have not been encountered, and for most personnel sexual orientation is a 'non-issue.'"

Army: "The general message from COs [commanding officers] is that there appears to have been no real change since the new policy was announced."

Air Force: "All COs agreed that there had been no tangible impact on operational effectiveness, team cohesion, or Service life generally."

As to the alleged "spate of resignations," what the Ministry report actually says is that "there remains some disquiet in the Senior Ratings' Messes concerning the policy on homosexuality within the Service. This has manifested itself in a number of personnel electing to leave the Service, although *in only one case* was the policy change cited as the only reason for going. Nonetheless, homosexuality is not a major issue and, to put the effect of the policy change into context, the introduction of Pay 2000 and pay grading caused a far greater reaction."[22] It is also worth noting that the *Daily Mail* is a conservative newspaper.

Polling on Gays and Lesbians in the Military

The Palm Center publishes all of the data it uncovers, regardless of the conclusions that evidence sustains. One good example is a 2006 Zogby poll of 545
service members,[23] which the opposition often cites to support their arguments
while simultaneously criticizing methodology of the poll to cast doubt on the
results they don't like. A common criticism is that the Zogby poll did not use a
random sampling strategy. But without the Pentagon's cooperation, no scholars
have been able to draw a random sample of military personnel. Hence, scholars
on both sides of the debate must develop techniques that are considered "next-
best" sampling strategies to assemble respondents who will answer their surveys and will best reflect the population they wish to study.[24]

Zogby, for example, used statistical weights to approximate a representative
sample of military respondents—mostly male, mostly conservative, and mostly
enlisted. Using statistical weights to approximate a randomly drawn sample is
less compelling than random sampling itself, but is a commonly used technique
when random access is not available, and is a much more scientific approach
than that used by a *Military Times* poll which is often cited by proponents of
the gay ban. Some have cast further doubt on the Zogby findings because the
poll was administered to a sample drawn "from a purchased list of U.S. Military
Personnel," with skeptics wrongly assuming this cannot be true because "the
U.S. military does not sell or provide access to personnel lists."[25] This is a misunderstanding of how the polling process works. The list in question was not
purchased or obtained *from* the military but was obtained from vendors who
compile such lists. The panel of potential respondents included more than one
million Americans, some of whom were service members. Zogby then used
statistical weights to draw a sample of service members from that panel.

The context of any research is also key. Findings are only as good as the
methodology from which they are derived. The unscientific *Military Times* poll
found that 10 percent of service members said they would leave the military if
the gay ban were repealed. Ten percent is a large number, and if it's true that that
many people would leave the military if the ban were lifted, that statistic should
be taken seriously. But this inference is, again, a misunderstanding of how polls
work. In the 1990s, polls suggested that two-thirds of male Canadian and British service members would refuse to work with gays if bans in those countries
were lifted.[26] But when the changes were made, no more than a handful of service members left the forces in each case.[27] Any social psychologist can explain
why: polls measure attitudes; they do not predict behavior. Claims about likely
behavior do not always correlate with people's actual behavior, particularly when
there is institutional pressure to respond to a poll in a certain way. In a famous
1934 experiment, a white scholar accompanied a Chinese friend to over 250
hotels, motels, and restaurants throughout the United States. All but one served
both individuals without problem. The scholar later contacted all the establishments to ask if they would serve a Chinese patron. More than 90 percent said
they would not.[28] Just because 10 percent of service members say they will leave

the military when the gay ban is repealed does not mean that 10 percent will actually leave. If analogous situations revealed that such behavior had actually taken place, such as a mass exodus from the CIA, police or fire departments, or foreign militaries when they lifted their bans, this would lend some credibility to claims that the policy change might prompt a large personnel loss. We submit there isn't an expert anywhere in the world who believes that any foreign military, police force, or fire department has suffered an overall detriment to cohesion, readiness, or morale as a result of a decision to lift a gay ban.

Final Thought

Earlier this year, a political group organized a statement signed by more than 1,000 retired general and flag officers warning President Obama that repealing "don't ask, don't tell" could "break the All Volunteer Force."[29] Shortly thereafter, General Shalikashvili scolded the signatories for speaking on the basis of emotion rather than evidence, writing that, "Not only is there no evidence to support these conclusions, but research shows conclusively that openly gay service members would not undermine military readiness."[30]

Despite how easy it is to make extreme claims based on exaggerated fears, the time has come for those who believe that lifting the gay ban would harm the military to take an honest look at the evidence.

Notes

1. See, e.g., DODI 1332.14, *Enlisted Administrative Separations*, 28 August 2008; AR 635-200, *Personnel Separations: Active Duty Enlisted Separations*, 6 June 2005; AFI 36-3208, *Administrative Separation of Airmen*, 9 July 2004; MILPERSMAN 1910-148, *Separation by Reason of Homosexual Conduct*, 16 June 2008, Ch. 23; MCO P1900.16F, *Marine Corps Separation and Retirement Manual* (MARCORSEPMAN), 6 June 2007, § 6207; COMDTINST M1000.6A, *Coast Guard Personnel Manual*, 8 January 1988, Ch. 12.E, "Homosexual Conduct"; AR 600-20, *Army Command Policy*, Ch. 4–19; and AR 635-200, Ch. 15. The other uniformed services, uniformed corps of the US Public Health Service, and the National Oceanographic and Aeronautic Administration do not currently have homosexual conduct policies based on 10 U.S.C. § 654, as the statute only applies to the "armed" forces. Their members may be subject to these policies when serving with the armed forces or in Naval or Coast Guard operations.

2. See, e.g., DODI 5505.8, *Defense Criminal Investigative Organizations and Other DoD Law Enforcement Organizations Investigations of Sexual Misconduct*, 24 January 2005.

3. Although UCMJ Articles 125 (Sodomy), 133 (Conduct unbecoming an officer and a gentleman), and 134 (General article) have been used to punish homosexual conduct, current statutory language, judicial opinions, and executive guidance do not expressly target same-sex sexual misconduct as compared to opposite-sex sexual misconduct. Military courts are already in the process of both construing the new provisions of Article 120 (effective since 1 October 2007), which transformed the UCMJ's rape statute into a series of specified sex crimes, and reinterpreting the prosecution of sodomy in light of the Supreme Court's decriminalization of civilian sodomy in *Lawrence v. Texas*, 539 U.S. 558 (2003), and the US Court of Appeals for the Armed Forces post-*Lawrence* decision, *United States v. Marcum*, 60 M.J. 198 (C.A.A.F. 2004).

4. Each of the services in the Department of Defense Direction has personnel management publications which include the homosexual conduct provisions that set the policy, prescribe how it is to be applied, and describe the procedures for separating enlisted members and officers. Similarly,

the US Coast Guard, which operates as a component of the Department of Homeland Security unless placed under the Department of Defense during time of conflict, enforces 10 U.S.C. §654 under its own regulations. For example, AR 600-200, *Army Command Policy*, lays out a detailed discussion of "don't ask don't tell."

5. These guidelines, for example, appear in materials used to train service members on the homosexual conduct policy itself. Furthermore, since 2000, in addition to the mandated training offered to service members from the point of enlistment and throughout their service, the services have prohibited the harassment of service members based on their sexual orientation or perceived sexual orientation and tasked the inspectors general with investigating harassment based on sexual orientation. See, for example, the US Navy Inspector General Web page addressing homosexual conduct issues including harassment: www.ig.navy.mil/complaints/Complaints(homosexuality).htm.

6. See, e.g., MARADMIN 451/02, "Homosexual Conduct Policy Tasks and Responsibilities," 22 August 2002, setting out a detailed schedule for training.

7. Department of the Army, *Improving Race Relations in the Army: Handbook for Leaders*, Pamphlet Number 600-16, 1973, 2.

8. Ibid.

9. The following appendix did not appear in the original May 2009 publication.

10. Om Prakash, "The Efficacy of 'Don't Ask, Don't Tell,'" *Joint Force Quarterly* 55 (4th Quarter, 2009): 88–94.

11. Elaine Donnelly, "Constructing the Co-Ed Military," *Duke Journal of Gender Law & Policy* 14 (2007): 815–952.

12. Jeanne Scheper, Nathaniel Frank, Aaron Belkin, and Gary J. Gates, "'The Importance of Objective Analysis' on Gays in the Military: A Response to Elaine Donnelly's *Constructing the Co-Ed Military*," *Duke Journal of Gender Law & Policy* 15 (2008): 419–448.

13. John M. Shalikashvili, "Gays in the Military: Let the Evidence Speak," *Washington Post*, 19 June 2009.

14. Randy Shilts, "Army Discharges Lesbian Who Challenged Ban," *San Francisco Chronicle*, 19 January 1991; Wade Lambert and Stephanie Simon, "US Military Moves to Discharge Some Gay Veterans of Gulf War," *Wall Street Journal*, 30 July 1991; Doug Grow, "Captain Did Her Duty, Despite Military's Mixed Messages," *Minneapolis Star Tribune*, 16 March 1993; Randy Shilts, "Gay Troops in the Gulf War Can't Come Out," *San Francisco Chronicle*, 18 February 1991; Randy Shilts, *Conduct Unbecoming: Gays and Lesbians in the U.S. Military* (New York: Ballentine Books, 1994), 735–8; and Nathaniel Frank, *Unfriendly Fire: How the Gay Ban Undermines the Military and Weakens America* (New York: St. Martin's Press, 2009), 227–30.

15. Bryan Bender, "Military Retaining More Gays," *Boston Globe*, 19 March 2006.

16. Joseph Giordono, "Discharged Gay Sailor Is Called Back to Active Duty," *Stars and Stripes*, 6 May 2007; and Joseph Giordono, "Navy Bars Outed Gay Sailor from Return to Service," *Stars and Stripes*, 10 June 2007.

17. David F. Burrelli and Charles V. Dale, *Homosexuals and U.S. Military Policy: Current Issues*, Congressional Research Services Report, 13 March 2006, 8.

18. Army Forces Command (FORSCOM) Regulation 500-3-3, vol. III, *Reserve Component Unit Commanders Handbook*, 1999, table 2.1: "Personnel actions during the mobilization process"; and Lou Chibbaro Jr., "Out Gay Soldiers Sent to Iraq," *Washington Blade*, 23 September 2005.

19. Allan Berube, *Coming Out Under Fire: The History of Gay Men and Women in World War Two* (New York: Free Press, 1990), 172, 262; and Randy Shilts, *Conduct Unbecoming: Gays and Lesbians in the U.S. Military* (New York: Ballentine Books, 1994), 70; for recent discharge figures, see www.sldn.org.

20. Aaron Belkin and Jason McNichol, "Effects of the 1992 Lifting of Restrictions on Gay and Lesbian Service in the Canadian Forces: Appraising the Evidence" (Santa Barbara, CA: Center for the Study of Sexual Minorities in the Military, 2000); Aaron Belkin and Melissa Levitt, "The Effects of Including Gay and Lesbian Soldiers in the Israeli Defense Forces: Appraising the Evidence" (Santa Barbara, CA: Center for the Study of Sexual Minorities in the Military, 2000); Aaron Belkin and Jason McNichol, "The Effects of Including Gay and Lesbian Soldiers in the Australian Defence Forces: Appraising the Evidence" (Santa Barbara, CA: Center for the Study of Sexual Minorities in the Military, 2000); and Aaron Belkin and R. L. Evans, "The Effects of In-

cluding Gay and Lesbian Soldiers in the British Armed Forces: Appraising the Evidence" (Santa Barbara, CA: Center for the Study of Sexual Minorities in the Military, 2000); all studies are available at www.palmcenter.ucsb.edu.

21. Donnelly, "Constructing the Co-Ed Military," 926, n. 545.

22. Ministry of Defence, Service Personnel Board, *Tri-Service Review of the Armed Forces Policy on Homosexuality and Code of Social Conduct,* December 2002, 2.

23. Sam Rodgers, *Opinions of Military Personnel on Sexual Minorities in the Military,* Zogby International, December 2006.

24. Language in this paragraph and the next is adapted from "'The Importance of Objective Analysis' on Gays in the Military: A Response to Elaine Donnelly's *Constructing the Co-Ed Military.*"

25. Donnelly, "Constructing the Co-ed Military," 918.

26. Aaron Belkin, "Don't Ask, Don't Tell: Is the Gay Ban Based on Military Necessity?" *Parameters,* Summer 2003, 108–19.

27. Ibid.

28. Lapiere, R. T. "Attitudes versus Actions," *Social Forces* 13 (1934): 230–37.

29. *PBS NewsHour with Jim Lehrer* contacted the four-star officers on the list and discovered that not all of them had, in fact, signed the statement (29 June 2009).

30. John Shalikashvili, "Gays in the Military: Let the Evidence Speak," *Washington Post,* 19 June 2009.

Relevant Sources

On Military Readiness and Decisions to Allow Service by Openly Gay Personnel

Belkin, Aaron. "Is the Gay Ban Based on Military Necessity?" *Parameters* 33 (2003): 108–19.

Belkin, Aaron, and Jason McNichol. "Homosexual Personnel Policy of the Canadian Forces: Did Lifting the Gay Ban Undermine Military Performance?" *International Journal* 56 (2001): 73–88.

Belkin, Aaron, and Jason McNichol. "Pink and Blue: Outcomes Associated with the Integration of Open Gay and Lesbian Personnel in the San Diego Police Department." *Police Quarterly* 5 (2002): 63–95.

Belkin Aaron, and Melissa Levitt. "Homosexuality and the Israel Defense Forces: Did Lifting the Gay Ban Undermine Military Performance?" *Armed Forces and Society* 27 (2001): 541–66.

Belkin, Aaron, and Melissa Sheridan Embser-Herbert. "A Modest Proposal: Privacy as a Rationale for Excluding Gays and Lesbians from the US Military." *International Security* 27 (2002): 178–97.

Canadian Forces. Briefing Note for Director of Public Policy. Ottawa, Canada, 25 August 1995.

Frank, Nathaniel. *Unfriendly Fire: How the Gay Ban Undermines the Military and Weakens America.* New York: St. Martin's Press, 2009.

General Accounting Office (GAO). *DOD's Policy on Homosexuality.* Washington, DC: GAO, 12 June 1992.

———. *Homosexuals in the Military: Policies and Practices of Foreign Countries.* Washington, DC: GAO, 1993.

Herek, Gregory M., Jared B. Jobe, and Ralph M. Carney, eds. *Out in Force: Sexual Orientation and the Military.* Chicago: University of Chicago Press, 1996.

Kier, Elizabeth. "Rights and Fights: Sexual Orientation and Military Effectiveness." *International Security* 24 (1999): 194–201.

MacCoun, Robert J., Elizabeth Kier, and Aaron Belkin. "Does Social Cohesion Determine Motivation in Combat? An Old Question with an Old Answer." *Armed Forces and Society* 32 (2006): 646–54.

Ministry of Defence (MOD). *A Review of the Armed Forces Policy on Homosexuality.* London: British MOD, 31 October 2000.

Moradi, Bonnie, and Laura Miller. "Attitudes of Iraq and Afghanistan War Veterans toward Gay and Lesbian Service Members." Manuscript under review at *Armed Forces and Society,* 2009.

National Defense Research Institute. *Sexual Orientation and U.S. Military Personnel Policy: Options and Assessment.* Santa Monica, CA: RAND, 1993.

Pinch, Franklin. *Perspectives on Organizational Change in the Canadian Forces.* US Army Research Institute for the Behavioral and Social Sciences, 1994.

Pond, Frank. "A Comparative Survey and Analysis of Military Policies with Regard to Service by Gay Persons." In *Policy Concerning Homosexuality in the Armed Forces,* Hearing Held by Senate Armed Services Committee. 103d Cong., 2d Sess. Washington, DC: Government Printing Office, 1993.

Prakash, Om. "The Efficacy of 'Don't Ask, Don't Tell.'" *Joint Force Quarterly* 55 (4th Quarter 2009): 88–94.

Report of the Board Appointed to Prepare and Submit Recommendations to the Secretary of the Navy for the Revision of Policies, Procedures and Directions Dealing with Homosexuality, 15 March 1957 (Crittenden Report).

Report of the General/Flag Officers' Study Group. Santa Barbara, CA: Palm Center, 2008. Reprinted as chapter 7 in this volume.

Sarbin, Theodore R., and Kenneth E. Karols. *Nonconforming Sexual Orientation and Military Suitability.* Monterey, CA: Defense Personnel Security Research and Education Center, 1988.

On Sexual Orientation and Organizational Change

Department of the Army. *Improving Race Relations in the Army: Handbook for Leaders.* Pamphlet Number 600–16. Washington, DC: Department of the Army, 1973.

Herek, Gregory M. "Sexual Orientation and Military Service: A Social Science Perspective." *American Psychologist* 48 (1993): 538–49.

Herek, Gregory M., and Aaron Belkin. "Sexual Orientation and Military Service: Prospects for Organizational and Individual Change in the United States." In *Military Life: The Psychology of Serving in Peace and Combat,* edited by T. W. Britt, A. B. Adler, and C. A. Castro, 119–42. Vol. 4, *Military Culture.* Westport, CT: Praeger Security International, 2005.

Kauth, Michael R., and Dan Landis. "Applying Lessons Learned from Minority Integration in the Military." In *Out in Force: Sexual Orientation and the*

Military, edited by Gregory M. Herek, Jared B. Jobe, and Ralph Carney, 86–105. Chicago, IL: University of Chicago Press, 1996.

Koegel, Paul. "Lessons Learned from the Experience of Domestic Police and Fire Departments." In *Out in Force: Sexual Orientation and the Military*, edited by Gregory M. Herek, Jared B. Jobe, and Ralph Carney, 131–53. Chicago, IL: University of Chicago Press, 1996.

Landis, Dan, Richard O. Hope, and Harry R. Day. "Training for Desegregation in the Military." In *Groups in Contact: The Psychology of Desegregation*, edited by Norman N. Miller and Marilynn B. Brewer, 257–78. Orlando, FL: Academic Press, Inc., 1984.

Sarbin, Theodore R. "The Deconstruction of Stereotypes: Homosexuals and Military Policy." In *Out in Force: Sexual Orientation and the Military*, edited by Gregory M. Herek, Jared B. Jobe, and Ralph Carney, 177–96. Chicago, IL: University of Chicago Press, 1996.

Thomas, Patricia J., and Marie D. Thomas. "Integration of Women in the Military: Parallels to the Progress of Homosexuals?" In *Out in Force: Sexual Orientation and the Military*, edited by Gregory M. Herek, Jared B. Jobe, and Ralph Carney, 65–85. Chicago, IL: University of Chicago Press, 1996.

Zellman, Gail L. "Implementing Policy Changes in Large Organizations: The Case of Gays and Lesbians in the Military." In *Out in Force: Sexual Orientation and the Military*, edited by Gregory M. Herek, Jared B. Jobe, and Ralph Carney, 266–89. Chicago, IL: University of Chicago Press, 1996.

Contributors

Dr. Aaron Belkin is director of the Palm Center and associate professor of political science at the University of California, Santa Barbara. For over a decade, he has played a leading role in the national conversation about "don't ask, don't tell." In that role, he and his colleagues at the Palm Center have broken major stories that have been covered widely by every major television network and newspaper in the United States. He has published peer-reviewed studies on gays in the military in leading journals, including official military journals such as *Parameters*, the official journal of the Army War College. Every year for the past six years, he has delivered lectures on "don't ask, don't tell" at West Point, the Army War College, and the Air Force Academy.

Dr. Nathaniel Frank is the author of *Unfriendly Fire: How the Gay Ban Undermines the Military and Weakens America*. He is a senior research fellow at the Palm Center at the University of California, Santa Barbara, and teaches history on the adjunct faculty at New York University's Gallatin School. His research and publications on gays in the military have appeared in the *New York Times, Washington Post, The New Republic, Slate, USA Today, Los Angeles Times, Huffington Post, Philadelphia Inquirer, Lingua Franca*, and others, and he is a frequent guest on national radio and television shows. Dr. Frank earned his doctorate and master's from Brown University.

Dr. Gregory M. Herek is a professor of psychology at the University of California at Davis. His edited books include *Out in Force: Sexual Orientation and the Military* and *Stigma and Sexual Orientation*. He is a fellow of the American Psychological Association and the Association for Psychological Science. He was the recipient of the 2006 Kurt Lewin Memorial Award for "outstanding contributions to the development and integration of psychological research and social action," presented by the Society for the Psychological Study of Social Issues.

Dr. Elizabeth L. Hillman, JD, is professor of law at the University of California, Hastings College of the Law and a veteran of the US Air Force. She holds both a PhD and JD from Yale University. Her work focuses on US military law and history since the mid twentieth century and the impact of gender and sexual norms on military culture. A veteran of the US Air Force, she taught history at the Air Force Academy and at Yale University before joining the faculty at Rutgers University School of Law. She is the author of *Defending America: Military Culture and the Cold War Court-Martial* and coauthor of *Military Justice Cases and Materials*.

Diane H. Mazur, JD, is professor of law at the University of Florida, a former Bigelow Fellow at the University of Chicago Law School, and a graduate of the University of Texas School of Law. Professor Mazur is a former aircraft and munitions maintenance officer in the US Air Force. She is the author of a forthcoming book, *A More Perfect Military: How the Constitution Can Make Our Military Stronger*.

Bridget J. Wilson, JD, practices law at Rosenstein Wilson & Dean in San Diego and is a veteran of the US Army Reserve. She is a graduate of Creighton University and the University of San Diego School of Law.

CHAPTER 11

FLAG & GENERAL OFFICERS FOR THE MILITARY

30 March 2009

Statement to: President Barack H. Obama and Members of Congress

Subject: Support for the 1993 Law Regarding Homosexuals in the Military (Section 654, Title 10, U.S.C.)

Dear Mr. President and Members of Congress:

In 1993 Congress passed a law (Section 654, Title 10), affirming that homosexuality is incompatible with military service. The law passed with bipartisan, veto-proof majorities in both houses, and federal courts have upheld it as constitutional several times. We believe strongly that this law, which Congress passed to protect good order, discipline, and morale in the unique environment of the military, deserves continued support.

The 111th Congress is likely to take up legislation to repeal the law (Section 654, Title 10) early in 2009. Our past experience as military leaders leads us to be greatly concerned about the impact of repeal on morale, discipline, unit cohesion, and overall military readiness. We believe that imposing this burden on our men and women in uniform would undermine recruiting and retention, impact leadership at all echelons, have adverse effects on the willingness of parents who lend their sons and daughters to military service, and eventually break the All-Volunteer Force.

As a matter of national security, we urge you to support the 1993 law regarding homosexuals in the military (Section 654, Title 10), and to oppose any legislative, judicial, or administrative effort to repeal or invalidate the law.

Very respectfully,

The Undersigned Flag & General Officers

This statement was delivered to Pres. Barack Obama, Pentagon officials, and senior members of Congress on 31 March 2009. Personal signatures are on file with the Center of Military Readiness. As of 4 February 2010, there were 1,163 signatories to the Flag & General Officers for the Military (FGOM) Statement. For further information, see *www.flagandgeneralofficersforthemilitary.com*.

Flag & General Officers for the Military

The following retired Flag & General Officers have signed a statement to the President of the United States and Members of Congress in support for the 1993 law regarding homosexuals in the military (Section 654, Title 10, U.S.C.).

4-Star Rank (51)

Gen E. E. Anderson, USMC *(ret.)*[1]
Gen Robert W. Bazley, USAF *(ret.)*[2]
Gen Walter E. Boomer, USMC *(ret.)*[3]
Gen Arthur E. Brown, Jr., USA *(ret.)*[4]
Gen Edwin H. Burba Jr., USA *(ret.)*[5]
Gen Paul K. Carlton, Sr., USAF *(ret.)*[6]
Gen John R. Dailey, USMC *(ret.)*[7]
Gen Terrence R. Dake, USMC *(ret.)*[8]
Gen James B. Davis, USAF *(ret.)*
Gen John K. Davis, USMC *(ret.)*[9]
Gen John R. Deane, Jr., USA *(ret.)*
Gen Michael J. Dugan, USAF *(ret.)*[10]
Gen Ronald R. Fogleman, USAF *(ret.)*[11]
Gen John W. Foss, USA *(ret.)*
Gen Carlton W. Fulford, Jr., USMC *(ret.)*
Gen Paul F. Gorman, USA *(ret.)*[12]
Gen Richard E. Hawley, USAF *(ret.)*[13]
Adm Ronald J. Hays, USN *(ret.)*[14]
Adm Thomas B. Hayward, USN *(ret.)*[15]
Gen C. A. Horner, USAF *(ret.)*[16]
Adm Jerome L. Johnson, USN *(ret.)*[17]
Gen P. X. Kelley, USMC *(ret.)*[18]
Gen William F. Kernan, USA *(ret.)*[19]
Gen William L. Kirk, USAF *(ret.)*[20]
Gen Frederick J. Kroesen, USA *(ret.)*[21]
Gen James J. Lindsay, USA *(ret.)*[22]
Adm James A. "Ace" Lyons, Jr., USN *(ret.)*[23]
Gen Robert Magnus, USMC *(ret.)*[24]
Adm Henry H. Mauz, Jr., USN *(ret.)*[25]
Gen Louis C. Menetrey, USA *(ret.)*[26]
Gen Edward C. Meyer, USA *(ret.)*[27]
Gen Thomas R. Morgan, USMC *(ret.)*[28]
Gen Carl E. Mundy, Jr., USMC *(ret.)*[29]
Gen Wallace H. Nutting, USA *(ret.)*[30]
Gen Glenn K. Otis, USA *(ret.)*[31]
Gen Joseph T. Palastra, Jr., USA *(ret.)*
Gen Crosbie E. Saint, USA *(ret.)*[32]
Gen Henry H. Shelton, USA *(ret.)*[33]
Gen Robert M. Shoemaker, USA *(ret.)*[34]
Gen Lawrence A. Skantze, USAF *(ret.)*[35]

Adm Leighton W. "Snuffy" Smith, USN *(ret.)*[36]
Gen Carl W. Stiner, USA *(ret.)*[37]
Gen Richard H. Thompson, USA *(ret.)*
Gen John W. Vessey, Jr., USA *(ret.)*[38]
Gen John W. Vogt, USAF *(ret.)*[39]
Gen Louis C. Wagner, Jr., USA *(ret.)*
Gen William S. Wallace, USA *(ret.)*[40]
Gen Volney F. Warner, USA *(ret.)*[41]
Gen Joseph J. Went, USMC *(ret.)*[42]
Gen John A. Wickham, Jr., USA *(ret.)*[43]
Gen Charles E. Wilhelm, USMC *(ret.)*[44]

3-Star Rank (193)

Lt Gen Teddy G. Allen, USA *(ret.)*
Lt Gen Edgar R. Anderson, Jr., USAF *(ret.)*
Lt Gen Edward G. Anderson III, USA *(ret.)*
Lt Gen Marcus A. Anderson, USAF *(ret.)*
Lt Gen Spence M. Armstrong, USAF *(ret.)*
Lt Gen George C. Axtell, USMC *(ret.)*
Lt Gen Donald M. Babers, USA *(ret.)*
Vice Adm Albert J. Baciocco, USN *(ret.)*
Lt Gen Robert J. Baer, USA *(ret.)*
Lt Gen Charles W. Bagnal, USA *(ret.)*
Vice Adm Robert B. Baldwin, USN *(ret.)*
Vice Adm John A. Baldwin, USN *(ret.)*
Lt Gen John L. Ballantyne III, USA *(ret.)*
Lt Gen Jared L. Bates, USA *(ret.)*
Lt Gen Emil R. Bedard, USMC *(ret.)*
Lt Gen Dennis L. Benchoff, USA *(ret.)*
Lt Gen Robert R. Blackman, Jr., USMC *(ret.)*
Lt Gen Paul E. Blackwell, USA *(ret.)*
Lt Gen Arthur C. Blades, USMC *(ret.)*
Lt Gen Harold W. Blot, USMC *(ret.)*
Lt Gen John B. Blount, USA *(ret.)*
Lt Gen Lawrence E. Boese, USAF *(ret.)*
Lt Gen James A. Brabham, USMC *(ret.)*
Lt Gen John N. Brandenburg, USA *(ret.)*
Lt Gen Martin L. Brandtner, USMC *(ret.)*
Lt Gen Devol Brett, USAF *(ret.)*
Vice Adm Edward S. Briggs, USN *(ret.)*
Lt Gen George M. Browning, Jr., USAF *(ret.)*

Lt Gen John D. Bruen, USA *(ret.)*
Lt Gen Peter G. Burbules, USA *(ret.)*
Vice Adm E. A. Burkhalter, Jr., USN *(ret.)*
Lt Gen Richard A. Burpee, USAF *(ret.)*
Lt Gen Tony Burshnick, USAF *(ret.)*
Lt Gen John S. Caldwell, Jr., USA *(ret.)*
Vice Adm James F. Calvert, USN *(ret.)*
Lt Gen William J. Campbell, USAF *(ret.)*
Lt Gen Richard E. Carey, USMC *(ret.)*
Lt Gen Paul K. Carlton, Jr., USAF *(ret.)*
Lt Gen Thomas P. Carney, USA *(ret.)*
Vice Adm Kenneth M. Carr, USN *(ret.)*
Vice Adm K. J. Carroll, USN *(ret.)*
Lt Gen William G. Carter III, USA *(ret.)*
Lt Gen Patrick P. Caruana, USAF *(ret.)*
Lt Gen Carmen J. Cavezza, USA *(ret.)*
Lt Gen Dennis D. Cavin, USA *(ret.)*
Lt Gen Paul G. Cerjan, USA *(ret.)*
Lt Gen Ernest C. Cheatham, USMC *(ret.)*
Lt Gen Richard A. Chilcoat, USA *(ret.)*
Lt Gen George R. Christmas, USMC *(ret.)*
Lt Gen Marc A. Cisneros, USA *(ret.)*
Lt Gen Charles G. Cleveland, USAF *(ret.)*
Lt Gen Charles G. Cooper, USMC *(ret.)*
Lt Gen George A. Crocker, USA *(ret.)*
Lt Gen John S. Crosby, USA *(ret.)*
Lt Gen James W. Crysel, USA *(ret.)*
Lt Gen John M. Curran, USA *(ret.)*
Lt Gen John J. Cusick, USA *(ret.)*
Vice Adm George Davis, USN *(ret.)*
Lt Gen David K. Doyle, USA *(ret.)*
Vice Adm James H. Doyle, USN *(ret.)*
Lt Gen Brett M. Dula, USAF *(ret.)*
Lt Gen Leo J. Dulacki, USMC *(ret.)*
Lt Gen Charles B. Eichelberger, USA *(ret.)*
Lt Gen James R. Ellis, USA *(ret.)*
Lt Gen Robert M. Elton, USA *(ret.)*
Lt Gen William R. Etnyre, USMC *(ret.)*
Lt Gen Bruce L. Fister, USAF *(ret.)*
Lt Gen William Harold Fitch, USMC *(ret.)*
Lt Gen Merle Freitag, USA *(ret.)*
Lt Gen Edward S. Fris, USMC *(ret.)*
Vice Adm Richard C. Gentz, USN *(ret.)*
Lt Gen Alvan C. Gillem II, USAF *(ret.)*
Lt Gen William H. Ginn, Jr., USAF *(ret.)*
Lt Gen Charles P. Graham, USA *(ret.)*
Vice Adm Howard E. Greer, USN *(ret.)*

Lt Gen Wallace C. Gregson, USMC *(ret.)*
Lt Gen Thomas N. Griffin, Jr., USA *(ret.)*
Lt Gen Earl B. Hailston, USMC *(ret.)*
Lt Gen James R. Hall, Jr., USA *(ret.)*
Vice Adm Patrick J. Hannifin, USN *(ret.)*
Lt Gen Edgar S. Harris, Jr., USAF *(ret.)*
Lt Gen Bruce R. Harris, USA *(ret.)*
Lt Gen Henry J. Hatch, USA *(ret.)*
Vice Adm Peter M. Hekman, USN *(ret.)*
Lt Gen Samuel T. Helland, USMC *(ret.)*
Lt Gen Richard C. Henry, USAF *(ret.)*
Lt Gen Fred Hissong, Jr., USA *(ret.)*
Lt Gen Jefferson D. Howell, Jr., USMC *(ret.)*
Lt Gen John I. Hudson, USMC *(ret.)*
Lt Gen Jan C. Huly, USMC *(ret.)*
Lt Gen Neal T. Jaco, USA *(ret.)*
Lt Gen Theodore G. Jenes, Jr., USA *(ret.)*
Lt Gen James H. Johnson, Jr., USA *(ret.)*
Lt Gen Johnny J. Johnston, USA *(ret.)*
Lt Gen Robert B. Johnston, USMC *(ret.)*
Lt Gen James M. Keck, USAF *(ret.)*
Lt Gen Robert Prescott Keller, USMC *(ret.)*
Lt Gen David J. Kelley, USA *(ret.)*
Lt Gen William M. Keys, USMC *(ret.)*
Lt Gen Joseph W. Kinzer, USA *(ret.)*
Lt Gen Jack W. Klimp, USMC *(ret.)*
Lt Gen Bruce B. Knutson, Jr., USMC *(ret.)*
Vice Adm E. R. Kohn, Jr., USN *(ret.)*
Lt Gen Alcide M. La Noue, USA *(ret.)*
Lt Gen Richard D. Lawrence, USA *(ret.)*
Lt Gen James M. Lee, USA *(ret.)*
Vice Adm Tony Less, USN *(ret.)*
Lt Gen Kenneth E. Lewi, USA *(ret.)*
Lt Gen Bennett Lewis, USA *(ret.)*
Lt Gen Frank Libutti, USMC *(ret.)*
Lt Gen James M. Link, USA *(ret.)*
Lt Gen Anthony Lukeman, USMC *(ret.)*
Lt Gen Robert J. Lunn, USA *(ret.)*
Lt Gen Lawson W. Magruder III, USA *(ret.)*
Lt Gen Charles S. Mahan, USA *(ret.)*
Lt Gen William R. Maloney, USMC *(ret.)*
Lt Gen Caryl G. Marsh, USA *(ret.)*
Lt Gen Charles A. May, Jr., USAF *(ret.)*
Lt Gen Frederick McCorkle, USMC *(ret.)*
Lt Gen Gary McKissock, USMC *(ret.)*
Lt Gen Clarence E. McKnight, Jr., USA *(ret.)*
Lt Gen Gary H. Mears, USAF *(ret.)*

Lt Gen John H. Miller, USMC *(ret.)*
Vice Adm Gerald E. Miller, USN *(ret.)*
Lt Gen Robert F. Milligan, USMC *(ret.)*
Lt Gen Harold G. Moore, Jr., USA *(ret.)*
Vice Adm J. P. Moorer, USN *(ret.)*
Lt Gen G. S. Newbold, USMC *(ret.)*
Lt Gen Jack P. Nix, Jr., USA *(ret.)*
Vice Adm John W. Nyquist, USN *(ret.)*
Lt Gen Edmund F. O'Connor, USAF *(ret.)*
Lt Gen David H. Ohle, USA *(ret.)*
Lt Gen Stephen G. Olmstead, USMC *(ret.)*
Lt Gen Allen K. Ono, USA *(ret.)*
Lt Gen Robert L. Ord III, USA *(ret.)*
Lt Gen John P. Otjen, USA *(ret.)*
Lt Gen Dave R. Palmer, USA *(ret.)*
Lt Gen Anthony L. Palumbo, USA *(ret.)*
Vice Adm Jimmy Pappas, USN *(ret.)*
Vice Adm John T. Parker, Jr., USN *(ret.)*
Lt Gen Garry L. Parks, USMC *(ret.)*
Lt Gen Burton D. Patrick, USA *(ret.)*
Lt Gen Ernest D. Peixotto, USA *(ret.)*
Lt Gen John Phillips, USMC *(ret.)*
Lt Gen Charles H. Pitman, USMC *(ret.)*
Lt Gen Benjamin F. Register, Jr., USA *(ret.)*
Lt Gen John H. Rhodes, USMC *(ret.)*
Vice Adm David C. Richardson, USN *(ret.)*
Lt Gen Thomas M. Rienzi, USA *(ret.)*
Lt Gen Randall L. Rigby, USA *(ret.)*
Lt Gen James C. Riley, USA *(ret.)*
Lt Gen Thurman D. Rodgers, USA *(ret.)*
Lt Gen Craven C. Rogers, USAF *(ret.)*
Lt Gen Donald E. Rosenblum, USA *(ret.)*
Lt Gen John B. Sams, USAF *(ret.)*
Vice Adm James R. Sanderson, USN *(ret.)*
Lt Gen Daniel R. Schroeder, USA *(ret.)*
Lt Gen James T. Scott, USA *(ret.)*
Vice Adm James E. Service, USN *(ret.)*
Lt Gen Wilson A. Shoffner, USA *(ret.)*
Vice Adm Robert F. "Dutch" Shoultz, USN *(ret.)*
Lt Gen E. G. Shuler, Jr., USAF *(ret.)*
Lt Gen Jeffrey G. Smith, USA *(ret.)*
Lt Gen Norman H. Smith, USMC *(ret.)*
Lt Gen Lawrence F. Snowden, USMC *(ret.)*
Lt Gen Michael F. Spigelmire, USA *(ret.)*
Lt Gen H. C. Stackpoke III, USMC *(ret.)*
Lt Gen William M. Steele, USA *(ret.)*

Lt Gen Howard F. Stone, USA *(ret.)*
Lt Gen George R. Stotser, USA *(ret.)*
Lt Gen John B. Sylvester, USA *(ret.)*
Lt Gen Billy M. Thomas, USA *(ret.)*
Lt Gen Nathaniel J. Thompson, Jr., USA *(ret.)*
Lt Gen James M. Thompson, USA *(ret.)*
Vice Adm Nils R. Thunman, USN *(ret.)*
Lt Gen Robert A. Tiebout, USMC *(ret.)*
Lt Gen Richard F. Timmons, USA *(ret.)*
Lt Gen Richard G. Trefry, USA *(ret.)*
Vice Adm Frederick C. Turner, USN *(ret.)*
Lt Gen Paul K. Van Riper, USMC *(ret.)*
Lt Gen John F. Wall, USA *(ret.)*
Lt Gen Claudius E. Watts III, USAF *(ret.)*
Lt Gen Ronald L. Watts, USA *(ret.)*
Lt Gen Joseph F. Weber, USMC *(ret.)*
Lt Gen Robert L. Wetzel, USA *(ret.)*
Lt Gen Alexander M. Weyand, USA *(ret.)*
Lt Gen Orren R. Whiddon, USA *(ret.)*
Lt Gen William J. White, USMC *(ret.)*
Lt Gen Robert J. Winglass, USMC *(ret.)*
Lt Gen Leonard P. Wishart III, USA *(ret.)*
Lt Gen Jack D. Woodall, USA *(ret.)*
Lt Gen John J. Yeosock, USA *(ret.)*
Vice Adm Lando W. Zech, Jr., USN *(ret.)*

2-Star Rank (512)

Rear Adm J. L. Abbot, Jr., USN *(ret.)*
Maj Gen William P. Acker, USAF *(ret.)*
Maj Gen Christopher S. Adams, Jr., USAF *(ret.)*
Rear Adm John W. Adams, USN *(ret.)*
Maj Gen Edwin M. Aguanno, USA *(ret.)*
Maj Gen Jere H. Akin, USA *(ret.)*
Maj Gen Willie A. Alexander, USA *(ret.)*
Maj Gen Gary M. Alkire, USAF *(ret.)*
Maj Gen James B. Allen, Jr., USA *(ret.)*
Maj Gen Phillip R. Anderson, USA *(ret.)*
Maj Gen Ronald K. Andreson, USA *(ret.)*
Rear Adm Philip Anselmo, USN *(ret.)*
Maj Gen Richard W. Anson, USA *(ret.)*
Maj Gen Joseph W. Arbuckle, USA *(ret.)*
Maj Gen Victor A. Armstrong, USMC *(ret.)*
Maj Gen Wallace C. Arnold, USA *(ret.)*
Maj Gen John C. Atkinson, USA *(ret.)*
Maj Gen Marvin G. Back, USA *(ret.)*
Maj Gen Donald M. Bagley, Jr., USA *(ret.)*
Maj Gen Darrel P. Baker, USA *(ret.)*

Maj Gen Charles Baldwin, USAF *(ret.)*
Maj Gen Thomas P. Ball, Jr., USAF *(ret.)*
Maj Gen Craig Bambrough, USA *(ret.)*
Maj Gen David J. Baratto, USA *(ret.)*
Maj Gen Eldon A. Bargewell, USA *(ret.)*
Rear Adm J. M. Barr, USN *(ret.)*
Maj Gen Raymond D. Barrett, Jr., USA *(ret.)*
Rear Adm John R. Batzler, USN *(ret.)*
Maj Gen George V. Bauer, AUS *(ret.)*
Maj Gen James B. Baylor, USA *(ret.)*
Maj Gen James E. Beal, USA *(ret.)*
Maj Gen Ronald L. Beckwith, USMC *(ret.)*
Maj Gen Richard D. Beltson, USA *(ret.)*
Maj Gen Calvert P. Benedict, USA *(ret.)*
Rear Adm James B. Best, USN *(ret.)*
Maj Gen Gerald H. Bethke, USA *(ret.)*
Rear Adm Thomas C. Betterton, USN *(ret.)*
Maj Gen John Bianchi, CSMR *(ret.)*
Maj Gen David F. Bice, USMC *(ret.)*
Maj Gen Charles S. Bishop, Jr., USMC *(ret.)*
Maj Gen John E. Blair, USA *(ret.)*
Maj Gen William Bland, Jr., USAF *(ret.)*
Maj Gen Jonas L. Blank, USAF *(ret.)*
Maj Gen Buford C. Blount III, USA *(ret.)*
Maj Gen William M. Boice, USA *(ret.)*
Maj Gen William L. Bond, USA *(ret.)*
Rear Adm Peter B. Booth, USN *(ret.)*
Maj Gen Richard T. Boverie, USAF *(ret.)*
Maj Gen Albert J. Bowley, USAF *(ret.)*
Maj Gen Edward R. Bracken, USAF *(ret.)*
Maj Gen Patrick H. Brady, USA *(ret.)*[45]
Maj Gen Robert J. Brandt, USA *(ret.)*
Maj Gen John A. Brashear, USAF *(ret.)*
Maj Gen Bobby F. Brashears, USA *(ret.)*
Maj Gen James A. Brooke, USA *(ret.)*
Maj Gen Ronald E. Brooks, USA *(ret.)*
Maj Gen James G. Browder, Jr., USA *(ret.)*
Rear Adm Thomas F. Brown III, USN *(ret.)*
Rear Adm D. Earl Brown, Jr., USN *(ret.)*
Maj Gen John M. Brown, USA *(ret.)*
Maj Gen Edward M. Browne, USA *(ret.)*
Maj Gen Robert O. Bugg, USA *(ret.)*
Maj Gen Robert H. Buker, USA *(ret.)*
Rear Adm Lyle F. Bull, USN *(ret.)*
Maj Gen James W. Bunting, USA *(ret.)*
Rear Adm Lawrence Burkhardt III, USN *(ret.)*
Maj Gen William F. Burns, USA *(ret.)*

Maj Gen Bobby G. Butcher, USMC *(ret.)*
Rear Adm William Callaghan, USN *(ret.)*
Maj Gen Colin C. Campbell, USA *(ret.)*
Maj Gen Henry D. Canterbury, USAF *(ret.)*
Rear Adm Walter H. Cantrell, USN *(ret.)*
Maj Gen John H. Capalbo, USA *(ret.)*
Rear Adm William C. Carlson, USN *(ret.)*
Maj Gen Fred H. Casey, USA *(ret.)*
Maj Gen John T. D. Casey, USA *(ret.)*
Maj Gen Frank A Catalano, Jr., USA *(ret.)*
Maj Gen George L. Cates, USMC *(ret.)*
Maj Gen James C. Cercy, USA *(ret.)*
Rear Adm Stephen K. Chadwick, USN *(ret.)*
Maj Gen Richard L. Chastain, USA *(ret.)*
Rear Adm Robert W. Chewning, USN *(ret.)*
Maj Gen Vernon Chong, USAF *(ret.)*
Rear Adm Albert H. Clancy, USN *(ret.)*
Maj Gen Peter W. Clegg, USA *(ret.)*
Maj Gen John R. D. Cleland, Jr., USA *(ret.)*
Maj Gen Reginal G. Clemmons, USA *(ret.)*
Maj Gen Fletcher C. Coker, USA *(ret.)*
Maj Gen Thomas F. Cole, USA *(ret.)*
Maj Gen Richard E. Coleman, USA *(ret.)*
Rear Adm Joseph L. Coleman, USN *(ret.)*
Maj Gen Richard E. Collier, USA *(ret.)*
Maj Gen Paul G. Collins, USA *(ret.)*
Maj Gen Anthony H. Conrad, Jr., USA *(ret.)*
Maj Gen Louis Conti, USMC *(ret.)*
Maj Gen Richard M. Cooke, USMC *(ret.)*
Maj Gen Andrew L. Cooley, USA *(ret.)*
Maj Gen J. Gary Cooper, USMCR *(ret.)*
Maj Gen Gregory A. Corliss, USMC *(ret.)*
Maj Gen Edward L. Correa, Jr., USA *(ret.)*
Maj Gen John V. Cox, USMC *(ret.)*
Rear Adm Michael Coyle, USN *(ret.)*
Maj Gen J. T. (Mike) Coyne, USMCR *(ret.)*
Maj Gen Wesley E. Craig, Jr., USA *(ret.)*
Maj Gen W. D. Crittenberger, USA *(ret.)*
Maj Gen Robert E. Crosser, USA *(ret.)*
Maj Gen John J. Cuddy, USA *(ret.)*
Rear Adm Richard E. Curtis, USN *(ret.)*
Rear Adm William D. Daniels, USN *(ret.)*
Maj Gen John R. D'Araujo, USA *(ret.)*
Maj Gen Thomas G. Darling, USAF *(ret.)*
Maj Gen William J. Davies, USA *(ret.)*
Maj Gen Harley C. Davis, USA *(ret.)*
Maj Gen Richard E. Davis, USA *(ret.)*

Maj Gen Jack A. Davis, USMC *(ret.)*

Maj Gen Hollis Davison, USMC *(ret.)*

Maj Gen William B. Davitte, USAF *(ret.)*

Maj Gen Gene A. Deegan, USMC *(ret.)*

Maj Gen David P. Delavergne, USA *(ret.)*

Maj Gen Frank M. Denton, USA *(ret.)*

Maj Gen Kenneth E. Dohleman, USA *(ret.)*

Maj Gen Ralph O. Doughty, USA *(ret.)*

Maj Gen George Douglas, USAF *(ret.)*

Maj Gen James W. Duffy, USA *(ret.)*

Maj Gen Travis N. Dyer, USA *(ret.)*

Maj Gen David B. Easson, USAF *(ret.)*

Rear Adm L. F. Eggert, USN *(ret.)*

Rear Adm J. J. Ekelund, USN *(ret.)*

Maj Gen Billy J. Ellis, USAF *(ret.)*

Rear Adm George Ellis, USN *(ret.)*

Maj Gen James W. Emerson, USA *(ret.)*

Rear Adm Thomas R. M. Emery, USN *(ret.)*

Rear Adm Paul H. Engel, USN *(ret.)*

Rear Adm Robert B. Erly, USN *(ret.)*

Maj Gen W. P. Eshelman, USMC *(ret.)*

Maj Gen Harry Falls, Jr., USAF *(ret.)*

Maj Gen Vincent E. Falter, USA *(ret.)*

Rear Adm Eugene H. Farrell, USN *(ret.)*

Maj Gen John R. Farrington, USAF *(ret.)*

Rear Adm Edward L. Feightner, USN *(ret.)*

Rear Adm D. L. Felt, USN *(ret.)*

Maj Gen Charles J. Fiala, USA *(ret.)*

Maj Gen Philip B. Finley, USA *(ret.)*

Maj Gen Jackson L. Flake, USA *(ret.)*

Maj Gen Robert M. Flanagan, USMC *(ret.)*

Rear Adm G. J. "Rod" Flannery, USN *(ret.)*

Rear Adm James H. Flatley III, USN *(ret.)*

Maj Gen Thomas C. Foley, USA *(ret.)*

Rear Adm Harry J. P. Foley, USN *(ret.)*

Rear Adm Arthur Fort, CEC, USN *(ret.)*

Maj Gen Larry D. Fortner, USAF *(ret.)*

Rear Adm Robert R. Fountain, USN *(ret.)*

Maj Gen Joseph P. Franklin, USA *(ret.)*

Maj Gen Ray Franklin, USMC *(ret.)*

Maj Gen Paul Fratarangelo, USMC *(ret.)*

Maj Gen Stuart French, USAF *(ret.)*

Rear Adm Richard D. Friichtenicht, USN *(ret.)*

Rear Adm S. David Frost, USN *(ret.)*

Maj Gen John L. Fugh, USA *(ret.)*

Maj Gen Martin C. Fulcher, USAF *(ret.)*

Maj Gen Donald J. Fulham, USMC *(ret.)*

Rear Adm Skip Furlong, USN *(ret.)*

Maj Gen Jon A. Gallinetti, USMC *(ret.)*

Rear Adm Albert A. Gallotta, Jr., USN *(ret.)*

Maj Gen Bradley D. Gambill, USA *(ret.)*

Maj Gen Peter A. Gannon, USA *(ret.)*

Maj Gen James H. Garner, USA *(ret.)*

Maj Gen George T. Garrett, USA *(ret.)*

Maj Gen William F. Garrison, USA *(ret.)*

Rear Adm Richard T. Gaskill, USN *(ret.)*

Rear Adm John D. Gavan, USN *(ret.)*

Rear Adm H. E. Gerhard, USN *(ret.)*

Maj Gen Timothy Ghormley, USMC *(ret.)*

Maj Gen Greg L. Gile, USA *(ret.)*

Maj Gen Louis H. Ginn III, USA *(ret.)*

Maj Gen Harold G. Glasgow, USMC *(ret.)*

Rear Adm James M. Gleim, USN *(ret.)*

Maj Gen Richard N. Goddard, USAF *(ret.)*

Maj Gen Robert A. Goodbary, USA *(ret.)*

Maj Gen Fred A. Gorden, USA *(ret.)*

Maj Gen Robert L. Gordon, USA *(ret.)*

Rear Adm John "Ted" Gordon, USN *(ret.)*

Rear Adm Robert H. Gormley, USN *(ret.)*

Maj Gen Albert E. Gorsky, USA *(ret.)*

Maj Gen William H. Gossell, USMCR *(ret.)*

Maj Gen Todd P. Graham, USA *(ret.)*

Maj Gen Roy C. Gray, Jr., USA *(ret.)*

Rear Adm James V. Grealish, USN *(ret.)*

Maj Gen Lee V. Greer, USAF *(ret.)*

Maj Gen Robert H. Griffin, USA *(ret.)*

Maj Gen John S. Grinalds, USMC *(ret.)*

Maj Gen W. C. Groeniger III, USMCR *(ret.)*

Maj Gen William J. Grove, Jr., USAF *(ret.)*

Maj Gen James A. Guest, USA *(ret.)*

Maj Gen Robert K. Guest, USA *(ret.)*

Maj Gen George L. Gunderman, USA *(ret.)*

Maj Gen Gaylord T. Gunhus, USA *(ret.)*

Rear Adm William A. Gureck, USN *(ret.)*

Maj Gen David R. Gust, USA *(ret.)*

Maj Gen Richard A. Gustafson, USMC *(ret.)*

Rear Adm Frank S. Haak, USN *(ret.)*

Maj Gen Timothy M. Haake, USA *(ret.)*

Maj Gen Robert E. Haerel, USMC *(ret.)*

Maj Gen Craig A. Hagan, USA *(ret.)*

Maj Gen Kenneth L. Hagemann, Jr., USAF *(ret.)*

Maj Gen Henry M. Hagwood, Jr., USA *(ret.)*

Maj Gen Raphael J. Hallada, USA *(ret.)*
Maj Gen Francis X. Hamilton, USMC *(ret.)*
Maj Gen Rudolph E. Hammond, USA *(ret.)*
Maj Gen Gus L. Hargett, Jr., USA *(ret.)*
Maj Gen William E. Harmon, USA *(ret.)*
Maj Gen Gary L. Harrell, USA *(ret.)*
Maj Gen James E. Harrell, USA *(ret.)*
Rear Adm William H. Harris, USN *(ret.)*
Maj Gen Ronald O. Harrison, USA *(ret.)*
Rear Adm Donald P. Harvey, USN *(ret.)*
Maj Gen James E. Haught, USA *(ret.)*
Maj Gen Ralph L. Haynes, USA *(ret.)*
Maj Gen Richard E. Haynes, USA *(ret.)*
Maj Gen Fred Haynes, USMC *(ret.)*
Rear Adm Kenneth G. Haynes, USN *(ret.)*
Maj Gen Guy L. Hecker, Jr., USAF *(ret.)*
Maj Gen Frank F. Henderson, USA *(ret.)*
Maj Gen Clyde A. Hennies, USA *(ret.)*
Maj Gen Curtis B. Herbert III, USA *(ret.)*
Maj Gen G. B. Higginbotham, USMC *(ret.)*
Maj Gen Donald C. Hilbert, USA *(ret.)*
Maj Gen John W. Hill, USMC *(ret.)*
Maj Gen William B. Hobgood, USA *(ret.)*
Maj Gen Carl W. Hoffman, USMC *(ret.)*
Rear Adm Lowell J. Holloway, USN *(ret.)*
Maj Gen Jerry D. Holmes, USAF *(ret.)*
Maj Gen Charles E. Honore, USA *(ret.)*
Rear Adm J. T. Hood, USN *(ret.)*
Maj Gen Marvin T. Hopgood, Jr., USMC *(ret.)*
Maj Gen Patrick G. Howard, USMC *(ret.)*
Maj Gen Richard A. Huck, USMC *(ret.)*
Maj Gen Jerry Humble, USMC *(ret.)*
Maj Gen Donald R. Infante, USA *(ret.)*
Maj Gen Dewitt T. Irby, Jr., USA *(ret.)*
Maj Gen James T. Jackson, USA *(ret.)*
Maj Gen Wayne P. Jackson, USA *(ret.)*
Rear Adm Grady L. Jackson, USN *(ret.)*
Maj Gen Billy F. Jester, USA *(ret.)*
Rear Adm C. A. E. Johnson, Jr., USN *(ret.)*
Maj Gen Alan D. Johnson, USA *(ret.)*
Maj Gen Stephen T. Johnson, USMC *(ret.)*
Maj Gen Warren R. Johnson, USMC *(ret.)*
Maj Gen Kenneth A. Jolemore, USA *(ret.)*
Maj Gen Alvin W. Jones, USA *(ret.)*
Maj Gen William G. Joslyn, USMC *(ret.)*
Maj Gen Jerry J. Josten, USA *(ret.)*
Maj Gen Angelo D. Juarez, USA *(ret.)*

Rear Adm Thomas A. Kamm, USNR *(ret.)*
Maj Gen John F. Kane, USA *(ret.)*
Maj Gen Harry G. Karegeannes, USA *(ret.)*
Maj Gen Jerry M. Keeton, USA *(ret.)*
Maj Gen Maurice W. Kendall, USA *(ret.)*
Rear Adm John M. Kersh, USN *(ret.)*
Maj Gen Thomas D. Kinley, USA *(ret.)*
Maj Gen Eugene P. Klynoot, USA *(ret.)*
Rear Adm J. Weldon Koenig, USN *(ret.)*
Maj Gen Herbert Koger, Jr., USA *(ret.)*
Maj Gen Joseph Koler, Jr., USMC *(ret.)*
Rear Adm L. S. Kollmorgen, USN *(ret.)*
Maj Gen Charles H. Kone, AUS *(ret.)*
Maj Gen Glenn H. Kothmann, USA *(ret.)*
Maj Gen Lloyd E. Krase, USA *(ret.)*
Maj Gen Richard A. Kuci, USMC *(ret.)*
Maj Gen Kevin B. Kuklok, USMC *(ret.)*
Maj Gen Robert A. Lame, USA *(ret.)*
Rear Adm Lee E. Landes, SC, USN *(ret.)*
Maj Gen Frank C. Lang, USMC *(ret.)*
Rear Adm James R. Lang, USN *(ret.)*
Maj Gen Leo J. LeBlanc, USMC *(ret.)*
Maj Gen Paul M. Lee, Jr., USMC *(ret.)*
Maj Gen Larry E. Lee, USA *(ret.)*
Maj Gen Kenneth C. Leuer, USA *(ret.)*
Rear Adm Frederick L. Lewis, USN *(ret.)*
Maj Gen Thomas G. Lightner, USA *(ret.)*
Maj Gen Charles D. Link, USAF *(ret.)*
Maj Gen John H. Little, USA *(ret.)*
Maj Gen James E. Livingston, USMC *(ret.)*[46]
Maj Gen Donald A. Logeais, USAF *(ret.)*
Maj Gen Homer S. Long, Jr., USA *(ret.)*
Maj Gen John E. Longhouser, USA *(ret.)*
Maj Gen Federico Lopez III, USA *(ret.)*
Maj Gen Bernard F. Losekamp, USA *(ret.)*
Maj Gen Bradley M. Lott, USMC *(ret.)*
Maj Gen J. D. Lynch, Jr., USMC *(ret.)*
Maj Gen Robert G. Lynn, USA *(ret.)*
Rear Adm Malcolm MacKinnon III, USN *(ret.)*
Maj Gen William G. MacLaren, Jr., USAF *(ret.)*
Maj Gen Richard H. MacMillan, Jr., USA *(ret.)*
Maj Gen Philip H. Mallory, USA *(ret.)*
Maj Gen Donald L. Marks, USAF *(ret.)*
Rear Adm John L. Marocchi, USN *(ret.)*

Rear Adm Larry R. Marsh, USN *(ret.)*
Maj Gen Wayne D. Marty, USA *(ret.)*
Maj Gen Michael R. Mazzucchi, USA *(ret.)*
Maj Gen Charles E. McCartney, USA *(ret.)*
Rear Adm Robert B. McClinton, USN *(ret.)*
Rear Adm Dan McCormick, USN *(ret.)*
Maj Gen Ray E. McCoy, USA *(ret.)*
Maj Gen Darrel W. McDaniel, USA *(ret.)*
Rear Adm William J. McDaniel, USN *(ret.)*
Maj Gen James M. McDougal, USA *(ret.)*
Maj Gen James C. McElroy, Jr., USA *(ret.)*
Rear Adm E. S. "Skip" McGinley II, USN *(ret.)*
Maj Gen Chester M. McKeen, Jr., USA *(ret.)*
Maj Gen James J. McMonagle, USMC *(ret.)*
Maj Gen John R. McWaters, USA *(ret.)*
Maj Gen David C. Meade, USA *(ret.)*
Maj Gen Guy S. Meloy III, USA *(ret.)*
Maj Gen Robert E. Messerli, USAF *(ret.)*
Rear Adm Frederick Metz, USN *(ret.)*
Rear Adm Floyd H. Miller, Jr., USN *(ret.)*
Maj Gen Geoffrey D. Miller, USA *(ret.)*
Rear Adm Robert G. Mills, USN *(ret.)*
Maj Gen Gerald P. Minetti, USA *(ret.)*
Rear Adm Riley D. Mixson, USN *(ret.)*
Maj Gen John P. Monahan, USMC *(ret.)*
Rear Adm A. J. Monger, USN *(ret.)*
Maj Gen Mario F. Montero, Jr., USA *(ret.)*
Rear Adm James W. Montgomery, USN *(ret.)*
Maj Gen William L. Moore, Jr., USA *(ret.)*
Maj Gen Royal N. Moore, Jr., USMC *(ret.)*
Maj Gen Thomas L. Moore, Jr., USMC *(ret.)*
Rear Adm Douglas M. Moore, Jr., USN *(ret.)*
Maj Gen James E. Moore, USA *(ret.)*
Maj Gen William C. Moore, USA *(ret.)*
Maj Gen Burton R. Moore, USAF *(ret.)*
Maj Gen Marc A. Moore, USMC *(ret.)*
Maj Gen Robert D. Morgan, USA *(ret.)*
Rear Adm Jack Moriarty, USN *(ret.)*
Rear Adm James B. Morin, USN *(ret.)*
Maj Gen Richard Mulberry, USMCR *(ret.)*
Maj Gen Mark B. Mullin, USA *(ret.)*
Maj Gen Thomas B. Murchie, USA *(ret.)*
Maj Gen Dennis J. Murphy, USMC *(ret.)*
Maj Gen James A. Musselman, USA *(ret.)*
Maj Gen Willie B. Nance, Jr., USA *(ret.)*
Maj Gen Thomas H. Needham, USA *(ret.)*
Maj Gen George W. Norwood, USAF *(ret.)*

Rear Adm James K. Nunneley, USN *(ret.)*
Maj Gen John M. O'Connell, USA *(ret.)*
Maj Gen Thomas R. Olsen, USAF *(ret.)*
Maj Gen Ray E. O'Mara, USAF *(ret.)*
Maj Gen G. R. Omrod, USMCR *(ret.)*
Maj Gen Daniel J. O'Neill, USA *(ret.)*
Maj Gen Rudolph Ostovich III, USA *(ret.)*
Rear Adm Robert S. Owens, USN *(ret.)*
Maj Gen William C. Page, Jr., USA *(ret.)*
Maj Gen James W. Parker, USA *(ret.)*
Maj Gen John R. Paulk, USAF *(ret.)*
Maj Gen Donald A. Pearson, USA *(ret.)*
Maj Gen Earl G. Peck, USAF *(ret.)*
Maj Gen Robert F. Pennycuick, USA *(ret.)*
Maj Gen Harry D. Penzler, USA *(ret.)*
Maj Gen John S. Peppers, USA *(ret.)*
Maj Gen Charles H. Perenick, Sr., USA *(ret.)*
Maj Gen Elbert N. Perkins, USA *(ret.)*
Maj Gen Richard L. Phillips, USMC *(ret.)*
Maj Gen John R. Piatak, USA *(ret.)*
Maj Gen Ross Plasterer, USMC *(ret.)*
Maj Gen Arthur J. Poillon, USMC *(ret.)*
Maj Gen Gerald L. Prather, USAF *(ret.)*
Rear Adm Don G. Primeau, USN *(ret.)*
Maj Gen Gerald H. Putman, USA *(ret.)*
Maj Gen James I. Pylant, USA *(ret.)*
Maj Gen Kenneth J. Quinlan, Jr., USA *(ret.)*
Maj Gen Hugh J. Quinn, USA *(ret.)*
Maj Gen W. R. Quinn, USMC *(ret.)*
Maj Gen Richard J. Quirk III, USA *(ret.)*
Maj Gen David C. Ralston, USA *(ret.)*
Maj Gen John B. Ramey, USA *(ret.)*
Maj Gen Norbert J. Rappl, USA *(ret.)*
Maj Gen Bentley B. Rayburn, USAF *(ret.)*
Rear Adm Robert T. Reimann, USN *(ret.)*
Maj Gen Claude Reinke, USMC *(ret.)*
Rear Adm Thomas H. Replogle, USN *(ret.)*
Rear Adm William A. Retz, USN *(ret.)*
Maj Gen W. H. Rice, USMC *(ret.)*
Maj Gen R. G. Richard, USMC *(ret.)*
Maj Gen D. A. Richwine, USMC *(ret.)*
Maj Gen John Ricottilli, Jr., USA *(ret.)*
Rear Adm G. L. Riendeau, USN *(ret.)*
Rear Adm Roland Rieve, USN *(ret.)*
Maj Gen William H. Riley, Jr., USA *(ret.)*
Maj Gen Claude J. Roberts, Jr., USA *(ret.)*
Maj Gen George R. Robertson, USA *(ret.)*

Maj Gen Henry D. Robertson, USA *(ret.)*
Maj Gen Mastin Robeson, USMC *(ret.)*
Maj Gen Kenneth L. Robinson, USMC *(ret.)*
Maj Gen Wayne E. Rollings, USMC *(ret.)*
Maj Gen William A. Roosma, USA *(ret.)*
Rear Adm C. J. Rorie, USN *(ret.)*
Maj Gen Robert R. Rose, USA *(ret.)*
Maj Gen Charles L. Rosenfeld, USA *(ret.)*
Maj Gen Robert B. Rosenkranz, USA *(ret.)*
Maj Gen William H. Russ, USA *(ret.)*
Maj Gen James A. Ryan, USA *(ret.)*
Maj Gen Michael D. Ryan, USMC *(ret.)*
Maj Gen Thomas M. Sadler, USAF *(ret.)*
Maj Gen Reymaldo Sanchez, USA *(ret.)*
Maj Gen C. Dean Sangalis, USMCR *(ret.)*
Rear Adm Louis R. Sarosdy, USN *(ret.)*
Maj Gen John W. Schaeffer, Jr., USA *(ret.)*
Maj Gen Richard S. Schneider, USA *(ret.)*
Maj Gen Edison E. Scholes, USA *(ret.)*
Rear Adm Hugh P. Scott, (MC) USN *(ret.)*
Maj Gen Charles E. Scott, USA *(ret.)*
Rear Adm Robert H. Shumaker, USN *(ret.)*
Maj Gen Richard S. Siegfried, USA *(ret.)*
Maj Gen Thomas F. Sikora, USA *(ret.)*
Maj Gen Stephen Silvasy, Jr., USA *(ret.)*
Maj Gen Wilbur F. Simlik, USMC *(ret.)*
Maj Gen James E. Simmons, USA *(ret.)*
Maj Gen Frank J. Simokaitis, USAF *(ret.)*
Maj Gen Darwin H. Simpson, USA *(ret.)*
Maj Gen John K. Singlaub, USA *(ret.)*
Maj Gen Mark J. Sisinyak, USA *(ret.)*
Maj Gen James D. Smith, USA *(ret.)*
Maj Gen Monroe T. Smith, USAF *(ret.)*
Maj Gen Ray L. Smith, USMC *(ret.)*
Rear Adm C. Bruce Smith, USN *(ret.)*
Maj Gen James R. Snider, USA *(ret.)*
Maj Gen John F. Sobke, USA *(ret.)*
Rear Adm Robert H. Spiro, Jr., USNR *(ret.)*
Maj Gen Richard E. Stearney, USA *(ret.)*
Maj Gen Harry V. Steel, Jr., USA *(ret.)*
Maj Gen Orlo K. Steele, USMC *(ret.)*
Maj Gen Elmer L. Stephens, USA *(ret.)*
Maj Gen Richard E. Stephenson, USA *(ret.)*
Maj Gen Pat M. Stevens IV, USA *(ret.)*
Maj Gen Lynn H. Stevens, USA *(ret.)*
Maj Gen John F. Stewart, Jr., USA *(ret.)*
Maj Gen Joseph D. Stewart, USMC *(ret.)*

Maj Gen Eugene L. Stillions, Jr., USA *(ret.)*
Maj Gen James B. Stodart, Jr., USA *(ret.)*
Rear Adm F. Bradford Stone, USN *(ret.)*
Maj Gen Henry W. Stratman, USA *(ret.)*
Maj Gen Michael D. Strong III, USA *(ret.)*
Maj Gen Jack Strukel, Jr., USA *(ret.)*
Maj Gen Duane H. Stubbs, USA *(ret.)*
Maj Gen John Anthony Studds, USMC *(ret.)*
Rear Adm Donald L. Sturtz, MC, USN *(ret.)*
Maj Gen Leroy N. Suddath, Jr., USA *(ret.)*
Maj Gen Lawrence F. Sullivan, USMC *(ret.)*
Maj Gen Michael P. Sullivan, USMC *(ret.)*
Rear Adm Paul E. Sutherland, USN *(ret.)*
Maj Gen Charles H. Swannack, Jr., USA *(ret.)*
Maj Gen Samuel H. Swart, Jr., USAF *(ret.)*
Rear Adm John J. Sweeney, USN *(ret.)*
Maj Gen Will Hill Tankersley, USA *(ret.)*
Maj Gen James R. Taylor, USA *(ret.)*
Maj Gen Larry S. Taylor, USMCR *(ret.)*
Maj Gen Mark W. Tenney, USA *(ret.)*
Maj Gen Melvin C. Thrash, USA *(ret.)*
Maj Gen Robert C. Thrasher, USA *(ret.)*
Maj Gen Larry N. Tibbetts, USAF *(ret.)*
Maj Gen Harold L. Timboe, USA *(ret.)*
Rear Adm W. D. Toole, Jr., USN *(ret.)*
Maj Gen Richard W. Tragemann, USA *(ret.)*
Maj Gen Terry L. Tucker, USA *(ret.)*
Rear Adm Merton Dick Van Orden, USN *(ret.)*
Rear Adm Lloyd R. Vasey, USN *(ret.)*
Maj Gen Clyde L. Vermilyea, USMC *(ret.)*
Maj Gen John M. Vest, USA *(ret.)*
Maj Gen Hal W. Vincent, USMC *(ret.)*
Maj Gen James E. Wagner, USA *(ret.)*
Maj Gen Robert E. Wagner, USA *(ret.)*
Maj Gen Wayne F. Wagner, USA *(ret.)*
Rear Adm E. K. Walker, Jr., USN *(ret.)*
Maj Gen Stewart W. Wallace, USA *(ret.)*
Maj Gen William F. Ward, Jr., USA *(ret.)*
Maj Gen Gerald G. Watson, USA *(ret.)*
Maj Gen Robert H. Waudby, USA *(ret.)*
Rear Adm Donald Weatherson, USN *(ret.)*
Rear Adm John C. Weaver, USN *(ret.)*
Maj Gen William L. Webb, Jr., USA *(ret.)*
Rear Adm Hugh L. Webster, USN *(ret.)*
Maj Gen Kenneth W. Weir, USMC *(ret.)*
Maj Gen Barclay O. Wellman, AUS *(ret.)*

Maj Gen Billy G. Wellman, USA *(ret.)*
Rear Adm R. S. Wentworth, USN *(ret.)*
Maj Gen Albin G. Wheeler, USA *(ret.)*
Maj Gen Gary J. Whipple, USA *(ret.)*
Maj Gen David E. White, USA *(ret.)*
Maj Gen Jerry A. White, USA *(ret.)*
Maj Gen Richard O. Wightman, Jr.,
    USA *(ret.)*
Maj Gen Claude A. Williams, USA *(ret.)*
Maj Gen Norman E. Williams, USA *(ret.)*
Maj Gen Peter D. Williams, USMC *(ret.)*
Rear Adm Allen D. Williams, USN *(ret.)*
Maj Gen Guilford J. Wilson, Jr., USA *(ret.)*
Maj Gen Charles L. Wilson, USAF *(ret.)*
Rear Adm John R. Wilson, USN *(ret.)*
Maj Gen W. Montague Winfield, USA *(ret.)*
Maj Gen Charles J. Wing, USA *(ret.)*
Rear Adm Dennis Wisely, USN *(ret.)*
Maj Gen George K. Withers, Jr., USA *(ret.)*
Maj Gen Walter Wojdakowski, USA *(ret.)*
Maj Gen John J. Womack, USA *(ret.)*
Maj Gen Stephen R. Woods, Jr., USA *(ret.)*
Rear Adm George R. Worthington,
    USN *(ret.)*
Maj Gen Edwin H. Wright, USA *(ret.)*
Rear Adm William C. Wyatt, USN *(ret.)*
Rear Adm Earl P. Yates, USN *(ret.)*
Maj Gen Walter H. Yates, Jr., USA *(ret.)*
Rear Adm H. L. Young, USN *(ret.)*
Rear Adm W. M. Zobel, USN *(ret.)*

1-Star Rank (407)

Brig Gen Norris P. Abts, USA *(ret.)*
Brig Gen W. T. Adams, USMC *(ret.)*
Brig Gen David M. Adamson, USA *(ret.)*
Brig Gen Michael J. Aguilar, USMC *(ret.)*
Brig Gen Thomas H. Alexander, USA *(ret.)*
Brig Gen John R. Allen, Jr., USAF *(ret.)*
Brig Gen Carroll G. Allen, USA *(ret.)*
Brig Gen David J. Allen, USA *(ret.)*
Brig Gen George L. Allen, USA *(ret.)*
Brig Gen Richard F. Allen, USA *(ret.)*
Brig Gen Benny P. Anderson, USA *(ret.)*
Brig Gen Charles H. Anderson, USA *(ret.)*
Brig Gen Dorian T. Anderson, USA *(ret.)*
Brig Gen Dale F. Andres, USA *(ret.)*
Brig Gen William S. Anthony, USA *(ret.)*

Brig Gen John C. Arick, USMC *(ret.)*
Brig Gen Terrence L. Arndt, USA *(ret.)*
Brig Gen Edwin J. Arnold, Jr., USA *(ret.)*
Brig Gen Maurice C. Ashley, USMC *(ret.)*
Brig Gen Loring R. Astorino, USAF *(ret.)*
Brig Gen James M. AuBuchon, USA *(ret.)*
Brig Gen Richard W. Averitt, USA *(ret.)*
Brig Gen Terry O. Ballard, USA *(ret.)*
Brig Gen Naman X. Barnes, USA *(ret.)*
Brig Gen Thomas P. Barrett, USA *(ret.)*
Brig Gen D. Joseph Bartlett, USMCR *(ret.)*
Brig Gen George L. Bartlett, USMC *(ret.)*
Brig Gen Hugh J. Bartley, USA *(ret.)*
Brig Gen Dana D. Batey, USA *(ret.)*
Brig Gen James L. Bauerle, USA *(ret.)*
Brig Gen Sheila R. Baxter, USA *(ret.)*
Brig Gen Robert H. Beahm, USA *(ret.)*
Brig Gen James D. Beans, USMC *(ret.)*
Brig Gen Floyd E. Bell, Jr., USA *(ret.)*
Brig Gen Julius L. Berthold, USA *(ret.)*
Brig Gen William C. Bilo, USA *(ret.)*
Brig Gen Harry E. Bivens, USA *(ret.)*
Brig Gen Darrel E. Bjorklund, USMC *(ret.)*
Brig Gen Richard A. Black, USA *(ret.)*
Brig Gen Vincent T. Blaz, USMC *(ret.)*
Brig Gen William A. Bloomer, USMC *(ret.)*
Brig Gen Spessard Boatright, USA *(ret.)*
Brig Gen James W. Boddie, Jr., USA *(ret.)*
Brig Gen Furman P. Bodenheimer, Jr.,
    USA *(ret.)*
Brig Gen David D. Boland, USA *(ret.)*
Brig Gen Stephen C. Boone, USA *(ret.)*
Brig Gen Ronald I. Botz, USA *(ret.)*
Brig Gen Guy M. Bourn, USA *(ret.)*
Brig Gen Darden J. Bourne, USA *(ret.)*
Brig Gen Gary D. Bray, USANG
Brig Gen A. E. Brewster, USMC *(ret.)*
Brig Gen John P. Brickley, USMC *(ret.)*
Brig Gen James F. Brickman, USA *(ret.)*
Brig Gen John C. Bridges, USA *(ret.)*
Brig Gen George R. Brier, USMC *(ret.)*
Brig Gen Ernest D. Brockman, Jr., USA *(ret.)*
Brig Gen Matthew E. Broderick, USMC *(ret.)*
Brig Gen Donald H. Brooks, USMC *(ret.)*
Brig Gen William R. Brooksher, USAF *(ret.)*
Brig Gen Jeremiah Brophy, USA *(ret.)*
Brig Gen Harvey E. Brown, USA *(ret.)*

Brig Gen J. Royston Brown, USA *(ret.)*

Brig Gen Lewis E. Brown, USA *(ret.)*

Brig Gen Ralph H. Brown, USA *(ret.)*

Brig Gen Stanford E. Brown, USAF *(ret.)*

Brig Gen Gary E. Brown, USMC *(ret.)*

Brig Gen Thomas J. Bruner, USA *(ret.)*

Brig Gen James M. Bullock, Jr., USA *(ret.)*

Brig Gen Edward R. Burka, USA *(ret.)*

Brig Gen John C. Burney, USA *(ret.)*

Brig Gen James H. Burns, USA *(ret.)*

Brig Gen Walter L. Busbee, USA *(ret.)*

Brig Gen Michael E. Byrne, USA *(ret.)*

Brig Gen Bruce B. Byrum, USMC *(ret.)*

Brig Gen Sherian G. Cadoria, USA *(ret.)*

Brig Gen Eddie Cain, USA *(ret.)*

Brig Gen James E. Caldwell III, USA *(ret.)*

Brig Gen Danny B. Callahan, USA *(ret.)*

Brig Gen Joseph W. Camp, Jr., USA *(ret.)*

Brig Gen James R. Carpenter, USA *(ret.)*

Brig Gen James T. Carper, USA *(ret.)*

Brig Gen Jimmy L. Cash, USAF *(ret.)*

Brig Gen Lomer R. Chambers, USA *(ret.)*

Brig Gen Paul Y. Chinen, USA *(ret.)*

Brig Gen G. Wesley Clark, USAF *(ret.)*

Brig Gen Robert V. Clements, USAF *(ret.)*

Brig Gen Samuel G. Cockerham, USA *(ret.)*

Brig Gen William P. Cody, USA *(ret.)*

Brig Gen George P. Cole, Jr., USAF *(ret.)*

Brig Gen Dan M. Colglazier, USA *(ret.)*

Brig Gen James P. Combs, USA *(ret.)*

Brig Gen Augustus L. Collins, USA *(ret.)*

Brig Gen Joseph F. Conlon III, USA *(ret.)*

Brig Gen Vernon L. Conner, USA *(ret.)*

Brig Gen William M. Constantine, USAF *(ret.)*

Brig Gen David E. K. Cooper, USA *(ret.)*

Brig Gen Paul D. Costilow, USA *(ret.)*

Brig Gen Christian B. Cowdrey, USMC *(ret.)*

Brig Gen Carroll E. Crawford, ARNG

Brig Gen Stephen J. Curry, USA *(ret.)*

Brig Gen Thomas S. Cushing, USA *(ret.)*

Brig Gen Robert J. Dacey, USA *(ret.)*

Brig Gen John N. Dailey, USA *(ret.)*

Brig Gen Robert L. Davis, USA *(ret.)*

Brig Gen Robert S. Davis, USA *(ret.)*

Brig Gen Benjamin W. Day, Jr., USA *(ret.)*

Brig Gen Richard D. Dean, AUS *(ret.)*

Brig Gen Alan E. Deegan, USA *(ret.)*

Brig Gen Thomas J. DeGraw, USA *(ret.)*

Brig Gen Arnaldo J. Dejesus, USA *(ret.)*

Brig Gen Richard D. DeMara, USA *(ret.)*

Brig Gen John F. DePue, AUS *(ret.)*

Brig Gen Ralph O. DeWitt, Jr., USA *(ret.)*

Brig Gen Charles O. Dillard, USA *(ret.)*

Brig Gen Francis R. Dillon, USAF *(ret.)*

Brig Gen Lyle C. Doerr, USA *(ret.)*

Brig Gen Walter Donovan, USMC *(ret.)*

Brig Gen Wilson T. Dreger III, USA *(ret.)*

Brig Gen Robert A. Drolet, USA *(ret.)*

Brig Gen Clifford A. Druit, USA *(ret.)*

Brig Gen Charles M. Duke, USAF *(ret.)*

Brig Gen James T. Dunn, USA *(ret.)*

Brig Gen Douglas B. Earhart, USA *(ret.)*

Brig Gen Raymond W. Edwards, USMC *(ret.)*

Brig Gen Randy J. Ence, USA *(ret.)*

Brig Gen John L. Enright, USA *(ret.)*

Brig Gen Burney H. Enzor, USA *(ret.)*

Brig Gen Frederick H. Essig, USA *(ret.)*

Brig Gen Donald M. Ewing, USA *(ret.)*

Brig Gen Gerald G. Fall, Jr., USAF *(ret.)*

Brig Gen Andrew N. Farley, USA *(ret.)*

Brig Gen Thomas D. Farmer, USA *(ret.)*

Brig Gen Anthony J. Farrington, Jr., USAF *(ret.)*

Brig Gen James M. Feigley, USMC *(ret.)*

Brig Gen W. Daniel Fillmore, USMC *(ret.)*

Brig Gen Arvid M. Flanum, USA *(ret.)*

Brig Gen Fred R. Flint, USA *(ret.)*

Brig Gen Robert L. Floyd II, USA *(ret.)*

Brig Gen Melvin V. Frandsen, USA *(ret.)*

Brig Gen Joe N. Frazar III, USA *(ret.)*

Brig Gen Uri S. French III, USA *(ret.)*

Brig Gen H. J. Fruchtnicht, USMC *(ret.)*

Brig Gen David L. Funk, USA *(ret.)*

Brig Gen Benard W. Gann, USAF *(ret.)*

Brig Gen Larry Garrett, USMC *(ret.)*

Brig Gen Augusto R. Gautier, USA *(ret.)*

Brig Gen David W. Gay, USA *(ret.)*

Brig Gen Gordon D. Gayle, USMC *(ret.)*

Brig Gen Stuart W. Gerald, USA *(ret.)*

Brig Gen Charles E. Getz, USA *(ret.)*

Brig Gen Jacob E. Glick, USMC *(ret.)*

Brig Gen Bryghte D. Godbold, USMC *(ret.)*

Brig Gen Joseph W. Godwin, Jr., USA *(ret.)*
Brig Gen Harold M. Goldstein, USA *(ret.)*
Brig Gen William W. Goodwin, USA *(ret.)*
Brig Gen David L. Grange, USA *(ret.)*
Brig Gen Roger H. Greenwood, USA *(ret.)*
Brig Gen Tommy F. Grier, Jr., USA *(ret.)*
Brig Gen Robert F. Griffin, USA *(ret.)*
Brig Gen Bruce G. Grover, AUS *(ret.)*
Brig Gen Clyde E. Gutzwiller, USA *(ret.)*
Brig Gen Harvey M. Haakenson, USA *(ret.)*
Brig Gen Harry T. Hagaman, USMC *(ret.)*
Brig Gen Max G. Halliday, USMC *(ret.)*
Brig Gen Donald W. Hansen, USA *(ret.)*
Brig Gen Gary G. Harber, USA *(ret.)*
Brig Gen Keith L. Hargrove, USA *(ret.)*
Brig Gen Michael H. Harris, USA *(ret.)*
Brig Gen Larry D. Haub, USA *(ret.)*
Brig Gen Donald F. Hawkins, USA *(ret.)*
Brig Gen Edison O. Hayes, USA *(ret.)*
Brig Gen John A. Hays, USA *(ret.)*
Brig Gen Lewis M. Helm, USA *(ret.)*
Brig Gen Leif Hendrickson, USMC *(ret.)*
Brig Gen Terence M. Henry, USA *(ret.)*
Brig Gen James A. Herbert, USA *(ret.)*
Brig Gen Ralph E. Hickman, USA *(ret.)*
Brig Gen Keith T. Holcomb, USMC *(ret.)*
Brig Gen Terry L. Holden, USA *(ret.)*
Brig Gen William A. Holland, USA *(ret.)*
Brig Gen Bob Hollingsworth, USMC *(ret.)*
Brig Gen William S. Hollis, USA *(ret.)*
Brig Gen Alben N. Hopkins, USA *(ret.)*
Brig Gen Ronald A. Hoppes, USA *(ret.)*
Brig Gen John D. Howard, USA *(ret.)*
Brig Gen Frank A. Huey, USMC *(ret.)*
Brig Gen Francis A. Hughes, USA *(ret.)*
Brig Gen Joseph C. Hurteau, USA *(ret.)*
Brig Gen Thomas R. Ice, USA *(ret.)*
Brig Gen Roderick J. Isler, USA *(ret.)*
Brig Gen Wesley V. Jacobs, USA *(ret.)*
Brig Gen Delbert H. Jacobs, USAF *(ret.)*
Brig Gen Gerald F. Janelle, USA *(ret.)*
Brig Gen Manning T. Jannell, USMC *(ret.)*
Brig Gen James M. Jellett, USA *(ret.)*
Brig Gen James C. Johnson, USA *(ret.)*
Brig Gen Julius F. Johnson, USA *(ret.)*
Brig Gen M. A. Johnson, USMC *(ret.)*
Brig Gen Alan D. Jones, USA *(ret.)*

Brig Gen John L. Jones, USA *(ret.)*
Brig Gen Thomas C. Jones, USA *(ret.)*
Brig Gen Thomas M. Jordan, USA *(ret.)*
Brig Gen Robert R. Jorgensen, USA *(ret.)*
Brig Gen James R. Joy, USMC *(ret.)*
Brig Gen Frederick J. Karch, USMC *(ret.)*
Brig Gen Kenneth J. Kavanaugh, USA *(ret.)*
Brig Gen Paul F. Kavanaugh, USA *(ret.)*
Brig Gen John H. Kern, USA *(ret.)*
Brig Gen Hugh T. Kerr, USMC *(ret.)*
Brig Gen Ronald K. Kerwood, AUS *(ret.)*
Brig Gen Boyd E. King, Jr., USA *(ret.)*
Brig Gen Roy L. Kline, USMC *(ret.)*
Brig Gen Jack H. Kotter, USA *(ret.)*
Brig Gen J. J. Krasovich, USMC *(ret.)*
Brig Gen Charles E. Kruse, USA *(ret.)*
Brig Gen John G. Kulhavi, USA *(ret.)*
Brig Gen Alan J. Kunschner, USA *(ret.)*
Brig Gen Stanley Kwieciak, Jr., USA *(ret.)*
Brig Gen Joseph G. Labrie, USA *(ret.)*
Brig Gen William H. Lanagan, USMC *(ret.)*
Brig Gen George A. Landis, USA *(ret.)*
Brig Gen Harvey T. Landwermeyer, Jr., USA *(ret.)*
Brig Gen Victor C. Langford III, USA *(ret.)*
Brig Gen Peter W. Lash, USA *(ret.)*
Brig Gen Richard M. Laskey, USA *(ret.)*
Brig Gen Jerry L. Laws, USA *(ret.)*
Brig Gen Dennis A. Leach, USA *(ret.)*
Brig Gen G. Dennis Leadbetter, USAF *(ret.)*
Brig Gen Gary E. LeBlanc, USA *(ret.)*
Brig Gen Douglas E. Lee, USA *(ret.)*
Brig Gen Robert C. Lee, USA *(ret.)*
Brig Gen Ward M. LeHardy, USA *(ret.)*
Brig Gen Samuel K. Lessey, Jr., USAF *(ret.)*
Brig Gen James H. Lewis, USA *(ret.)*
Brig Gen William Lindeman, USAF *(ret.)*
Brig Gen Roscoe Lindsay, Jr., USA *(ret.)*
Brig Gen Frederick R. Lopez, USMC *(ret.)*
Brig Gen Jay M. Lotz, USA *(ret.)*
Brig Gen Thomas P. Luczynski, USA *(ret.)*
Brig Gen James E. Mace, USA *(ret.)*
Brig Gen Pasquale J. Macrone, Jr., USA *(ret.)*
Brig Gen Peter T. Madsen, USA *(ret.)*
Brig Gen Paul M. Majerick, USA *(ret.)*
Brig Gen Wayne C. Majors, USA *(ret.)*
Brig Gen James G. Martin, USA *(ret.)*

Brig Gen Walter E. Mather, Jr., USA *(ret.)*

Brig Gen Philip M. Mattox, USA *(ret.)*

Brig Gen Paul A. Maye, USAF *(ret.)*

Brig Gen Bain McClintock, USMC *(ret.)*

Brig Gen Gerald B. McConnell, USA *(ret.)*

Brig Gen William L. McCulloch,
USMC *(ret.)*

Brig Gen Ronald V. McDougal, USA *(ret.)*

Brig Gen Robert P. McFarlin, USA *(ret.)*

Brig Gen Allan F. McGilbra, USA *(ret.)*

Brig Gen William F. McIntosh, USA *(ret.)*

Brig Gen Robert H. McInvale, Jr., USA *(ret.)*

Brig Gen Gerald L. McKay, USMC *(ret.)*

Brig Gen Max V. McLaughlin, USA *(ret.)*

Brig Gen Claude H. McLeod, USA *(ret.)*

Brig Gen A. P. McMillan, USMC *(ret.)*

Brig Gen Keith E. McWilliams, AUS *(ret.)*

Brig Gen James M. Mead, USMC *(ret.)*

Brig Gen Clayton E. Melton, USA *(ret.)*

Brig Gen Bruce T. Miketinac, USA *(ret.)*

Brig Gen Leonard D. Miller, USA *(ret.)*

Brig Gen Gerald L. Miller, USMC *(ret.)*

Brig Gen James E. Mitchell, USA *(ret.)*

Brig Gen Lawrence A. Mitchell, USAF *(ret.)*

Brig Gen Marvin E. Mitchiner, Jr., USA *(ret.)*

Brig Gen James M. Morris, USA *(ret.)*

Brig Gen Harry J. Mott III, USA *(ret.)*

Brig Gen John W. Mountcastle, USA *(ret.)*

Brig Gen Orlin L. Mullen, USA *(ret.)*

Brig Gen Benton D. Murdock, USA *(ret.)*

Brig Gen Joseph T. Murphy, USA *(ret.)*

Brig Gen Cecil Neely, USA *(ret.)*

Brig Gen Michael I. Neil, USMCR *(ret.)*

Brig Gen Harold J. Nevin, Jr., USA *(ret.)*

Brig Gen Joseph O. Nixon, USA *(ret.)*

Brig Gen J. W. Noles, USA *(ret.)*

Brig Gen Curtis D. Norenberg, USA *(ret.)*

Brig Gen Thomas P. O'Brien, Jr., USA *(ret.)*

Brig Gen Joseph E. Oder, USA *(ret.)*

Brig Gen George C. Ogden, Jr., USA *(ret.)*

Brig Gen James P. O'Neal, USA *(ret.)*

Brig Gen Michael B. Pace, USA *(ret.)*

Brig Gen Charles R. Painter, USA *(ret.)*

Brig Gen Peter J. Palmer, USA *(ret.)*

Brig Gen Ralph E. Parker, Jr., USMC *(ret.)*

Brig Gen Roland L. Parkhill, USA *(ret.)*

Brig Gen Edward A. Parnell, USMC *(ret.)*

Brig Gen Robert V. Paschon, USAF *(ret.)*

Brig Gen Terry L. Paul, USMC *(ret.)*

Brig Gen Frederick R. Payne, USMC *(ret.)*

Brig Gen Gary H. Pendleton, USA *(ret.)*

Brig Gen Michael J. Pepe, USA *(ret.)*

Brig Gen Mark V. Phelan, USA *(ret.)*

Brig Gen James H. Phillips, USA *(ret.)*

Brig Gen Bruce W. Pieratt, USA *(ret.)*

Brig Gen Jeffrey L. Pierson, USA *(ret.)*

Brig Gen Lloyd G. Pool, USMC *(ret.)*

Brig Gen Guido J. Portante, Jr., USA *(ret.)*

Brig Gen Robert Russell Porter, USMC *(ret.)*

Brig Gen Joseph N. Pouliot, USA *(ret.)*

Brig Gen Darryl H. Powell, USA *(ret.)*

Brig Gen Larry G. Powell, USA *(ret.)*

Brig Gen Richard O. Proctor, USA *(ret.)*

Brig Gen James D. Randall, Jr., AUS *(ret.)*

Brig Gen Richard D. Read, USA *(ret.)*

Brig Gen Stanley E. Reinhart, Jr., USA *(ret.)*

Brig Gen Thomas W. Reynolds, USA *(ret.)*

Brig Gen James C. Rinaman, USA *(ret.)*

Brig Gen Harold E. Roberts, USA *(ret.)*

Brig Gen Charles S. Robertson, USMC *(ret.)*

Brig Gen Domenic P. Rocco, Jr., USA *(ret.)*

Brig Gen Jose M. Rosado, USA *(ret.)*

Brig Gen Roswell E. Round, Jr., USA *(ret.)*

Brig Gen Roger E. Rowe, USA *(ret.)*

Brig Gen Floyd L. Runyon, USA *(ret.)*

Brig Gen Theodore R. Sadler, USA *(ret.)*

Brig Gen Walter R. Schellhase, USA *(ret.)*

Brig Gen John R. Schmader, USA *(ret.)*

Brig Gen Eugene W. Schmidt, USA *(ret.)*

Brig Gen John K. Schmitt, USA *(ret.)*

Brig Gen Joseph D. Schott, USA *(ret.)*

Brig Gen Joseph Schroedel, USA *(ret.)*

Brig Gen Lawrence R. Seamon, USMC *(ret.)*

Brig Gen Michael L. Seely, USA *(ret.)*

Brig Gen Robert L. Sentman, USA *(ret.)*

Brig Gen James E. Shane, Jr., USA *(ret.)*

Brig Gen Jerome M. Shinaver, Jr., USA *(ret.)*

Brig Gen Denis L. Shortal, USMC *(ret.)*

Brig Gen David V. Shuter, USMC *(ret.)*

Brig Gen Elmer O. Simonson, USA *(ret.)*

Brig Gen Paul D. Slack, USMC *(ret.)*

Brig Gen James D. Slavin, Jr., USA *(ret.)*

Brig Gen John W. Smith, USA *(ret.)*

Brig Gen Joseph A. Smith, USA *(ret.)*

Brig Gen Jerry C. Smithers, USA *(ret.)*
Brig Gen Charles E. St. Arnaud, USA *(ret.)*
Brig Gen Vincent E. Stahl, USA *(ret.)*
Brig Gen Jon A. Standridge, USA *(ret.)*
Brig Gen Jackie D. Stephenson, USA *(ret.)*
Brig Gen Velton R. Stevens, USA *(ret.)*
Brig Gen Robert L. Stewart, USA *(ret.)*
Brig Gen Herbert G. Stocking, USA *(ret.)*
Brig Gen Thomas G. Stone, USA *(ret.)*
Brig Gen Robert J. Strader, Sr., USA *(ret.)*
Brig Gen Joseph S. Stringham, USA *(ret.)*
Brig Gen James J. Sullivan, USA *(ret.)*
Brig Gen Thomas J. Sullivan, USA *(ret.)*
Brig Gen Russell E. Summerlin, USA *(ret.)*
Brig Gen Russell H. Sutton, USMC *(ret.)*
Brig Gen Thomas E. Swain, USA *(ret.)*
Brig Gen Burt S. Tackaberry, USA *(ret.)*
Brig Gen Lance A. Talmage, Sr., USA *(ret.)*
Brig Gen Hugh B. Tant III, USA *(ret.)*
Brig Gen Robert W. Taylor, USMC *(ret.)*
Brig Gen James A. Teal, Jr., USA *(ret.)*
Brig Gen Samuel S. Thompson III, USA *(ret.)*
Brig Gen Hoyt E. Thompson, USA *(ret.)*
Brig Gen Stanley R. Thompson, USA *(ret.)*
Brig Gen W. H. J. Tiernan, USMC *(ret.)*
Brig Gen Alfred E. Tobin, USA *(ret.)*
Brig Gen Warren A. Todd, Jr., USA *(ret.)*
Brig Gen William T. Tolbert, USAF *(ret.)*
Brig Gen Frank J. Toney, Jr., USA *(ret.)*
Brig Gen Peter D. Tosi, Jr., USA *(ret.)*
Brig Gen Floyd H. Trogdon, USAF *(ret.)*
Brig Gen Gary R. Truex, USA *(ret.)*
Brig Gen John S. Tuck, USA *(ret.)*
Brig Gen Richard J. Valente, USA *(ret.)*
Brig Gen Jose M. Vallejo, USA *(ret.)*
Brig Gen James R. Van Den Elzen,
  USMC *(ret.)*

Brig Gen Sharon K. Vander Zyl, USA *(ret.)*
Brig Gen R. L. Vogel, USMC *(ret.)*
Brig Gen Rudolph F. Wacker, USAF *(ret.)*
Brig Gen John D. Wakelin, USA *(ret.)*
Brig Gen George H. Walker, Jr., USA *(ret.)*
Brig Gen James E. Walker, USA *(ret.)*
Brig Gen Joseph M. Waller, USAF *(ret.)*
Brig Gen John R. Walsh, USA *(ret.)*
Brig Gen Floyd J. Walters, Jr., USA *(ret.)*
Brig Gen Larry Ware, USA *(ret.)*
Brig Gen William B. Watson, USA *(ret.)*
Brig Gen Clark C. Watts, USA *(ret.)*
Brig Gen Robert H. Wedinger, USA *(ret.)*
Brig Gen William Weise, USMC *(ret.)*
Brig Gen Arthur A. Weller, USA *(ret.)*
Brig Gen Arvid E. West, Jr., USA *(ret.)*
Brig Gen William A. West, USA *(ret.)*
Brig Gen Richard W. Wharton, Jr., USA *(ret.)*
Brig Gen William A. Whitlow, USMC *(ret.)*
Brig Gen Stanley J. Wilk, USA *(ret.)*
Brig Gen Dennis A. Wilkie, USA *(ret.)*
Brig Gen Teddy E. Williams, USA *(ret.)*
Brig Gen Sherman H. Williford, USA *(ret.)*
Brig Gen Mary C. Willis, USA *(ret.)*
Brig Gen James L. Wilson, USA *(ret.)*
Brig Gen Paul G. Wisley, USA *(ret.)*
Brig Gen Rodney D. Wolfe, USA *(ret.)*
Brig Gen Columbus M. Womble, USA *(ret.)*
Brig Gen Kenneth F. Wondrack, USA *(ret.)*
Brig Gen Edward H. Wulgaert, USA *(ret.)*
Brig Gen Mitchell M. Zais, USA *(ret.)*
Brig Gen Frederick A. Zehrer, USAF *(ret.)*
Brig Gen John G. Zierdt, Jr., USA *(ret.)*

Notes

1. Former Assistant Commandant, US Marine Corps.
2. Former Commander in Chief (CINC), Pacific Air Forces.
3. Former Assistant Commandant, US Marine Corps.
4. Former Vice Chief of Staff, US Army.
5. Former CINC, US Army Forces Command (FORSCOM).
6. Former CINC, USAF Military Airlift Command.
7. Former Assistant Commandant, US Marine Corps.
8. Former Assistant Commandant, US Marine Corps.
9. Former Assistant Commandant, US Marine Corps.
10. Former Chief of Staff, US Air Force.
11. Former Chief of Staff, US Air Force.
12. Former CINC, Southern Command.
13. Former Commander, Air Combat Command.
14. Former Commander, US Pacific Command.
15. Former Chief of Naval Operations, US Navy.
16. Former CINC, North American Aerospace Command and Commander, Air Force Space Command.
17. Former Vice Chief of Naval Operations, US Navy.
18. Former Commandant, US Marine Corps.
19. Former CINC, Joint Forces Command.
20. Former CINC, US Air Forces in Europe and Commander, Allied Air Forces Central Europe.
21. Former Commander, US Army Europe.
22. Former CINC, Readiness Command and CINC, Special Operations Command.
23. Former CINC, US Pacific Fleet and Deputy Chief of Naval Operations, where he was principal advisor on all Joint Chiefs of Staff matters.
24. Former Assistant Commandant, US Marine Corps.
25. Former CINC, US Atlantic Fleet.
26. Commanded Sub-unified Command in Korea. Deceased, 14 January 2009. Mrs. Menetrey signed (with Power of Attorney) and stated that her husband "believed strongly in this law and would want his name added."
27. Former Chief of Staff, US Army.
28. Former Assistant Commandant, US Marine Corps.
29. Former Commandant, US Marine Corps.
30. Former CINC, US Southern Command.
31. Former Commander, US Army Europe.
32. Former Commander, US Army Europe.
33. Former Chairman of the Joint Chiefs of Staff.
34. Former Commander, US Army Forces Command (FORSCOM).
35. Former Commander, Air Force Systems Command.
36. Former CINC, Southern Command.
37. Former CINC, Special Operations Command.
38. Former Chairman Joint Chiefs of Staff.
39. Former CINC, US Air Forces in Europe and Commander, Allied Air Forces Central Europe.
40. Former Commanding General, US Army Training and Doctrine Command (TRADOC).
41. Former CINC, Readiness Command.
42. Former Assistant Commandant, US Marine Corps.
43. Former Chief of Staff, US Army.
44. Former Assistant Commandant, US Marine Corps.
45. Medal of Honor recipient.
46. Medal of Honor recipient.

CHAPTER 12

DEFENDING THE CULTURE
OF THE MILITARY

Elaine Donnelly

Statement of Priorities and Overview

Any discussion of the issue of gays in the military should begin with a statement of priorities. In the formulation of personnel policies, equal opportunity is important, but if there is a conflict between equal opportunity and military necessity, the needs of the military must come first.

Assigning higher priority to equal opportunity, at the expense of military necessity, opens the door to a wide range of problematic social policies. The campaign to repeal Section 654, Title 10, U.S.C., the 1993 law regarding homosexuals in the military, which is usually mislabeled "Don't Ask, Don't Tell," is a prime example of misplaced priorities.

Members of Congress should ask a basic question: Would repeal of the law Section 654, Title 10 improve or undermine discipline, morale, and overall readiness in the all-volunteer force? In 2009 more than 1,160 high-ranking retired flag and general officers—51 of them retired four-star officers—personally signed a public statement expressing great concern that repeal of the law would weaken unit cohesion, discipline, and combat effectiveness:

> We believe that imposing this burden on our men and women in uniform would undermine recruiting and retention, impact leadership at all levels, have adverse effects on the willingness of parents who lend their sons and daughters to military service, and eventually break the All-Volunteer Force.[1]

Some advocates argue that this statement reflects only the views of a previous generation, which are not relevant to young people today. But there are reasons why twenty-somethings do not make policies for an institution that puts men and women into harm's way. Experience matters. The counsel of leaders with invaluable experience should not be dismissed so lightly. Nor should younger counterparts—the flag and general officers of tomorrow—be punished and forced out of the military if they hold similar views.

The armed forces are organizationally strong. All branches and communities of the military have proud histories, cultural traditions, and members motivated by patriotism as well as personal career goals. The institutional strength of the military, however, makes it vulnerable to political pressures that can undermine its culture. Military tradition requires obedience to lawful authority that is—as it should be—subject to civilian control.

Controversy occurs when civilian officials attempt to impose questionable policies and practices on the armed forces in pursuit of misplaced priorities. Such policies, designed to put egalitarian goals to the ultimate test, frequently conflict with classic elements of military culture. Because the armed forces differ from the civilian world in many respects, an inherent tension exists between sociological goals and the needs of the military.

Unit cohesion, for example, is essential for a strong military force. Cohesion is more than being liked by others; it is a willingness to die for someone else. Horizontal cohesion within a given unit involves mutual dependence for survival in combat.[2] Vertical cohesion is the bond of trust that must exist between the commander in chief, subordinate leaders, and the troops they lead.[3]

Both types of cohesion develop from strong bonds of mutual confidence, trust, and discipline that make survival possible under chaotic wartime conditions. Military discipline does not just happen—it must be taught by leaders who have the trust of people who will live, and sometimes die, under their command. Essential elements of military culture foster qualities that are not duplicated anywhere in the civilian world, including selfless courage under fire during war far from home.

Without essential factors such as unit cohesion, discipline, and high morale, the armed forces would degrade into disorganized cohorts of self-interested and leaderless young people armed with lethal weapons. This is why morale and the culture of the military, defined most simply as "how things are done," must be guarded at all times and never taken for granted. As columnist Thomas Sowell wrote, "Military morale is an intangible, but it is one of those intangibles without which the tangibles do not work."[4]

Legislative History of Section 654, Title 10, U.S.C.

In 1993 Pres. Bill Clinton attempted to lift the ban on homosexuals in the military. It was one of the most contentious efforts of his administration, sparking months of intense debate. Following 12 legislative hearings and field trips, Congress passed a law codifying the pre-Clinton policy. That statute, technically named Section 654, Title 10, U.S.C.,[5] frequently is mislabeled "Don't Ask, Don't Tell." The statute clearly states that homosexuals are not eligible for military service, and federal courts have upheld it as constitutional several times.[6]

Members of Congress seriously considered a concept known as "Don't Ask, Don't Tell," which Pres. Bill Clinton formally proposed on 19 July 1993. The proposal suggested that homosexuals could serve in the military as long as they didn't *say* they were homosexual. Congress wisely rejected the convoluted concept and did not write it into law.[7]

Members recognized an inherent inconsistency that would render the proposed "Don't Ask, Don't Tell" policy unworkable and indefensible in court: If homosexuality is not a disqualifying characteristic, how could the armed forces justify dismissal of a person who merely reveals the presence of such a characteristic? Instead of approving such a legally questionable concept, Congress chose to codify Department of Defense (DOD) regulations that were in place long before Bill Clinton took office.[8]

The resulting law, Section 654, Title 10, U.S.C., codified the long-standing DOD policy stating that homosexuals are not eligible for military service. Following extensive debate in both houses, the legislation passed with overwhelming, veto-proof bipartisan majority votes.[9] In writing this law, members wisely chose statutory language almost identical to the 1981 DOD directives regarding homosexual conduct, which stated "homosexuality is incompatible with military service." Those regulations had already been challenged and upheld as constitutional by the federal courts.[10]

The 1993 statute was designed to encourage good order and discipline, not the situational dishonesty inherent in "Don't Ask, Don't Tell." Having rejected that concept, Congress chose instead to codify unambiguous findings and statements that were understandable, enforceable, consistent with the unique requirements of the military, and devoid of the First Amendment conundrums that were obvious in President Clinton's 19 July proposal.

Among other things, the law states that "military life is fundamentally different from civilian life," and standards of conduct apply "whether the member is on base or off base, and whether the member is on duty or off duty." It further notes that members of the armed forces must "involuntarily . . . accept living conditions and working conditions that are . . . characterized by forced intimacy with little or no privacy." Therefore, "the prohibition against homosexual conduct is *a long-standing element of military law that continues to be necessary* in the unique conditions of military service" (emphasis added).

These findings and statements are very different from the language proposed by Bill Clinton on 19 July 1993, which Congress did not write into law: "Sexual orientation is considered a personal and private matter, and homosexual orientation is not a bar to service entry or continued service unless manifested by homosexual conduct."[11]

A thorough search of media reports at the time reveals that there were few news stories reporting passage of the law, and those that did appear in print failed to report its language and meaning with accuracy. Those news accounts and contradictory DOD statements since then have confused the issue by erroneously suggesting that Congress voted for Pres. Bill Clinton's flawed proposal, known by the catch-phrase "Don't Ask, Don't Tell."[12] The situation brings to mind a statement of Oliver Wendell Holmes, quoted by *National Review* editor Rich Lowry and others: "A good catchword can obscure analysis for 50 years."

Describing the law as a "compromise" and referring to it as "Don't Ask, Don't Tell" gave political cover to President Clinton, who had promised to lift the ban shortly after his election in 1992. In fact, due to overwhelming public opposition, President Clinton failed to deliver on his promise. The only compromise involved allowed the Clinton administration to continue its interim policy of not asking "the question" regarding homosexuality that used to appear on routine induction forms.[13]

This politically expedient concession on a matter of process was ill-advised, but it did not nullify the language and substance of the actual law. The statute also includes language that authorizes the secretary of defense to reinstate the question about homosexuality at any time, without additional legislation.[14]

Differences between the Law and "Don't Ask, Don't Tell"

It is no accident that the vague phrase "sexual orientation," the key to Bill Clinton's original "Don't Ask, Don't Tell" proposal, does not appear anywhere in the law that Congress actually passed. Members of Congress recognized that the phrase would be difficult to define or enforce. Instead, the law is firmly based on conduct, evidenced by actions or statements.

Absent unusual circumstances, a person who says that he is homosexual is presumed to engage in the conduct that defines what homosexuality is. Using the same logic, a person who says he is a philanthropist is presumed to give away money—the conduct that defines what a philanthropist is. It is not necessary for an individual to be "caught in the act" for the eligibility law to apply.

The law should have been given a name of its own, such as the "Military Personnel Eligibility Act of 1993." Differences between the law and the Clinton administrative policy explain why opposing factions are critical of "Don't Ask, Don't Tell." Even though Congress rejected the concept in 1993, with good reason, the Clinton administration imposed it on the military anyway in the form of enforcement regulations that were announced in December 1993. Those expendable regulations, unfortunately, remain in effect today.[15]

In 1996 the US Court of Appeals for the Fourth Circuit said in a ruling upholding the constitutionality of the law that the Clinton administration's enforcement policies ("Don't Ask, Don't Tell") were not consistent with the statute that Congress actually passed (Section 654, Title 10, U.S.C.).[16] The Clinton administration disregarded the Court of Appeals and perpetuated deliberate confusion by retaining the inconsistent "Don't Ask, Don't Tell" policy in DOD enforcement regulations.[17]

Problems with the "Don't Ask, Don't Tell" Administrative Policy

President Clinton's convoluted "Don't Ask, Don't Tell" regulations were and still are inefficient and contrary to sound policy. In the civilian world it would be tantamount to a state law forbidding store and bar owners to check ID before selling liquor to younger customers. Such a law would force the proprietor of a bar to assume the risk that if an underage customer drives and accidentally kills someone on the way home, the proprietor will be held liable. That risk is reduced by the posting and enforcement of signs stating "We Check ID."[18]

Properly enforced liquor control laws protect the public interest even if some 18-year-olds successfully conceal or lie about their age and some adults do not ask for proof. It would not be accurate to claim, however, that the age of customers is "personal and private," and state law allows 18-year-olds to drink alcohol as long as they do not *say* they are underage.

This is, however, how the "Don't Ask, Don't Tell" policy works. It forbids the Department of Defense to include on induction forms a routine inquiry regarding homosexuality that would help to determine eligibility for military service.

The omission of that question and the lack of consistent, accurate information regarding the law mislead potential recruits about their eligibility to serve. Homosexualist leaders,[19] who want government power to impose their agenda on the military, are well aware of what the law actually says and are a large part of this problem.

Groups such as the Servicemembers Legal Defense Network (SLDN) and the Human Rights Campaign (HRC) constantly attack the wrong target—an administrative policy that Congress did not inscribe in law. Their multimillion-dollar public relations campaign exploits human interest stories demonstrating problems that members of Congress predicted when they rejected Bill Clinton's 19 July 1993 "Don't Ask, Don't Tell" proposal. Many personal dilemmas could have been avoided if the Department of Defense clearly explained to potential inductees the meaning of the 1993 Eligibility Law.

Many well-meaning people who may not understand the issues involved are opposed to the convoluted "Don't Ask, Don't Tell" policy or think it needs to be reviewed. They are correct—Congress did not vote for the Clinton "sexual orientation" policy and the secretary of defense should have exercised the option to drop it long ago. "Don't Ask, Don't Tell" diversions, however, should not preclude objective discussion of the consequences of repealing the 1993 Eligibility Law.

Consequences of Repealing the 1993 Eligibility Law, Section 654, Title 10, U.S.C.

Legislation to repeal the 1993 law, H.R.1283, was introduced in the 111th Congress by Rep. Ellen Tauscher (D-CA), who has been replaced as primary sponsor by Rep. Patrick Murphy (D-PA). The Murphy bill, which would apply retroactively, would forbid discrimination based on "homosexuality or bisexuality, whether the orientation is real or perceived."

If Congress approves Congressman Murphy's new lesbian, gay, bisexual, transgender (LGBT) law, commanders, mid-level career officers, and noncommissioned officers (NCOs) would be required to determine how the open-ended "real or perceived" legislative language would apply. Federal courts asked to interpret the new "nondiscrimination" paradigm are likely to extend it to all sexual minorities, including transgendered individuals perceiving themselves to be persons of the opposite sex.[20]

"Forced Intimacy" Unlike the Civilian World

The new LGBT law would govern the lives of men and women in all military branches and communities, including Army and Marine infantry battalions, special operations forces, Navy SEALS, and submarines. Unlike civilians, in these communities military personnel do not return home at night after work. They must accept living conditions involving what the 1993 Eligibility Law describes as "forced intimacy," offering little or no privacy.

A law mandating the inclusion of professed (not just discreet) homosexuals and bisexuals in this high-pressure environment, 24/7, would be tantamount to forcing female soldiers to share private living quarters with men. Such a situation would be unacceptable to the majority of military women even if actual assaults never occurred. Stated in gender-neutral terms, the military would require military *persons* to accept exposure to *persons* who may be sexually attracted to them.

We want and need women in our military, and personnel policies work best when they encourage discipline rather than indiscipline. This is why the military separates men from women in close quarters where there is little or no privacy, to the greatest extent possible. Sexual tension or misconduct of any kind is inherently disruptive whether it occurs on the romantic end of the behavioral spectrum or on the other end where harassment or sexual assaults occur.

The new nondiscrimination law requiring cohabitation with homosexuals or bisexuals, "whether the orientation is real or perceived," would disregard what we know about men and women in the military. The imagined "gender-free" culture desired by theorists exists nowhere on Earth, except in Hollywood's social science fiction movies.

Some advocates of gays in the military argue that modern military facilities provide more privacy than older ones, and even if people are exposed to sexual minorities in the field, younger people are used to it, and this is not a big deal.[21]

But the armed forces are not a *Will & Grace* world, created by television sitcom writers for laughs. The issue involves sexuality and the normal human desire for personal privacy and modesty in sexual matters. Elitist arguments equating sexual differences with skin-deep, irrelevant racial differences stand in stark contrast with commonsense customs that are culturally routine.[22]

Consider, for example, a typical family-oriented community recreation center that has separate locker rooms for men and women. Inside the entrance of the women's locker room, a sign clearly states that boys of any age are not permitted. A similar sign regarding girls is posted in the men's locker room.

The signs are there not as an affront to young boys (or girls). They are there because the community respects the desire for sexual modesty in conditions involving personal exposure to others using the same facility. This is so even though people using the recreation center visit for only an hour or two; they do not live and sleep there for months at a time.

Signs mandating racial segregation in the same community center would never be acceptable. Racial segregation has no rational basis; separation by gender does. Military volunteers deserve the same consideration.

Predictable Sexual Misconduct

If repeal of the law forces the military to disregard basic human psychology, risks of demoralizing misconduct will escalate to include male/male and female/female incidents, in addition to those that already occur. Predictable tensions ensuing from this unprecedented and provocative social experiment would constantly increase the stress of daily life and generate the full range of emotional turmoil, accusations, and legal jeopardy that undermines individual and unit morale.[23]

Some advocates of repeal try to end objective debate by accusing anyone concerned about these issues of somehow insulting the troops. The attempt at intimidation fails due to logic. Various types of sexual misconduct occur in the military because men and women are human and therefore imperfect. It is not an affront to anyone to state a simple fact: Human beings are not perfect, and homosexuals are no more perfect than anyone else.

Equality in Elevated Risks

Activists demanding repeal of the law dismiss concerns about sexual misconduct by claiming that existing regulations against heterosexual misconduct can and will be equally applied to misconduct involving openly gay personnel. This is an unrealistic, elitist argument, which was addressed in a House Armed Services Committee Report:

> The committee . . . heard a recommendation that the department should, as a matter of policy, enforce the Uniform Code of Military Justice [UCMJ] equally on heterosexuals and homosexuals. . . . The committee believes that such an eventuality is neither conducive to justice nor discipline. Violations of the [UCMJ] ought to be prosecuted on their individual merits, without an effort to compel the department to equalize prosecutions among groups of people, offenses, or artificially comparative categories.[24]

Reliance on "equal" prosecutions after the fact of harassment or worse would be small comfort to personnel forced to live in conditions that encourage inappropriate, passive/aggressive behavior conveying an unwelcome sexual message. Many women, both civilian and military (including this author), have experienced such behaviors, which are disturbing but do not involve physical assault that would spark disciplinary intervention or prosecution.

Members of Congress who have investigated and expressed outrage about such behavior when it involves women in the military should be among the first to anticipate and try to prevent predictable problems. Despite constant professional training and "leadership," unwelcome sexual tension occurs and causes division in groups that need to be cohesive in order to be effective.

Brian Maue, PhD, an Air Force major and instructor at the Air Force Academy, addressed this issue in the *New York Times*. Dr. Maue pointed out that a sexual preference–mixed atmosphere in the military would create conditions comparable to what feminists describe as a "hostile work environment":

Consider that the U.S. military does not allow swimsuit calendars in its workplaces because they can negatively affect the morale of female military members. . . . For example, if a female soldier was sexually uncomfortable with the way a male soldier looked at her, she or anyone who witnessed the situation could file a complaint, even if the man thought that his glance was not done in a sexually aggressive manner. . . .

Thus, if the morale of a heterosexual female military member can be negatively affected by a swimsuit calendar or by the behavior of a male soldier with no sexual interest in her, could she lodge a similar "hostile environment" complaint if she was forced to share a bathroom, a locker room or a bedroom (say, in a tent or in the barracks) with a lesbian soldier who has no sexual interest in her?

The military has traditionally prevented unnecessary privacy violations and complaints by separating men and women wherever privacy issues could arise. . . .

. . . Combining sexual preferences (i.e., lesbians with heterosexual women) would challenge American military commanders with privacy violations and dignity infractions that would reduce unit effectiveness.[25]

Any attempt to "equalize" regulations between heterosexuals and sexual minorities would lead to constant inconsistencies, persistent doubts about appropriate sexual expression, and an incremental erosion of personal discipline standards.

Equal Enforcement and the Lt Col Victor Fehrenbach Case

It is significant to note that many of the most outspoken advocates of gays in the military also demand the repeal of what they call "antiquated" provisions of the UCMJ that impose higher standards of personal conduct than exist in the civilian world.[26] The highly publicized case of Air Force Lt Col Victor Fehrenbach, an 18-year F-15 weapons systems officer,[27] demonstrates how "equality" might work to erode and eventually lead to the repeal of personal conduct sections of the UCMJ.

Colonel Fehrenbach became a public figure when he protested an honorable discharge resulting from his admission of homosexual conduct, which had been revealed by someone else. An investigative report in the 23 August 2009 *Idaho Statesman* revealed a more distasteful story relevant to the national debate.[28]

Prior to the *Statesman* report, supporters tried to generate sympathy for Fehrenbach because he had been "outed" by a third party. That person turned out to be Cameron Shaner, a criminal justice student who told the Boise police that he met Victor Fehrenbach through a gay Web site. Shaner reportedly went to the aviator's home on 12 May 2008, after Fehrenbach invited him with a text message and "stud" photographs.

According to the *Statesman*, Shaner did not explain why he "got naked" with Fehrenbach in a hot tub, but at 3:00 a.m. he called Boise police to report a sexual assault. Fehrenbach asserted that the encounter was consensual and was cleared of the rape charge, but his admission of homosexual conduct triggered discharge proceedings. Under the 1993 Eligibility Law, persons who engage in homosexual conduct at any time, on- or off-base, are not eligible for military service.

Colonel Fehrenbach deserves respect for participating in the 2003 liberation of Baghdad. The fact remains that despite provisions of the UCMJ (Article

131) that impose higher standards for "officers and gentlemen," Fehrenbach showed very poor judgment.

One of Fehrenbach's lawyers claimed that if his accuser had been a woman, "he'd have gone back to work with no further issue." Dozens of former naval aviators whose careers were ruined by the 1991 Tailhook scandal, some even without evidence of misconduct, certainly would disagree.[29]

Consider what would happen if a military officer posted nude photographs of himself and used Craigslist to obtain sex from an unknown woman who subsequently accused him of rape. Even if assault never happened, under the UCMJ that man's career would be over. Fehrenbach and his allies are demanding special treatment just because his conduct was homosexual rather than heterosexual. "Equal" enforcement would lower standards, weaken discipline, and vitiate the culture of the military.

If Rep. Barney Frank (D-MA) and other homosexualists successfully repeal what they call "antiquated" rules governing personal sexual conduct and make the UCMJ consistent with the proposed LGBT law, a wide range of personal conduct regulations would become a thing of the past. Special treatment for Fehrenbach, effectively permitting admitted misconduct if it is consensual, would define discipline down.

Regulations do not allow unmarried heterosexuals to live and sleep with persons of the opposite sex in military close quarters. How would it work if gays and lesbians get to share close quarters with "significant others," but heterosexual colleagues are denied the same comforts? Unit cohesion weakens when people pair off in sexual relationships, causing others to wonder where their primary allegiance lies.

Personal Reluctance to Report Sexual Tension or Physical Abuse

When a female soldier reports an incident of sexual harassment or abuse, she enjoys the presumption of truthfulness. But under the new LGBT law, if a male soldier reports an incident of homosexual harassment or abuse, he will face the suspicion, if not the presumption, of unacceptable attitudes toward fellow soldiers who are homosexual.

Both male and female heterosexuals whose sexual privacy and values are violated by the new LGBT law will hesitate to file complaints, lest they be suspected or accused of prejudiced attitudes that violate the new "zero tolerance" policy favoring homosexuals in the military. Having no recourse, many will leave the all-volunteer force.

When problems occur, commanders will face the thankless burden of trying to find out what happened and who was responsible for what. Regardless of the he-said or she-said details, in emotionally charged disputes such as this, the consequences would be the same, tearing individual units apart.

There are many personal reasons why women hesitate to file complaints when unwanted sexual approaches occur—embarrassment, intimidation by a superior, fear of not being believed, and so forth. Heterosexual men confronted with the same type of approaches from other men would face all of the factors that deter women, plus the additional concern that a complaint might lead to questions about their own sexuality. Among men, such insinuations are considered "fighting words."

A March 2008 story in *Clinical Psychiatry News*, quoting speakers at an annual meeting of the International Society for Traumatic Stress Studies, reported that "male veterans who have a history of military sexual trauma often fail to disclose their condition until well into treatment for post-traumatic stress disorder, and have many motivations for covering up their problems."[30]

According to a special report in the *Florida Times* quoting Veterans Affairs psychologists, a unique program designed to counsel veterans, particularly men who were raped or sexually assaulted in the military, found that men are even more reluctant to report such incidents and subsequent problems than women are. "Military men do not report the attacks because they fear no one will believe them, their careers will be damaged, they will be labeled homosexuals or they will suffer retribution from the attackers or their commanders."[31]

In an article about male military sexual trauma (MST), Harvard Medical School psychology instructor Jim Hopper commented, "When they get assaulted, they're unprepared to deal with their vulnerable emotions. They resist seeking help. They believe that their hard-earned soldier-based masculinity has been shattered." Gay activists writing on favorite Web sites frequently deride or ridicule such concerns about personal privacy, berating anyone who even mentions the subject.[32]

Institutional Barriers to Full Disclosure of Problems

A *Navy Times* editorial reported that incidents of male sexual assault often are underreported and may be more prevalent in the military than in other parts of society. *Navy Times* further reported that unlike the civilian judicial system, military courts do not offer a publicly accessible docket of pending court-martial cases. As a result, "military commanders release that information at will, giving them unmatched control over information that should be out in the open."[33]

Two cases summarized below demonstrate the risks of sexual abuse that could occur, with little or no public notice, if the 1993 Eligibility Law is repealed.

Navy Lt Cmdr John Thomas Lee. Lt Cmdr J. T. Lee, a 42-year-old Catholic priest, was a Navy chaplain who tested positive for HIV, an indicator of AIDS, in 2005. Between 2003 and 2007, Chaplain Lee was assigned to counsel midshipmen at the US Naval Academy and Marines at Quantico, VA. According to court testimony and factual stipulations signed by Lee and Navy prosecutors, Lee committed numerous sexual offenses with a young midshipman, an Air Force lieutenant colonel, and a Marine corporal. His conduct was all the more reprehensible due to his undisclosed HIV-positive status and the betrayal of trust associated with his role as a priest and chaplain.[34]

The *Washington Post* reported on 7 December 2007 that Lieutenant Commander Lee pleaded guilty to several serious charges, but nevertheless got off with a 12-year prison sentence reduced to two, with only 18 months to be served. The plea bargain effectively swept the case under the rug with little public awareness that the scandal even happened.

A surprisingly candid article in *Newsweek* stated that according to a 2007 report, up to 60 military chaplains were convicted or strongly suspected of committing sexual abuse over the past four decades, sometimes against the children of military personnel.[35] Studies suggest that sexual assault among military men is most prevalent among junior enlisted ranks.[36]

According to a recent *Navy Times* article about sexual misconduct, a Navy Department online survey of about 85,000 Sailors and Marines found that reports of male-on-male sexual assaults have increased sharply, up to about 7 percent from 4 percent in 2004. Navy official Jill Loftus indicated that reasons for the increased reports were unclear, but resources for men experiencing sexual assault are few in comparison to those available to women. She added that some commanders of all-male units told Navy officials that they didn't need sexual assault training or coordinators because they assumed they were not needed with only men in their units. The required inclusion of openly gay and bisexual personnel in all-male and mixed-gender units would worsen the underlying problem, not improve it.[37] Chief of Naval Operations Adm Gary Roughead, who had previously dismissed such reports as "anecdotal," should order a full investigation and a detailed report on all alleged male-on-male assaults. Absent such a review, claims that there have been no problems with discreet gays in the military should not be considered reliable.

Pfc Johnny Lamar Dalton. In 2007 Pfc Johnny Lamar Dalton, 25, was charged with assault with a deadly weapon—the HIV virus.[38] Dalton reportedly disobeyed orders by having unprotected, consensual sex with an 18-year-old, who became HIV-positive shortly after the encounter with Dalton. The Associated Press reported that Dalton pleaded guilty to assault for unprotected sex and was sentenced to 40 months in prison, reduction in rank, and a dishonorable discharge.[39]

In answer to an inquiry from the Center for Military Readiness (CMR), an Army spokesman confirmed that Dalton's records would show only his criminal violations, not the lesser offense of homosexual conduct. This is standard practice, especially when authorities are mindful of the impact of charges on innocent family members.[40] For this reason, discharges that involve homosexual conduct may not be reported to the public or to members of Congress—now or in the future if Congress votes to repeal the 1993 law.

Nondeployability of HIV-Positive Personnel

Advocates of gays in the military consider concerns about the nondeployability of HIV-positive personnel to be a taboo subject.[41] Nevertheless, as this author stated in testimony before the House Armed Services Personnel Subcommittee, responsible officials who make policy for the military should give serious consideration to all consequences of repealing the 1993 law.[42]

To the greatest extent possible, the armed forces try to reduce or eliminate any behavior, or the propensity for behavior, which elevates risks of survival for any service member. Congress has recognized that all personnel fighting in a close combat environment may be exposed to the blood of their colleagues, and all are potential blood donors for each other. Persons found to be HIV-positive, therefore, are not eligible for induction into the military.

If serving members are diagnosed as HIV-positive, regulations require that they be retained for as long as they are physically able. The military provides appropriate medical care, but HIV-positive personnel are not eligible for deployment overseas.[43]

An examination of military HIV nondeployability cases shows that since the passage of Section 654, Title 10, the incidence of HIV servicewide has trended downward.[44] Reasons for the trend are not clear, but it is reasonable to expect that if the law is repealed and great numbers of men having sex with men are inducted into the military,[45] the line indicating nondeployable personnel who are HIV-positive probably would trend upward.

Given the officially recognized correlation between homosexual conduct and HIV infection, it is reasonable to expect that repeal of the law could increase the number of troops who require medical benefits for many years but cannot be deployed. At a time when multiple deployments are putting great stress on the volunteer force, Congress should not make a major change in policy that could increase the number of nondeployable personnel.

Military Families and Children

In Britain, one of the countries hailed as a role model for homosexual equality, same-sex couples live in military family housing.[46] Before voting to repeal the 1993 Eligibility Law, members of Congress should consider whether a similar "nondiscriminatory" housing policy would have negative effects on family retention in our military.

The British Ministry of Defence also meets regularly with LGBT activist groups to promote "anti-gay-bullying" programs, similar to controversial programs adopted in some American public school systems.[47]

Our military is likely to follow these examples, mandating programs to teach everyone how to get along with incoming homosexuals of all ages. If military parents are unable to opt out or change their children's schools, how would they react? No one should expect public protests against official intolerance in the name of "tolerance." Because our military is an all-volunteer force, families will simply leave.

Since the Department of Defense runs the largest school and childcare systems in the world, this would be a huge victory for homosexuals who want the military to become the cutting edge of radical cultural change. New, unprecedented practices ultimately would affect all institutions of American life, far beyond what is already happening today.

The Intolerance of "Zero Tolerance"

Once the military establishes an issue as a matter of "civil rights," it does not do things halfway. Passage of the new LGBT law would introduce corollary "zero tolerance" policies that would punish anyone who disagrees. Any military man or woman who expresses concerns about professed (not discreet) homosexuals in the military, for any reason, will be assumed "intolerant" and suspected of harassment, bad attitudes, or worse. Attitudes judged to be unacceptable will require disciplinary action and denials of promotions—penalties that end military careers.

Enforcement of the gay agenda in the military would be particularly divisive among men and women whose personal feelings and convictions are thrown into direct conflict with the new LGBT law and corollary zero tolerance policy. Among the first to be affected would be chaplains of major religions that disapprove of homosexuality for doctrinal or moral reasons. These would include major denominations of the Jewish, Christian (Catholic, Protestant, and Orthodox), and Muslim faiths. Likely issues of conscience would include personal counseling of same-sex couples and requests to perform marriages or to bless civil unions between same-sex couples.

The language of Section 654, Title 10 is completely secular, but individual service members who are practicing members of the religions mentioned above also would face choices involving matters of conscience. These would include the accommodation of same-sex couples in married/family housing and the introduction of personnel and curricula that promote the homosexual agenda in military base schools and childcare centers.

Even those who do not see this as a moral issue could be affected by cultural changes and mandates associated with official zero tolerance of dissent. At the House Armed Services Personnel Subcommittee hearing on 23 July 2008, a member of the committee asked retired Army Sgt Maj Brian Jones, who was testifying in support of the 1993 law, whether he saw the issue as a matter of religious conviction.

Jones, a former Ranger and Delta Force soldier who rescued fallen colleagues in the 1994 "Black Hawk Down" incident in Somalia, said that readiness for combat was his most important concern.[48] Mid-career and non-commissioned officers who are key leaders in combat-oriented communities could be hit with severe zero tolerance penalties just for expressing opinions similar to those of Brian Jones. Among these would be potential four-stars and senior NCOs who are needed to lead the military of tomorrow.

Carrots, Sticks, and Zero Tolerance Taken to Extremes

In a May 2009 report promoting a road map for repealing the 1993 law, the Palm Center provided insight into social difficulties that the activist group expects the military to overcome with conscious coercion.[49] In a three-page section of that report, subtitled "Organizational Changes that Should Accompany Policy Change," the authors used variations of the word "implementation," "enforcement," or "compliance," often in tandem with the word "problems," no less than 35 times.[50]

The largely civilian leaders of the Palm Center based their recommendations not on military history or experience, but on "social science research that has focused specifically on sexual orientation and on the open service of gays and lesbians in militaries abroad." Recommendations proceed from an erroneous premise, suggesting that military organizational culture is essentially a "theme" related to successful inclusion of racial minorities.[51] The inapt comparison underlies an apparent plan to *redefine* military culture as a means to advance social goals, not to achieve military objectives—that is, deterring or winning wars.

In this paragraph of the Road Map Report, the Palm Center confirmed consequences of zero tolerance that would have devastating effects on the culture of the military:

> Compliance with the new policy will be facilitated to the extent that personnel understand that *enforcement will be strict and that noncompliance will carry high costs*, and thus perceive that *their own self-interest* lies in supporting the new policy. Consequently, the implementation plan should include clear *enforcement mechanisms* and *strong sanctions for noncompliance*, as well as support for effective implementation in the form of adequate resources, allowances for input from unit leaders for improving the implementation process, and *rewards* for effective implementation. *Toward this end, the Defense Department should work to identify the most potent "carrots" and "sticks" for implementing the new policy.*[52] (emphasis added)

Under such a regime, the "most potent" career "carrots" would reward commanders who embrace the new law enthusiastically. Civilian and military commanders would be required to interpret and apply the law in all stages of training, education, and deployment and to do so under threat of career penalties if they fail to make it "work."

Career incentives for superior officers—recommended by the Palm Center as "carrots," "self-interest," or "rewards for effective implementation"—could create conflict with the expectation of "accurate information about implementation problems."[53] Human nature being what it is, some officers might be tempted to advance their own careers by reporting no issues of concern under the new law, even if they are aware that subordinates are experiencing demoralizing problems.

Other commanders might fear that accusations of unacceptable attitudes and poor leadership could sink their careers if they take the side of a heterosexual person over a homosexual one. The appearance of self-interest in the decisions of superior officers—an element that the Palm Center considers a *positive* thing—would undermine the bond of vertical cohesion and trust that must exist between commanders and the troops they lead.

Disciplinary "sticks," described as "strong sanctions for noncompliance," would deny promotions and end the military career of anyone who disagrees for any reason. This would force out of the military thousands of junior officers and enlisted personnel who are the land, sea, and air combat commanders, chiefs of staff, and senior enlisted advisors of tomorrow.

Involuntary losses of good people would compound the harmful effects of shortages caused when others decline reenlistment or avoid military service in the first place. It is impossible to justify the potential loss of valued future leaders such as this, incurred just to satisfy the demands of determined homosexualists and their civilian allies in academia and the media.

"Diversity" Training and Education

The Palm Center recommends that "military leaders must signal clearly that they expect all members of the armed forces to adhere to the new policy, regardless of their personal beliefs."[54] Coercive implementation would require what the Palm Center described as "surveillance and monitoring of compliance" combined with mandatory training programs to change attitudes and make the new gay-friendly policy work.[55]

Absent current law, the DOD will "salute smartly" and proceed to implement all-encompassing, "nondiscriminatory" training and education programs to enforce acceptance—even among mid-level commanders who would be forced to set aside their own objections in order to teach others. Success for such training would be far more difficult than historic programs designed to end discrimination and irrational prejudice against racial minorities. Mandatory sensitivity sessions will attempt to overcome the normal human desire for modesty and privacy in sexual matters—a quest that is inappropriate for the military and unlikely to succeed.

With the exception of lawyers needed to defend military personnel accused of "bad attitudes," the only people likely to benefit from the mandatory implementation of such programs would be LGBT advocates and professional diversity trainers that the Department of Defense invites to participate.

None of the time or expense involved in these activities would improve morale, discipline, or readiness in the all-volunteer force. Our military respects women and does not expect them to accept constant exposure to passive/aggressive approaches of a sexual nature. It should not be ordered to change personal feelings and beliefs about human sexuality.

Special Events and Sexual Expression

Gay activists expect special events and occasions to celebrate homosexual service members, in the same way that special days or months are scheduled to recognize minority groups and women in the military. Early in the Clinton administration, the Department of Defense sponsored a day-long "Diversity Day Training Event" in an Arlington, Virginia, Crystal City building near the Pentagon. Programs cosponsored with 18 other government agencies featured lectures, anti-Christian panel discussions, exhibits, workshops, and a controversial video titled "On Being Gay."[56]

In 2009, Pres. Barack Obama signed a statement proclaiming June to be "LGBT Pride Month." The Department of State and NASA followed with similar gay and lesbian pride proclamations and activities posted on their Web sites.[57]

Social events can have consequences. According to the *Washington Post*, in May 2009 employees of the American Embassy in Baghdad celebrated gay rights by sponsoring a "Pink Zone" theme party event at a pub called BagDaddy's.[58] Guests were invited to attend dressed in drag as their favorite gay icon. An embassy spokesman explained that social events are permitted there because there are no gathering places elsewhere in Baghdad. The same rationale could apply to military people serving on remote bases in war zones.

Consistency in gay-friendly social events would create a new *inconsistency* with policies requiring Americans to avoid practices considered offensive to the Muslim civilians and soldiers that Americans are supposed to train in combat or local security skills. The problem was presaged in July 2009, when the State Department came to regret an incident involving male security contractors in Kabul, Afghanistan.[59] The alcohol-besotted men partied wildly around a bonfire in a state of near-nudity—bacchanalian behavior that rivaled the most offensive abuses of Abu Ghraib.

Public nudity will not become acceptable in the military, but if the Pentagon follows the State Department's lead in equating consensual heterosexual and homosexual behavior, where will local commanders be able to draw the line? It is difficult to put one's foot down when there is no visible floor on which to place one's foot.

Advocacy not Evidence—Five Flawed Arguments

The Gays in the Military Campaign (GIMC) rarely addresses any of the consequences listed above. Instead, Rep. Patrick Murphy and the Human Rights Campaign, the nation's largest LGBT activist group, have been coordinating a multimillion-dollar campaign of media events in cities around the country, which began 8 July 2009 at the National Press Club in Washington, DC. The campaign has focused on the human interest stories of homosexuals who were honorably discharged due to the 1993 Eligibility Law.

Ensuing media reports rarely explain the eligibility issue or put the matter of discharges into perspective. Virtually all repeat standard arguments that sound plausible but do not withstand closer scrutiny. There are at least five flawed arguments that Congress should analyze critically before it votes to repeal the 1993 Eligibility Law.

1. The Civil Rights Argument

Advocates for repeal of the 1993 law constantly wrap their cause in the honored banner of "civil rights." The argument, however, is among the weakest. There is no constitutional right to serve in the military. Sometimes there is an obligation, as in times of war when conscription is imposed. But there is no "right" to serve; the military is not just another equal opportunity employer.

The Male/Female Analogy. Pres. Harry Truman's executive order to end racial discrimination in 1948 advanced civil rights, but its primary purpose was military necessity. Title VII of the 1964 Civil Rights Act does not apply to uniformed military personnel because its provisions might make it harder to confront enemies that are not subject to similar rules.[60]

The military's "can do" efforts to implement zero tolerance for racial prejudice have succeeded faster than in the civilian world because there is no rational justification for racial discrimination. Separation of men and women in circumstances affecting sexual privacy, however, is rational, reasonable, and usually appropriate in the civilian world as well as in the military. The late Charles Moskos, a respected military sociologist and former enlisted draftee, rejected the "black/white analogy" during his testimony before the Senate on 29 April 1993. Moskos asserted that it really is a "male/female analogy."[61]

Gen Colin Powell, who was chairman of the Joint Chiefs of Staff early in the Clinton administration, wrote a classic letter addressing the subject to then-Rep. Patricia Schroeder (D-CO) in 1993. Dismissing Schroeder's argument that his position reminded her of arguments used in the 1950s against desegregating the military, General Powell replied: "I know you are a history major but I can assure you I need no reminders concerning the history of African-Americans in the defense of their nation and the tribulations they faced. I am part of that history. . . . Skin color is a benign, non-behavioral characteristic. Sexual orientation is perhaps the most profound of human behavioral characteristics. Comparison of the two is a convenient but invalid argument."[62]

Columnist Charles Krauthammer agreed: "Powell's case does not just rest on tradition or fear. It rests on the distinct difference between men and women. Because the cramped and intimate quarters of the military afford no privacy, the military sensibly and non-controversially does not force men and women to share barracks."[63]

Dr. Brian Maue points out that the military policy regarding homosexuals is not arbitrary. When the introduction of large numbers of women changed the "sexual preference" makeup of the military, women were accommodated with an infrastructure of separate facilities: "When it comes to open homosexuality, however, another sexual preference would be added that cannot be accommodated separately, even if the military possessed a limitless budget. Homosexual advocates are not asking for equal rights, they are asking for an exception to the universal principle of separate sexual preferences in areas of close body proximity."[64]

Affirmative Action and Retroactive Consequences of a Civil Rights Standard. Campbell University law professor William A. Woodruff has expressed concern about the likely extension of the civil rights standard to logical extremes. Legislation (H.R. 1283) to repeal the 1993 law (Section 654, Title 10) would apply the civil rights model in all matters involving homosexuals on a retroactive basis. Professor Woodruff wrote:

We all know that the military has used various "affirmative action" measures to promote women and minorities. Every selection board instruction by the secretary of the service tells the promotion board to look specifically at minorities and women and make sure they are given fair consideration for promotion because they may not have had the best assignments or gotten the best OERs [officer evaluation recommendations]—evaluations that need to be considered in that context.

Several successful court cases have resulted in back pay for officers non-selected for promotion or who have been forced into selective early retirement because women and minorities were given special consideration in the board's instructions.... But, in affirmative action-land a history of institutional discrimination is one of the factors that courts look to in determining whether quotas or other preferential policies are warranted. I suggest that in context, homosexuals will have a stronger argument for affirmative action recruiting than women and minorities. Will application of the affirmative action efforts require the service to ask about sexual orientation? ... How else can you identify the people entitled to special consideration? This opens a can of worms that most folks won't want to deal with.[65]

In addition to the offer of enlistment to persons previously denied, such policies could mean retroactive promotions, which would be extremely disruptive if forced on existing units. Recruiting quotas for gay personnel and financial settlements for persons claiming past discrimination also would be within the realm of possibility.

2. Alleged National Security Argument: Discharges of Homosexuals

The ongoing campaign for homosexuals in the military keeps focusing on almost 13,000 discharges for homosexuality that have occurred since 1994, implying that such losses—over a period of 15 years—have nearly crippled the all-volunteer force. Under closer examination, the argument falls apart.

Newly released DOD figures documenting military discharges in the past five fiscal years (2004–2008) show the same pattern evident in the previous decade: Discharges due to homosexuality affect a minuscule number of troops and represent less than one percent of personnel losses that occur for other legitimate reasons.[66]

According to numbers provided to the Congressional Research Service by the Department of Defense, discharges due to homosexuality, averaged over five years, accounted for only 0.32 percent of all losses—only 0.73 percent if departures due to retirement or completion of service are excluded.[67]

The Department of Defense first put the issue into perspective in 2005, when the under secretary for personnel and readiness provided figures on discharges for homosexuality compared to losses in general for the years 1994–2003. The average percentage of discharges due to homosexuality during those 10 years, as calculated by the Department of Defense, was 0.37.[68]

In 2005 the Department of Defense also provided figures comparing discharges for six reasons, including homosexuality, for 10 years (1994–2003). Highlighting the same categories for the subsequent five years (2004–2008), it is easy to see that proportionate losses for the six reasons noted have not changed significantly (figure 12-1).

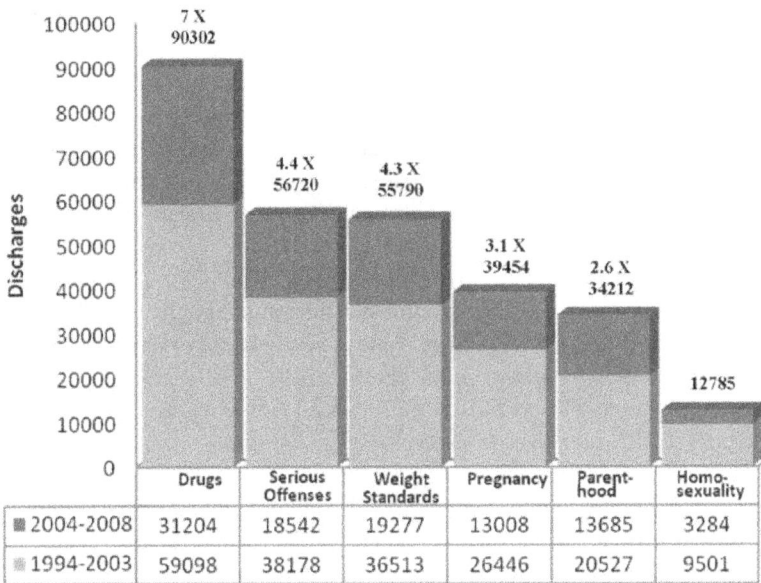

	Drugs	Serious Offenses	Weight Standards	Pregnancy	Parent-hood	Homo-sexuality
▤ 2004-2008	31204	18542	19277	13008	13685	3284
▤ 1994-2003	59098	38178	36513	26446	20527	9501

Figure 12-1. Number of discharges by reason, 1994–2008. (Based on data from GAO, *Military Personnel: Financial Costs and Loss of Critical Skills Due to DOD's Homosexual Conduct Policy Cannot Be Completely Estimated,* GAO-05-299, February 2005, 42–43.)

The report produced by the Congressional Research Service (CRS) on 14 August 2009 confirmed that the small numbers and percentages of discharges for homosexuality do not threaten military readiness. A table in that report showing both the numbers and percentages of homosexual discharges compared to the total active force over a period of 28 years (1980–2008) indicated that the percentage of losses ranged between a high of 0.095 in 1982 to a low of 0.038 in 1994, with the average being 0.063.[69]

The 14 August 2009 CRS report also refuted the legend that discharges declined during the 1991 Persian Gulf War, and the gay discharge process was suspended to retain openly homosexual troops to fight.[70] The Palm Center made the claim without citation in its Road Map Report, and it was repeated in a *Washington Post* op-ed signed by Gen John M. Shalikashvili.

But according to a number of high-ranking generals in a position to know, there was no suspension of homosexual discharges (under DOD regulations) during the Persian Gulf War.[71] According to CRS, a review of discharges during that time indicated that "such a pattern is not evident in these data." As in a previous February 2005 report, CRS noted that personnel not subject to stop-loss orders include "soldiers eligible for disability retirement or separation, dependency, hardship, pregnancy, misconduct, punitive actions, unsatisfactory performance and homosexuality."

Some activists who complain about too many discharges nevertheless claim that there are too few, due to alleged suspension of regulations regarding homosexuals when units are deployed in the current war.[72] On the contrary, the CRS has confirmed that if a person claims to be homosexual just prior to deployment, an investigation taking as long as eight weeks still must take place. During that time he may be deployed, but if his claim is confirmed, he is returned home and honorably discharged. These rules discourage the possibility of "false claim[s] of same-sex behavior being used as a means of avoiding a mobilization." CRS added that retention of individuals who are not eligible for service is a "violation of federal law."[73]

On page six of its 2009 analysis, CRS quoted an April 1998 DOD report that confirmed that most losses due to homosexuality occur among "junior personnel with very little time in the military ... [and] the number of cases involving career service members is relatively small." Furthermore, "the great majority of discharges for homosexual conduct are uncontested and processed administratively. ... [In FY 1997] more than 98 percent received honorable discharges."

The secretary of defense could reduce these numbers to near-zero by complying with language in the 1993 law directing that all personnel receive required briefings on the meaning and effect of the law. The secretary also could repeal the administrative "Don't Ask, Don't Tell" policy/regulations that former Pres. Bill Clinton imposed on the military and exercise his legally authorized option to restore the question about homosexuality that used to appear on induction forms. Homosexuals can serve our country in many ways, but they are not eligible to serve in the military.

3. Foreign Countries as Role Models

Activist groups promoting the cause of gays in the military frequently cite as role models for the United States 25 mostly Western European countries that have no restrictions on professed homosexuals in their militaries.[74] The number is small compared to approximately 200 nations in the world, and comparisons by sheer numbers put the picture into clearer perspective.

Cultural differences between America's military and the forces of other countries, to include potential adversaries such as North Korea, Iran, and China, also are important. For four basic reasons, nothing in the experiences of other nations justifies repeal of the 1993 law, Section 654, Title 10, U.S.C.:

1. There are vast differences in the culture and missions of the American military in comparison to much smaller forces maintained by countries that depend on America for defense.

2. Foreign military authorities do not provide independent, objective information about the effects of gay integration on the majority of personnel—not just those who are homosexual.

3. Official or self-imposed restrictions on homosexual behavior in the militaries of foreign countries, which are comparable to the "Don't Ask, Don't Tell" policy in this country, would not be acceptable to American gay activists whose definition of nondiscrimination is far more extreme.

4. Our superior military is a role model for other countries, not the other way around.

With all due respect to Austria, Belgium, the Czech Republic, Denmark, Estonia, Finland, France (excepting the elite Foreign Legion), Ireland, Italy, Lithuania, Luxembourg, New Zealand, Norway, Slovenia, South Africa, Spain, Sweden, Switzerland, and Uruguay—none of these 19 nations' small militaries bear burdens and responsibilities comparable to ours.

The American Army, Navy, Air Force, and Marines accept far-away, months-long deployments, and our direct ground combat battalions, special operations forces, and submarines require living conditions offering little or no privacy.

Israel. Israel's situation differs from the United States because all able-bodied citizens, including women, are compelled to serve in the military. In addition, deployments do not involve long distances, close quarters, or other conditions comparable to those common in our military, which elevate the potential for sexual tension.

Israeli popular culture is somewhat accepting of homosexuality, but most homosexuals in the Israel Defense Forces are discreet.[75] Israeli soldiers usually do not reveal their homosexuality, and used to be barred from elite combat positions if they did.[76]

In the United States, gay activists are not asking for the right to be discreet in the military. The goal is to celebrate and expand that status into every military occupation and eventually into the civilian world. The limited experiences of homosexuals in the Israel Defense Forces do not recommend implementation of this goal.

Germany. The late Prof. Charles Moskos noted that nations without official restrictions on gays in the military were also very restrictive in actual practice. Germany, for example, dropped criminal sanctions against homosexual conduct in 1969, but also imposed many restrictions on open homosexual behavior and career penalties such as denial of promotions and access to classified information.[77]

According to veteran foreign correspondent Dr. Uwe Siemon-Netto, Germany has conscription for both civilian and military duties. About one-fifth of Bundeswehr soldiers are draftees who are not subject to deployment overseas. Homosexuals used to be exempt from conscription but are now subject to it. Due to strong feelings in the ranks, there are few homosexuals in German elite combat units that are subject to deployment in war zones such as Afghanistan.

There are few complaints about the treatment of homosexuals in the German military because young homosexuals of draft age tend to choose alternative forms of civilian national service, including hospital, hospice, or ecology-related assignments.[78] According to the chairman of their own advocacy group, few of the gays in the German military choose elite combat units that are subject to deployment in war zones.[79]

In 2009 Germany had some 7,700 troops stationed abroad, with 4,000–4,500 in northern Afghanistan and Uzbekistan. Because draftees are not deployed, and because there are strong feelings of opposition to gays in close combat units, these troops do not provide a model for American forces or for the type of force envisioned by homosexualists in this country.[80]

In his correspondence, Dr. Siemon-Netto added a comment about the French Foreign Legion, which consists primarily but not exclusively of foreign volunteers. Considered to be one of the toughest fighting forces in the world, the French Foreign Legion's corps of nine regiments has been deployed to international crises in Afghanistan, Africa, and the Middle East. Dr. Siemon-Netto wrote, "I have mentioned the Foreign Legion only in support of the assertion that Continental European forces, to wit the German airborne elite units, are not a happy venue for homosexuals to 'out' themselves."

Australia. Australian forces represent one of several nations with civilian and military social cultures far more liberal than the United States. A Web site of the Australian Defense Force has created a romantic image for gays in the ranks, who are described as a "largely invisible" minority.[81] This may reflect the culture of liberal Australian society, but a recent report in the *Sydney Morning Herald* suggests that the nation has priorities for its military vastly different from our own.

On 17 November 2008, the *Herald* reported that personnel shortages were so severe, the Australian navy found it necessary to shut down for a two-month Christmas break.[82] The stand-down period was scheduled to run from 3 December to 3 February 2009 and will be a permanent arrangement every year. (If Australia is part of an allied naval force in the Pacific, the best time for an enemy to attack would be during the Christmas break.)

The *Herald* reported that the plan was announced to make the Aussie navy more "family friendly," in order to improve retention and remedy personnel shortages. Their navy loses 11 percent of its personnel every year and achieved only 74 percent of its full-time recruitment goals in the previous fiscal year.

The Netherlands and Canada. The Netherlands and Canada have civilian and military cultures quite different from the United States,[83] and both countries enjoy the protection of American forces. Dutch and Canadian forces primarily deploy for support or peacekeeping missions that depend on the nearby presence of American forces. In these militaries most homosexuals are discreet, but American gay activists are demanding far more than that.

Dutch society, known worldwide for socially liberal policies regarding sexual matters, is not a model suitable for the US military to follow. Deployments normally do not involve offensive combat or conditions comparable to those experienced by American troops.

Canada chose to include homosexuals in the Canadian Forces in 1992, after the conclusion of the Persian Gulf War. Some Canadian troops have been deployed in supportive roles in the current war, but not under conditions comparable to American forces. Canadian society is more culturally liberal than the United States, becoming one of the first countries to legalize same-sex marriage.

Canada's policy has made it necessary for officials to establish protocols for chaplains asked to perform same-sex marriages. If a chaplain cannot participate as a matter of conscience, a referral to a colleague or civilian officiating clergy (COC) can be made.[84] Comparable regulations in the American military would not shield a chaplain from disciplinary measures, such as career-ending denial of promotions for refusing to perform same-sex marriages or to bless civil unions. Nor would chaplains or other military officials be protected from predictable litigation claiming discrimination against same-sex couples.

Britain and the United Kingdom. In September 1999, the European Court of Human Rights ordered the United Kingdom to open its military ranks to homosexuals. Instead of exercising its option to resist, Britain complied with the order. This unnecessary capitulation, in itself, demonstrated profound differences in British and American governments and the cultures of their respective militaries.[85]

Independent information about what is happening in Britain is difficult to obtain, since the Ministry of Defence (MOD) no longer releases objective reports on the integration of gays in the military.[86] A 2002 MOD report on the subject was kept secret, but in 2007 the *London Daily Mail* obtained a copy by means of a Freedom of Information request. According to a *Daily Mail* article about the 2002 report, Britain's armed forces faced significant protest when the government lifted the ban on homosexuals serving. The Royal Navy, in particular, suffered a loss of experienced senior rates and warrant officers who preferred to quit.[87]

Eight years later, homosexual service members have told activists in this country that the integration process, from their perspective, has been a complete success in Britain. This is not surprising, since they have no reason to complain. Same-sex couples live in married and family housing, dance at social events, and march in gay pride parades.[88]

The Ministry of Defence meets regularly with LGBT activist groups to discuss even more progress for their agenda.[89] A multicolored "rainbow" version of the official seal appears on the MOD Web site,[90] which posts newsletters and other documents of interest on the Web site of the MOD LGBT Forum. The forum is looking at issues such as future acceptance of transgenders in the military, and the gay activist group Stonewall praised the Ministry of Defence for working with them on "homophobic bullying."[91] (This is an interesting comment since activists claim that the British experience has been completely positive.)

Britain is often held out as a model for the United States on social change, but the Ministry of Defence has not cooperated by allowing independent interviews. In 2007, the *New York Times* included this in a story promoting the success of gays in the British military: "For this article, the Defense Ministry refused to give permission for any member of the forces to be interviewed, either on or off the record. Those who spoke did so before the ministry made its position clear." Instead of questioning why the restrictions on interviews were so tight, the *Times* headlined the article as if the British experience were an unqualified success.[92]

Britain is an ally of the United States, and the efforts of its men and women in uniform are admirable and appreciated. Still, there have been indications that all is not well with British forces. European newspapers have reported recruiting and disciplinary problems in the British military.[93] When Royal Navy officials stood by and allowed 15 of their sailors and marines to be taken hostage by Iranians in 2007, many observers wondered if the culture of the service had changed, and not for the better.[94]

In January 2009, the British military's top commander agreed with American Secretary of Defense Robert Gates that the British military had been less than effective in carrying out counterinsurgency operations against the Taliban in southern Afghanistan when they first deployed to Helmand Province in 2006.[95] It is impossible to determine the effect of changes in military culture caused by liberal social policies, but the British military should not be a role model for the American all-volunteer force.

Middle East and Muslim Allies. In this debate there has been little discussion about the cultural values of some of our allies, which could present problems in military situations. In Iraq and Afghanistan, American forces are training Muslim forces in small units in the field. Nine- to 11-man military training teams in Iraq, called embedded training teams in Afghanistan, live, sleep, and train together constantly.

Reportedly, under Sharia law homosexual conduct is a crime in many countries in the Middle East, punishable by imprisonment, flogging, or primitive, violent death. The US military cannot change such attitudes, but it does try to avoid offending Muslim allies whenever possible.[96] The challenge of training Iraqi and Afghan troops already is difficult enough. If our military creates a serious cultural problem and then "solves" it by exempting openly gay soldiers from close-combat training and deployments with Muslim troops, how would that affect military readiness and the morale of everyone else? Modern history provides few answers to such questions, but members of Congress should consider them before voting to repeal the 1993 law.

Potential Adversaries. Conspicuously missing from the list of 25 gay-friendly militaries are potential adversaries China, North Korea, and Iran. Their combined forces (3.8 million, not counting reserves) are more than two times greater than the active-duty forces of the 25 foreign countries with gays in their militaries (1.7 million).[97]

Congress is being asked to impose a risky military social experiment that is duplicated nowhere in the world. Instead, members of Congress should assign priority to national security, putting the needs of our military first.

4. Civilian Surveys and Polls

The Zogby/Palm Poll. In January 2007, retired Army Gen John M. Shalikashvili, chairman of the Joint Chiefs of Staff from 1993 to 1997, joined the gays-in-the-military cause by writing an op-ed for publication in the *New York Times*.[98] The general's article, and a second one published in 2009 in the

Washington Post, drew attention to a December 2006 poll of 545 service members conducted by Zogby International, indicating that 73 percent of the respondents said they were "comfortable interacting with gay people."[99]

The only surprising thing about this innocuous question was that the favorable percentage was not closer to 100 percent. Virtually everyone knows and likes at least one person who is gay—but this is not the most relevant issue.

The Zogby poll asked another, more important question that was not even mentioned in the news release announcing the poll's results: "Do you agree or disagree with allowing gays and lesbians to serve openly in the military?" On that question, 26 percent of those surveyed agreed, but 37 percent disagreed. The Zogby poll also found that 32 percent of respondents were "neutral" and only 5 percent were "not sure."[100]

If this poll were considered representative of military personnel, the 26 percent of respondents who wanted the law repealed were far fewer than the combined 69 percent of people who were opposed to or neutral on repeal. This minority opinion was hardly a mandate for radical change, but the poll has been spun and trumpeted for years as if it were.

A closer look at the Zogby poll reveals more interesting details that should have been recognized by news media people reporting on it:

a. The news release announcing results stated, "The Zogby Interactive poll of 545 troops who served in Iraq and Afghanistan was designed in conjunction with the Michael D. Palm Center at the University of California, Santa Barbara." Since the Palm Center paid for the survey, it is appropriate to refer to it as the Zogby/Palm poll.

b. The methodology page stated, "Zogby International conducted interviews of 545 US Military Personnel online from a purchased list of US Military personnel [*sic*]." However, the US military does not sell or provide access to personnel lists to civilian pollsters or anyone else.[101] The authors of a separate report analyzing the Zogby/Palm poll undermined its credibility with an honest comment: "Initial attempts to secure a list of military personnel from the Department of Defense in order to draw a random sample for this survey were unsuccessful."[102]

c. The Zogby/Palm poll further weakened its own credibility with this overstatement: "The panel used for this survey is composed of over 1 million members and correlates closely with the U.S. population on all key profiles." If this was a reference to the US military, it was not credible for reasons stated above. If a "million-man" polling sample existed, why did it locate only 545 respondents? This sample was only slightly more than one-quarter of the number used by the *Military Times* poll described below.

d. The Zogby/Palm poll's description of methodology referred to a "double opt-in format through an invitation only method."[103] The obfuscation was no substitute for the plain and conspicuously-missing word *random*. Respondents, apparently, self-selected themselves to answer a survey on gays in the military, which might have led to a disproportionately large sample of gay or liberal participants.[104]

e. Activists frequently claim that since greater numbers of younger people are more comfortable with homosexuals, this is evidence enough to justify changing the 1993 law.[105] However, personal relationships among younger people do not seem to be decisive when voters actually decide matters of policy. In 30 states (increased in 2009 to 31), voters (as opposed to courts or legislatures) have approved referenda or other measures banning same-sex marriage, often with comfortable majorities.[106]

Civilian Polls. Some civilian polls, such as the *Washington Post*/ABC News poll released on 19 July 2008,[107] have asked respondents whether gays should serve in the military "openly" or "undisclosed."[108] These questions are not on point because they focus on elements of the "Don't Ask, Don't Tell" administrative policy, not the consequences of repealing the law.

Such surveys also measure opinions among people who generally know as much about the military as they do about remote issues currently being debated by the Canadian Parliament. The results, therefore, are less relevant to members of Congress considering legislation to repeal the actual 1993 law.

Polling organizations recognize that respondents who believe a policy already exists are more likely to favor that policy, while those who know otherwise are less likely.[109] Constant news reports suggesting that homosexuals already are in the military probably skew civilian surveys to the positive side. This is especially so when a poll asks innocuous questions about knowing or liking individual people who are gay.

Military Times **Polls.** The annual *Military Times* poll of almost 2,000 active-duty subscriber/respondents found that 58 percent opposed repeal of the 1993 law, described as "Don't Ask, Don't Tell," for four years in a row.[110] Contrary to some criticisms from activist groups, the *Military Times* editors did not imply that the survey reflected military demographics perfectly. Nor did the editors of *Military Times*, a Gannet-owned publication that has supported efforts to repeal the 1993 law, try to inflate the survey's credibility in the same way that the Zogby/Palm poll did.

As in previous years, the *Military Times* mailed surveys to subscribers at random, but they counted only the responses from almost 2,000 active-duty military. Unlike the Zogby/Palm poll, questions on the survey covered a wide range of topics, not the gays-in-the-military issue alone.

The 2008 *Military Times* poll asked a new question that produced significant results: "If the 'don't ask, don't tell' policy is overturned and gays are allowed to serve openly, how would you respond?" The article emphasized that 71 percent of respondents said they would continue to serve. But almost 10 percent said, "I would not re-enlist or extend my service," and 14 percent said, "I would consider not re-enlisting or extending my service." Only 6 percent responded "No Opinion."

Absent unusual circumstances, the military cannot force anyone to enlist or reenlist in the volunteer force. Such results indicate potential recruiting and retention problems that could become even more difficult during a time of intense warfare or during times of economic prosperity, when a recruiter's job is more difficult.[111]

Military professionals follow orders and honor induction contracts that do not allow them to end their military careers overnight. The gradual but persistent loss of even a few thousand careerists in grades and skills that are not quickly or easily replaceable would be devastating to the all-volunteer force.

The Military Officers Association of America (MOAA) Online Survey. In October 2008, MOAA invited readers of their magazine *Military Officer* to participate in an online opinion survey on gays in the military. No one claimed that it was "scientific" or random, (civilian polling companies cannot reach military people directly). Still, the professionally designed online survey, which tabulated the ages and military background of respondents, provided useful insights more relevant than "scientific" surveys of uninformed civilians.

In July 2009, the *Washington Times* reported that the MOAA survey revealed strong support for current policy (16 percent) or an even stronger law excluding homosexuals from the military (52 percent). The same combined percentage, 68 percent, expressed the belief that repeal of the 1993 law would have a very negative effect (48 percent) or a moderately negative effect (20 percent) on troop morale and military readiness.

The MOAA survey of 1,664 respondents included a significant number of younger, active-duty or drilling reserve/guard personnel, many of whom responded to the survey weeks after it was announced in the organization's publicly available magazine. By July 2009, 64 percent of MOAA survey respondents were under the age of 45, and the percentage of currently active-duty or reserve/guard military personnel was 51 percent.

Shattering the usual presumptions, by two-to-one margins these younger, closer-to-active-duty respondents came down in support of current law and opposed to harmful consequences of repeal. Contrary to stereotype, a combined 35 percent of MOAA respondents simultaneously indicated that today's service members are "much more" or "moderately more" tolerant toward homosexuals in the military, while 45 percent said that attitudes were "no different" from those who served in the 1980s and earlier.[112]

There was no time limitation on the survey, but a MOAA spokesman said the group was scuttling the poll because there were only 500 responses in the first 11 days. Revising an earlier statement, MOAA officials belatedly described the survey as "statistically invalid" because "some non-members" may have passed the survey around to friends in order to "skew results." No evidence of the alleged activity, on either side, was provided or evident to anyone.

Prior to withdrawing all data that the *Washington Times* had reported, there were 1,664 responses—a significant sample that tripled the size of the initial 500 who responded in the first 11 days. The incident brought to mind an Andy

Rooney aphorism, "To ignore the facts does not change the facts." The online survey was not invalid—but it was inconvenient.

5. Human Interest Stories

The Gays in the Military Campaign (GIMC). For many years gay activists have been pushing hard for repeal of the 1993 Eligibility Law with a multimillion-dollar public relations campaign focusing on the human interest stories of former military personnel who were discharged because of homosexual conduct, usually evidenced by voluntary statements.

Special attention has been given to linguists who speak Arabic—an important skill in the current war. In 2005, activists decried the loss of "fifty-four Arabic linguists" who were discharged from the military due to homosexuality. The number appeared in a column of personnel losses included in a 2005 Government Accountability Office (GAO) report, but details about the type and level of proficiency of the language trainees, which varied considerably, put the matter into perspective.[113]

In 2002 authorities discharged 12 homosexual language trainees at the Army's Defense Language Institute (DLI) in Monterey, California. Two of the students were found in bed together, and the others voluntarily admitted their homosexuality. When the language trainees were honorably discharged, gay activist groups protested the dismissals as a loss for national security.[114]

The true loss occurred, however, when 12 students who were not eligible to serve occupied the spaces of other language trainees who could be participating in the current war. This loss of time and resources was a direct result of President Clinton's calculated action to accommodate homosexuals with his "Don't Ask, Don't Tell" administrative regulations, despite prohibitions in the law.[115] The Pentagon should clarify the meaning of the 1993 Eligibility Law and pursue other ways to recruit and train qualified language trainees.[116]

The Servicemembers Legal Defense Network (SLDN) and allied groups such as the Human Rights Campaign (HRC) and Servicemembers United have played up emotional stories of several young men and women who were trained and served in the military but received an honorable discharge when they disclosed they were gay. Even an article in the *Joint Forces Quarterly* appealed to emotions with this: "Several homosexuals interviewed were in tears as they described the enormous personal compromise in integrity they had been making, and the pain felt in serving in an organization they wholly believed in, yet that did not accept them."[117]

In most cases it is appropriate to assume good faith on the part of these individuals who want to serve in uniform. The problem is the Department of Defense, which keeps issuing contradictory statements regarding the eligibility of homosexuals to serve. Gay activist groups also aggravate the problem by misinforming young people about the meaning of the 1993 Eligibility Law.

Many people who are patriots and willing to serve are not eligible for reasons such as age, health, personal violations of law, and the like. It makes no sense to recruit, train, and deploy people who are not eligible to serve. This is the problem with Bill Clinton's convoluted "Don't Ask, Don't Tell" policy, which has created human interest problems that members of Congress predicted and tried to avoid by rejecting it. Criticism of "Don't Ask, Don't Tell," however, does not justify repeal of the 1993 Eligibility Law.

Speculation About Willingness to Serve. A July 2008 Palm Center report advocating repeal of the 1993 law, signed by a study group of four retired general and flag officers, suggested that possible personnel losses were not the group's primary concern.[118] In Finding Five of the document, the panel conceded that an estimated 4,000 military personnel would be lost to the service if the law were repealed. The report also claimed, with no credible support, that the loss would be "a wash in terms of recruiting and retention" because 4,000 gays and lesbians would enlist to take their places.

The study group's estimates were based on responses to a survey question in the same Zogby poll that the Palm Center commissioned and paid for in late 2006. Finding Five cited responses to Zogby question 27 suggesting that if gays and lesbians had been allowed to serve openly in the military, 2 percent of potential recruits—about 4,000 presumably heterosexual military men and women—probably would have declined enlistment in the past 14 years. Then the study group claimed without support that the 4,000 losses would be "canceled out" by 4,000 gays and lesbians likely to enlist in their places.

The estimate of potential losses, however, was miscalculated. The percentage of military people identified by Zogby in survey question 27 was not 2 percent; it was 10 percent, five times greater, with 13 percent undecided. Taking those percentages and estimates at face value, that means 20,000 people would have declined to join the military since 1994, or 32,000 men and women if half of Zogby's undecided group was factored in.[119]

Yet another estimate came from Dr. Aaron Belkin, director of the Palm Center, who submitted to the House Armed Services Committee a brief statement claiming that if the law were repealed, 41,000 new recruits would join the military.[120] If 10 times more than the Palm Center's own study group's 4,000 figure was good, why not pick another number—any number—to make the estimate even better?

Belkin's statement quoted Gary D. Gates, PhD, whose statement filed with the House Armed Services Committee used the same 41,000 figure and cited his own speculative claim that 65,000 homosexuals are currently serving in the military.[121] Gates was the author of a September 2004 report published by the Urban Institute, titled *Gay Men and Lesbians in the U.S. Military: Estimates from Census 2000*.[122]

The 24-page Gates Report included several tables of numbers regarding military service rates, age, gender, and other factors. It concluded, "Estimates suggest that more than 36,000 gay men and lesbians are serving in active duty, representing 2.5 percent of active-duty personnel. When the guard and reserve are included, nearly 65,000 men and women in uniform are likely gay or lesbian, accounting for 2.8 percent of military personnel."[123]

The Gates Report was widely described as definitive, even though many of the numbers used to calculate percentages of gays in the military were based on speculation derived from several social science sources as well as the 2000 Decennial Census. The document stated that same-sex couples living in the same household are "commonly understood to be primarily gay and lesbian couples even though the census does not ask any questions about sexual orientation, sexual behavior, or sexual attraction (three common ways used to identify gay men and lesbians in surveys)."[124]

This is one of several caveats in the Gates Report, including this observation: "Prevalence estimates of the proportion of men and women in the United States who are gay or lesbian drawn from samples that can be used to make nationally representative estimates are rare."[125]

Using a statistical method called the Bayes Rule, author Gates added up speculative figures regarding different military communities (active duty, guard, and reserve) to come up with the 65,000 figure. Paul Winfree, a policy analyst at the Heritage Foundation Center for Data Analysis, has described the Bayes Rule or Bayes Theorem as "basically a calculation of the probability of an event occurring subject to certain known priors."[126] Statisticians use the Bayes Rule as a formula to determine probability when relevant factors are known with certainty.[127]

The Gates Report calculated the number of gays and lesbians in the military by using estimated figures derived from the 2000 Census. Winfree noted that the Bayes Rule methodology used in the Gates Report was standard, but the resulting estimate was only as good as other estimates made using the 2000 Decennial Census. (A judgment on those figures was beyond the scope of his review.)

Given the element of speculation throughout, it is an overstatement to describe it as an objective presentation of "real numbers."[128] It is not possible to determine the accuracy of estimates used in the Gates Report, which was prepared in consultation with the Center for the Study of Sexual Minorities in the Military (now the Palm Center) and the Servicemembers Legal Defense Network.

Even if there were 65,000 homosexuals serving discreetly in the military, it would not follow that the time has come to repeal the 1993 law. Homosexualists are not seeking the right to serve discreetly in the military. The goal is unrestricted acceptance of professed sexual minorities in the military, regardless of the consequences.

Road Map or Railroad?

"Stop-Loss" Authority for National Security Only

In May 2009 the Palm Center issued a 29-page Road Map Report claiming that President Obama can and should suspend enforcement of Section 654, Title 10 by signing an executive order.[129]

Under the terms of 10 U.S.C. 12305 the president may suspend any law regarding "separation" of military personnel in time of a declared national emergency—defined as a period when reservists are serving on involuntary active duty, as they are now. But according to law professor William A. Woodruff, the purpose of the stop loss authority is to benefit national security, not to achieve political objectives:

> The authority under the stop loss law (10 USC 12305) is quite broad, but the real issue is whether permitting homosexuals to serve is vital to national security. This is where the Palm Center takes things off track. They are urging the President to use his national emergency authority to create an environment that will eventually lead to the repeal of 10 USC 654. However, Congress passed the law, 10 USC 12305, to allow the President flexibility to respond to national emergencies, not to give him political cover to socially engineer the military.

> If, as the Palm Center apparently believes, service by homosexuals is beneficial and not detrimental to national security, with no adverse impact on unit cohesion and combat effectiveness, the issue should be debated on that basis and not by using statutory authority enacted for other purposes.

> Even though President Obama has promised to work to repeal 10 USC 654, he seems to understand and appreciate that such unilateral political decisions in an area the Constitution specifically vests in the Congress would show profound disrespect for a coordinate branch of government. The situation is very similar to the deference the courts show the Executive and Legislative branches in areas the Constitution assigns to the political branches. Likewise, the political branches must be sensitive to their respective areas of constitutional authority and not try to usurp each others' legitimate areas of responsibility.

If President Obama yields to gay activist pressure and unilaterally suspends or stops enforcement of the law, the troops would perceive that action as an evasion of his oath to "faithfully execute the office of the President of the United States." Even the *Washington Post,* a strong proponent of gays in the military, questioned the Palm Center's plan to "get around existing policy."[130]

Preempting the Joint Chiefs

Palm Center reports have twice suggested strategies to coerce senior members of the military to go along with their agenda. The May Road Map Report set the tone by suggesting that "the President should not ask military leaders if they support lifting the ban. . . . Any consultation with uniformed leaders should take the form of a clear mandate to give the President input about how, not whether, to make this transition."[131]

In a subsequent report, the Palm Center went even further in advocating a strategy that would short-circuit the political system. The document confirmed insufficient votes to repeal the law and criticized fellow activists for not having a plan to overcome the resistance of reluctant members of Congress, including Democrats from fairly conservative districts.[132] Claiming that military leaders consider repeal of the 1993 law to be inevitable, Palm disrespected military leaders with this: "In terms of their capacity to make trouble, it is the legislative process that would open a can of worms by allowing military leaders to testify at hearings and forge alliances with opponents on the Hill. *A swift executive order would eliminate opportunities for them to resist*" (emphasis added).[133]

This two-step plan to box in members of the Joint Chiefs of Staff with a presidential executive order reveals an attitude of arrogance and elitism that should not be allowed to prevail.

Conclusion

Proposed radical change demands a heavy burden of proof. Advocates of repealing the 1993 Eligibility Law have not carried that burden or made a convincing case. Lofty civil rights rhetoric cannot erase the normal desire for sexual privacy in the real world of the military. Consistently small numbers and percentages of people discharged due to homosexuality contradict any claim that a national security emergency justifies repeal of the law.

It is not convincing to hold up as role models for America's forces the small, dissimilar militaries of foreign nations—none of which have adopted the extreme agenda being proposed for our military. Nor does it help to ignore the stated opinions of experienced and current military personnel of all ranks or to advocate zero tolerance and punishment for anyone who disagrees with the gay agenda.

Some advocates cavalierly argue that intimate living conditions in infantry battalions and aboard submarines should be made *more* uncomfortable to accommodate at least four different gender and sexual orientation groups. Others have admitted that some units may become dysfunctional if Congress repeals the 1993 law.[134] No one has justified these costs in terms of personnel disruptions, operational distractions, or scarce defense dollars.

As stated at the beginning of this chapter, the rationale for social policy changes begins with a choice of priorities. Advocates of this cause assign higher priority to career considerations than they do to the needs of the military. Arguments for repeal of the 1993 Eligibility Law center on self interest, not concerns about morale and readiness required for a strong military culture and an effective national defense. Professor Woodruff has noted:

> The military is not popular culture. It is very different and must remain so to defend the freedoms that advance our popular culture. Those who favor personnel policies grounded in notions of fairness to the individual must be required to demonstrate beyond any doubt that military discipline, unit cohesion, and combat effectiveness will not be diminished one iota by adoption of their preferred policy. Otherwise, it elevates the individual over the mission and that is the antithesis of military service.

Policy changes involving political coercion, compromised standards, and elevated risks of social disruption would undermine the culture of the military, and complicate the lives of thousands of good men and women in our military whose voices rarely are heard. For their sake as well as the nation's, we have to get this right. We need to maintain our military as the strongest in the world—it is the only one we have.

Appendix: 10 USC 654

§ 654. Policy concerning homosexuality in the armed forces

(a) **Findings.**— Congress makes the following findings:

(1) Section 8 of article I of the Constitution of the United States commits exclusively to the Congress the powers to raise and support armies, provide and maintain a Navy, and make rules for the government and regulation of the land and naval forces.

(2) There is no constitutional right to serve in the armed forces.

(3) Pursuant to the powers conferred by section 8 of article I of the Constitution of the United States, it lies within the discretion of the Congress to establish qualifications for and conditions of service in the armed forces.

(4) The primary purpose of the armed forces is to prepare for and to prevail in combat should the need arise.

(5) The conduct of military operations requires members of the armed forces to make extraordinary sacrifices, including the ultimate sacrifice, in order to provide for the common defense.

(6) Success in combat requires military units that are characterized by high morale, good order and discipline, and unit cohesion.

(7) One of the most critical elements in combat capability is unit cohesion, that is, the bonds of trust among individual service members that make the combat effectiveness of a military unit greater than the sum of the combat effectiveness of the individual unit members.

(8) Military life is fundamentally different from civilian life in that—

(A) the extraordinary responsibilities of the armed forces, the unique conditions of military service, and the critical role of unit cohesion, require that the military community, while subject to civilian control, exist as a specialized society; and

(B) the military society is characterized by its own laws, rules, customs, and traditions, including numerous restrictions on personal behavior, that would not be acceptable in civilian society.

(9) The standards of conduct for members of the armed forces regulate a member's life for 24 hours each day beginning at the moment the member enters military status and not ending until that person is discharged or otherwise separated from the armed forces.

(10) Those standards of conduct, including the Uniform Code of Military Justice, apply to a member of the armed forces at all times that the member has a military status, whether the member is on base or off base, and whether the member is on duty or off duty.

(11) The pervasive application of the standards of conduct is necessary because members of the armed forces must be ready at all times for worldwide deployment to a combat environment.

(12) The worldwide deployment of United States military forces, the international responsibilities of the United States, and the potential for involvement of the armed forces in actual combat routinely make it necessary for members of the armed forces involuntarily to accept living conditions and working conditions that are often spartan, primitive, and characterized by forced intimacy with little or no privacy.

(13) The prohibition against homosexual conduct is a longstanding element of military law that continues to be necessary in the unique circumstances of military service.

(14) The armed forces must maintain personnel policies that exclude persons whose presence in the armed forces would create an unacceptable risk to the armed forces' high standards of morale, good order and discipline, and unit cohesion that are the essence of military capability.

(15) The presence in the armed forces of persons who demonstrate a propensity or intent to engage in homosexual acts would create an unacceptable risk to the high standards of morale, good order and discipline, and unit cohesion that are the essence of military capability.

(b) **Policy.**— A member of the armed forces shall be separated from the armed forces under regulations prescribed by the Secretary of Defense if one or more of the following findings is made and approved in accordance with procedures set forth in such regulations:

(1) That the member has engaged in, attempted to engage in, or solicited another to engage in a homosexual act or acts unless there are further findings, made and approved in accordance with procedures set forth in such regulations, that the member has demonstrated that—

    (A) such conduct is a departure from the member's usual and customary behavior;

    (B) such conduct, under all the circumstances, is unlikely to recur;

    (C) such conduct was not accomplished by use of force, coercion, or intimidation;

    (D) under the particular circumstances of the case, the member's continued presence in the armed forces is consistent with the interests of the armed forces in proper discipline, good order, and morale; and

    (E) the member does not have a propensity or intent to engage in homosexual acts.

(2) That the member has stated that he or she is a homosexual or bisexual, or words to that effect, unless there is a further finding, made and approved in accordance with procedures set forth in the regulations, that the member has demonstrated that he or she is not a person who engages in, attempts to engage in, has a propensity to engage in, or intends to engage in homosexual acts.

(3) That the member has married or attempted to marry a person known to be of the same biological sex.

(c) **Entry Standards and Documents.**—

(1) The Secretary of Defense shall ensure that the standards for enlistment and appointment of members of the armed forces reflect the policies set forth in subsection (b).

(2) The documents used to effectuate the enlistment or appointment of a person as a member of the armed forces shall set forth the provisions of subsection (b).

(d) **Required Briefings.**— The briefings that members of the armed forces receive upon entry into the armed forces and periodically thereafter under section 937 of this title (article 137 of the Uniform Code of Military Justice) shall include a detailed explanation of the applicable laws and regulations governing sexual conduct by members of the armed forces, including the policies prescribed under subsection (b).

(e) **Rule of Construction.**— Nothing in subsection (b) shall be construed to require that a member of the armed forces be processed for separation from the armed forces when a determination is made in accordance with regulations prescribed by the Secretary of Defense that—

(1) the member engaged in conduct or made statements for the purpose of avoiding or terminating military service; and

(2) separation of the member would not be in the best interest of the armed forces.

(f) **Definitions.**— In this section:

(1) The term "homosexual" means a person, regardless of sex, who engages in, attempts to engage in, has a propensity to engage in, or intends to engage in homosexual acts, and includes the terms "gay" and "lesbian".

(2) The term "bisexual" means a person who engages in, attempts to engage in, has a propensity to engage in, or intends to engage in homosexual and heterosexual acts.

(3) The term "homosexual act" means—

    (A) any bodily contact, actively undertaken or passively permitted, between members of the same sex for the purpose of satisfying sexual desires; and

    (B) any bodily contact which a reasonable person would understand to demonstrate a propensity or intent to engage in an act described in subparagraph (A).

(b) Regulations.—Not later than 90 days after the date of enactment of this Act [Nov. 30, 1993], the Secretary of Defense shall revise Department of Defense regulations, and issue such new regulations as may be necessary, to implement section 654 of title 10, United States Code, as added by subsection (a).

(c) Savings provision.—Nothing in this section or section 654 of title 10, United States Code, as added by subsection (a), may be construed to invalidate any inquiry, investigation, administrative action or proceeding, court-martial, or judicial proceeding conducted before the effective date of regulations issued by the Secretary of Defense to implement such section 654.

(d) Sense of Congress.--It is the sense of Congress that—

(1) the suspension of questioning concerning homosexuality as part of the processing of individuals for accession into the Armed Forces under the interim policy of January 29, 1993, should be continued, but the Secretary of Defense may reinstate that questioning with such questions or such revised questions as he considers appropriate if the Secretary determines that it is necessary to do so in order to effectuate the policy set forth in section 654 of title 10, United States Code, as added by subsection (a) and

(2) the Secretary of Defense should consider issuing guidance governing the circumstances under which members of the Armed Forces questioned about homosexuality for administrative purposes should be afforded warnings similar to the warnings under section 831(b) of title 10, United States Code (article 31(b) of the Uniform Code of Military Justice).

Notes

1. The Flag & General Officers for the Military (FGOM) statement, dated 30 March 2009, was delivered to Pres. Barack Obama, Pentagon officials, and senior members of Congress on 31 March 2009. The statement and a brief issue overview are available at www.flagandgeneralofficers forthemilitary.com. Personal signatures were requested by a Steering Committee of seven four- and three-star retired officers via regular postal mail (not e-mail). A clear majority of the Army officers listed, for whom retirement dates are available, were in command after 1994, when the 1993 Eligibility Law was in effect. As of 4 February 2010, 1,163 handwritten signatures were received and are on file with the Center for Military Readiness, which provided administrative support for the FGOM project. The full statement appears as chapter 11 in this volume.

2. In his address to the "Boys of Pointe du Hoc" on the 40th anniversary of D-Day 1984, Pres. Ronald Reagan described the force of military cohesion: "You were young the day you took these cliffs; some of you were hardly more than boys, with the deepest joys of life before you. Yet, you risked everything here. Why? Why did you do it? What impelled you to put aside the instinct for self-preservation and risk your lives to take these cliffs? . . . We look at you, and somehow we know the answer. It was faith and belief; it was loyalty and love."

3. William Darryl Henderson, PhD, testimony before the Presidential Commission on the Assignment of Women in the Armed Forces, 26 June 1992, and Commission Report to the President, 15 November 1992, Finding 2.5.1, page C-80-81, quoting Dr. Henderson's book, *Cohesion: The Human Element*, National Defense University Press, 1985.

4. Thomas Sowell, "The Anointed and Those Who Aren't," *Washington Times*, 8 February 1993, E3.

5. National Defense Authorization Act (NDAA) for Fiscal Year (FY) 1994, Pub. L. no. 103–60, § 571, 107 Stat. 1547, 1670, (1993), codified at 10 U.S.C. § 654, reprinted in appendix. The 1993 law codified long-standing DOD regulations adopted in January 1981. See Elaine Donnelly, "Constructing the Co-Ed Military," *Journal of Gender Law & Policy* (Duke University) 14 (May 2007), hereafter cited as Duke *Law Journal*, 906–10. The article is available in full at http://www .law.duke.edu/shell/cite.pl?14+Duke+J.+Gender+L.+&+Pol%27y+815l.

6. Jody Feder. "'Don't Ask, Don't Tell': A Legal Analysis," *Congressional Research Service Report* 7-5700, R40795, 2 September 2009.

7. Legislative history clearly shows that members of Congress did not intend to accommodate professed homosexuals in the military. See 103rd Congress, House Report 103-200, NDAA for

FY 1994, 287. Rep. Steve Buyer (R-IN), then-chairman of the House Armed Services Committee (HASC) Personnel Subcommittee, underscored the point in a 16 December 1999 Memorandum for Members of the Republican Conference, "Policy Regarding the Present Ban on Homosexuals in the Military": "Although some would assert that section 654 of Title 10, US Code . . . embodied the compromise now referred to as 'Don't Ask, Don't Tell,' there is no evidence to suggest that the Congress believed the new law to be anything other than a continuation of a firm prohibition against military service for homosexuals that had been the historical policy." See document available at http://cmrlink.org/problemgays.asp, and Duke *Law Journal*, 905–8.

8. Section 654, Title 10, U.S.C., and Duke *Law Journal*, 904–7.

9. The FY 1994 NDAA codified language almost identical to that in DOD directives promulgated in 1981. An amendment offered by Sen. Barbara Boxer (D-Cal.), which would have allowed the president to decide policy regarding gays in the military, was defeated on 9 September 1993, on a bipartisan 63 to 33 vote. On 28 September the House rejected a similar amendment, sponsored by Rep. Martin Meehan (D-MA) and Rep. Patricia Schroeder (D-CO), which would have stricken the Senate-approved language and expressed the sense that the issue should be decided by the president and his advisors. The Meehan/Schroeder amendment was defeated on a bipartisan roll-call vote, 264 to 169.

10. See extensive analysis in the University of Missouri-Kansas City *Law Review* article by Campbell University law professor William A. Woodruff, "Homosexuality and Military Service," 64 UMKC L. Rev. 121, 123–24 (Fall 1995). Also Section 654, Title 10, U.S.C., and Duke *Law Journal*, 903–10. Professor Woodruff is a leading expert on the issue of homosexuals in the military.

11. Duke *Law Journal*, 908–11.

12. Official spokesmen continue to suggest, erroneously, that homosexuals are eligible for military service if they do not *say* they are homosexual. Statutory language requires briefings and educational materials to clarify the meaning and intent of the law, but the Department of Defense has failed to comply with this provision. See Section 654, Title 10, U.S.C., "Required Briefings." Also Duke *Law Journal*, 907–8.

13. David E. Burrelli. "'Don't Ask, Don't Tell': The Law and Military Policy on Same-Sex Behavior," *Congressional Research Service Report* 7-5700, R40782, 14 August 2009, hereafter cited as *CRS Report*, 14 August 2009, 1–5.

14. Duke *Law Journal*, 900–908.

15. See DOD Release No. 605-93, Dec. 22, 1993. The DOD News Release announcing enforcement regulations primarily referred to the "Don't Ask, Don't Tell" policy announced by President Clinton on 19 July 1993, not the language and meaning of Section 654, Title 10. The unnoticed discrepancy has been the source of confusion and controversy ever since.

16. In a 9-4 decision that denied the appeal of Navy Lt Paul G. Thomasson, a professed homosexual who wanted to stay in the Navy, US Circuit Judge Michael Luttig wrote about the exclusion law: "Like the pre-1993 [policy] it codifies, [the statute] unambiguously prohibits all known homosexuals from serving in the military." Judge Luttig added that the Clinton administration "fully understands" that the law and DOD enforcement regulations are inconsistent and has engaged in "repeated mischaracterization of the statute itself."

17. Pres. George W. Bush could have rescinded the Clinton regulations with a stroke of the pen, but did not do so, for reasons unknown.

18. CDR Wayne L. Johnson, Judge Advocate General's Corps, Navy, retired, notes that the same principle of effective enforcement applies even if there are some underage individuals who are more mature and trustworthy than some 21-year-olds who are legally permitted to purchase alcohol.

19. The term is taken from a July 2008 Dutch newspaper commenting on gay issues in that country. There are homosexuals who are not homosexualists and activist heterosexuals who are.

20. According to its own *LGBT News* Web site and newsletters, the British Ministry of Defence meets regularly with LGBT groups advocating transgender rights. See www.lgbt.mod.uk. The Behavioral Sciences and Leadership Department at the US Military Academy at West Point invited a formerly male graduate and transgender activist to address classes on 4 November 2008. The Michael D. Palm Center, formerly the Center for the Study of Sexual Minorities in the Military, has posted on its Web site an article titled "Transgender People in the U.S. Military." In

another July 2009 article titled "Self-Inflicted Wound," the Palm Center complains that proposed legislation, H. R. 1283, would "do nothing for transgender service members," signaling an intent to expand that agenda during or after the current legislative process (27 July 2009, p. 6). The list of expectations from the transgender faction would include military housing access and medical coverage for pre– and post–gender reassignment surgery.

21. Aaron Belkin and Melissa Sheridan Embser-Herbert, "A Modest Proposal," *International Security*, 27 (Fall 2002): 178.

22. In a presentation opposite Nathaniel Frank, PhD, of the Palm Center, in Chicago on 17 June 2009, Air Force Academy instructor Brian Maue, PhD, speaking for himself only, noted that in the Air Force, body-touching measurements to determine waist size and personal fitness are done only by persons of the same sex. Respect for sexual privacy also is apparent at every commercial airport, where female security workers perform more extensive body searches of women. Fleeting risks of dignity discomforts are minimized by reasonable practices that respect sexual differences and sensitivities. Maue added that men and women in the military, who must share close quarters on a constant basis, deserve the same respect.

23. Andrew Tilgham, *Navy Times*, "Why So Many Skippers Get Fired," 14 September 2009, 18. The article reports that "personal misconduct is by far the most significant cause of CO firings. Some 45, or 35 percent of the firings during the past 10 years, were due to misbehavior rather than a significant mishap, command performance, or a troubled command climate."

24. See House Report 103-200, 103rd Cong., 1st sess., NDAA for FY 1994, Report of the Committee on Armed Services on H.R. 2401, 30 July 1993, 290.

25. Brian E. A. Maue, PhD, "The Locker Room Issue," in "In the Barracks, Out of the Closet," Room for Debate, *New York Times*, 3 May 2009. Dr. Maue's opinions were identified as his own.

26. William H. McMichael, "Report: Outdated Sodomy Law Should Be Repealed," *Navy Times*, 16 November 2009, 12. Previous reports by this private commission, headed by retired military judge Walter T. Cox III and by a 1998 Task Force on Good Order and Discipline that was appointed by then-Defense Secretary William S. Cohen in 1997, have issued several proposals for revising manuals for courts-martial on several sexual offenses, including adultery.

27. Some reports described Lt Col Fehrenbach, a WSO, as an F-15 pilot whose training cost $25 million. DOD figures provided to the 1992 Presidential Commission on the Assignment of Women in the Armed Forces estimated training costs for fighter or bomber pilots to be $3.1 million. See Commission Report, Finding #2.6.1GH, p. C-93.

28. Dan Popkey, "Gay Boise Air Force Pilot 'Outed' by False Accusation," *Idaho Statesman*, 23 August 2009. SLDN lawyers representing Fehrenbach did not contest the Boise Police Report, DR#813-786.

29. Col W. Hays Parks. "Tailhook: What Happened, Why, and What's to be Learned," US Naval Institute *Proceedings*, September 1994, 89–102.

30. Jeff Evans, "Men with Military Sexual Trauma Often Resist Disclosure," *Adult Psychiatry*, March 2008, 21.

31. Alan Snel. "Male (and Female) Rape in the Military," *Florida Times Special Report*, 17 January 2003. This article included graphic descriptions of some of the assaults suffered by men seeking treatment for military sexual trauma.

32. Bill Sizemore, "Military Men Are Silent Victims of Sexual Assault," *Virginian Pilot*, PilotOnline.com, 5 October 2009.

33. Editorial. "Corps Puts Spin Control Ahead of Victims' Health," *Navy Times*, 17 December 2007, 44.

34. Ernesto Londono, "Navy Chaplain Pleads Guilty: HIV-Positive Priest Is Sentenced in Sex Case," *Washington Post*, 7 December 2007, B-1. In one of the pornographic photos obtained by the *Post*, Lieutenant Commander Lee was sitting nude on a sofa in his office flanked by an image of the Virgin Mary and a framed photo of Marine Gen Peter Pace, former chairman of the Joint Chiefs of Staff.

35. Dan Ephron, "Questionable Conduct," *Newsweek*, 15 December 2007.

36. Andrew Tilghman, "Military among Settings in Which Assault 'Most Likely,'" *Navy Times*, 17 December 2007, 9. This article quotes Mic Hunter, a psychologist and author of *Honor*

Betrayed: Sexual Abuse in America's Military. "The military, boarding schools, sports teams and prison—these are the settings where a male is most likely to be assaulted."

37. Philip Ewing, *Navy Times*, "Male-on-Male Sex Assaults Increase," *Navy Times*, 7 December 2009, 22.

38. Michael Moore, "Soldier at Bragg Charged with HIV Assault," *Raleigh News & Observer*, 18 July 2007.

39. Associated Press (AP), "US: HIV-Positive Paratrooper Pleads Guilty to Assault for Unprotected Sex," *Washington Post*, 1 November 2007.

40. Maj Thomas Earnhardt, US Army Forces Command (FORSCOM), to the author, e-mail, 28 January 2008. Major Earnhardt wrote that Private First Class Dalton was not charged with homosexual conduct because "it's not in the Army's interest to pursue an additional charge that imposes no criminal penalty."

41. Rep. Vic Snyder (D-AR) nearly went "bonkers" (quoting his word) in protest against a mention of this subject by this author as part of her 23 July 2008 testimony. Snyder's intemperate language betrayed an apparent inability to comprehend or discuss a serious subject affecting the health and readiness of deployable units.

42. See testimony of Elaine Donnelly, House Armed Services Personnel Subcommittee, 23 July 2008, available at http://armedservices.house.gov/pdfs/MilPers072308/Donnelly_Testimony072308 .pdf., 15–16.

43. DOD Instruction 6485.01, 17 October 2006, Subject: Human Immunodeficiency Virus.

44. See analysis and graph prepared and posted by the Center for Military Readiness at http:// cmrlink.org/cmrnotes/HIV_Statistics100107.pdf.

45. See Centers for Disease Control HIV/AIDS Fact Sheet, "HIV/AIDS Among Men Who Have Sex with Men," June 2007. "In the United States, HIV infection and AIDS have had a tremendous effect on men who have sex with men (MSM). MSM accounted for 71 percent of all HIV infections among male adults and adolescents in 2005"; and Sarah Kershaw, "New H.I.V. Cases Drop but Rise in Young Gay Men," *New York Times*, 2 January 2008.

46. Chris Johnston, "Navy to Advertise for Homosexual Sailors," *London Times Online*, 21 February 2005.

47. In May 2009, an Alameda County, CA, school district mandated an LGBT curriculum for all students that denied parents the right to opt out. One activist reportedly said that the children of parents who would opt-out of such education were the ones who need it most. "Gay Curriculum Proposal Riles Elementary School Parents," *Fox News*, 22 May 2009; and "Compulsory LGBT Curriculum Pushes 'Political Agenda' on Schoolkids, California Parents Charge," *Catholic News Agency*, 24 May 2009.

48. USA retired Sgt Maj Brian Jones testimony. House Armed Services Committee, Subcommittee on Personnel, 110th Cong., 23 July 2008, available at http://cmrlink.org/fileuploads/HASC 072308JonesTestimony.pdf.

49. "How to End 'Don't Ask, Don't Tell': A Roadmap of Political, Legal, Regulatory, and Organizations Steps to Equal Treatment," Michael D. Palm Center, University of California Santa Barbara, CA, May 2009, hereafter referred to as the Palm Roadmap Report. The report appears in full text as chapter 10 in this volume.

50. Palm Roadmap Report, 19–21.

51. The Palm Roadmap Report suggests that "a new policy will work best if personnel are persuaded that it will not be harmful to the armed forces or to themselves, and may even result in gains. Toward this end, explanations of the new policy should be framed using themes reflecting military culture, such as the military's pride in professional conduct, its priority of mission over individual preferences, its culture of hierarchy and obedience, its norms of inclusion and equality, and its traditional 'can do' attitude," 19.

52. Ibid., 20.

53. Ibid., 21.

54. Ibid., 6.

55. Ibid., 20.

56. Rowan Scarborough, "Navy Officers Balk at Pro-Gay Seminar," *Washington Times*, 8 September 1994, A-1.

57. Office of the White House Press Secretary, "Lesbian, Gay, Bisexual and Transgendered Month," News Release, 1 June 2009. Also see Department of State, http://www.state.gov/r/pa/ei/pix/lgbt/, and NASA Equal Opportunity Programs Office, http://eeo.gsfc.nasa.gov/.

58. Al Kamen, "For One Night, Baghdad Gets a Pink Zone," *Washington Post*, 22 May 2009.

59. "14 from U.S. Embassy Security Staff in Afghanistan Fired," *CNN.com*, 5 September 2009, available at http://edition.cnn.com/2009/WORLD/asiapcf/09/04/afghanistan.contractors/index.html. Indiscipline is cumulative and progressive. Date-stamped photos taken at Abu Ghraib prison indicated that the Soldiers debased themselves before they abused Iraqi prisoners. See Duke *Law Journal*, 886.

60. Presidential Commission Report, Findings 1.32, 1.33, and 1.33A, p. C-40.

61. Senate Hearing 103-845, 1993, p. 424.

62. Karen DeYoung, *Soldier: The Life of Colin Powell* (New York: Knopf, 2006): 230–33. Army Gen Colin Powell and other members of the Joint Chiefs of Staff resisted President Clinton's move to lift the ban on gays in the military. Powell was frustrated that the issue was overtaking every other issue. "He had never been attacked by liberals before, particularly as a bigot; it bothered him far more than he had anticipated." This intimidation factor is relevant to the current debate.

63. Charles Krauthammer, "Powell Needs No Lectures," *Washington Post*, 29 January 1993, A23.

64. Brian Maue, PhD, an Air Force major and professor at the Air Force Academy, e-mail to the author, 17 July 2009. He was expressing personal views only.

65. July 2008 communication with the Center for Military Readiness. Following the civil rights model, in 2007 the British Ministry of Defence issued an open apology to all servicemen and servicewomen who were not admitted to or retained in the military before the ban on homosexuality was lifted (by order of the European Court of Human Rights) in 1999. See Damian Barr and Lucy Bannerman, "Soldiers Can Wear Their Uniforms with Pride at Gay Parade, says MoD," *London Times Online*, 14 June 2008.

66. Additional graphs and tables are displayed in "False 'National Security' Argument for Gays in the Military," available at http://cmrlink.org/CMRDocuments/DoDDischarges-090809.pdf.

67. David F. Burrelli, *CRS Report*, 14 August 2009, 5–10.

68. Dr. David Chu, under secretary of defense for personnel and readiness, letter, 7 February 2005, published in "Military Personnel Financial Cost and Loss of Critical Skills Due to DoD's Homosexual Conduct Policy Cannot be Completely Estimated," GAO Report GAO-05-299, February 2005, 42–43.

69. *CRS Report*, 14 August 2009, 9–10.

70. Ibid.

71. Palm Roadmap Report, 7; and Gen John M. Shalikashvili, USA, retired, "Gays in the Military: Let the Evidence Speak," 19 June 2009. In addition to the CRS, several officers in command during the Persian Gulf War have refuted this assertion. Lt Gen Robert B. Johnston, USMC, retired, who served as chief of staff, US Central Command (CENTCOM) at the time, would have been privy to any accommodation to that effect or conversation about it between Joint Chiefs chairman Gen Colin Powell and Gulf War commander Gen Norman Schwartzkopf. According to General Johnston and several former commanders at the time, the alleged suspension of discharges never happened.

72. Michael D. Palm Center, "Researchers Locate Army Document Ordering Commanders Not to Fire Gays," 13 September 2005; and "Pentagon Acknowledges Sending Openly Gay Service Members to War," 23 September 2005. The Palm Center claimed to have found an Army handbook from FORSCOM, Fort McPherson, GA, stating that homosexuals could be retained during deployments.

73. *CRS Report*, 14 August 2009, 6–8.

74. The Michael D. Palm Center, "Nations Allowing Gays to Serve Openly in Military," June 2009; David Crary, AP, "Allies Stance Cited in US Gays-in-Military Debate," 13 July 2009; and Otto Kreisher, "Few Armies Accept Homosexuals," *Sacramento Union*, 7 June 1993.

75. Charles Moskos, "Services Will Suffer If Used for Social Experiments," *Richmond-Times Dispatch*, 28 February 1993, F1; Susan Taylor Martin, "Israeli Experience May Sway U.S. Army Policy on Gays," *Israel 21c.com*, 10 January 2007; and author's e-mail correspondence with Israeli policy analyst Ethan Dor-Shav, May 2009.

76. Charles Moskos, "Services Will Suffer If Used for Social Experiments," *Richmond-Times Dispatch*, 28 February 1993, P. F1; and e-mail correspondence with Israeli policy analyst Ethan Dor-Shav, May 2009.

77. Otto Kreisher, "Few Armies Accept Homosexuals," *Sacramento Union*, 7 June 1993, A5.

78. Caucus of Homosexual Members of the Bundeswehr (AHsAB e.V.). Uwe Siemon-Netto, PhD, a veteran German foreign correspondent, translated this information and other German documents relevant to this subject. E-mail correspondence on file with author, May 2009.

79. Ibid.

80. On 22 January 2008, the national German wire service *Deutsche Presseagentur* reported that gay activist and AHsAB e.V chairman Jan Trautmann, a chief petty officer, said that though he has personally never had "negative experiences" since "coming out" in the navy, "most homosexuals prefer to stay away from elite units such as paratroopers."

81. Australian Defence Force. "Understanding Homosexuality," http://www.defence.gov.au/fr/education/Understanding%20Homosexuality%202003/index.html.

82. Cynthia Banham, "Navy Closes for Christmas, Families First in New Year," *Sydney Morning Herald*, 18 November 2008.

83. Kate Monaghan, "Dutch Political Party Wants to Normalize Pedophilia," *CNSNews.com*, 26 July 2006, http://www.cnsnews.com/ViewSpecialReports.asp?Page=/SpecialReports/archive/200607/SPE20060726a.html.

84. Interfaith Committee on Canadian Military Chaplaincy, "Same Sex Marriage/Blessing of a Relationship: Guidelines for Canadian Forces Chaplains," 25 September 2007.

85. *Lustig-Prean and Beckett v. United Kingdom*, 29 Euro. Ct. H.R. 548, 587 (1999); Human Rights Watch: Uniform Discrimination, 38; and *BBC News*, "Delight and Despair at Gay Ban Ruling," 27 September 1999, http://news.bbc.co.uk/2/hi/uk_news/458842.stm (reporting that the ruling of the European Court of Human Rights was "not binding on the UK Government").

86. Lawrence Korb, Sean E. Duggan, Laura Conley. "Ending 'Don't Ask, Don't Tell'." The Center for American Progress, June 2009, 17. This report cites a report due six months after the United Kingdom capitulated to the European Court order to accomodate gays in the military, but nothing more recent.

87. "Lifting Ban on Gays in Armed Forces Caused Resignations Report Reveals," *Daily Mail Online*, 15 October 2007, http://www.dailymail.co.uk/news/article-487750/Lifting-ban-gays-armed-forces-caused-resignations-report-reveals.html.

88. Chris Gourley, "Armed Forces March United for Gay Rights at Pride London," *London Times Online*, 5 July 2008, http://www.timesonline.co.uk/tol/news/uk/article4276099.ece; and Chris Johnston, "Navy to Advertise for Homosexual Sailors," *London Times Online*, 21 February 2005.

89. Nicholas Hellen, "Navy Signals for Help to Recruit Gay Sailors," *London Times Online,* 20 February 2005; and Chris Gourley, "Armed Forces March United for Gay Rights at Pride London," *London Times Online*, 5 July 2008, http://www.timesonline.co.uk/tol/news/uk/article4276099.ece.

90. Ministry of Defence, *LGBT News* Web site, available at http://www.mod.uk/NR/rdonlyres/370C9F8D-4728-4805-BE98-27A0207C2271/0/LGBTNewsletterMay08.pdf.

91. Richard Hatfield, "A Few Words from Our Diversity Champion," in Ministry of Defence, *LGBT News,* May 2008, http://www.mod.uk/NR/rdonlyres/370C9F8D-4728-4805-BE98-27A0207C2271/0/LGBTNewsletterMay08.pdf; and "LGBT Definitions – Transexuality," Ministry of Defence Web site, http://www.mod.uk/DefenceInternet/AboutDefence/WhatWeDo/Personnel/EqualityAndDiversity/LGBT/LgbtDefinitionsTranssexuality.htm.

92. Sarah Lyall, "Gay Britons Serve in Military With Little Fuss, as Predicted Discord Does Not Occur," *New York Times*, 21 May 2007, 8.

93. Tony Czuczka, "British Soldier Admits to Assault on Captive," *Washington Times*, 19 January 2005, available at http://www.buzztracker.org/2005//01/19/cache/441692.html; and Glenda Cooper. "Photos Indicating Abuse Renew British Debate," *Washington Post*, 20 January 2005, A18.

The reported abuse of male Iraqi soldiers with a forklift involved forced sexual acts, but details are not known because of court-ordered gag orders.

94. Mary Jordan and Robin Wright, "Iran Seizes 15 British Seamen," *Washington Post*, A-11; also see US Naval Institute *Proceedings*, May 2007, 10, which ran an editorial cartoon comparing the British navy of 1982 that sailed immediately to free the Falklands to a sailor of 2007. The first panel (1982) read, "Britannia Rules the Waves!" The second one (2007), read "Er, I say, Britannia Let Iran Waive the Rules!"

95. Michael Evans, Defence Editor, "British Were Complacent in Afghanistan, Says Sir Jock Stirrup," *London Times Online*, 30 January 2009.

96. "Bringing Serenity to Soldiers," *Army Times*, 6. This article, about the Army's first Buddhist chaplain, reports that "the military is trying to find chaplains who can minister to American troops without offending Muslim allies."

97. *Time/Encylopaedia Britannica Almanac* 2009.

98. John M. Shalikashvili, op-ed, "Second Thoughts on Gays in the Military," *New York Times*, 2 January 2007, 17; and "Gays in the Military: Let the Evidence Speak," *Washington Post*, 19 June 2009.

99. Zogby International, Opinions of Military Personnel on Gays in the Military, December 2006, submitted to Aaron Belkin, director, Michael D. Palm Center (hereafter Zogby/Palm poll), available at http://www.zogby.com/news/ReadNews.dbm?ID=1222.

100. See Zogby/Palm poll, 14–15, question 13.

101. Due to security rules that were tightened in the aftermath of 9/11, personal details and even general information about the location of individual personnel is highly restricted. Memorandum from Deputy Secretary of Defense Paul Wolfowitz to Secretaries of the Military Departments et al., 18 October 2001, addressing "Operations Security throughout the Department of Defense."

102. Bonnie Moradi, PhD, and Laura Miller, PhD, "Attitudes of Iraq and Afghanistan War Veterans toward Gay and Lesbian Service Members," *Armed Forces & Society* OnlineFirst, 29 October 2009, hereafter referred to as Moradi and Miller report, http://afs.sagepub.com/cgi/rapidpdf/0095327X09352960v1. The *Boston Globe* and other major media misrepresented this commissioned paper as if it were a genuine research report of the RAND Corporation. See Bryan Bender, "Study Builds Case for Repealing Don't Ask," *Boston Globe*, 9 November, 2009, http://www.boston.com/news/politics/politicalintelligence/2009/11/study_builds_ca.html.

These reports disregarded RAND's news release, which indicated that the paper "was the product of a contract directly with the researchers and not through RAND" (http://www.rand.org/news/press/2009/11/09/index.html).

In e-mail correspondence with the RAND Media Relations Department, the Center for Military Readiness determined that RAND employee Dr. Miller produced the paper on her own time, together with an academic associate at the University of Florida. Survey results discussed in the paper were from the 2006 Zogby International Poll, which also was commissioned by the Palm Center.

103. Zogby/Palm poll methodology, 2.

104. The Moradi and Miller report acknowledged that the Zogby/Palm poll and other studies shared "the limitation of being unable to distinguish responses by sexual orientation, as asking for sexual orientation disclosure on a survey would pose a substantial risk to participants under 'Don't Ask, Don't Tell,'" 6.

105. Zogby/Palm poll, 2. Zogby's 2006 polling sample was somewhat questionable, but if it were to be considered credible, internal data in the poll revealed interesting insights. The poll seemed to indicate that opinions on this issue have more to do with military occupation than they do with age. Active-duty people in the younger and older ranks were more favorable to the idea, but the ones in the middle age and experience group, who were more likely to be involved in close combat situations, were more strongly opposed.

106. Defense of Marriage Act (DOMA) Watch, Alliance Defense Fund, Marriage Amendment Summary, updated Fall 2008, http://www.domawatch.org/amendments/amendmentsummary.html.

107. Kyle Dropp and Jon Cohen, "Acceptance of Gay People in Military Grows Dramatically," *Washington Post*, 19 July 2008, A03.

108. While 71 percent of self-identified veterans in the *Washington Post* poll said gay people who do not declare themselves as such should be allowed to serve, that number dropped sharply to 50 percent for those who are open about their sexuality.

109. *Presidential Commission Report*, 15 November 1992, Commissioner-Generated Finding 14, p. C-135, citing Roper Organization, Inc., "Attitudes Regarding the Assignment of Women in the Armed Forces: The Public Perspective," September 1992.

110. Brendan McGarry, "Troops Oppose Repeal of 'Don't Ask,'" *Navy Times*, 5 January 2009, 16. Annual *Military Times* surveys are done by mailing questionnaires randomly to subscribers to the affiliated newspapers *Air Force Times, Army Times, Navy Times,* and *Marine Corps Times.* The polls tabulate only responses from active-duty personnel. Results are published in all four affiliated newspapers.

111. Michelle Tan, "No More Felony Waivers," *Army Times*, 4 May 2009, 28. In 2004, the US Army felt compelled to adjust waiver policies to allow some recruits to join despite previous run-ins with the law. Due to the economic downturn of 2009, however, recruiting results improved and felony waivers were suspended.

112. Grace Vuoto, "Is Obama Administration Listening to the Troops?" *Washington Times*, 30 July 2009, B2.

113. General Accountability Office, "Military Personnel: Financial Costs and Loss of Critical Skills Due to DoD's Homosexual Conduct Policy Cannot be Completely Estimated," February 2005, 21; cited in Duke *Law Journal*, 923.

114. Nathaniel Frank, "Don't Ask, Don't Tell' v. the War on Terrorism," *The New Republic*, 18 November 2002, 18; also Alistair Gamble, op-ed, "A Military at War Needs Its Gay Soldiers," *New York Times*, 29 November 2002.

115. On 11 December 2002, CMR filed a formal request for assistance with the Army inspector general, asking for an investigation of this waste of educational resources at DLI, followed by a 17 November 2003, Freedom of Information (FOIA) request that did not ask for personal information. The request was initially denied and later "answered" with largely blank pages marked with the FOIA exemption code that is used when government officials refuse to confirm or deny that disciplinary proceedings have taken place.

116. See Donnelly, HASC Testimony, 23–25.

117. Col Om Prakash, USAF, "The Efficacy of 'Don't Ask, Don't Tell,'" *Joint Forces Quarterly* 55 (4th Quarter, 2009), 92. Publication of Colonel Prakash's essay in the *JFQ* was the automatic result of a writing competition judged by officials at the National Defense University, which also awarded a generous Amazon.com gift certificate to the first-place winner. A spokesman for Adm Mike Mullen confirmed that the chairman had nothing to do with Colonel Prakash's essay prior to its winning the competition and its automatic publication in the *JFQ*. Colonel Prakash's article primarily drew upon sources favoring repeal of the law and failed to draw distinctions between "Don't Ask, Don't Tell" and the 1993 Eligibility Law.

118. *Report of the General/Flag Officers' Study Group*, Michael D. Palm Center, July 2008. The document claimed that the four retired officers "devoted particular and extensive effort" to the study of published works submitted by named "invited experts" who disagree with the Palm Center's views. There are no footnotes referring to opposing views that this author and others gave or recommended to the panel in response to a letter from the project co-coordinator, Brant Shalikashvili, whose father served as chairman of the Joint Chiefs of Staff. The report appears as chapter 7 in this volume.

119. Ibid., Finding 5, 8. The prospect of losing thousands of personnel apparently did not disturb the Palm Center study group because, they said, the military would become more "diverse" as a result. So much for concerns about recruiting, retention, or other factors associated with military necessity.

120. Dr. Aaron Belkin, written testimony submitted to the Military Personnel Subcommittee, Committee on the Armed Services, US House of Representatives, Hearing on "Don't Ask, Don't Tell" Review, 23 July 2008, 2.

121. Gary J. Gates, testimony on "Don't Ask, Don't Tell," submitted to US House of Representatives, Armed Services Committee, Military Personnel Subcommittee, 23 July 2008, 2.

122. Gary J. Gates, PhD, the Urban Institute, *Gay Men and Lesbians in the US Military: Estimates from Census 2000*, 26 September 2004. Dr. Gates is now a senior research fellow with the progressive Williams Institute at the UCLA Law School, Los Angeles, CA.

123. Ibid., iii.

124. Ibid., 1.

125. Ibid., 2.

126. Paul Winfree, policy analyst, Center for Data Analysis, Heritage Foundation, e-mail correspondence with the author, 19 February 2009.

127. Mr. Winfree agreed that proper use of the formula could be demonstrated with a bag of multicolored M&M candies. If a researcher knows the total number of candies and the number in each color group, the Bayes Rule can be applied as a formula to determine the probability that a child reaching into the bag will pick out a particular color.

128. Joanne Kimberlin, "Study Finds 65,000 Gay Men, Women in the Military," *Virginian-Pilot*, 21 October 2004.

129. Palm Roadmap Report, 11–12.

130. "Do Tell," editorial, *Washington Post*, 27 June 2009.

131. Palm Road Map Report, 4–5.

132. Aaron Belkin, "Self-Inflicted Wound: How and Why Gays Give the White House a Free Pass on 'Don't Ask, Don't Tell,'" Michael D. Palm Center, 27 July 2009, 4–5.

133. Ibid., 11.

134. Col Om Prakash, USAF, "The Efficacy of 'Don't Ask, Don't Tell,'" *JFQ* 55 (4th Quarter 2009), 93.

About the Author

Elaine Donnelly is president of the Center for Military Readiness, an independent, nonpartisan public policy organization that specializes in military/social issues. Founded in 1993, CMR advocates high, single standards in all forms of military training and sound priorities in the making of military/social policies.

Secretary of Defense Caspar Weinberger appointed Mrs. Donnelly to be a member of the Defense Advisory Committee on Women in the Services (DACOWITS) for a three-year term (1984–1986). In 1992, Pres. George H. W. Bush appointed her to the Presidential Commission on the Assignment of Women in the Armed Forces.

In May 2007 the Duke University *Journal of Gender Law & Policy* published her comprehensive, peer-reviewed article titled "Constructing the Co-Ed Military." In July 2008 she presented testimony on the issue of homosexuals in the military before the House Armed Services Personnel Subcommittee.

Mrs. Donnelly has published articles on military personnel issues in many newspapers and magazines nationwide, including the *Washington Post*, *USA Today*, the *Boston Globe*, *Congressional Quarterly Researcher*, *U.S. News & World Report*, the *Washington Times*, the Naval Institute's *Proceedings*, and *Human Events*, and has appeared on most network and cable network discussion programs. Formerly active in volunteer political and issue activities, in 2002 she was the recipient of the American Conservative Union's Ronald Reagan Award.

RACE AND GENDER

★★★★★★★★★ ★ ★★★★★★★★★

My fundamental belief is that we, as a military, must represent our country. We must represent the demographics of it. It is our greatest strength.
— Adm Mike Mullen, Chairman, Joint Chiefs of Staff

The United States' masculine military model has produced the most powerful military in the world with the most well-trained personnel and the world's most powerful and accurate weapons. . . . If the United States wants to maintain the world's best military, it should not focus on the feminine as weakness—it should instead focus on the possibilities of the feminine as a force multiplier.
— Dr. Edie Disler

Consider fitness testing: Why does the notion of being built differently become rational only when males have the advantage and not when females have an advantage?
— Dr. Steve and Dena Samuels

Women are in the military, women are serving in combat roles, and women should be required to register for Selective Service.
— Maj Maurleen Cobb, JD, USAF

I urge all of you . . . to talk with your friends and co-workers on the other side of the divide about racial matters. In this way we can hasten the day when we truly become one America.
— US Attorney General Eric Holder

RACE AND GENDER

When military members hear the word "diversity" these days, it is not uncommon to see a look of "here we go again" come across their faces. Some people think of diversity as nothing more than a politically correct social program. Others see diversity as just another online training module to "click through" once a year. Sadly, few people perk up and say, "Diversity, hey, that's a great idea! Let's talk about it." Instead, most would prefer to stop talking about it altogether. This isn't surprising. After all, the military has been studying, tracking, reporting, and discussing diversity for as long as most people can remember. By now, everyone understands that diversity is important. It is part of the military culture.

But do they? Do people really understand the importance of diversity, and have the correct lessons been assimilated into the culture of the military? There has been, and continues to be, "much ado" about diversity in the military, but the fundamental question still remains: what should the US military look like demographically?

At one time, military service in the United States was believed to be a predominantly white male endeavor. Blacks and Hispanics were eventually integrated through the course of history, followed by women. The process of integration was not easy, nor is it complete. Even in 2010, demographic data look very different when comparing the officer and enlisted ranks across the services. But merely observing demographic statistics that may appear out of balance does not necessarily indicate racial or gender discrimination. When the imbalances are not immediately obvious to those in charge, they are prone to inadvertently create systems which are less conducive to diversity over time if diversity is not an explicit and primary goal.

Embracing diversity has not always been a forte of the armed services. *Separate but equal* was a widely endorsed philosophy in the US military prior to 1948. It would seem from a resource perspective that combining blacks and whites into integrated units would have made a lot of sense. There would have been a reduced need for redundant facilities (barracks, showers, chow halls, etc.). And yet, when Pres. Harry S. Truman integrated the US armed forces with the stroke of a pen, dissent persisted. Many arguments and justifications were offered as to why it would have been contrary to good order and discipline to have blacks and whites serving in integrated units. But the data we have gathered over the last six decades makes a pretty compelling case that the only real barrier to immediately successful integration was bigotry.

By the time the Vietnam War wound to a close, the political debates began to shift and address the future role of women in the military. Once again, arguments emerged claiming that full integration of a minority class would impact the good order and discipline of the military. There would be an increased need for additional facilities (bathrooms and barracks), and, of course, many expressed concern about women in combat. This was just not an issue people wanted to consider on the heels of a failed Vietnam experience. Nevertheless, Congress acted, and by the 1980s, the military had expanded the diversity of its talent pool.

As we move forward in the twenty-first century, despite the progress we have made in the context of a broader society electing our first black president in 2008, race remains a contentious issue. Take for instance the following comments made in an interview on 15 September 2009 by former Pres. Jimmy Carter on the issue of racism:

> I think an overwhelming portion of the intensely demonstrated animosity toward President Barack Obama is based on the fact that he is a black man. I live in the South, and I've seen the South come a long way, and I've seen the rest of the country that share the South's attitude toward minority groups at that time, particularly African Americans. And that racism inclination still exists. And I think it's bubbled up to the surface because of the belief among many white people, not just in the South, but around the country, that African-Americans are not qualified to lead this great country. It's an abominable circumstance, and it grieves me and concerns me very deeply.

Naturally, the current administration downplayed these comments. After all, it was their only prudent course of action. One can only imagine what might have happened if the Obama administration had come out in support of the former president's remarks. The point of this example is not to judge the merits of the perspective, but rather to illustrate that in 2010, perceptions of racism between majority and minority groups are still *very real*. Senior leaders and policy makers must remain mindful that regardless of their own perspectives on the issue of race, there are many within their organizations who likely see the world very differently.

One can also imagine what kinds of political debates might be occurring now if, instead of Barack Obama being elected as the 43rd president of the United States in 2008, former US senator and current secretary of state Hillary Clinton had won. Senior military leaders must not ponder the question of "if," but rather of "when" the military will be ultimately directed by a female commander in chief. One needs only to look at the current wording of Article 133 of the Uniform Code of Military Justice (UCMJ)[1] where it deals with "conduct becoming an officer and a gentleman." With the full-scale integration of women in the military, the UCMJ was changed, but instead of removing the word "gentleman," the UCMJ merely redefined gentlemen to include women. Thus, we should not be surprised in the years to come when we collectively return our primary attention to gender issues.

The lesson to be drawn here is that racial and gender diversity has, is, and will remain a contentious topic facing military leaders into the foreseeable future. If history has taught us anything, the precursor to any successful resolution is transparent and respectful dialogue about the issues from a breadth of perspectives. In the following chapters, this is precisely what you will find.

Dr. Michael Allsep, a professor with Air Command and Staff College, introduces the section by telling a story of a former US Marine, turned secretary of the Navy and presently a US senator, James Webb. As he analyzes Senator Webb's publicly espoused opinions of the integration of women into the military from the mid-1970s through the present, Professor Allsep provides us with a lens through which to view the adaptive nature of the military when its internal culture and traditions are threatened by externally forced change. He argues the resistance is less about defending the military culture and more about defending martial masculinity in the face of evolving societal gender norms.

Leaders do not always understand that their own advantaged statuses can affect their decisions and behavior. The husband and wife research team of **Steve and Dena Samuels**, professors with the US Air Force Academy and the University of Colorado, respectively, discuss the concept of "privilege" and argue that leaders must look beyond their own frames of reference and perspectives when they desire to embark on cultural change. Whether looking at claims of sexual harassment or addressing accusations of religious intolerance, leaders must embrace privilege as a framework to enable effective change. Specifically, they highlight the necessity for every group member to internalize the idea of "equal human value" amongst colleagues.

With the transition of women into most aspects of military service over the past 30 years, the United States finds itself in a dramatically different context facing two very different types of wars than experienced in previous eras. **Maj Maurleen Cobb**, a US Air Force Reserve judge advocate, reviews the post-Vietnam history of the Selective Service System and argues the time has come to relax the male-only requirement for registration. In light of the expressed desire by the current presidential administration to potentially expand the selective service concept to "national service" beyond the realm of the Department of Defense, she argues that taking the initiative now could alleviate challenges the Department of Defense would inevitably face if it maintained a reactive posture to the policy.

Maj Valarie Long, a USAF career intelligence officer, reports on her study of dual-military couples and discusses the unique challenges they face as compared to couples where only one person is a service member. She finds that dual-military members begin their careers highly motivated, but by the 10-year point, they are comparatively less motivated to remain on active duty as compared to their peers. Children, deployments, and perceptions of promotion opportunity remain a hindrance to dual-military member retention. She recommends reforms in deployment scheduling, assignments, and return-to-service opportunities for all service members to improve retention of mid-grade officers.

As a retired officer and linguistics professor from the US Air Force Academy, **Dr. Edie Disler** argues that the hyper-masculine environment of the military favors overly masculine approaches to armed conflict. She provides several historical and contemporary examples to illustrate the value of the feminine in woman-centered cognition, tactics, and strategies—value which could easily translate to more effective ways of approaching military missions with greater success.

In January 2009, the Department of Defense established the Military Leadership Diversity Commission to conduct a comprehensive evaluation and assessment of policies that provide opportunities for the promotion and advancement of minority members of the armed forces, including minority members who are senior officers. In September 2009, the **chairman of the Joint Chiefs of Staff, Adm Mike Mullen,** addressed the commission, and his comments are reprinted here to frame the current senior military leadership perspective on diversity.

Charles V. Bush, the first African-American to graduate from the US Air Force Academy (Class of 1963), and colleagues argue a lack of diversity in the Department of Defense and related intelligence communities threatens the morale, welfare, and effectiveness of our military and risks compromising the safety of our nation. They assert the military has come up short in accomplishing any real measurable or sustainable progress in the diversity of its senior ranks and recommend recruiting and development efforts to change course.

Maj AaBram Marsh, a USAF maintenance squadron commander, continues along the same line of reasoning as Bush et al., providing the history of the integration of blacks into the Air Force officer corps. He argues that although the Air Force has made significant progress in overcoming overt segregation and bias, black officers have continued to lag behind their white counterparts, particularly at the most senior ranks. He recommends immediate and comprehensive reform to mentorship programs for minority officers. Given that the majority of Americans are likely to be today's minority in the foreseeable future, maintaining the status quo of predominantly white senior military leaders could lead to a repeat of history fraught with racial tension.

Finally, we end the section with a speech given by **US Attorney General Eric Holder** during African-American History Month in February 2009 at the Department of Justice. In what has since become known as the "Nation of Cowards" speech, he reflects on the racial challenges of both the past and present. Mr. Holder's message is clear and foreshadows the later comments of former President Carter: we do not talk amongst one another nearly enough about racial issues, which continue to dominate political discussions. Although the focus of his speech is specifically directed at race relations, his argument can and should be extended to every matter identified in this volume—if we are to make progress as a diverse nation, we must get comfortable with having frank conversations about the social matters that divide us.

Notes

1. Consistent with Article 1, Chapter 1, Section 1 of the US Code where Congress declared that "words importing the masculine gender include the feminine as well."

CHAPTER 13

THE ODYSSEY OF JAMES WEBB
AN ADAPTIVE GENDER PERSPECTIVE

Michael Allsep

If an odyssey describes an adventurous and hazardous journey with many twists and turns, then the story of James Webb's relationship with women in the military certainly qualifies. More importantly, his journey highlights the complexity of an issue that the Department of Defense grapples with to this day. Webb's odyssey had its origins in Public Law 94-106, passed in 1975, which required the service academies to open their doors to women beginning with the classes entering in July 1976. Opposition to the law among the armed services was immediate, but muted. The official line was that little was changing. "Today we are inducting 1,274 midshipmen," a Naval Academy spokesman was quoted as saying on induction day. "Eighty-one of them happen to be women, but they're all midshipmen to us."[1] For Webb, a Naval Academy graduate and highly decorated Marine who was grievously wounded in Vietnam, his opposition was nurtured in private during a semester as a visiting professor at the academy in 1979. It blossomed into public view with his publication of "Women Can't Fight" in the *Washingtonian* magazine in November of that year.[2]

Still carrying the whiff of cordite, his article was one of the first shots of the culture wars that began in earnest with the election of Ronald Reagan the following year. An excellent writer whose first novel, *Fields of Fire*, was loosely based on his experiences in Vietnam, Webb's article was meant to shock. "We became vicious and aggressive and debased, and reveled in it," he wrote about his time as a Marine in Vietnam. "I woke up one night to the sounds of one of my machinegunners stabbing an already-dead enemy soldier, emptying his fear and frustrations into the corpse's chest." He described seeing another soldier, "a wholesome Midwest boy, yank the trousers off a dead woman while under fire, just to see if he really remembered what it looked like."[3]

Webb believed that his best preparation for that environment had been his plebe year at the Naval Academy. The plebe system was "harsh and cruel," but it took him deep inside himself in a way that prepared him to withstand the multiple assaults of combat and its inevitable casualties. He specifically recalled one instance at the academy when, after being run to exhaustion, he was set upon by four upperclassmen who "took turns beating me with a cricket bat, telling me they would stop if I admitted it hurt. Finally, they broke the bat on my ass. I returned to my room and stuck my head inside my laundry bag and cried for fifteen minutes, standing in the closet so my roommates wouldn't see me." He admitted that it may "seem a sadistic and brutal way to learn self-truths," but he insisted it was essential preparation for combat leadership, and the presence of women at the academies was "poisoning that preparation."[4]

The Emergence of Martial Masculinity

Webb was opposed to women at the service academies because the primary mission of the academies was to train combat leaders, and he believed women were unfit for combat. "I have never met a woman," he declared, "including the dozens of female midshipmen I encountered during my recent semester as a professor at the Naval Academy, whom I would trust to provide those men with combat leadership. Furthermore, men fight better without women around."[5] This opposition ultimately arose from masculinity, especially the relationship between masculinity and war, and the relationship between masculinity and nationalism.

Throughout history, men have dominated the ranks of warriors, almost to the complete exclusion of women. This began to change during the industrial era as the nature of war changed. As R. W. Connell observed, "Violence was now combined with rationality, with bureaucratic techniques of organization and constant technological advance in weaponry and transport."[6] In military terms, this resulted in the first general staffs and what Walter Millis called the "organizational revolution" in warfare.[7] In gender terms, it meant a split in hegemonic masculinity between dominance and technical expertise. Dominance behavior had to make room for expertise, which not only was incompatible with traditional notions of dominant masculinity, but also eventually opened doors for women. Neither form of masculinity displaced the other, but their coexistence was uneasy and often competitive, creating a battle between expertise "on tap or on top."[8] The creation of nationalism sharpened the distinctions between "us" and "them" in a way that also empowered women by giving them space within the national community.[9] Inside the military, however, this space was caught in the contest between the competing versions of masculinity. As a result, a form of hegemonic masculinity that privileged direct dominance, excluded women, and felt challenged by claims of expertise also claimed the mantle of nationalism. From it emerged a kind of masculinity that is most accurately characterized as "martial masculinity."

Webb's rejection of women in the service academies was matched in vehemence by his disdain for politicians. "Civilian political control over the military is a good principle," he allowed, but "civilian arrogance permeates our government." The military had become "a politician's toy," and under the banner of equality, politicians were using the military as "a test tube for social experimentation." "Nowhere is this more of a problem than in the area of women's political issues," he asserted. He went so far as to claim that the civil-military relationship in the

United States was "dangerously close" to becoming like that of Nazi Germany, "a military system so paralyzed in every detail by the political process that it ceases to be able to control even its internal policies." This was more than a decade before Rush Limbaugh popularized the term "feminazis." Webb believed there was a war being waged against the traditional military culture and it included a "realignment of sexual roles," which was ultimately destructive to masculinity and national security—two concepts that merged in his mind.[10] As the image of the soldier fell into disrepute as a result of Vietnam, veterans like Webb lashed back from a deep sense of frustration and betrayal.[11]

A Sense of Honor

If the magazine article was a *cri de coeur*, Webb quickly retreated to home ground and produced a novel that more thoughtfully developed his argument. *A Sense of Honor*, published in 1981, was a thinly veiled polemic against the assault on martial masculinity. Set at the Naval Academy, the story unfolded over five days in February 1968, with the Tet Offensive raging constantly in the background. The novel was essentially the story of a troubled freshman, the senior who instilled in him the martial masculine virtues, and the changing academy environment that punished them both. Webb highlighted the connection to the Vietnam War by having the senior's plebe-year mentor killed in combat and by giving the two midshipmen a Marine company officer who was a decorated and wounded Vietnam combat veteran. At the end of the novel, the Marine hero was also victimized by the new "politically correct" academy regime and volunteered to return to Vietnam, but before he left, he visited a badly wounded comrade at Bethesda Medical Center and vented his rage. "It's a sad state of affairs when a few candy-ass lawyers and a couple of congressmen who wouldn't know a come-around from a walk in the woods can knock a whole institution on its ass."[12] The final challenge was delivered by the expelled senior, also on his way to join the Marines in Vietnam, who confronted the young MIT-educated professor who was the principal author of their misfortune. Hanging his full dress-blue uniform on the professor's classroom podium, with its insignia of a high-ranking midshipman officer in the Brigade of Midshipman, he explained all that the uniform signified to him and then faced down the unmilitary, uncomprehending professor. "So who should wear these, Professor?" the now-disgraced senior demands. "You decide."[13] Webb's message was powerful and succinct. If military leadership was no longer to be instilled on the basis of traditional martial masculinity, but taught instead through standards set by meddling politicians, candy-ass lawyers, and clueless liberal professors, then what could we expect from our future combat leaders?

The book quickly became a cult classic not only among males at Annapolis, but at each of the service academies. Webb's despair was in part a symptom of the political and cultural malaise of the late 1970s. The liberal activism of those years, especially in congressional legislation, flowed directly from the political activism and turmoil of the 1960s, and a loss of belief in the legitimacy of established political traditions. At the time it seemed an unstoppable wave of the future—a realization of many of the better dreams of that controversial decade. Though it was fated to be a short-lived moment before a much more powerful conservative political movement swept to power, during that moment feminism achieved perhaps its most lasting victories, though in the shadow of its most prominent defeat.

The Seventies

A new wave of feminist awareness and grassroots activity began in the 1960s, but amid the tumult of that decade, the feminist movement did not become politically powerful in Washington until the early 1970s. Its success in breaking down historic barriers to equal opportunities for women led some to call the 1970s the "She Decade."[14] The centerpiece of that political activism was an initiative to amend the Constitution to establish clear equality before the law for all women.[15] The same year Congress sent the Equal Rights Amendment (ERA) to the states, it also passed legislation adding Title IX to the existing civil rights laws.[16] Title IX barred all educational institutions that received federal funds from discriminating on the basis of sex. While not written to bar single-sex education, it did accelerate the demise of all-male colleges and contributed to coeducation becoming the almost exclusive model for higher education in the United States.[17]

The year after Congress passed the ERA and Title IX, the US Supreme Court decided the case of *Roe v. Wade*, declaring that the Constitution granted a right of privacy that extended to the doctor-patient relationship. The result of applying that right in this case was to protect a woman's right to choose to medically abort a pregnancy without government interference, therefore striking down laws criminalizing abortion.[18] Over the previous several years, individual state legislatures had gradually been removing those laws. At the time *Roe v. Wade* was announced, it seemed to many only the capstone to that movement. It would take time to see that the Court had overreached and that by tying its agenda so tightly to the pro-abortion cause, the feminist movement had doomed itself to decades of decline as it came to be seen largely in the light of that single issue.

It was in this atmosphere that Congress turned its attention to the question of admitting women to the service academies. As early as the 1960s, Rep. Robert B. Duncan had nominated a woman to West Point, but there was little public interest in the issue and no movement in Congress to force the service academies to admit women. Then Senator Jacob Javits of New York nominated Barbra J. Brimmer to the Naval Academy in 1972 and publicized the academy's refusal to consider her nomination because of her sex. In September 1973, two women and four members of Congress brought lawsuits against the Naval and Air Force Academies on equal protection grounds. In October of that year, Rep. Pierre du Pont introduced legislation calling for the admittance of women. Other bills followed. In December, the Senate by a voice vote passed an amendment allowing women admission to the service academies. The House struck the amendment, arguing that the issue should be considered separately, but agreed to schedule hearings before a subcommittee of the House Armed Services Committee. During the months leading up to the hearings, the position of the services solidified against admitting women, with the Navy echoing the failed Southern segregationist strategy by opening its Naval Reserve Officer Training Corps (NROTC) programs to women in an attempt to stave off admitting them to Annapolis. The hearings began on 29 May 1974. For the only time in Congress, both sides were given an opportunity to make their case during nine days of hearings spread over four months, concluding on 8 August 1974.[19]

The first witness before the committee, Rep. Patricia Schroeder of Colorado, began her testimony by reminding her colleagues of the overwhelming support Congress gave the ERA and the inevitability of its ultimate ratification. "The eventuality of the admission of women to the service academies is clear," she asserted, arguing that this legislation was needed mostly because bureaucracies "often aren't too responsive to changing circumstances." "The point is," she continued, "that when Congress passed the equal rights amendment, it did make a very clear statement as to whether this kind of exclusion should be permitted."[20] When asked later if her position on combat training at the academies was that "the law should treat all midshipmen and cadets alike," she responded, "that is what the equal rights amendment says, and I think that will soon be the law."[21] Over the course of the hearings, numerous arguments were made for and against admitting women to the service academies, but dominating all of those discussions was the assumption that the ERA would eventually become law and require their admission in any case. What nobody seemed to consider was what would happen if the service academies were required to admit women and the ERA failed.

Women and the Military Academies

Webb's indictment was that allowing women into the service academies destroyed the environment necessary to train officers properly for war. "By attempting to sexually sterilize the Naval Academy environment in the name of equality," he lamented, "this country has sterilized the whole process of combat leadership training, and our military forces are doomed to suffer the consequences."[22] Yet comparatively little testimony was offered or elicited during the nine days of hearings that directly addressed the ramifications of destroying the all-male culture at the academies as it related to successful training for combat. Only Secretary of the Army Howard H. Callaway, a West Point graduate and Korean War combat veteran from Georgia, made the argument that women would destroy "the Spartan atmosphere" that was necessary for producing combat leaders. "Let there be no doubt in anyone's mind about one thing," he testified. "Admitting women to West Point would irrevocably change the Academy. And all the evidence seems to say that the change could only be for the worse."[23] Like Webb, his argument was largely based on the evidence of history. "We should take a very long look indeed before we tamper with something that has proven over so long a time to be so successful." Also like Webb, he fell back on the oft-cited speech of Gen Douglas MacArthur at West Point in 1962, which famously ended with a benediction to "The Corps, and The Corps, and The Corps," inserting the speech in its entirety into his testimony.[24] Callaway's view was echoed in the short statement of a West Point senior, Stephen Townes, who accompanied him to the hearings. "I think by injecting women into this last bastion of military puritanism that West Point truly is, I think you are going to start a weakening process . . . of the Army." He went on to claim that many of the cadets who had not yet incurred their full military commitment would resign rather than continue at a coed West Point. "True, there is a bit of chauvinism in all this," he admitted. "We're just saying that West Point is our school."[25]

None of the other witnesses appealed to either the Spartans or the Puritans. One committee member even recalled that on his last visit to the Air Force Academy, he posed the question of admitting women to a dozen cadets, and most of them "thought it would be the greatest thing in the world." One even said, "Gee, it would clean up the language around here."[26] More importantly, Vice Adm William P. Mack, the superintendent of the Naval Academy, after reading a statement opposing admitting women, testified under questioning that his opposition rested solely on the current exclusion of women from combat. "In my estimation," he testified, "women could serve in any role in the US Navy at any time if this law were changed. They could come to the Naval Academy; they could pass the course in large numbers, and do all that's required of them physically, mentally, professionally, and in any other way, and there would be little requirements for change in our course curriculum, physical facilities, or anything of that sort." "If the law were changed," he concluded, "in my mind, women could do anything that men could do, and in some cases, perhaps even better."[27] Calling on a more recent history than Callaway or Webb, Admiral Mack argued that "having seen summer Olympics on television, having seen Billie Jean King on television, there are many women who can do all sorts of things that they are prepared for, and it would be a question, sir, of taking the training, passing it successfully, and demonstrating that that particular person, man or woman, could do the job."[28]

This then was the fundamental clash between those unalterably opposed to admitting women on an equal basis and those who were willing to accept at least the possibility. Those whose arguments were grounded in martial masculinity already had all the evidence they needed from military history and tradition; never mind that it was a history that treated women as second-class citizens and denied them opportunities to prove their equal worth. As long as men like Webb understood masculinity as a gender reality ultimately determined by nature and confirmed by historical experience, there could be only respect for traditions built on the natural gender order, or a perversion of it. Those who believed that individual women might prove themselves equal to the task if given the chance, but who were opposed to special treatment or a separate track for women, were, whether they recognized it or not, admitting that gender was a social construct. Theirs was a masculinity that respected expertise over direct dominance, and the evidence of their eyes rather than the traditions of their forebears was their standard of reference. The appeal of women to be given an equal chance to succeed or fail was appealing to officers like Mack partly because they could ultimately judge on the observed evidence, rather than relying on the nostrums of a sacred past. For Webb and others who shared his masculine values, men or women, the call of the sacred past was altogether too strong to be so easily dismissed.

While the House hearings were underway, Senators William Hathaway, Strom Thurmond, John Stennis, Mike Mansfield, and Jacob Javits reintroduced legislation in the Senate requiring the admission of women to the service academies. Though no action was taken that year, sponsorship by such a diverse group of senators sent a powerful message of support. In May 1975, the House finally passed legislation opening the service academies to women, and the Senate passed similar legislation in June. In October, Pres. Gerald Ford signed Public Law 94-106 (Title VIII), which included the provision requiring equal treatment for women at the service academies. The pending court cases were subsequently dismissed as moot. The feminists seemingly had their victory, and the She Decade could chalk up another achievement. The services acted quickly to comply with the law, and the Class of 1980 at each of the service academies for the first time in history included cadets and midshipmen of the female sex when they stood on induction day to take the oath.

In the wave of publicity that followed these women into the ranks, little space was given to the notion that their very presence was destructive of the martial masculinity vital for defense of the nation. Instead, reporters generally celebrated their courage while marveling at the obstacles they still faced. Despite the official line that sex was irrelevant, "the women were looked at with eyes ranging from disdain to skepticism to outright sexist appraisal of their legs. Everyone gawked." At the Naval Academy, recently graduated Ensign Kevin Cheatham remarked to a reporter that passing Midshipman Patricia Thudium was "certainly a well put-together girl," before realizing that he had traversed official lines of conduct. It did not bode well for a smooth integration when he quickly retreated into a warning that women midshipmen were already courting trouble by "talking sweet to first class (senior) midshipmen," neatly side-stepping his own inappropriate remarks.[29]

Sex was certainly an issue from day one, but the deeper kind of gender issues that would soon consume Webb were still largely hidden from view behind a gauzy fascination with the novelty and prurient interest of the thing. To Webb, such incidents were only symptoms of a larger problem, a problem caused by dangerous cultural trends and an ever-expanding revolution in civil rights.

The Feminist Movement

The wave of feminist legislation that opened the doors of the service academies and paved the way for Ensign Cheatham's encounter with Midshipman

Thudium was part of a larger rights revolution. This revolution not only produced a comprehensive set of individual rights enforceable by law, but it also
constituted what legal scholar Samuel Walker called "a new rights consciousness, a way of thinking about ourselves and society." Americans at every level of
society learned to think of their relationships to each other and to government
in terms of individual rights, defining all problems in terms of their rights and
talking in terms of what they had a "right" to do.[30] It was in part a victory of the
individual over the collective, but because this new rights culture depended on
government to define and enforce these newly recognized rights, it also meant
an expansion of government power even as a panoply of new rights attempted
to limit the reach of government power into the personal—especially sexual
lives of individuals. The rights revolution quickly found its reactionary response
in the form a "culture war" waged from the political right in defense of what
conservatives considered traditional values. "The crisis in American society today," thundered Supreme Court Justice Clarence Thomas, "is a result of the
'judicial rights revolution.'"[31] At the center of that response was a fiery critique
of the politics of feminism and the cultural revolution of the 1960s that gave it
new political life in the 1970s.

The turn against the feminist agenda was as sharp as it was sudden. In response to *Roe v. Wade*, Congress passed the Hyde Amendment in 1976 that
limited federal funding for abortions paid for by Medicaid, except where necessary to preserve the life of the mother or where the pregnancy was the result of
rape or incest. The Supreme Court upheld these restrictions by a five-to-four
vote in 1980, opening the door to a persistent effort to limit the result in *Roe v.
Wade* by chipping away at its margins that continued well into the following
century. At the same time, conservative grassroots activists in the states that
had not yet ratified the ERA began to mobilize in opposition, claiming that the
amendment, first proposed in the early 1920s, was an effort by feminists and
New Left radicals to remake the culture to fit their secular, unisex vision. No
new states ratified the amendment after 1977, but five would attempt to rescind their earlier ratification votes. Though Congress extended the time for
ratification in an attempt to save the amendment, the ratification effort finally
died in 1982.[32] What Schroeder and her allies had taken for granted had not
come to pass after all. Women were in the service academies and rising through
the ranks, but the political movement that put them there was dead.

A Warrior Is Born

Webb's response to the arrival of women at Annapolis can be seen as part of
that larger reaction against the feminist movement, and he certainly allied himself with that movement and adopted some of its rhetoric, but his argument
was more pointed and precise. Though his opposition to women at the service
academies was in many respects the issue that launched his political career, his
reaction was rooted in beliefs more firmly held than mere political convictions.
He saw the feminist advocacy of women in the military as part of a larger and

in many ways cynical attack on a tradition of martial masculinity that had deep roots not only in his personal identity, but also in the identity of the country he loved so much. It was a tradition that largely defined his sense of self, his sense of country, and even his sense of ethnic identity. In "Women Can't Fight," Webb quoted a midshipman, a former enlisted Marine, "who watched the Naval Academy change with the addition of women, believes the institution is dying, and with it a part of our culture as a whole." The military once "provided a ritualistic rite of passage into manhood," claimed Midshipman Jeff McFadden, but with women infiltrating that world, "Where in this country can someone go to find out if he is a man? And where can someone who knows he is a man go to celebrate his masculinity?"[33]

From his childhood, Webb had imbibed deep in a warrior ethos that was inseparable to him from true manhood. "My most memorable childhood moments were the ones spent at the outer edges of what other cultures might call the tribal circle," he wrote, "listening to my father and his long-time friends swap tales." In that circle, he learned that for a man

> to be recognized as a leader, he must know how to fight and be willing to do so, even in the face of certain defeat. He must be willing to compete in games of skill, whether they are something as traditional as organized athletics, as specialized as motorcycle or stock car racing, or as esoteric as billiards or video games. He must know how to use a weapon to defend himself, his family, and his friends. He should know how to hunt and fish and camp, and thus survive. And throughout his young life he should observe and learn from the strong men in his midst, so that he can take their lessons with him into adulthood and pass them on to the next generation. Perhaps, as some claim, the advance of civilization and the sophistication of our society have made many of these lessons irrelevant. But to me, the attitudes they ingrained have been the most consistent sustaining forces in my life.[34]

He sometimes described his father, an Air Force colonel, as "a Great Santini dad," referring to the novel *The Great Santini* by Pat Conroy.[35] Loosely based on his own childhood experiences, Conroy's novel described a household dominated by a Marine Corps fighter pilot who beat his wife and children through tears of mingled anger and love. While Webb makes no such allegations of physical cruelty against his father, he admitted that he "had not been an easy man to grow up with. He did not spare the lash."[36]

His father was also "given to making taunts and impossible challenges." When Webb was a small boy, his father used to ask him if he was tough, then hold out his large fist and order the youngster to "hit it again and again, telling me I could stop if I admitted I wasn't tough. My small fist would crumple against his and I would be unable to stop my tears, but I would never admit I wasn't tough."[37] Fast forward to Webb's plebe year at the Naval Academy, and the connection between the child and the young man comes full circle when four upperclassmen try to break him by beating him with a cricket bat until he admits it hurts. Small wonder that they broke the bat before they broke the plebe, though Webb was once again reduced to uncontrollable tears that he found shameful and tried to hide inside a canvas laundry bag.[38] The connection

Midshipman McFadden made between the rites of passage into manhood and the rigors of Annapolis was one that he shared with Webb and countless others. It was not just a feature of academy training, but a seamless web linking the traditions of masculinity with the traditions of war, made more sacred by a patina of patriotism. The demands of masculinity were entwined with training for war in a way that made it seem to those steeped in its traditions that to pull on one thread was to unravel the entire skein.

Vietnam's Influence

More than any other influence, Webb was shaped by Vietnam. His understandings of nationalism and masculinity were indelibly hardened in that fiery furnace. After graduating from the Naval Academy in June 1968, Webb completed the grueling Marine Basic School at Quantico, emerging first in his class. He was posted to the Fifth Marine Regiment in Vietnam in March 1969. The Tet Offensive was more than a year in the past, and peace talks had begun in Paris, but in Vietnam an average of 440 Americans were dying every week. Webb was ready to live the dream of combat he had fantasized about since childhood and trained for since first arriving at the Academy, but that dream was already out of sync with the emerging mood of the country. The antiwar movement, already strong on college campuses before Tet, metastasized afterwards into a groundswell of opposition that soon dominated the country's political discourse. The discord only seemed highlighted when Webb flew to war not in the rough confines of a military transport plane as he had anticipated, but in the plush seating of a Continental Airlines passenger jet. As the marines deplaned, the attractive German stewardess sent them off with, "Well, gentlemen, have a good war."[39]

The Fifth Marines was in its third year of continuous combat operations in the An Hoa Basin west of Danang, an area the Americans called the Arizona Valley, when Webb arrived to command a rifle platoon.[40] Combat seemed to come naturally to Webb; he had spent much of his life preparing for it as the ultimate test of his manhood and worth. He was courageous, adept at tactics, and popular with his men as a leader who would not get them unnecessarily killed. One of them, Mac McGarvey, became a "blood brother" to Webb over the two months he served as Webb's radio operator. One night they discussed how they would react if they were ever seriously wounded. The next day a booby trap blew off McGarvey's right arm. As Webb leaned over his bleeding friend, tears streaming down his grimy face, McGarvey looked him in the eye and said, "Knock that shit off, sir, it's only an arm."[41] In moments like that, Webb's measure of manhood became firmly fixed in his mind, and with it his understanding of war and the camaraderie that was combat's fuel and byproduct.

During his time in Vietnam, he was no stranger to the moral complexities of guerilla warfare. Most of the villages in the An Hoa Basin were considered enemy-controlled free-fire zones, areas where the Marines and Air Force could unleash firepower with little or no restraint. "Air strikes and artillery missions on populated areas were the order of the day." He later recalled walking the moonscape of what had once been a village after a B-52 "arc light" strike left "craters twenty feet deep in places where a day before there had been thatch-roofed homes." "The An Hoa Basin," he concluded, "was a bloody, morally conflicted mess." He resolved his own moral dilemmas in the way soldiers of all cultures usually do. "You are walking in the real world, with Marines around you and a weapon in your hand," he recalled, "not sitting in a college class on moral philosophy." Your bond is with your fellow soldiers, so you "take care of your Marines" even if it means firing into a tree line studded with populated hamlets.[42] It was the age-old equation that soldiers count more than civilians, but Webb didn't simply bury his guilt. "The villages in these contested areas paid a horrible price," he wrote years later. "No matter one's feelings about the war itself, or which side they might have been on, we owe them."[43] Since the war ended, Vietnam has called him back many times, and it is fair to say that the bonds created there have never loosened. Vietnam in all its complexities and contradictions haunts him still.

Webb had been told at Quantico that Marine officers in the rifle companies had an 85 percent probability of being killed or wounded, and commanding a platoon in a company nicknamed "Dying Delta" seemed unlikely to improve the odds. Webb's company commander was wounded, and every other platoon leader was either killed or wounded during his time in combat.[44] For a while, it seemed Webb might just buck the odds, but courage has its price, and Webb was nothing if not courageous in battle. During his tour, he earned two Bronze Star medals, two Silver Star medals, two Purple Hearts, and the Navy Cross, the highest award for valor the Marine Corps can give. Only the Congressional Medal of Honor rates higher.[45]

In July 1969 his luck ran out as two grenades hit him as he led his men in clearing a series of bunkers. Even then he was not out of the war, as he discharged himself from the hospital and returned to his company before the wounds had properly healed. He continued to lead his men with skill and courage, but the wounds to his left leg became infected, and despite two years of surgery after he left Vietnam, the damage was permanent. His determined efforts at rehabilitation were unavailing, and he was finally forced to leave the Marine Corps in 1972.[46]

Civilian Transition

By the standards of martial masculinity, he had a "good war," but his return from Vietnam and the bitter disappointment of the ending of his Marine career required Webb to consider other life choices. His sense of still wanting to serve led him to the Georgetown Law Center, but his experience there led him away from the courtroom and towards a life of writing and politics. To say that Webb felt alienated in the Georgetown environment is to do a great disservice to the word *alienated*. It would be more accurate to say that he felt his Vietnam experience had simply found another battlefield. "Few things in life have come as naturally to me as combat, however difficult those days proved to be," he later wrote. "And conversely, few things have surprised me so completely as the other world I entered a few years later when I arrived at the Georgetown University Law Center."[47]

Webb quickly saw himself at odds with the students and the faculty. "Years of intellectual conditioning had taught them that the government was corrupt, the capitalist system was rapacious, that the military was incompetent and even invidious, and that the WASP culture that had largely built America had done so at the expense of other ethnic and racial groups." As for those like Webb who had volunteered to serve, they "were either criminal or stupid."[48] He felt that one of his professors had targeted him directly when the final exam was based on a fact pattern about a platoon sergeant named "Jack Webb" who smuggled black market jade in the bodies of his dead comrades. Webb found it difficult to finish the exam, but despite his protest to the dean, in the topsy-turvy world of Georgetown the professor won tenure, and Webb's concerns were brushed aside.[49] However murky the moral picture sometimes looked to him in Vietnam, he found a different kind of moral quagmire in the elite universities.

On graduating from Georgetown, Webb began his twin careers of writing and government service. He worked as counsel to the House Committee on Veterans Affairs from 1977 to 1981, while publishing *Fields of Fire* in 1978, quickly followed by *A Sense of Honor* in 1981, and another Vietnam novel, *A Country Such As This*, in 1983.[50] His service in Washington and his introduction to the literary scene only served to reinforce his sense of isolation from the liberal orthodoxy that seemed to dominate elite politics and culture.

Arriving at the Naval Academy for the spring semester of 1979 as that institution's first writer in residence, and finding it seemingly so different from the institution he left in 1968, only served to make his growing sense of political outrage personal. In the women midshipmen he now saw in the classrooms, on the drill fields, and in the dormitory rooms of Bancroft Hall, Webb found the personification of a feminist-driven attack on the values that he had been taught to revere and defend. The result was not only a series of articles and books arguing against the erosion of the warrior culture, but a crusade against the encroachment of feminism into what he considered the masculine sphere. Webb took that crusading spirit with him when he entered the Reagan administration, but as he worked with a military that was becoming increasingly integrated with competent women officers, his adversarial position against them slowly began to change into a grudging support and admiration.

The Journey into Politics

In 1984 he became the first-ever assistant secretary of defense for reserve affairs, and in 1987 he became secretary of the Navy. One of his first acts as Navy secretary was to order that Naval Academy graduates who selected Marine Corps service be sent to "Bulldog" training, a process of toughening from which academy graduates had once been exempt.[51] The implication was that the rigors of academy training, especially plebe year, had been so softened since the introduction of women that it now resembled a civilian campus. One of his first messages to the fleet directed that the promotion process give new emphasis to "demanding assignments," especially combat assignments. Women in the naval services understood that they were barred from what Webb would consider "demanding assignments" and feared for their future promotion prospects.[52] Nonetheless, when the results of a comprehensive review Webb ordered of the progress of women in the Navy recommended expanding their opportunities for service at sea and in naval aviation, he opened more types of ships and aircraft to women and crucially changed the definition of combatant to include only those units that were actually meant to engage the enemy. When a similar review he ordered of women in the Marine Corps was considered by his successor, many new jobs were opened to women in the Marine Corps as well.[53]

Webb also "directed vigorous corrective and preventive actions in areas of sexual harassment, fraternization, treatment, and morale, and approved a number of other recommendations to improve career opportunities" for women.[54] Whatever his views on proper masculine and feminine roles or the state of the culture wars, Webb proved in office that he was as willing as Admiral Mack to act on evidence of competence even if it flew against his preconceived notions of gender.

If people thought these initiatives indicated a lessening of his passion over the issue of women in the military, his 1997 front-page article in the conservative magazine *The Weekly Standard* surely disabused them of the notion. Webb believed the issue of women in the military was part of a larger cultural assault on the bastions of the traditional gender order, and however he might have shifted his views on aspects of the issue, his opposition to the assault on martial masculinity came from his core identity. Echoing his earlier "Women Can't Fight," he returned to the issue of women at the service academies. He argued that many of those who opened the doors of the service academies to women did so not just to advance the cause of equality for women in the military, but because they were "quasi-revolutionaries who took delight in the chaos into which our country had fallen." Sounding again the tocsin of the culture wars, he railed against the usual suspects of feminists, liberal politicians, the media, and academic elites, accusing them of seeking to undermine the military's "historic culture." Where they spoke of rights, he spoke of values. "Next to the clergy," he wrote, "the military is the most values-driven culture in our society."[55] How deep his belief in those values ran was not fully apparent until 2004, when he published the book he had spent his whole life writing.

Born Fighting: How the Scots-Irish Shaped America was itself born from two impulses. The first was Webb's reverence for his own particular ethnic heritage, but the most powerful was his response to the cultural and political upheaval of the 1960s and 1970s. While "left-wing activists" looked on those years as "a time of victory, and even of affirmation," for traditionalists like Webb "it was a time of confusion and defeat."[56] To Webb and many others, the rights revolution put self-interest over national interest and created a new "reverse discrimination."

These initiatives appeared to "cast even the lowliest Euro-American as a privileged oppressor."[57] As a proud descendant of hardscrabble Appalachia, Webb resented that his Scots-Irish heritage had been lumped in with a generic WASP culture that was not only distinct from his own, but had in many ways oppressed his forebears as surely as it had any other.[58] *Born Fighting*, he wrote, was an attempt "to defend myself in this new world of hyphenated Americans. . . . And in a society obsessed with multicultural jealousies, those who cannot articulate their ethnic origins are doomed to a form of social and political isolation."[59]

As is always the case when history is yoked to the task of defining ethnic identity, more myth than truth emerges. Webb's book was no exception. In his survey of American history, Reconstruction appears as "The Mess the Yankees Made," while the genocidal Andrew Jackson is raised to the level of a Scots-Irish deity. Yet his history is doubtless a reasonably accurate reflection of what other descendants of that tradition might accept, and as a context for the martial masculinity that is the engine of Webb's history, it adds much to understanding his conception of it. It was that martial masculinity that had always motivated Webb's reaction to women in the military, and because it was a privileged part of his ethnic identity, it was not an idea to be examined but a belief through which all other ideas were examined. "The warrior ethic has always been the culture's strong suit," he wrote. "The Scots-Irish emphasis on soldiering builds military leaders with the same focus and intensity that Talmudic tradition creates legal scholars."[60]

The largely missing and silent part of that culture was the role of women. They appear rarely in his narrative and would have disappeared altogether had he not felt compelled to say some good words about the women in his own family. Even here, women appear only in supporting roles. They either keep the home while the warrior is away, hold it together after the warrior has fallen, or fade into the background when the warrior returns, all the while passing down the warrior stories from one generation to the next.[61] More to the point, this near absence of women from his history seems to pass unnoticed by Webb. In a world defined by the values of martial masculinity, there was no place for women in the main story.

Amartya Sen gave his book *Identity and Violence* the subtitle *The Illusion of Destiny*. All people have a fear of losing their past and their historical identity in the melting pot of the present, but identities are "robustly plural," and the importance of one identity, no matter how deeply felt, "need not obliterate the importance of the others."[62]

"There is a critically important need to see the role of *choice* in determining the cogency and relevance of particular identities which are inescapably diverse."[63] History, in other words, is not destiny, but it can become a trap when one particular identity is given such priority and privilege that it obscures the others. "The military culture emphatically views itself as part of an historical continuum," Webb argued, and for that reason the denigration of the military that began in the Vietnam era still resonates in the twenty-first century.[64] The implication is that destiny has decreed the divide that exists between those who champion women's equality in the armed services and those who still resist full acceptance of women in the military. The warrior ethos is masculine, Webb insists, and women only serve to dilute its essence. The martial masculine identity, grounded in military history and ethnic tradition, transcends all others when considering the identity of the warrior. As there was no place for women in military history or Scots-Irish history, there can be no place for them in the front ranks of the military.

Senator Webb

The unexpected turns of history can sometimes offer a way out of what at times seems the trap of history, however, and no turn in recent US history was more unexpected than the events of 9/11. By chance, Webb was at the Pentagon when the first plane hit the World Trade Center towers for a scheduled book signing for *Born Fighting*. He quickly left to watch CNN from his office across the way, and from there he heard the plane that hit the Pentagon "almost like the tinny sound of outgoing artillery." That afternoon an editor from the *Wall Street Journal* called and asked him to write a piece for the paper. Entitled "Where Do We Go from Here?" it argued for a robust response to the terrorists but warned against deploying major forces to the Persian Gulf region. "Do not occupy territory," he warned. It remains the only thing he ever wrote for that paper that they never published.[65] From the very first day of the "War on Terror," a divide was opening between Webb and his putative allies in the Republican camp. As events progressed, Webb widened the gulf by warning repeatedly of the folly of invading Iraq and the consequences of an American occupation of that country. Six months before the invasion of Iraq, Webb published a piece in the *Washington Post* whose argument was in its title, "Heading for Trouble: Do We Really Want to Occupy Iraq for the Next Thirty Years?" As his former political associates ignored his warnings, he slowly became alienated from the Republican Party.

Like many professional officers, Webb had gravitated towards the GOP during the late 1960s and 1970s in response to the antiwar movement and the liberal activism of those years that challenged military culture and values. By the time George W. Bush was awarded victory in the disputed 2000 election, Republican officers outnumbered Democrats by a ratio of 8 to 1, and the Republican Party had come to treat the military as its political constituency. The Iraq War has threatened to change the political landscape for the military. As Webb declared, "The historical tables have turned."

> It is now the Republican Party that populated the Defense Department with a cast of unseemly true believers who propelled America into an unnecessary and strategically unsound war; the Republican Party that insisted in distorting the integrity of the military's officer corps by rewarding sycophancy and punishing honesty; the Republican Party that has most glaringly violated its stewardship of those in uniform; and the Republican Party that continually seeks to politicize military service for its own ends even as it uses their sacrifices as a political shield against criticism for its failed policies. And in that sense, it is the Republican Party that most glaringly does not understand the true nature of military service.[66]

On 8 February 2006, Webb made the break with the Republicans complete by announcing his candidacy for the US Senate from Virginia, taking on incumbent Republican Senator George Allen, a man many political insiders had already tabbed as a favorite for the 2008 presidential campaign. Despite a resume that made him look more naturally Allen's successor than opponent, the circumstances of a new foreign war were powerful enough to force Webb to realign his political identity and assume the mantle of the Democratic Party, still the party of the "political radicals, peaceniks, Black Power activists, [and] flower children" that had once stirred his rhetorical passion.[67]

His role as polemicist against women at the service academies could not go unnoticed, just as it had dogged him during his earlier confirmation hearings during the 1980s. It remained a minor irritant during his primary campaign, but when Allen's repeated racial gaffes and the rising unpopularity of the Iraq War and the Bush administration propelled Webb into a real battle, the article in all its rhetorical excesses came back to bite.[68]

On 13 September 2006, five women graduates of the Naval Academy appeared at a news conference at the Richmond Marriott hotel to finally fire back at the author of "Women Can't Fight." "Jim Webb didn't create harassment against women at the academy, but I truly believe he increased its intensity," said retired CDR Jennifer Brooks. Particularly offensive was Webb's allegation that the academy was "a horny woman's dream." "My mother read that," remembered Brooks. "I joined the Navy to serve. It was unbelievably demoralizing to be painted as a pampered slut."[69] Another graduate recalled that Webb's article "was like throwing gasoline on a fire." It gave the misogynist attitudes of some male midshipmen a hero and a rally point. "The article was brandished repeatedly," the women recalled, and used as justification for sexually harassing them.[70] Some male midshipmen even took to wearing "Jim Webb Fan Club" t-shirts.[71] One woman said that after the article came out, she was forced to memorize passages and shout them at the top of her lungs at noon meal.[72] In that room, the real impact of Webb's angry tirade in defense of martial masculinity became apparent, and the victims weren't the meddling politicians, candy-ass lawyers, or clueless liberal professors of his imagination; they were young women who were willing to test themselves against the best the academy could throw at them in order to serve their country. They were the real victims of martial masculinity.[73]

The old Webb had proved combative against such attacks. When Webb was nominated for his first Pentagon post in 1984, Defense Secretary Caspar Weinberger submitted a letter to the Senate committee announcing that Webb had "reversed" his previous hostile views on women in the naval service and was now fully in line with official policy. In private, Webb chafed at the letter, resenting being treated "like a reformed smoker" and claiming only to have promised not to try to "turn back the clock."[74] This time he offered a full and complete apology. Within two hours of the news conference, Webb released a statement distancing himself from the article. "To the extent that my writing subjected women at the academy or the active armed forces to undue hardship, I remain profoundly sorry." He argued that the article was written at a time when his emotions over Vietnam were still raw, and "during a time of great emotional debate over a wide range of social issues and the tone of this article was no exception." "I am completely comfortable with the roles of women in today's military," he concluded, "and I fully support the advancements that have taken place."[75] Publicly at least, Webb abandoned his former opposition to women in the service, though he reserved the right to oppose full removal of the restrictions on women in combat. His dogged defense of martial masculinity was at least muted in service of what he now saw as greater national priorities.

For at least one of the women who spoke out against Webb, his apology was enough. "I would like to thank Jim Webb for recognizing that an apology was needed and for issuing one," Linda Postenriender wrote in a statement that also endorsed him for election.[76] While the rest of the women remained firm in their opposition to Webb, and while Allen campaign ads featured them on what seemed a continuously running reel, an even more unlikely woman stepped forward to his defense.

At a fundraising luncheon in Old Town Alexandria, Senator Hillary Clinton endorsed Jim Webb for the US Senate. Just writing those words makes the mind boggle. "I've watched and analyzed what Jim Webb has done when he was in the Pentagon, opening up positions—thousands of them—to women when previously they had not been available," she said. "We have women serving with valor and distinction because the battlefield has changed." Webb had once called the Clinton administration the most "corrupt" in history, but now he owed his Senate seat largely to the endorsement of the woman he and his former culture war allies always considered their *bête noire*.[77] When it was his turn to speak, Webb said, "probably the most important thing I can say about that entire episode is that there's a term in law, *res ipsa loquitur*, the thing speaks for itself." He then introduced his senior campaign staff, five of whom were women. It seemed that Webb was not a masculine purist after all and that he could still be the combat warrior of old while respecting the role of expertise, in whatever gender it appeared.

Clinton and Webb were both on message that day, and there was plenty of room for political cynicism, but Webb's public record is extensive in words and deeds, and if anything stands out it is his lack of cynicism. He has been authentic throughout. "The books we write march alongside us," he wrote. "If we have written directly or even tangentially about social policy, our books speak to our judgment and intellectual breadth at one particular moment, as well as to the events that were then dominating our consciousness. And yet in our books that moment survives, vulnerable to rebuttal in different social and national circumstances as the years wear on, forever frozen on the pages we have written." That was not written in response to the campaign attacks on "Women Can't Fight"; it was written two years earlier in the foreword to *Born Fighting*.[78] As Clinton observed, the battlefield had changed from the liberal activism of the 1960s and 1970s and from the conservative backlash of the 1980s and 1990s, and

Webb had changed with it. The new priorities created by the Iraq War caused Webb to shift his priorities in a way that required him to give less emphasis to the masculine values that once seemed nonnegotiable.

Drawing Conclusions

As the hegemonic masculinity that has long characterized military culture split over the contest between direct dominance and technical expertise, the military has always been of two minds on the issue of women serving equally alongside men. On the one hand, military history and tradition supported the concept of a martial masculinity that excluded women from all but the safest supporting roles. On the other hand, military experts understood that a volunteer military could not possibly work without women playing increasingly more prominent roles in an ever-increasing number of military specialties.

Admiral Mack and many of the others who testified on the issue of admitting women to the service academies were mostly concerned with personnel issues and the costs and benefits of integrating women. Though most of them were also motivated by a masculinity that made them uncomfortable to varying degrees with the idea of women in combat, they were susceptible to arguments based on the proven ability of women and the needs of the service. Only some like Webb were masculine purists, and few of them possessed his gifts as a writer or his insights into history and culture. His ability to occasionally rise above the level of most partisan rhetoric made it possible to see the inherently honorable and legitimate concerns of those who opposed women in the military on the basis of profoundly held and deeply rooted cultural beliefs.

Since 9/11, the demands placed on the military and the steady rise of women in the ranks have combined to create the one thing that an exclusionary history never could—solid evidence that Mack was right when he testified that there were many women who could do all sorts of things; it was just a matter of "taking the training, passing it successfully, and demonstrating that that particular person, man or woman, could do the job."[79] The last remaining question for the military is whether the cultural demands of martial masculinity can ultimately accept women as warriors. Only once that question is resolved can the military finally be at peace with itself as a fully integrated fighting force.

The answer lies with culture. Just as martial masculinity is a cultural construct that creates a powerful personal identity, that identity can be changed over time, and more importantly it can be reprioritized in recognition of other equally important identities. "In our normal lives," Sen wrote, "we see ourselves as members of a variety of groups—we belong to all of them." That plurality of identities forces us to continually decide on their priority in any particular context. Under normal circumstances, this occurs naturally and without much conscious thought. It's not difficult for most people to recognize when they need to be a parent, when they need to be a friend, when they need to be a commander, or when they need to be a fellow human being helping another human being in distress.

Even when identities conflict, most people can get past the contradictions. "Do I contradict myself?" Walt Whitman wrote. "Very well then I contradict myself, (I am large, I contain multitudes.)" Conflict is promoted "by the cultivation of a sense of inevitability about some allegedly unique—often belligerent—identity that we are supposed to have and that makes extensive demands on us."[80] Martial masculinity is a socially constructed identity that excludes and devalues women by making unique and belligerent claims on the identity of those who wish to serve their country in uniform. It is as harmful to men as to women, for by blurring the things all Soldiers, Sailors, Airmen, and Marines have in common, it weakens what they might accomplish together. Fortunately, it is not the only masculinity the military recognizes, and as Webb proved, it can adjust to the greater demands of national security.

Explaining "Women Can't Fight," Webb claimed that he had not anticipated the firestorm the article would create, but surely this was disingenuous. It wasn't that he didn't want a firestorm; it was that his passionate defense of martial masculinity blinded him to the more important identities he shared with the women midshipmen he so cavalierly disparaged. He could have seen them as fellow officers, as patriots, as daughters of courageous veterans, as our country's future leaders; in any of those guises, it is hard to see him describing their motivation as a "horny woman's dream."

It is too easy to dismiss Webb's assaults on women in the military as merely rhetorical excesses during times of passionate political debate. The depth of his judgment and his intellectual breadth argue against such a facile response. Webb was, in his own way, as much a victim of the martial masculinity he embraced as the women against whom he brandished it as a weapon in the culture wars. That he proved able to move past the masculine straightjacket of the past and accept, if not actually embrace, the advances women have made in the military, especially since 9/11, indicates hope for the survival of a healthy military culture, adaptable to necessary change. It also offers an example for the military establishment in general as it continues to face the challenges of a changing gender order in American culture.

Like the story of Odysseus, Webb's story finally winds its way back home, but a home much changed from the one he first knew as a young midshipman at the United States Naval Academy. As a member of the Democratic Party, he now sits as the junior senator from Virginia in the US Senate, once again directly responsible for this nation's defense. He has been changed by his journey and has adapted in recognition of higher priorities and a greater calling, but the military still struggles with the demons of martial masculinity. It is too much to ask or expect more from Webb that he disown an identity that he has been at such pains to explain, justify, and defend. Nor is it necessary. This country's military past can be fairly interpreted as largely a triumph of the martial masculine virtues, but the past need not confine the future.

Notes

1. Karen DeYoung, "Academy Girls-uh, Guys-Blend In," *The Washington Post*, 7 July 1976, C1.

2. James Webb, "Women Can't Fight," *Washingtonian*, November 1979.

3. Ibid.

4. Ibid.

5. Ibid.

6. R. W. Connell, *Masculinities* (Berkeley, CA: University of California Press, 1995), 192.

7. Walter Millis, *Arms and Men: A Study in American Military History* (New Brunswick, NJ: Rutgers University Press, 1981).

8. Connell, *Masculinities,* 191–92.

9. Cynthia Enloe, *Bananas, Beaches and Bases: Making Feminist Sense of International Politics* (Berkeley, CA: University of California Press, 1989, 1990 and 2000), 61–62.

10. Webb, "Women Can't Fight,"

11. Michael S. Kimmel, *Manhood in America: A Cultural History* (New York: Oxford University Press, 2006), 174 and 197.

12. James Webb, *A Sense of Honor* (Englewood Cliffs, NJ: Prentice-Hall, 1981), 302. For those who don't know a come-around from a walk in the woods, it's a part of the plebe (first year) indoctrination system that requires a plebe when so ordered to come-around to an upperclassman's room, usually just before meal formations or in the evening, and recite mandatory rates (memorized items of naval and academy knowledge) and undergo other military instruction. This was once a great source of hazing, some of which Webb describes and defends, but is now tightly controlled. Though professing his opposition to hazing, it was this control that most incensed Webb.

13. Ibid., 308.

14. James T. Patterson, *Restless Giant: The United States from Watergate to Bush v. Gore"* (Oxford: Oxford University Press, 2005), 52.

15. In 1972, Congress sent the Equal Rights Amendment (ERA) to the states for ratification. Few doubted it would soon become part of the constitution. The vote in the Congress was overwhelmingly in favor, with only 23 representatives and 8 senators voting against it. Within minutes of final Congressional action, the Hawaiian legislature began the ratification process, unanimously ratifying the amendment the same day. Delaware, Nebraska, and New Hampshire ratified the next day. As early as 1970, the first year pollsters asked the question, 56 percent of the American public favored the amendment. Both political parties had already endorsed the ERA, the Republican Party being on record in support since 1944, and President Richard Nixon publicly supported its passage. Besides, the simple affirmation in the wording of the amendment, "Equality of rights under the law shall not be denied or abridged by the United States or by any State on account of sex," seemed beyond serious dispute outside those areas of the South that were already stigmatized by the violent racism so recently overcome. Even in South Carolina, the lower house of the General Assembly voted 83 to 0 in favor of ratification within days of final Congressional action before the measure failed in the state senate, and Tennessee became the tenth state to ratify in April. By 1974, final ratification seemed all but assured, with 33 of the required 38 states already having ratified the amendment and political momentum still seemingly favoring final approval.

16. Ruth Rosen, *The World Split Open: How the Modern Women's Movement Changed America* (New York: Viking, 2000), 89.

17. Though it didn't specifically target athletics, it's most important and contentious effect was to require equal access to athletic opportunities, placing sacred football programs under apparent threat at universities whose athletic departments were little more than semi-professional developmental programs for the professional football league. The wide popularity of college football as entertainment, a phenomenon spurred by television, and the cynical manipulation of the issue by university athletic directors, virtually all of them male, popularized the notion of Title IX as affirmative action that pushed women into sports programs largely to meet the desires of feminists for a unisex culture.

18. *Roe v. Wade*, 410 U.S. 113, 158 (1973).

19. Maj Gen Jeanne Holm, *Women in the Military: An Unfinished Revolution* (Novato, CA: Presidio Press, 1982; *Women in Higher Education*, ed. Aleman and Kristen A. Renn 1992), 305–307.

20. House, *HR 9832, Hearings before Subcommittee No. 2 of the Committee on Armed Services* 93rd Cong., 2nd sess. (Washington, DC: US Government Printing Office, 1975), 25–26.

21. Ibid., 28.

22. Webb, "Women Can't Fight,"

23. House, *HR 9832*, 162–165 and 175.

24. Ibid., 163–166.

25. Ibid., 203–204.

26. Ibid., 204.

27. Ibid., 99.

28. Ibid., 100.

29. DeYoung, "Academy Girls-uh, Guys-Blend In," C1.

30. Samuel Walker, *The Rights Revolution: Rights and Community in Modern America* (New York: Oxford University Press, 1998), vii.

31. Quoted in Walker, *Rights Revolution*, 3.

32. Patterson, *Restless Giant*, 42.

33. Webb, "Women Can't Fight."

34. James Webb, *Born Fighting: How the Scots-Irish Shaped America* (New York: Broadway Books, 2005), 329–31.

35. Pat Conroy, *The Great Santini* (New York: Houghton Mifflin, 1976); James Webb, *A Time to Fight* (New York: Broadway Books, 2008), 200.

36. Webb, *Born Fighting,* 334.

37. Ibid.

38. Webb, "Women Can't Fight," 22.

39. Robert Timberg, *The Nightingale's Song* (New York: Simon & Schuster, 1995), 152.

40. Webb, *Born Fighting*, 313 and 345.

41. Timberg, *Nightingale's Song*, 157.

42. Webb, *Time to Fight*, 243–45.

43. Ibid., 248.

44. Webb, *Born Fighting*, 314.

45. Ibid., 345.

46. Ibid., 315–17.

47. Ibid., 312.

48. Ibid., 319.

49. Ibid., 320–21.

50. Ibid., 346.

51. Brian Mitchell, *Women in the Military: Flirting With Disaster* (Washington, DC: Regnery Press, 1998), 76.

52. Holm, *Women in the Military*, 412.

53. Mitchell, *Flirting With Disaster*, 133–34.

54. Holm, *Women in the Military*, 413.

55. James Webb, "The War on Military Culture," *The Weekly Standard* 2, no. 18 (20 January 1997).

56. Webb, *Born Fighting*, xvi–xvii.

57. Patterson, *Restless Giant*, 30.

58. Webb, *Born Fighting*, xvii and 13–14.

59. Ibid., xviii and 8.

60. Ibid., 253.

61. Ibid., 338–42.

62. Amartya Sen, *Identity and Violence: The Illusion of Destiny* (New York: W. W. Norton & Co., 2006), 18–19.

63. Ibid., 4.

64. Webb, *Born Fighting*, 202.

65. Webb, *Time to Fight*, 149–52.

66. Ibid., 209.

67. Webb, *Born Fighting*, 318.

68. Tyler Whitley and Jeff E. Schapiro, "Webb Backs Off Old Stance on Women; 1979 Article Sparked Criticism; Allen Blasts TV Ad on Armor," *Richmond Times Dispatch*, 14 September 2006, A-1.

69. Warren Fiske, "Jim Webb Assailed for Essay on Women in Combat," *The Virginian-Pilot*, 14 September 2006, A1.

70. Michael D. Shear and Tim Craig, "Va. Senate Race Goes Negative on 1979 Essay; Women Didn't Belong at Annapolis, Webb Said," *The Washington Post*, 14 September 2006, A01.

71. Fiske, "Jim Webb Assailed for Essay on Women in Combat," A1.

72. Bradley Olson, "Academy's Past is Present in Va. Senate Campaign; Webb Essay Polarizes Election," *The Baltimore Sun*, 3 October 2006, 1B.

73. Those prohibited by their service from directly confronting Webb, they did fight back with what weapons they had. When Webb came to the academy to be sworn in as secretary of the Navy, the women midshipmen festooned the trees around Tecumseh Court with their underwear. Olson, "Academy's Past is Present in Va. Senate Campaign; Webb Essay Polarizes Election," 1B.

74. Mitchell, *Women in the Military*, 132.

75. Fiske, "Jim Webb Assailed for Essay on Women in Combat," A1; Whitley and Schapiro, "Webb Backs Off Old Stance on Women; 1979 Article Sparked Criticism; Allen Blasts TV Ad on Armor," A-1.

76. Olson, "Academy's Past Is Present in Va. Senate Campaign," 1B.

77. Michael D. Shear and Tim Craig, "Sen. Clinton Supports Webb in Va. Campaign; Allen, Rival Compete For Women's Votes," *The Washington Post*, 4 October 2006, A01.

78. Webb, *Born Fighting*, viii.

79. House, *HR 9832*, 100.

80. Sen, *Identity and Violence*, xii–xiii.

About the Author

Dr. Michael Allsep, JD, is an assistant professor of comparative military history at Air University, where he teaches in the Department of Leadership and Strategy at the Air Command and Staff College. He earned a PhD in history at the University of North Carolina at Chapel Hill and was a visiting assistant professor in the History Department of Duke University before coming to Air University. He also holds a JD from the University of South Carolina School of Law, and practiced law in Charleston, South Carolina. He can be contacted at allsep@mac.com.

CHAPTER 14

Incorporating the Concept of Privilege into Policy and Practice

Guidance for Leaders Who Strive to Create Sustainable Change

Steven M. Samuels
Dena R. Samuels

At the US Air Force Academy, cadets must complete a physical fitness test each semester consisting of pull-ups, a long jump, sit-ups, push-ups, and a 600-yard run. To achieve the highest score, female cadets are not required to perform at the same level as males. For example, females must complete 48 push-ups while males must complete 72; females need to obtain a 1:53 time on the run while males need 1:30; and so forth. Many cadets and staff, mostly male, argue that this system is biased against male cadets since female cadets need less stringent outputs in each event to achieve the same score.[1]

Interestingly, neither males nor females seem to protest the fact that males have a *less* rigorous standard in the case of waist measurement for the Air Force Fitness Test, which all Air Force enlisted and officers must take twice a year. To achieve the highest score, males need waists that measure less than 32.5 inches, while females' waists must be less than 29.5 inches. In this fitness test, everyone seems to understand that females are built differently than males and thus different specific outcomes are needed to represent a single standard of "overall fitness." Why does this notion of being built differently become rational only when males have an advantage and not when females have an advantage? Is this reaction an anomaly? Or is it predictable given a systemic way of understanding identity in culture, especially in the military? If the latter, then understanding how such a system works can help create sustainable change that can, in fact, improve the institution's ability to accomplish its mission.

Policy is often created to fix a problem or to safeguard against the start of one. Leaders typically look at trends in an organization, and perhaps in related fields, to determine the critical elements that must be included in the policy. But how effective would a policy be, in its own right or in its implementation, if it were based only on a part of the whole? Say, for example, you were interested in creating a policy whose purpose was to increase the retention rates of students of color at a military academy. If you wanted to obtain useful information about the elements that should be included in the policy, would you only look at the demographic statistics of those who had accepted their appointments to the academy? You also would want to acquire information from the students themselves as to why they are persisting and, conversely, why former students have left. Moreover, you would want to obtain information that would provide insight into the climate from students, faculty, and staff of color as well as from those academy members in the dominant race. These additional sources of data could provide you with a far more comprehensive picture of what is occurring in terms of the retention of students of color. Too often, policy is created and implemented in ways that only focus on a part of the picture or one component of the population.

Further, policies are often rooted in an organization's values, mission, and vision. This, of course, makes sense. The integrity of an organization should be protected. Sometimes, however, the mission of the organization does not include an appreciation of its diverse membership. Policy, then, runs the risk of perpetuating the idea that organization members must always adjust to the organization, rather than the organization at non-mission-essential times adjusting to the diverse needs of its members. No organization, including the military, exists in a vacuum. Organizational members come to their positions with social identities and experiences that affect who they are, which in turn can affect how successful they are at implementing their organizational role. Without an awareness of the cultural diversity of one's organization and the needs of different cultural groups, it is difficult to achieve an inclusive culture where members feel like they belong and believe they can succeed. For example, the Air Force Academy long held a policy that there was no academic testing on Sundays. They recently added that any cadet whose faith named Saturday as a holy day could reschedule Saturday tests.

To better understand diverse cultures within an institution, we will discuss the concept of privilege at a general level and provide specific examples from many aspects of the military. Based on the research and experiences of the authors, the Air Force Academy will be utilized as the primary exemplar. Upon conclusion of this chapter, we hope that readers will be able to apply the concept of privilege to their own units and experiences.

The Need for a Framework

Without cultural awareness, opportunities for organizational change become abridged. Without diverse input, it becomes extremely difficult to discover functional alternatives to "the way things have always been done." It is easy, and therefore common, for leaders to miss opportunities and creative ideas because they hadn't thought to ask others with different perspectives for suggestions before proceeding. Often, this is due to a leader's lack of cultural awareness of others and a failure to recognize that leaders themselves also come to their positions with social identities and experiences that affect who they are and how they see the world. That is, leaders do not always understand that their own advantaged statuses can affect their decisions and behavior. Examining the inherent diversity of one's organization (or lack thereof) can create greater insight into sustainable and successful change.

Perhaps the most recent celebrated military example of creating success through diversity would be Gen David Petraeus' 2007 assumption of command of the war in Iraq. He took over what was largely assumed to be a failing mission: trying to create stability and security while simultaneously pulling American military forces out of the country. While his "brain trust" was renowned for the high percentage of military officers with doctorates from prestigious institutions (e.g., Stanford, Harvard, Princeton, Columbia, Oxford, etc.), what was not as well known was the dissident make-up of his handpicked group, as many had been privately (and publicly) against the invasion of Iraq. General Petraeus ensured the group included civilians and retired military, many of whom had been ignored (e.g., Stephen Biddle) or had previously quit due to frustration (e.g., Timothy Carney); Arab-Americans (e.g., Col Pete Mansoor and Sadi Othman); and even foreign nationals (e.g., David Kilcullen from Australia and Toby Dodge and Emma Sky from Great Britain). This diverse group created policies that literally redefined the goals of the conflict, reversed many failures, and turned around the security situation to the point that pulling out American troops without Iraqi collapse became possible.[2]

Many organizations have trouble creating and maintaining racial or gender diversity. Even with the best of intentions, it is common to make surface-level, often cosmetic, changes in the hope of alleviating the problem. In fact, the changes may even be predicated on the mistaken belief that there is not a problem at all, just an anomaly or a misunderstanding. Since leaders do not believe there is any underlying problem in situations like these, they see no need to make any underlying changes. Thus, they may release public statements pointing to successes they have accomplished in these domains, add a statement about being an equal-opportunity employer in their recruitment advertisements, or put women and people of color into their training films. While these may be positive steps, these minimal changes do not alter the underlying organizational culture.

It is also critical to understand why members of underrepresented groups are not applying for jobs in an organization, and if they do apply and get hired, why they are less likely to stay in their positions. If the culture of an organization is not critically examined, no significant changes will be made from minor, surface improvements. Lasting diversity in an organization is not something that can be created quickly or haphazardly, but rather requires policy and practice that are methodical and purposeful. Thus, a framework is needed to help leaders become more culturally aware of other organizational members' experiences and needs. This frame also needs to highlight the manner in which the statuses of leaders might serve as blinders and even inhibitors to creating a diverse and inclusive workplace.

In some organizations, policies have been created to rectify the unequal treatment of people based on social identities (e.g., race, gender, class, religion, etc.). Typically, the claims that have been made to demonstrate the inequalities that exist in these categories focus on those that have been traditionally discriminated against (people of color, women, poor people, non-Christians, etc.). Not to diminish in any way the abundant literature and policies that have been created in this realm, it is important to note that it is only part of the story. Often omitted from the analysis are the legacy of historical, institutional social inequalities and how those systems continue to create and perpetuate disparities, making it an ongoing challenge to successfully build a culturally inclusive environment.

Such treatment disparities can involve even military status. For example, the Air Force Academy (and West Point) faculty was exclusively military until Congress mandated integration of civilian professors in the fall of 1993. The inclusion of these obviously qualified teachers for the military (the Naval Academy has had civilian instructors nearly since its inception) met with *institutionally accepted* personal criticism from military faculty, staff, and even cadets. In fact, the first author, a civilian professor, experienced officers coming up to him nearly weekly to announce they did not think he should be teaching there. Even a leading colonel and department head continually and publicly commented on how this was the worst thing to happen to the academy (which strangely continued even after he retired and was given a job there as a civilian). No effort was made by superiors or peers to quell the dissent, and it was only through time that the commentary slowed and ultimately ceased (aided by the fact that within four years nearly all students graduated and most officers rotated out or retired, leaving few with a memory of an all-military faculty). In an organization that prided itself on discipline and following the chain of command, public refutation of a clear and lawful order should have been shocking, but instead was not only accepted but also perpetuated in a systemic manner.

Thus, systemic inequality can exist for any social identity, although race and gender often are the two most salient. This is extraordinarily difficult to understand and absorb as so many in America and other Western countries have been socialized on the myth of equality: that everyone has an equal and fair chance of advancement through effort and hard work. It surprises many people when told this is not the truth, and those people often have a backlash against the information. A most intriguing example of a backlash was when commentator Glenn Beck called Pres. Barack Obama "a racist" and "a guy who has a deep-seated hatred for white people or the white culture," despite the fact that he was brought up by his white mother and grandparents and nearly all of his staff is white.[3] While it is difficult to hear and understand concepts from alternate perspectives, learning them becomes necessary to overcome failures of the status quo. Just as the alternate perspectives of General Petraeus and his group reversed the failures in Iraq, new ideas and viewpoints are necessary when trying to change embedded culture.

Even when such changes breed success, it is still not easy to reconcile with previously held beliefs. Again, despite the clear success of General Petraeus' counterinsurgency strategies, many in the Army hierarchy still did not agree with his nontraditional methods. The fact that one of his primary advisors, Col H. R. McMaster, was passed over for brigadier general twice despite his considerable achievements (e.g., writing the Army required-reading book *Dereliction of Duty* and personally leading the 3rd Armored Cavalry Regiment in securing Tal Afar, one of the most dangerous areas in all of Iraq in 2005) demonstrates the resistance that often comes with new ideas that disrupt the status quo. Going against commonly accepted ways of thinking has costs, but shedding those preconceptions can allow for unexpected success.

In this chapter, we will discuss a crucial element that addresses existing systemic inequalities and provide a framework for understanding such issues. It is often left out of the creation and implementation of policy and thus left out of any subsequent decision-making in the practice of the policy. That element is the concept of *privilege*.

Understanding Privilege

Privilege in our society is usually considered a positive attribute or reward that anyone can attain if he or she works hard enough. But in her seminal work, Peggy McIntosh depicted privilege as something else entirely.[4] She focused on two unique perspectives of privilege. The first is "unearned entitlements": things that theoretically everyone has access to but in actuality belong only to some groups. For example, Americans believe that higher education is easily accessible to everyone. Children of wealthy parents, even if they lack intelligence, don't work hard, are addicts, and so forth, nearly always end up going to college. Even if they fail out, they usually are able to go back again (and again) until they eventually graduate. Their status in society guarantees them innumerable chances to finish their degree and thus get one of the most important predeterminants for success. Poor persons, however, do not have the same opportunities. Even when they are intelligent, driven, have a high work ethic, and so forth, they may not have the financial resources to attend or even access to information to learn how to obtain acceptance into schools. If they do gain acceptance, they may not know what they should expect once they enroll or how to deal with the radical personal changes that accompany such a change in environment. Their safety net is also reduced, if not completely absent—they usually get one and only one chance. That chance may be additionally compromised by financial issues at home if they are considered a primary supporter.

McIntosh's second focus flows from the above. "Conferred dominance" is the privilege that gives one group power over another.[5] Based on social group identities, systemic power inequalities work to privilege/include dominant statuses (e.g., white, male, wealthy people) at the expense/exclusion of others (e.g., people of color, females, poor people).[6] It is important to note that everyone is endowed with *some* type of privilege in society (e.g., heterosexuality, mental ability, Christianity, etc.). The notion of privilege, therefore, affects us all in some way, and therefore, we are all implicated in the systems of inequality.[7] For example, a white, heterosexual, Christian male cadet has race, sexual orientation, religious, and gender privileges. Thus, he is more likely to be viewed as someone who "belongs" at the Air Force Academy, rather than someone who is there only because of affirmative action (even though he might be if he comes from a state that has low numbers of people applying to the Academy). The identities that define the cadet in this example allow him access to resources and protections that other cadets may be denied simply because of their identities. Understanding these distinctions is essential to comprehending the power and inequalities that exist in our society.

It is important to note that privilege is often invisible to those who have it. Since dominant statuses are considered the norm, everyone else is measured against that norm, and named accordingly.[8] A useful example of the invisibility of privilege can be gleaned from sociologist Michael Kimmel's experience. As a white, heterosexual male, he tells the story of his first experience recognizing his privileged statuses. He was at a conference and heard two women having a conversation about what they saw when they looked in the mirror. The white woman said to the African-American woman, "I see a woman."

The African-American woman replied, "That's precisely the problem. . . . I see a *black* woman. To me, race is visible every day, because race is how I am *not* privileged in our culture. Race is invisible to you, because it's how you are privileged. It's why there will always be differences in our experience."[9]

Kimmel then shares his reaction: "I groaned—more audibly, perhaps, than I had intended. Since I was the only man in the room, someone asked what my response had meant. 'Well,' I said, 'when I look in the mirror, I see a human being. I'm universally generalizable. As a middle-class white man, I have no class, no race, no gender. I'm the generic person!'"[10]

Kimmel goes on to state this was the first time he clearly recognized that race, class, and gender refer to *everyone,* including himself, and that his own experiences are shaped by these social identities. Before this, he had not thought about the fact that he even had a race or gender and thus was unaware that he was privileged by any of these statuses. In other words, he, like others in privileged categories, was able to ignore not only the statuses but also the effects that these identities had on his experiences. He concludes, "Only white people in our society have the luxury not to think about race every minute in their lives. And only men have the luxury to pretend that gender does not matter."[11] Ultimately, when privilege is normalized, those in dominant positions tend not to see themselves as privileged and thus run the risk of ignoring their own role in perpetuating inequalities.

When this happens, well-meaning individuals can actually distance themselves from their goals. When a lieutenant general came to the Air Force Academy to speak on diversity, he did not intend to alienate people of color. In an attempt to reach out and support diversity, he thought he would establish a connection by pointing out that he (a white male) was a role model for all Airmen[12] regardless of race. Those last three words, though meant as a bridge to the cadets of color in the room, instead served only to discount them and their experiences. True, anyone could consider the lieutenant general a role model based on his dedication, work effort, and success. However, without recognizing his own privilege, he made it clear that he did not understand the challenges specific to people of color in the Air Force. Disregarding his own privileged status, and for that matter, race at all, he overlooked the impact that race has on all of our experiences, both the advantages given to white people and the disadvantages given to people of color. Again, the point is not that white people cannot serve as role models for people of color (or men serve as role models for women, etc.), but it is crucial to acknowledge and incorporate systemic inequalities into their interactions. Assumptions, oversights, and lack of understanding can lead well-meaning officers to create roadblocks where they had hoped to build bridges.

The Importance of Incorporating Privilege into Policy

Why would it be important to incorporate the notion of privilege into policy and practice? As Powell, Branscombe, and Schmitt[13] suggest, framing social inequalities only in the context of the disadvantaged outgroup encourages prejudicial attitudes by privileged group members. If we consider only those who are discriminated against in determining policy, we run the risk of replicating an unfair system. It is critical to the success of policy creation and implementation to understand how privilege operates so that we can lessen prejudicial attitudes and discriminatory practices and develop policies that are more equitable and thus more effective.

An example of this can be found in the way the Air Force Academy handled sexual assault allegations in 2003. At that time, several female cadets went public about their experiences, both their alleged sexual assaults and the way they felt they were treated afterward by their fellow cadets and the staff. They accused the Academy of supporting a culture where they were re-victimized by being ostracized, threatened, ignored, and/or punished for breaking other rules (e.g., drinking alcohol, being off the Academy grounds when they weren't authorized to be, fraternizing with older cadets, etc.). They claimed the Academy was biased against them: against their reporting sexual assaults in the first place and then once they reported the assault, shutting them down as quickly as possible.[14]

In an early policy evaluation of this situation, Samuels and Samuels argued that first efforts to deal with what had occurred were limited in scope and rooted in only one framework: studying those who were victimized.[15] Many of the staff and cadets blamed victims for the assaults, accused them of lying, and reprimanded them for any other behavior that could have contributed to their situations. Clearly, the focus was on those who were on the losing end of an unfair system that allowed assaults to continue and go largely unreported for so long: namely, women. The men who allegedly perpetrated the assaults were not ostracized or belittled; many in fact were actively protected by their colleagues. This is a fitting example of the benefit of having privileged status. That is, those who are valued by the system are given the benefit of the doubt, and their behaviors are more easily justified than those with lower power and status. It is not surprising any resulting changes made by the leadership at that time were cosmetic at best.

Thus, when cadets heard about a sexual assault, many were quick to accept the alleged perpetrator's story and just as quick to reject the sexual assault survivor's version. Additionally, wild stories that minimized crimes were rampant. Stories about how some female cadets would make rape accusations "just because she didn't like him and wanted him thrown out" were commonplace, as was the strange story about the male cadet "who was thrown out for winking at a female cadet on the track when he got dust in his eye." In the last example, the male cadet was never identified (and, in fact, did not exist), but nearly every cadet seemed to have heard the story, and most, if not all, believed it.

This becomes less humorous when sexual assault survivors who came forward were re-victimized by the group. Many survivors were invariably labeled as "unfit" cadets having loose morals or blamed for *leading on* their attackers. One startling story was when cadets came into the first author's class wanting to tell him what *really* happened in a recent rape case. They said that everyone knew the sexual assault survivor's father found out about her dating a black man and was furious. To save herself from her father's wrath, she reported their consensual sex of a few weeks ago as rape. The author slowly worked through the illogic of this with each class until the last class of the day, when a woman in back stood up and said, "That's not true. I was her roommate. She came home bleeding and hysterical that night and I helped take her to the hospital for a rape kit." The rest of class time was spent working through why stories that "everyone knows" are not always true.

When the stories of sexual assault survivors went public in late 2003 and denial was no longer an option, the Academy began to take the problem more seriously and changed course dramatically. It instituted sweeping changes, including firing four top officers and putting in a new leader as superintendent, Lt Gen John Rosa. General Rosa brought with him a reputation for revolutionary vision. As a graduate of The Citadel, he was the first Air Force Academy superintendent ever *not* to be an Air Force Academy or West Point graduate. He instituted a policy directed by the secretary and chief of staff of the Air Force called the *Agenda for Change*, a four-plus year plan to create and embed a new, more inclusive culture. In 2003, we analyzed this policy using the privilege framework. While the concept of privilege was not consciously used to create the *Agenda for Change*, we predicted the most long-term successes would be in the areas of the plan that most aligned with it. However, without privilege as a concept fully ingrained into the process, we predicted short-term gains could give way to long-term derailments. The next sections of this chapter will revisit predictions made in 2003 on the Air Force Academy's *Agenda for Change* and report on its successes and failures to achieve sustainable changes for a more inclusive environment. Survey data will be examined from 2003,[16] 2004, 2006,[17] and 2007.[18] We will then make some recommendations for sustainable cultural changes that can be applied to any organization.

Successful Change: Incidental Use of Privilege

Evaluating change at the Air Force Academy is somewhat difficult to accomplish. First of all, items in various surveys changed over time, which made it difficult to follow specific issues. In addition, to compare their findings with other institutions, the Air Force Academy moved toward using national surveys. Secondly, as reported by Samuels and Samuels, leadership became less transparent about climate data (e.g., 2005 data was not made available until more than 18 months after it was collected).[19] Fortunately, this latter trend appears to be reversing, and most data, at least at the descriptive level, was available at the time of the writing of this chapter.

Overall results show generally positive gains for a more equitable environment at the Academy and corroborate our 2003 prediction that there would be less resistance to cultural changes made over time. Resistance to the *Agenda for Change* was eased as the influence of several privileged groups was reduced. Due to the passage of time, there was a natural disconnection between those who were at the Academy at the time the *Agenda for Change* was first implemented and those currently surveyed. First, the four-degrees (first-year students) of the 2003 survey became the seniors of the latest survey; thus almost no cadet surveyed in 2007 was at the Academy before extreme changes were made in response to the scandal. Second, most officers at the Academy under the old system (pre-*Agenda for Change*) had departed as most were only assigned on three- or four-year tours of duty. Third, there had always been a tenuous relationship between graduates under the old system and the Academy. Graduates were often the most resistant to any change made, especially when they perceived the change made things easier for current cadets (this is known as WHITLY—we had it tougher last year—whereby every class believes the following classes are "weak" and have it much easier). Their negative influence diminished as more classes went through the new system and graduated, and tended to see previous systems as antiquated.

It is clear that strong leadership was instrumental in creating change at the Academy. Since privilege operates invisibly, the most important starting point in 2003 was the recognition that a problem existed. Both the superintendent and the dean made it clear that the system needed to be transformed, and the change would start at the top. The new leadership recognized that they had to examine the needs of the diverse members of their organization and that any changes would be challenging for the organization as a whole. Nevertheless, by 2006, significant cultural change had occurred due to the *Agenda for Change*. So much so, in fact, that 96 percent of those surveyed that year claimed they valued diversity in statements like "Women are an asset to organizational effectiveness," "It's not your gender, but what you do that counts," and "Female leaders are as effective as male leaders." Additionally, 91 percent claimed, "In my unit, discrimination based on gender is not tolerated."

Long-Term Derailment without Privilege Focus

Despite the many gains, the concept of privilege was not fundamental to the plan. Organizational members were enlightened on the importance of diversity, but without the focus on privilege, fewer substantive changes were made than could have been. For example, cultural improvements included changes in specific factual procedures (e.g., now over 98 percent stated they knew the reporting process for sexual harassment and assault) but not in interpretations of behaviors. Within two years, from 2004 to 2006, the belief that the Academy "has a culture and climate problem" dropped precipitously from 58 percent to 38 percent. Not surprisingly, these same members were less likely to "support the efforts being made to change the AFA's culture," were less likely to believe that "improving the acceptance of women in the military profession should be a critical element of culture change at USAFA," and had reduced trust in organizational leadership (i.e., the new superintendent who took command in October 2005). Without a focus on systemic, institutional inequalities, many people assumed these surface changes meant that the problem had been solved and was therefore something they no longer needed to worry about. Further, these superficial changes gave the impression that no institutional change was necessary, despite the fact that the problems were much more deeply rooted.

Without the recognition of privilege, comments made by those in underrepresented groups were now considered complaints, and any gains they experienced were considered unfair advantages. Interestingly, the majority were able to accurately see the extent of problematic behavior. That is, when asked "How often have you personally observed a person being given preferential (favored) treatment due to gender?" the responses dropped from 59 percent in 2004 to 52 percent in 2006. But the *perceptions* of "preferential treatment" for women went up dramatically. Positive responses more than doubled for the belief that "women are given preferential treatment" (26 percent, up from 12 percent) and increased by over a third for "women are less likely to be held accountable for poor performance" (19 percent, up from 14 percent). So even though people factually saw *less* preferential treatment, they interpreted it as *more* prevalent and as the fault of those in underrepresented populations. Once again, since they assumed the problems of inequality had already been solved, anyone complaining must be at fault in the new so-called fair system.

In terms of factual procedures, the Academy once again demonstrated some successful change. Overall, several survey items addressed trust and confidence that leaders would take appropriate actions in response to crises (e.g., investigate reports of sexual assault, provide appropriate care to victims, make sincere efforts to create a culture where unwanted sexual attention will not be tolerated, etc.). Responses to these items were quite high in 2004 (all above 90 percent), and still increased two years later (up 2–2.5 percent). But ultimately, despite changes in policy designed to help women report sexual assaults, those levels remained remarkably similar to even the 2003 data: approximately 80 percent of women claimed they would report if they were sexually assaulted. The reasons they gave for not reporting also appear similar, regardless of policy changes, although "fear of being blamed" and "fear of ostracism/harassment" were less common and moved down the list of most likely responses.

In sum, there were both gains and losses in this culture. While 7 percent of open-ended comments mentioned gender (and sexual assault and harassment training) was something the Academy was doing right, 6 percent mentioned that gender inequalities were something that hindered the development of a positive culture. Additionally, the question of what respondents meant by inequalities was not exactly clear. Were people saying that women were still not being treated as well as men? Or were they saying women were treated *better* than men? Without examining and incorporating a framework of privilege into the context, those with dominant statuses may be less likely to comprehend their own invisible benefits and thus may be less likely to understand their own contribution to the systemic inequalities.

Differing Perceptions of Favoritism

If there was confusion about which sex was being treated better, we would expect to see differences between men and women. And that is exactly what occurred in cadet data from 2007. Despite the fact that women reported coming to and staying at the Academy for selfless reasons more than men did (e.g., to serve one's country, do meaningful work, be an Air Force officer), both sexes agreed women were considered less socially accepted. Additionally, female cadets were perceived by male cadets as benefiting due to their gender in nearly every area. For example, male cadets believed female cadets were favored in military grading, while female cadets believed they received less favorable assessments than males. Objectively over the last six years, both received almost exactly the same grades. The possibility that females were given subjectively higher military grades than deserved seems unlikely given their above-mentioned lower social acceptance that all agreed upon. Not surprisingly, it followed that female cadets believed they received fewer leadership opportunities than males, and male cadets believed females received more such opportunities. Again, objectively over the last five years, representative percentages of both sexes who served at higher levels were almost identical.

These patterns are similar to other studies of military cadets. In their investigation of perceptions of men and women in the Texas A&M Corps of Cadets, Boldry, Wood, and Kashy found that although objective measures of performance between men and women were equivalent, for both perceptions and subjective evaluations of individual cadets, men were believed to have the qualities of leadership needed for effective military performance.[20] Women, on the other hand, were believed to possess attributes that were detrimental to successful military performance. Like our interpretations of the above data, Boldry et al. asserted that negative evaluation of female cadets was based on gender stereotyping rather than reality. Yet it is important to note in our data, women also were inaccurate in their perceptions of being unfavorably treated compared to the objective data. It may be that their responses were more influenced by their lower social acceptance.

In sum, the data from the two surveys above support the conclusion that fewer people believed there were still climate problems than in previous years, and more people believed that disadvantaged groups were receiving unfair advantages. Again, the lack of focus on privilege resulted in misconstruing the inequalities as the problem of discriminatory individuals, as opposed to the inherent consequences of an unequal system. Another consequence was a culture of blaming others for these inequalities rather than gaining a more comprehensive perspective of the institutional culture of privilege.

Identity: Different Focus, Same Problem

The leadership at the Air Force Academy is certainly aware of its cultural climate. However, we believe the wrong framework is being used for interpreting these results. The Academy leadership seriously and sincerely attempted to make change to create a more accepting environment for everyone. Yet, without the concept of privilege to understand the true underpinnings of the systemic inequalities, many of their attempts went awry.

This lack of framework still continues. For example, the 2007 survey analysts' first recommendation was to "focus on change of identity [from subgroup] to military member and stress that the military identity is respectful and accepting of all individual differences and abilities."[21] Although the goal here seems to be respect and acceptance, the underlying suggestion is to make the organizational members conform to one common identity. That is, the construal of what a "cadet" believes or how an "Airman" acts is more easily defined and controlled than defining and controlling each and every subgroup. This serves as an example of the organizational expectation that its members adjust to the organization without addressing the needs of its diverse members. If individuals do not believe the prototypical military identity respects other crucial aspects of their identity, they will resist assuming the military identity as central. For them, until the prototypical military identity is in harmony with the rest of their identities, they cannot be reconciled.

Further examples become obvious when examined at the individual level. Not surprisingly, women were more likely to self-identify by gender (the second most common response at 19 percent) compared to men (not even listed in the top five responses at 3 percent). The same is true for nearly every non-majority group (e.g., intercollegiate, racial minorities, etc.);[22] they tended to be more likely to identify as members of their subgroup rather than as a cadet first. It may have been that when they thought of what it meant to be a cadet, they thought of the majority group (i.e., white males). If the analysts' recommendation calls for everyone to identify with the majority group, they are missing the crux of the problem of inequalities, and in turn, blaming the victim for not identifying as a cadet first. Moreover, they are leaving out of this identity the plurality of differences that makes the Academy and the Air Force strong.

Ultimately, the unaddressed problem may be that while the perception the majority group has of itself is "we respect everyone," the responses of those in subgroups indicate they may not believe this to be true. They have seen the privilege that majority groups have—a privilege often born on the backs of those in underrepresented groups. Understanding the concept of privilege shifts the burden of this recommendation from those who choose to identify primarily with their subgroups to those in power. Now, the first step becomes convincing those people who *already* identify as "military members" that their privilege must be mediated, that respect for all must be an inherent part of the military role. They need to understand real respect is not merely tolerance and not merely lip service to a "we're all the same" ideal. Instead, they need to discontinue making external attributions about the success of people from other subgroups (e.g., "she only got that leadership position because she's a girl") while assuming that those who succeed from the majority groups earned it.

Additionally, if you hope to attract all members to a central identity, there must be successful examples of all subgroups as part of that central identity.[23] For example, it becomes easier for people of color to identify as an Airman if they see people of their own race succeeding at all levels of the Air Force. If nobody in the entire leadership looks like you, why should you identify as a member of that group? Wisely, the Academy used the first opportunity after 2003 to put highly qualified women in top-ranked positions. When Brig Gen David Wagie finished his tour as the dean, Brig Gen Dana Born replaced him, and when Brig Gen Johnny Weida finished his tour as commandant (the leader of the military part of the Academy), he was replaced by Brig Gen Susan Desjardins. As mentioned above, mentorship and role modeling can certainly come from those who have different social identities, but ultimately, there must be people from your own subgroup experiencing success for you to believe you can succeed as well.

It is crucial that the majority comes to understand that respect and responsibility are in everyone's job description. Only then will members of underrepresented groups be willing to reach for and identify as the central group identity. That is, underrepresented groups may resist assimilating into a dominant group if they do not see it respecting their subgroup. They will be more willing to adopt that identity if they see it as aligned with their subgroup instead of opposed to it. On the other hand, if widespread negative attributions are made about the successes of underrepresented groups (e.g., "Generals Born and Desjardins only got their positions because they were women"), members of these groups are less likely to adopt the central group identity.

Implications for the Future:
Using the Concept of Privilege for Change

Many recommendations for using privilege to instill successful long-term transformation have not changed. The issues are the same today as we discussed in 2003 and are still appropriate for policy changes:

1. "Currently, if the leadership wishes to create significant change, they must keep the focus (and impetus for change) on those who are privileged."[24] When leaders do not keep the focus squarely on the concept of privilege as a fundamental underpinning of the problem, empowered group members are more than happy to claim victory and slough off the responsibility, especially if it appears that the immediate crisis has dissipated or disappeared. Privilege is the lens that allows successful change to occur. It provides a set of guidelines for understanding not only those who currently experience privilege in the culture, but also those who are being oppressed by the system. In fact, without understanding privilege, the focus is likely to shift to one of discrimination, which is easier for most people to understand. As in our examples above, however, this reduces the likelihood that any resulting cultural changes will be sustainable. In other words, it is crucial to view inequalities and inequities as enduring challenges of identity, power, and conflict rather than simply short-term problems to be solved.

2. "Without a framework, change is created intuitively, which can lead to arbitrary and haphazard results."[25] This is almost certainly the focal point of this current chapter. Successes can and do come from well-meaning people in power who want to create change. But well-meaning people in power do not necessarily understand the experience of those without power. In 2003, differing official reports had contradictory recommendations: one advised the elimination of confidentiality for those who reported sexual assaults in order to increase the likelihood that perpetrators would be punished, while another advocated the reinstatement of confidentiality so that sexual assault survivors could access aid and support without having to start an investigation. Another example is that currently, students from West Point, the Naval Academy, and the Air Force Academy are gender-segregated in their dormitories as dictated since the sexual assault scandal. Across all three service academies, however, both women (96 percent) and men (80 percent) overwhelmingly disagreed that dormitories/barracks should be physically separated by gender.[26] Not surprisingly, segregation has been shown to strongly correlate to a lower status of women, so much so that Sanday argued it may be a necessary prerequisite for male dominance and sexual inequality.[27]

3. "Taking personal responsibility for one's role in perpetuating inequality is the first step in making a difference."[28] Every group member must internalize the idea that all their colleagues are of equal human value and must be treated accordingly. Each must understand he or she is personally responsible for creating a culture that protects and maintains that equality. As many report that the problem is past, it is clear that they are failing to name the problem, and therefore they are not likely to see a need for their own personal action in solving it. If, on the other hand, they were taught to understand the problem through a privilege

lens, they would be more likely to see not only the inequality that exists, but also their own role in challenging the unequal system. It is not about blaming the individual for the systemic inequalities that exist, but rather to be cognizant of the legacy of those inequalities and know that everyone has a personal responsibility to challenge them. Once the systemic consequences of privilege are understood, it is incumbent that action should follow. Without understanding the root of the problem, no corrective action should be expected. But with complete comprehension, *not* to act is truly to be complicit.

4. "Real conversations need to happen for the culture to change."[29] All members of the organization need to be engaged and valued to avoid resistance to the desired changes. Creating forums for cadets who understand these concepts to meet and discuss with those who do not could be useful as well. Of course, those who are facilitating these discussions must be taught how to do so effectively. There is a budding literature[30] and many training programs on how to talk about and teach privilege.[31] These resources should be made available to students and staff alike.

In addition to the recommendations we suggested in 2003, it is clear that incorporating a privilege framework into future policy and practice is still critical. Here, we add three more recommendations to our previous analysis:

5. Ensure all stakeholders have a voice during the recommendation and implementation stages of policy development. Once aware that different groups have different views on the status quo, it only makes sense to learn from them and to share the power of decision making to institute more sustainable change. One example is that two years after the sexual assault scandal, the Academy went through a severe religious intolerance scandal due to overt fundamentalist Christian evangelism. In reaction, the administration brought together lawyers, chaplains from all faiths, and civilian and military leaders of various invested groups (e.g., Freethinkers, Jews, Buddhists, Christians, etc.) to create a training entitled RSVP (Respecting the Spiritual Values of all People). This is the kind of positive response that can be replicated in other areas to build a more inclusive Academy.

Additionally, privilege highlights the need to follow through on this advice. If underrepresented group members are brought into the policy development process only to be ignored, any potential for change will invariably dissipate into cynicism. Such was the case in 2007 when the Academy initiated a reflection weekend where cadets were gathered and asked for input on character development. A long list of issues was brought up by cadets. However, since timely follow-through was poor (e.g., two years later, committees still exist considering action), most cadets simply assumed their voices were not heard and their suggestions were for naught. It is important for feedback to be expedited and stakeholders updated constantly on the status of their contributions.

6. Change will occur only if resources are allocated to promote the change. For example, despite the diverse group General Petraeus created, success would have been much less likely if Congress had not allocated the extra troops needed for the "surge." The RSVP training discussion groups mentioned above in the aftermath of the religious intolerance scandal never reached their full potential as the civilian and military leaders assigned to the task were either brought in temporarily or had their assignment in addition to their myriad regular duties. There were simply not enough resources invested to fully complete the recommended changes; while phase one and two of the program did occur, phase three was never fully implemented. Due to a lack of resources, the changes they had hoped to instill did not occur. In fact, instead of actually creating attitude changes in many fundamentalist evangelical Christian cadets, they simply drove these movements "underground," creating a more entrenched problem.[32] On a more positive note, though, the Academy is currently increasing resources for building a more inclusive environment by creating a new department to serve as a clearinghouse for all character and leadership issues: the Center for Character and Leadership Development. It also appears they will be recruiting a chief diversity officer for the institution.

7. Using the framework of privilege and confronting issues from a systemic perspective can allow for well-thought-out, long-term solutions. This is almost certainly the thesis of this current chapter, as without an understanding of how social identity interacts with privilege, success becomes arbitrary (see point 2 above). Is it surprising, then, that one of the leaders brought in to realign the culture on the sexual assault scandal ended up being part of the next scandal on religious intolerance? The commandant in 2003 hired to replace the previous leader who was fired did not equate the privilege he experienced as a male with the privilege he experienced as a fundamentalist evangelical Christian. Thus, he did not see any problem with creating a new system where such Christians were allowed to evangelize to their "non-saved" classmates.[33] Had he understood the concept of privilege and the fact that privilege exists for every social identity, he might have chosen a different path.

In contrast, a positive step toward incorporating privilege into policy and practice is in the recent work by the Committee on Respect and Human Dignity. The group, led by Col Gary Packard, a psychologist who is cognizant of the concept of privilege, has attempted to define exactly what respect means in practice and then to instill that in all members of the Academy. So that no misconstrual can take place, this definition includes an appreciation for the different social identities that each member brings to the institution. Thus, the committee hopes to create a culture where majority groups are equally responsible for change. If the above analysis of identity is correct, underrepresented groups may then feel comfortable assuming the central identity, knowing they will not be asked to ignore, hide, or minimize their personal identities.

Additionally, the committee has tried to focus attention on power differentials in general. They have created a developmental system whereby new cadets write on powerlessness during basic training and then reexamine their essays multiple times during the next four years. Samuels, Samuels, and Martínez have pointed out that people are more receptive to the concept of privilege when they are made aware of their own oppressed identities.[34] This writing assignment and review keep the focus on the inherent power differences of rank and remind each cadet of what it means to be without power.

Conclusion

We have argued the importance of using privilege as the underpinning for policy change, especially when that change must reach down into the very fabric of the culture. Despite our use of military examples, this is true for any organization. Without the lens of privilege, problems appear to be more simplistic, and thus their fixes are conceived as simplistic as well. The salient point is that deep-rooted problems need deep-rooted solutions. Superficial policy and/or implementation changes put forth to ameliorate a crisis in any organization are not an antidote to discrimination. And as we have seen, they may even make the cultural environment worse. Putting privilege in the foreground of policy creation allows organization members to respond accordingly; it gives them insight into the social inequalities that exist, helps them understand, appreciate, and make use of members' diverse experiences, and most importantly, empowers all members to be agents of real, sustainable, positive change.

Ultimately, any institution that wants to change its culture must do so not simply from the top down (although that is fundamental), but also from the bottom up. Thus, empowering every member becomes crucial. But as we specified above, first must come understanding and not blame. All people are products of systemic pressures, and those who benefit often do so without realizing it. As Kimmel stated, privilege is invisible.[35] Returning to our example at the beginning of the chapter, it becomes clear why women are considered advantaged in the cadet physical-fitness test, but men are not in the Air Force test. The largely invisible system expects women to not belong, especially in terms of physical fitness, so potential advantages become highlighted. Men do not have this negative expectation built for them, and thus they assume there must be a reason for any advantage they have. The hierarchy is so embedded in the culture, in fact, that neither cadets nor staff recognize the injustice their accusations create. Without an understanding of privilege, claims of reverse discrimination, tokenism, and feelings of even greater prejudice are inevitable as people who benefit from their invisible privilege are motivated both cognitively and emotionally to reject threats to their power. If the framework of privilege can be explained carefully and used appropriately, much of the resistance can be turned into support, and all members of the institution can become part of the solution.

Authors' Note

The authors wish to thank Gary Packard and Edie Disler for their suggestions and Kathleen O'Donnell for her help with accessing Air Force Academy survey data. The views expressed in this paper are those of the authors and do not necessarily represent the policy of the United States Air Force Academy or any other government agency.

Correspondence concerning this article may be sent to Steven Samuels at HQ USAFA/DFBL, 2354 Fairchild Dr., Suite 6L-101B, USAF Academy, CO 80840-6228. Email: Steven.Samuels@usafa.edu.

Notes

(All notes appear in shortened form. For full details, see the appropriate entry in the bibliography.)

1. Do, *Understanding Attitudes on Gender and Training.*
2. See Ricks, *The Gamble.*
3. Associated Press, "Glenn Beck."
4. McIntosh, "White Privilege and Male Privilege."
5. Ibid.
6. Samuels, "Understanding Oppression."
7. Ferber, Jiménez, O'Reilly Herrera, and Samuels, *Matrix Reader.*
8. Johnson, *Privilege, Power, and Difference.*
9. Kimmel, *Gendered Society,* 6.
10. Ibid., 7
11. Ibid.
12. *Airman* is the term for any member of the Air Force, similar to a Navy Sailor or an Army Soldier.
13. Powell, Branscombe, and Schmitt, "Inequality as Ingroup Privilege."
14. "Gallery of Comments," *Oprah.com.*
15. Samuels and Samuels, "Reconstructing Culture."
16. US Air Force Academy, *Fall 2003 Superintendent's Social Climate.*
17. Rast, *USAFA 2006 Permanent Party Survey.*
18. O'Donnell, *Fall 2007 Cadet Climate Survey.*
19. Samuels and Samuels, "Privilege and Cultural Reform."
20. Boldry, Wood, and Kashy, "Gender Stereotypes."
21. O'Donnell, *Fall 2007 Cadet Climate Survey,* 8.
22. Only majority Christians identify religion more than non-Christians: 15.2 percent versus 12.9 percent, the fourth highest ratings for both groups. This may be because their privilege was highlighted during the religious intolerance scandal and thus became more salient to them.
23. Zirkel, "Is There a Place for Me?"
24. Samuels and Samuels, "Reconstructing Culture," 138.
25. Ibid., 139.
26. Department of Defense, Office of the Inspector General, *Report on the Service Academy Survey.*
27. Sanday, *Female Power.*
28. Samuels and Samuels, "Reconstructing Culture," 139.
29. Ibid., 141.
30. For example, Kubal, Torres Stone, Meyler, and Mauney, "Teaching Diversity"; Samuels, "What's in Your Student's 'Invisible Knapsack'?"; Samuels, Ferber, and O'Reilly Herrera, "Introducing the Concepts of Oppression."
31. For example, University of Denver's Pedagogy of Privilege Conference, University of Colorado at Colorado Springs Knapsack Institute, and so forth.
32. Sharlet, "Jesus Killed Mohammed," 34.

33. Ibid., 32.
34. Samuels, Samuels, and Martinez, "Privilege Tag."
35. Kimmel, *Gendered Society*.

Bibliography

Associated Press. "Glenn Beck: Obama Is a Racist." *CBSNews.com*, 29 July 2009. http://www.cbsnews.com/stories/2009/07/29/politics/main5195604.shtml.

Boldry, Jennifer, Wendy Wood, and Deborah A. Kashy. "Gender Stereotypes and the Evaluation of Men and Women in Military Training." *Journal of Social Issues* 57, no. 4 (2001): 689–705.

Department of Defense, Office of the Inspector General. *Report on the Service Academy Sexual Assault and Leadership Survey*, 4 March 2005. http://www.dodig.mil/occl/pdfs/ExecSumFinal.pdf (accessed 1 May 2009).

Do, James J. *Understanding Attitudes on Gender and Training at the United States Air Force Academy*. Unpublished master's thesis, University of Colorado at Colorado Springs, 2005.

Ferber, Abby L., Christina M. Jiménez, Andrea O'Reilly Herrera, and Dena R. Samuels. "Introduction." In *The Matrix Reader: Examining the Dynamics of Oppression and Privilege*, edited by Abby L. Ferber, Christina M. Jiménez, Andrea O'Reilly Herrera, and Dena R. Samuels, 1–5. New York: McGraw-Hill, 2009.

"Gallery of Comments." *Oprah.com*, December 2003. http://www.oprah.com/tows/slide/200312/20031208/tows_slide_20031208_01.jhtml;jsessionid=MMQ3SQ2ZY4KHPLARAYHCFEQ (accessed 30 December 2003).

Johnson, Allan G. *Privilege, Power, and Difference*. Mountain View, CA: Mayfield Publishing Company, 2001.

Kimmel, Michael S. *The Gendered Society*. New York: Oxford University Press, 2000.

Kubal, Timothy, Rosalie Torres Stone, Deanna Meyler, and Teelyn Mauney. "Teaching Diversity and Learning Outcomes: Bringing Lived Experience into the Classroom." *Teaching Sociology* 31, no. 1 (January 2003): 441–55.

McIntosh, Peggy. "White Privilege and Male Privilege: A Personal Account of Coming to See Correspondences through Work in Women's Studies." Excerpted from Working Paper 189, Wellesley College Center for Research on Women, Wellesley, MA, 1988.

O'Donnell, Kathleen. *Fall 2007 Cadet Climate Survey Results*. Colorado Springs, CO: US Air Force Academy, Directorate of Plans and Programs—Institutional Assessment Division, September 2008.

Powell, Adam A., Nyla R. Branscombe, and Michael T. Schmitt. "Inequality as Ingroup Privilege or Outgroup Disadvantage: The Impact of Group Focus on Collective Guilt and Interracial Attitudes." *Personality and Social Psychology Bulletin* 31 (2005): 508–21.

Rast, Vicki. *USAFA 2006 Permanent Party Organizational Culture & Climate Survey.* Colorado Springs, CO: US Air Force Academy, Directorate of Plans and Programs—Institutional Assessment Division, 26 April 2007.

Ricks, Thomas E. *The Gamble: General David Petraeus and the American Military Adventure in Iraq.* New York: Penguin Press, 2009.

Samuels, Dena R. "Understanding Oppression and Privilege." In *The Matrix Reader: Examining the Dynamics of Oppression and Privilege*, edited by Abby L. Ferber, Christina M. Jiménez, Andrea O'Reilly Herrera, and Dena R. Samuels, 139–45. New York: McGraw-Hill, 2009.

———. "What's in Your Student's 'Invisible Knapsack'? Facilitating Their Connection with Oppression and Privilege." In *Race, Gender, and Class in Sociology: Toward an Inclusive Curriculum*, edited by Barbara Scott, Joya Misra, and Marcia Segal, 5–14. Washington, DC: American Sociological Association, 2003.

Samuels, Dena R., Abby L. Ferber, and Andrea O'Reilly Herrera. "Introducing the Concepts of Oppression and Privilege into the Classroom." *Race, Gender & Class, Special Edition on Privilege* 10 (2003): 5–21.

Samuels, Steven M., and Dena R. Samuels. "Privilege and Cultural Reform at the U.S. Air Force Academy." In *The Matrix Reader: Examining the Dynamics of Oppression and Privilege*, edited by Abby L. Ferber, Christina M. Jiménez, Andrea O'Reilly Herrera, and Dena R. Samuels. 579–83. New York: McGraw-Hill, 2009.

———. "Reconstructing Culture: Privilege and Change at the United States Air Force Academy." *Race, Gender, and Class: Special Edition on Privilege* 10 (2003): 120–44.

Samuels, Steven M., Dena R. Samuels, and Jaime Martínez. "Privilege Tag: Learning and Transmitting the Concept of Privilege." *Reflections: Special Edition on Issues of Privilege*, in press.

Sanday, Peggy Reeves. *Female Power and Male Dominance.* New York: Cambridge University Press, 1981.

Sharlet, Jeff. "Jesus Killed Mohammed: The Crusade for a Christian Military." *Harper's Magazine*, May 2009, 31–43.

US Air Force Academy. *Fall 2003 Superintendent's Social Climate Assessment*, September 2003. http://www.usafa.af.mil/supt/Supt%27s_Social_Climate _Assessment_USAFA.ppt (accessed 28 December 2003).

Zirkel, Sabrina. "Is There a Place for Me? Role Models and Academic Identity among White Students and Students of Color." *Teachers College Record* 104, no. 2 (2002): 357–76.

About the Authors

Dr. Steven Samuels is a professor in the Department of Behavioral Sciences and Leadership at the US Air Force Academy. He has been active publishing and working internally on issues of inclusion and pedagogy. As a fellow at The Center for the Study of Professional Military Ethics at the US Naval Academy, he focused on ethical behavior, investigating the problem of individual ethical accountability in a situationist world. He works on leadership modeling, experiential leadership, and communication issues. He has also investigated the effects of leadership efficacy in other areas such as freefall parachuting and combat physical education classes. Samuels received a bachelor's degree from Brandeis University and a PhD in social psychology from Stanford University. Samuels completed this chapter while he was a visiting scholar at the University of Colorado—Colorado Springs' Matrix Center for the Advancement of Social Equity and Inclusion.

Dena Samuels is a sociologist specializing in race, gender, sexuality, and social justice curriculum development and training. She is a senior instructor in Women's and Ethnic Studies at the University of Colorado—Colorado Springs (UCCS) and has received the university's Outstanding Instructor Award. She is also the recipient of the Student Multicultural Affairs' Honorary Award for Outstanding Achievement in Diversity. Among her many publications on the pedagogy of social justice, she co-edited the anthology *The Matrix Reader: Examining the Dynamics of Oppression and Privilege* and is author of *Teaching Race, Gender, Class, and Sexuality*, a teaching guide that accompanies that volume. Samuels has been interviewed by local news stations on issues of race, gender, and sexuality. She holds a bachelor's degree from Brandeis University and a master's degree in sociology from UCCS. In addition to teaching, she is currently completing her PhD in educational leadership, research, and policy.

CHAPTER 15

NOT SO SELECTIVE SERVICE

Maurleen W. Cobb

O n 1 December 1969, the nation watched, and listened, as Cong. Alexander Pirnie of the House Armed Services Committee pulled a small, blue plastic ball from a large glass container. In his hand he held a lot more than a plastic sphere with a piece of paper inside—he held the lives of thousands of young Americans. This lottery, the first of its kind since 1942, would determine the order in which young men between the ages of 18 and 26 would be drafted into combat service in Vietnam.[1]

The draft of 1969 differed considerably from the 1942 draft. The mood of the country had changed. There was no feeling of unity, of patriotism, or of pride that often comes with being asked to defend one's country against a foreign oppressor. Instead, there were emotions of dread and fear of sending young Americans to fight and die, in a country of which few Americans knew before then. In the 28 years between the two lotteries, the term "draft" changed from being associated with fighting for freedom to becoming for many a synonym for meaningless suffering and loss. World War II was easy to understand. Hitler was trying to conquer the world. We had to stop him, but it was not as easy to understand how Vietnam was comparable. For many, it was hard to fathom how this tiny country, an ocean away, posed a threat to American freedom necessitating the conscription of nearly two million of our young men.[2]

Now more than 50 years since those blue balls were pulled from the jar to overcome the manpower shortages of a prolonged conflict, the United States again finds itself facing similar prospects. Headlines of any major newspaper provide clear and convincing evidence that our military is stretched thin fighting long wars while maintaining its Cold War obligations across the globe. Congress is already beginning to offer proposed legislation to reinstitute a draft, which is certain to place the debate on the political center stage. The time has come for us to reconsider the resources available to satisfy the obligations we have to our allies and maximize the military's readiness.

The Selective Service and Draft Are Not the Same Thing

Laws are the foundation of a nation. They allow a government to achieve its goals. Thus, to change these laws and make them more applicable to current issues, it is a prerequisite to develop a comprehensive understanding of the laws. Most people assume that the Selective Service and "the draft" are synonymous, interchangeable terms; however, this is not the case. While related, they are in fact two very different pieces of legislation. The current Selective Service System is nothing more than a law requiring all males in the United States between the ages of 18 and 26 to register with a national database and then, after Congress has passed legislation implementing a draft, the agency that puts Congress's and the president's orders in motion. The draft itself is a process by which names are selected, in the manner designated by Congress,[3] from the Selective Service list for conscription into military service. Once this distinction is made, it is easy to understand how three generations of men have registered for Selective Service without performing any form of conscripted military service. However, the opposite is not true—there cannot be a draft without the list of eligible men to draw from. Historically, the only times the Selective Service System has been used is for a military draft, but there is no requirement that any future uses of the Selective Service list must be only for military conscription.

The purpose of this chapter is to expand on this distinction and highlight some alternative considerations for modifying the Selective Service System independent of the draft. Thus, it is critical for readers to keep these concepts separate. From a pragmatic perspective, these pieces of legislation are resource management tools, with American citizens being the resource. Given that it has been nearly half a century since these tools were used, it is time to reconsider them in the context of a dramatically different world.

The Issue

Arguably, the most contentious debate surrounding Selective Service has been the inclusion of women in the registration process. The post-Vietnam experience and the thought of sending our daughters to war are often cited as some of the primary reasons the Equal Rights Amendment failed.[4] It was argued that had the amendment been ratified, not only would women have been required to register for the Selective Service, but they also would have been eligible for the draft. This example helps illustrate why people often confuse the Selective Service with the draft, because up until this point, they were each necessary and sequential steps for conscription. However, this issue is not about women's rights or gender equality. It is a contemporary issue where the military is excluding a vital resource *de facto* because "that's the way it has always been done." Given the profound changes that have emerged with the manifest integration of women into most facets of the military structure, it is time to question why women should not be required to register with the Selective Service.

History of Selective Service Legislation

The current (male-only) Selective Service registration requirement was passed by Congress as the Military Selective Service Act (MSSA)[5] and upheld as nondiscriminatory by the US Supreme Court in 1981.[6] The Court was asked to determine if the MSSA violated the Fifth Amendment[7] in requiring only males to register for selective service. The Court determined the only possible use for the Selective Service was for combat. After analyzing the various laws[8] and policies of the day which disqualified women from combat service, the Court deemed that women and men were not considered equal when it came to combat service, and the logic followed that the MSSA was not discriminatory.

What Has Changed?

Three major developments have occurred in the last 29 years that challenge this logic. First, the laws the Supreme Court previously relied upon have since been repealed and amended to expand women's roles in the military to include traditional combat functions.[9] Presently, women serve on most combat-capable vessels and have been highly decorated for their achievements in aerial combat. Second, the way that America fights wars has changed. There is no "frontline" in contingency and counterinsurgency operations. The support chain, where women could safely serve in the rear echelons, no longer exists, and a supply clerk must be as proficient in combat techniques as the infantryman. The need for women to come in contact with the local population has also changed. Women have become essential resources in certain cases. For instance, women must be present at checkpoints in the Middle East, as it would be considered the greatest of sins for a female suspect to be checked for weapons and explosives by a male soldier. Finally, technology has dramatically reduced the impact of the physical differences between men and women in combat. Pictures of all-female cargo and refueling crews participating in combat operations make the headlines, and all services have female combat pilots in the air. These women have proved they are as effective as their male counterparts.

One example is that of Lt Col Kim Campbell. In 2003, then-Captain Campbell's A-10 Thunderbolt was hit by antiaircraft artillery during a combat flight over Baghdad. Although her plane was heavily damaged and she had lost all hydraulic support, she was able to land safely back at her deployed location. To date, Colonel Campbell is one of the few pilots to land an A-10 safely in manual mode, a feat that would require all the strength of most men.[10]

What Is Likely to Change?

Because the Selective Service is merely a list of personnel eligible for service as the government deems appropriate, how this list is used remains the privilege of Congress and the Department of Defense (DOD). President Barack Obama has suggested the United States continue on its journey as a world leader diplomatically, economically, and militarily by embracing a concept of "national service." National service could entail such manpower-intensive projects as expanding the Peace Corps and AmeriCorps or requiring volunteer hours as a part of receiving public benefits.[11] Such a concept suggests that a list of eligible Americans, such as that already in place with the Selective Service, would become necessary, particularly if the concept of a draft is expanded in scope beyond military conscription. To the extent such changes are introduced into future draft legislation in Congress, it would open a new world of possibilities for both men and women to serve the nation in venues other than the military.

The profoundly different context in which we find ourselves in 2010 in comparison to the past is at the heart of the issue of women being required to register for Selective Service. While the phrase "we live in a litigious society" has developed a negative connotation, there are aspects of the phrase which a country that lives by the rule of law should be proud to claim. Everyday citizens can challenge any law, at any time, if they have standing to do so. Imagine what would happen if a class-action lawsuit were brought before the Supreme Court this year asking the same question that was asked in 1981: "whether the [MSSA] violates the Fifth Amendment to the United States Constitution in authorizing the President to require the registration of males and not females."[12] Naturally, it is a legal question that only the Court could decide, but it does not require extensive legal training to deduce that the decision today would be very different than in 1981. This suggests the necessity for senior military leaders and officials to strongly consider the flexibility they now have to control the seemingly inevitable changes to the MSSA and their impact on DOD policy, organization, and missions before such changes are forced upon them.

There is no doubt that including both men and women of a certain age in the Selective Service would pose challenges for Congress and the DOD because the act of registration is the first step towards mobilizing our country from civilian life to government service. Both military service and a national service option create separate issues that would have many problems of their own. But speaking specifically from a military perspective of invoking a draft with both men and women on the list, there are many difficult questions that must first be answered:

- Do you send a single mother to war?
- What about the single father?
- What happens if both parents are called to service?
- Can legislation similar to the sole surviving child exemption[13] be drafted to ensure a child is never orphaned under these circumstances?
- What family hardship exemptions will be offered which are fair to both genders?
- Should drafted parents be allowed to choose who stays and who goes?
- Should young parents who are both subject to Selective Service be required to file a plan for the care of their children as part of registration?
- How much consideration should be given to religious groups who take issue with various roles taken on by the separate genders?

These are but a few of the many questions which must be addressed with any potential changes in draft legislation, and they must be considered with any discussion of Selective Service registration. Currently these types of issues are addressed in Title 32 of the Code of Federal Regulations as exemptions and limitations on service once an individual has been selected by lottery.[14] These issues should not be seen as oceans we cannot cross but rather as rivers we must overcome. Compromises can and will be reached. We must get beyond the traditional rebuttals that stop discussion short by negating the contributions women have made to our military. Painting images of women in combat that are based more in fear than in reality only serve to maintain the status quo of male-only Selective Service registration.

Recommendation: Require Women to Register for Selective Service

Women are in the military, women are serving in combat roles, and women should be required to register for Selective Service. In the words of President Obama, in today's world, excluding women from Selective Service makes as much sense as segregating African American soldiers did in World War II. Requiring both men and women to register for the Selective Service would emphasize to every American his or her duty to our nation.[15] If women were required to register for the Selective Service, Congress and the DOD would still retain the flexibility of choosing whom to call upon as well as controlling when, where, and how to use the conscripted manpower.

The only remaining question is why we continue to limit the pool of our most valuable national resource in a time of constrained resources. Requiring women to register for the Selective Service not only makes the list of our most valuable, most critical resources complete, but it also makes sense.

Notes

1. Selective Service System, "The Vietnam Lotteries," http://www.sss.gov/lotter1.htm.

2. International World History Project, "Vietnam War Statistics," http://history-world.org/vietnam_war_statistics.htm. The actual number of young men conscripted is 1,728,344.

3. In current legislation, this selection is done by lottery.

4. Roberta W. Francis. "Frequently Asked Questions," June 2009, Equal Rights Amendment, www.equalrightsamendment.org/faq.htm (accessed 14 September 2009).

5. Specifically, the registration language found in *Military Selective Service Act, US Code,* vol. 50, app 451, sec. 453 as amended.

6. *Rostker v. Goldberg,* in *United States Supreme Court Reporter,* vol. 453 U.S. 57, 59, 1981.

7. The case was brought before the Supreme Court under the due process clause of the Fifth Amendment to the Constitution of the United States.

8. *Armed Forces Act, US Code,* vol. 10, sec. 6015 (repealed 1991) and vol. 10, sec. 8549 (amended 1991).

9. The Army's rules against women in combat were never codified and currently retain the same language as in 1981. However, the interpretation of what constitutes a combat position has changed. For example, an attack helicopter pilot is not considered a "combat position." Comment by Army combat unit commanders in private conversations with the author.

10. Comment by more than one A-10 pilot in private conversations with the author.

11. Fox News, "Barack Obama Proposes New Expanded National Service Programs," 5 December 2007, http://www.foxnews.com/story/0,2933,315361.html.

12. *Rostker v. Goldberg*, 453 U.S. 57, 59.

13. Made popular by the movie *Saving Private Ryan*, this exemption allows a sole surviving child to end his military combat service. See DOD Instruction 1315.15, *Special Separation Policies for Survivorship*, 5 January 2007.

14. See 32 *C.F.R.* § 163.

15. Jerome L. Sherman, "Candidates Differ on Female Draft," *Pittsburgh Post-Gazette*, 13 October 2008. Comparison made by Pres. Barack Obama.

Bibliography

Department of Defense Instruction 1315.15. *Special Separation Policies for Survivorship*, 5 January 2007.

Francis, Roberta W. "Frequently Asked Questions," June 2009. Equal Rights Amendment. www.equalrightsamendment.org/faq.htm (accessed 14 September 2009).

Fox News. "Barack Obama Proposes New Expanded National Service Programs," 5 December 2007. http://www.foxnews.com/politics/2007/12/05/barack-obama-proposes-new-expanded-national-service-programs/ (accessed 14 September 2009).

International World History Project. "Vietnam War Statistics." http://history-world.org/vietnam_war_statistics.htm (accessed 14 September 2009).

Rostker v. Goldberg. In *United States Supreme Court Reports*, 453 U.S. 57, 1981.

Sherman, Jerome L. "Candidates Differ on Female Draft." *Pittsburgh Post-Gazette*, 13 October 2008. http://www.post-gazette.com/pg/08287/919582-470.stm?cmpid=elections.xml (accessed 26 October 2008).

About the Author

Maj Maurleen Cobb is a reserve judge advocate attached to the Air Force Judge Advocate General's School at Maxwell Air Force Base. She is licensed to practice law in Texas and North Carolina. She earned a bachelor's degree from Austin College and a juris doctorate from Southern Methodist University. She is also a graduate of Air Command and Staff College with a master's degree in military operational art. She is currently in private practice in San Antonio, Texas.

CHAPTER 16

RETENTION AND THE DUAL-MILITARY COUPLE
IMPLICATIONS FOR MILITARY READINESS

Valarie A. Long

Introduction

The ability of the services to retain highly trained personnel contributes, in large part, to military readiness. When a group within the military is retained at a lower rate than the majority of military members, readiness can be negatively affected. Such is the case of service members who are part of dual-military couples, that is, a couple consisting of two military members from the same or different services. During the later stages of military careers, members of dual-military couples often experience lower retention than officers who are either single or married to civilian spouses. While this chapter focuses on the Air Force, this problem can be generalized across the military services. After examining the current retention context, this chapter offers recommendations that should be considered to promote higher retention rates for mid-level career dual-military couples.

The Current Dual-Career and Dual-Military Literature

With the integration of women into the armed services over the past few decades, an increasing number of members have married within the service. In 1978, dual-military couples composed 6 percent of the active duty Air Force.[1] In 2006, slightly over 8 percent of the active duty Air Force officer corps were members of dual-military couples. It is important to emphasize that this is not a "woman's" issue *per se*—an issue seen to affect only women. It affects roughly an equal number of men and women. However, this issue affects a higher proportion of women than men in the Air Force because 23.3 percent of the female officers in the Air Force are dual-military, while only 4.7 percent of male Air Force officers are dual-military.[2]

Much of the dual-career literature centers on civilian couples and focuses on aspects of work-life conflict. The core of work-life conflict is captured well in this statement: "The traditional family operated with two jobs and two adults.

The husband had a full-time paid job in the world of work while the wife had a full-time unpaid job. . . . In today's two-career families, one more paid job has been added and nothing subtracted."[3] How a family deals with two adults working outside the home while trying to maintain a satisfactory family and home life has been the focus of dual-career studies.[4]

Recently, there has been a reconceptualization of work-life conflict through the acknowledgment of some of American society's unquestioned norms surrounding the world of work and family life. Those norms include domesticity, the ideal worker schedule, the expectation that executives must put in a substantial amount of overtime, marginalization of part-time workers, and the expectation that persons who are "executive material" will relocate their family to take a better job.[5] The marginalization of part-time workers is an important factor in explaining why many workers do not take advantage of "family friendly" policies. Remedies advocated in the civilian literature include institutional changes that work toward the "reunification of work and life," which might include decoupling "making a living" and participating in the labor market, cooperative living, and corporations relocating "families" versus individual workers.[6]

The literature on civilian dual-career couples identifies specific factors affecting their decisions to remain in the work force and includes the issues of family formation, career precedence, and occupational mobility. Family formation—marriage and its timing, whether to have children, how many, and when—as it relates to dual-career couples, is a topic of particular interest. Delaying marriage, as well as having fewer children, have been identified as strategies used by dual-career couples to coordinate work and family life.[7] Career precedence and occupational mobility are two issues that are closely linked. The question of career priority is usually answered once one spouse is offered a promotion that would require a geographic move. Early studies were almost unanimous in finding that the husband's career took precedence. However, follow-up studies indicate the pattern might be changing to allow for more creative solutions, such as "commuter marriage," scaling back, and trading off on career precedence.[8]

Compared to the literature covering civilian dual-career couples, much less research focuses specifically on dual-military couples. Much of the literature that discusses dual-military couples focuses primarily on issues that might affect women more than men ("women's issues"), such as equitable promotions, maternity leave policy, retention, quality of life, and unique stresses. The focus on women's retention is understandable in the context of the comparably recent integration of women and their full-time participation in the US military. Fortunately, many of the findings from civilian life apply to dual-military couples because they both face similar issues, although the challenges for dual-military couples may be intensified.

Unique factors affecting military members in general include unpredictability, unlimited commitment (as in, the service can legitimately ask a member to give his or her life), geographic mobility, long separations, residence in foreign countries, and isolation from social networks.[9] Just as in the civilian world, family and life factors play into a military member's decision to remain in the military or to leave. Key variables in determining military retention include retention intentions, level of spousal support for the member's career, member's and spouse's level of satisfaction with military life, and the member's and spouse's level of satisfaction with marital and family life.[10]

Like civilian dual-career couples, dual-military couples face struggles and stresses concerning relationships, parenting, and career mobility. There is also a strong perception that when a military person marries another in the military, "one or the other is committed to a 'tag along' job, versus a career."[11] As with civilian dual-career couples, career decisions are, in fact, joint decisions. The military treats each member of the dual-military couple as an independent entity; however, the decision-making process includes both members acting in concert.

Parenting issues are also top on the list of stresses cited by dual-military couples. A recent RAND survey shows that among the military community, childcare issues most negatively affect the retention intentions of dual-military members. RAND concludes that despite policies that favor single and dual-military parents in terms of enrollment in Department of Defense (DOD) Child Development Centers, these families still find it difficult to manage a military career and provide the appropriate level of care for their children.[12]

Deployments may also affect dual-military retention differently than the retention of more traditional service members. Traditional service members in 1995 were away from their families 15–20 percent of the time. The situation for dual-military couples is even more difficult because each member must perform the same types of missions away from home, resulting in the couple being away from each other approximately 33 percent of the time.[13] However, since Reeves' 1995 article, military personnel tempo (PERSTEMPO) has increased across the service. While common wisdom holds that an increase in deployments causes a decrease in retention, a recent RAND study advises caution in accepting that hypothesis. In general, for junior- and mid-grade officers, more deployments to non–war zone areas equated to higher retention.[14]

Much like their civilian counterparts, Air Force women report that they feel they must make a choice between work and family.[15] There were several coping strategies identified by the women interviewed for the study. Some choose work over having a family; some retire right at 20 years so they can "finally spend what they perceive as quality time with their family"; some opt for third-party childcare options; and a few have "stay at home husbands."[16] This study brings to light the fact that Air Force women are aware that the decisions they make with regard to work and family will significantly impact their lives and those of their family; being successful in one area is perceived to require sacrifices in the other area.[17]

Current Policy and Regulations
Affecting Dual-Military Couples

The Air Force provides specific guidance pertaining to dual-military couples within overarching regulations. Regulations dealing with assignments, permanent changes of station (PCS), and family care planning all include specific mention of rules applying to dual-military couples.

Assignments

The topic of assignments, in terms of both jobs and duty locations, is always at the forefront when discussing dual-military couples. With respect to equal opportunity, "the AF [Air Force] assigns members without regard to color, race, religious preference (except chaplains), national origin, ethnic background, age, marital status (except military couples), spouse's employment, educational or volunteer service activities of a spouse, or gender (except as provided for by statute or other policies)."[18] The exception for dual-military couples covered in Air Force Instruction (AFI) 36-2110, *Assignments*, is most informative for this study. Members of a dual-military couple serve in their own right; they each must fulfill their own personal obligations to the Air Force. There is no job sharing in the civilian sense. The military considers each spouse for assignment based on his or her individual training and skills and the needs of the Air Force. The term "join spouse" is used to refer to the assignment of military spouses in close enough proximity that they can establish joint domicile (usually within 50 miles).[19] During the assignment process, both members of a military couple indicate their join spouse preference; if it is in the best interest of the Air Force, the service will work to assign dual-military spouses together. A normal assignment within the United States used to be three years; however, recent budget considerations caused the USAF to increase assignment length to four years.[20] The Air Force recently changed its time on station requirement from 24 to 12 months for a funded join spouse PCS, which should alleviate some stress on dual-career families.[21]

Much like civilian anti-nepotism laws, the USAF prohibits family members from having supervisory or command position over each other. However, family members, defined as siblings, parents, children, or spouses, can be assigned to the same unit as long as there is not a command or supervisory relationship.[22] Some flexibility is given specifically to aircrew members within the same family, who can request reassignment to different units to avoid exposure to a common danger.[23]

Most assignments are "accompanied" assignments, meaning that the member's family moves with the member to the new assignment. Some assignments, usually shorter tours (12–18 months), are "unaccompanied" or "remote" tours, and, as the name implies, family members must remain behind. Examples of unaccompanied remote tours include one-year tours to Iraq, Afghanistan, and South Korea. An exception to the idea that members of dual-military couples "serve in their own right" is implied in the policy that they cannot be assigned to the same or nearby locations for concurrent unaccompanied short tours.[24] Some overseas short tour locations, however, include "command sponsored" billets. Command sponsorship "is approval . . . for dependents to reside with the member at the OS [overseas] duty station."[25] These billets are traditionally allocated among leadership positions within the unit. Dual-military couples (with or without dependents) can be assigned to: (1) concurrent unaccompanied short tours to different areas (e.g., one member is assigned to Iraq for a year and one member is assigned to South Korea during the same time period); and (2) the same hostile duty location such as Iraq or Afghanistan at the same time for shorter-term temporary duty (normally up to six months). However, couples cannot be assigned to the same or nearby overseas short tour locations without one member being assigned to a command sponsored billet. This appears to contradict the policy that members of the military serve in their own right.

Dependent Care

Dependent care is another topic affecting dual-military couples. The term "dependents" encompasses both children and possibly elderly parents within the household. For taxation and benefit purposes, only one member of a dual-military couple can claim dependents (for example the couple's wife might claim both children so she is credited with two dependents, while the husband has zero dependents on his record). By regulation, dual-military couples and single members with dependents must file a written family care plan.[26] The family care plan designates short- and long-term caregivers for a dual-military couple's dependents and is approved by a member's unit commander or first sergeant. The family care plan is put into action when both members are required to perform duty away from home. This includes temporary duty assignments (which can last from a couple of hours to approximately 179 days away from home) as well as overseas short tour assignments (as discussed above). Along with filing a family care plan, members must also make arrangements such as powers of attorney and base passes for caregivers.[27] Through family care plans, "the Air Force assures itself of an available force to meet all of its needs by making certain that each member has made adequate arrangements for the care of his/her family members."[28] Those who are unwilling or unable to make adequate and acceptable arrangements for their family are subject to discharge or separation.[29]

Findings and Discussion

While recent studies provide interesting and compelling qualitative evidence concerning dual-military couples and the stresses they face, few studies use quantitative data when examining dual-military member issues. For a complete discussion of the data and methodology, please refer to "Retention and the Dual-Military Couple: Implications for Military Readiness."[30] However, the summary findings from the study looking at the intent of various spouse status groups (dual-military, single, divorced, married to civilian, and married to National Guard/Reserve) from a database of 29,427 surveyed members to remain in the Air Force for a 20-year career are reported below.

Finding #1: Dual-military members begin their careers highly motivated to remain in the service. However, by the 10-year point dual-military members are comparatively less motivated to complete full careers than are their peers.

Figures 16-1 and 16-2 illuminate a distinct change near the 7.5-year point. Years one to seven in figure 16-1 show a positive relationship—as the years of service increase, so do the number of dual-military members. However, figure 16-2 indicates that after seven years of service, the number of dual-military members decreases each year.

When questioned during the first half of their careers, dual-military members have the highest intent to remain in the service for 20 years compared to all other spouse status groups. However, after spending at least 10 years in the Air Force, dual-military members' motivation lags behind those married to civilians and those married to Guard and Reserve members. A comparatively low level of retention for dual-military members during the second half of their careers alludes to some of the possible (de)motivating factors for a declining dual-military population.

Finding #2: Overall, the presence of children in the household is a motivating factor for members to remain in for 20 years. However, children are less of a motivating factor for dual-military members.

Of households with *no* children present, dual-military members are second only to those who are divorced in motivation to remain for a full 20-year career.

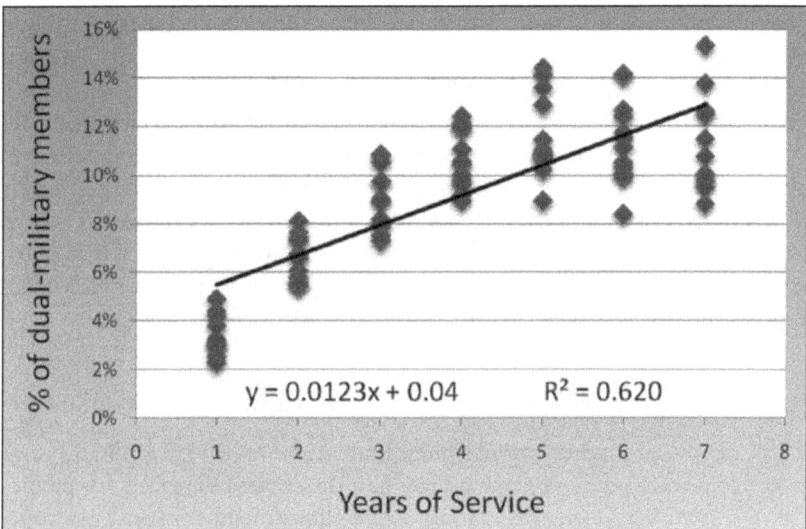

Figure 16-1. Percentage of USAF dual-military members from one to seven years of service.
(For more on the data, see Valarie Long, "Retention and the Dual-Military Couple: Implications for Military Readiness.")

Once children are present in a household, the most positive effect on intent to remain in the Air Force is on those who are married to civilians. The only group less motivated to remain in the service once children are present in the household are those who are divorced with children at home.

Having children present in the household, while a positive retention motivator for all but divorced officers, is not as positive an influence on dual-military members as it is for those married to civilians. Despite policies that favor single and dual-military parents in terms of enrollment in Department of Defense Child Development Centers, these families still find it difficult to manage a military career and provide the appropriate level of care for their children.[31]

Finding #3: Deployments have a more negative effect on dual-military member retention than on other groups.

Unacceptable amounts of time deployed, as judged by the individual military member, might affect dual-military couples differently than other service members. On average, satisfaction with the number and duration of deployments is a positive influence on a member's intent to remain. Satisfaction with the number and duration of deployments has a more positive effect on the intent of singles and those married to civilians to remain than on those married to active duty members.

Finding #4: Perceptions of promotion opportunity remain a hindrance to dual-military member retention.

When perceived promotion opportunity is held constant, those who are married to civilians and those who are divorced show a higher average intent to remain than those who are married to active duty. Singles were more positively influenced by promotion opportunity as compared to the other spouse status groups. This suggests that some officers married to active duty spouses feel they

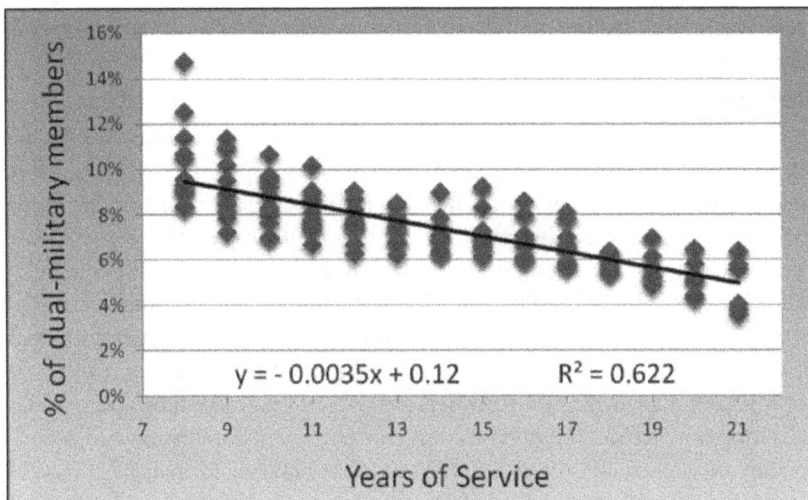

Figure 16-2. Percentage of USAF dual-military members from eight to 21 years of service.

are relegated to a "tag along" job and will have fewer command opportunities, thus reducing their potential for promotion and reducing their intent to remain.[32] This is also consistent with Evertson and Nesbitt's findings that many of those who are part of a dual-military couple feel that being married to a military spouse means that one of them has to make career sacrifices.[33]

Finding #5: Coordinating PCS moves takes a toll on dual-military member retention.

Frequent moves can be beneficial, but can also be a source of stress that can disrupt family life.[34] They can be especially disruptive in dual-military families because, while the Air Force attempts to keep military spouses together ("join spouse" assignments), sometimes it is not possible, and this can result in significant time apart.[35] Stress can also result from being unable to remain in the same geographic location as a spouse. The join spouse assignment process has been cited by Air Force women as a critical retention issue.[36]

Satisfaction with the number of PCS moves is a significantly more positive influence on single officers and those married to civilians than on dual-military members as a retention factor. This supports the contention that coordinating PCS moves in the Air Force can take a toll on dual-military couples.[37]

Policy Recommendations

The following recommendations are consistent with Segal's theory that military commitment to the family results in increased institutional commitment by service members,[38] and also consistent with Williams' theory that flexibility on the part of the employer increases workplace retention.[39] The idea that the Air Force and DOD are more likely to enact policy reforms once an issue transitions from a "women's issue" to an issue of "organizational concern"[40] is also very applicable as retention of highly dedicated dual-military members affects readiness in all areas of the Air Force.

Recommendation #1: Allow for more flexible deployment scheduling.

The Air Force is expeditionary in nature, and deployments away from home are a way of life. The analysis above suggests that deployments have a more negative effect on dual-military member retention than for other Air Force members. The Air Force has made strides in providing predictability to members concerning when they will deploy. The next step is to provide some flexibility to dual-military couples. Allowing dual-military couples some flexibility in choosing when they deploy, either by deconflicting deployments so that care-giving obligations can be met by ensuring one parent is at home or by allowing them to synchronize their deployments to minimize time apart, would go a long way in reducing family stress. Currently, synchronization and deconfliction are worked out informally when possible; however, formalizing a process would provide a framework for dual-military members and would aid those couples who work in different career fields and across different commands.

Recommendation #2: Allow dual-military members to deconflict or synchronize one-year remote assignments, and eliminate the policy that forbids married couples to serve concurrently at the same remote base.

Like coordinating deployments, coordinating remote assignments would reduce stress on the family by providing dual-military members some flexibility to cover care-giving responsibilities, if needed, or to reduce time apart, if they so choose.

Recommendation #3: Continue to increase opportunities for dual-career couples to be stationed together during the latter halves of their careers.

These opportunities can be increased in three ways. First, continue efforts at base consolidation. Base consolidation not only makes good fiscal sense, but it would also increase opportunities for dual-career couples to be stationed together. Larger "mega" bases and metro complexes would have a higher concentration of jobs for those who have served more than 10 years.[41] Second, encourage the philosophy that an officer is a leader first and a specialist in his or her career field second. Providing more opportunities to work and lead outside of an officer's career field not only provides the Air Force with officers who understand issues outside of their career field, but it also provides more flexibility for stationing dual-military couples together. Providing more flexibility in command opportunities could also help alleviate the outmoded idea that one dual-military spouse must be relegated to a "tag-along" job. Third, continue to strongly support the concept of "jointness," as providing more Air Force officers to combatant commands results in more leadership and job opportunities.

Recommendation #4: Maintain the current PCS policy of four-year instead of three-year tours.

The analysis above shows that the number of PCS moves more negatively affects dual-military members than single officers and officers married to civilians. The data for this analysis was collected before the most recent policy change that mandated less frequent PCS moves. However, there is reason to believe that this policy change will also increase retention among dual-military members.

Recommendation #5: Work to provide more flexible childcare options.

Childcare issues continue to negatively affect dual-military member retention. This supports RAND's 2006 findings that childcare issues not only affect military readiness, but also negatively affect dual-military and single parent retention.[42]

Recommendation #6: Look at potential lateral reentry and return-to-service opportunities for all service members.

This echoes Evertson and Nesbitt's recommendation from their study on women's retention.[43] Allowing members to have more control over their careers and giving them the opportunity to meet both their professional and care-giving responsibilities will increase retention. Thie et al. provide a fairly comprehensive list of viable options in their 2003 study.[44]

A Few Other Concerns

While the data for this chapter exclusively concerned the Air Force officer corps, these findings can be generalized to the rest of the military services because policies governing dual-military personnel are relatively consistent across the Department of Defense. The policies on dual-military personnel apply to both officers and enlisted, so the results can also be generalized to the enlisted population because the policies affect the two populations in the same ways. There may be some differences in the career calculations used by the enlisted population because career timing and enlistment terms are different from the officer population; but in general, one can expect to find lower retention rates for dual-military members in both populations.

A very deliberate effort was made during this analysis to remain impartial and to see where the numbers led. The analysis here only scratches the surface of some of the underlying issues causing retention problems for dual-military members. Many of the underlying issues are spawned from unaddressed institutional norms within both the military specifically and American society as a whole. Williams' work concerning the norms of domesticity, the ideal worker schedule, overtime, marginalization of part-time workers, and professional mobility was truly revolutionary and opens the door to some practical policy changes that address core issues.[45]

Conclusion

Retention of dual-military couples beyond mid-career is a question of policy and a choice to consider for senior leaders and policy makers. Just as flexibility is the key to air power, the concept of flexibility is inherent in increasing retention among dual-military members. Finding a way to balance work and family life remains one of the main obstacles for both dual-military men and women. By further integrating work and family life through innovative social policies, the Air Force specifically, and DOD generally, have the ability to improve retention levels for dual-military members. It's merely a matter of choice.

Notes

(All notes appear in shortened form. For full details, see appropriate entry in the bibliography.)

1. Segal, "The Military and Family as Greedy Institutions," 90.
2. Air Force Personnel Center, Interactive Demographic Analysis System (IDEAS) Web site, http://www.afpc.randolph.af.mil/demographics/ (accessed 21 January 2008).
3. Moen, *It's about Time*, vii.
4. Rapoport and Rapoport, "Three Generations of Dual-Career Family Research," 23–48; and Rapoport and Rapoport, *Dual Career Families Re-examined*, 9, 17.
5. Williams, *Unbending Gender*, 70.
6. Beck, *Risk Society*, 121–23.

7. Oppenheimer, "A Theory of Marriage Timing," 563–91; Altucher and Williams, "Family Clocks: Timing Parenthood," 49–59; and Rapoport and Rapoport, "Three Generations of Dual-Career Family Research," 24, 43.

8. Becker and Moen, "Scaling Back," 1002; Pixley and Moen, "Prioritizing Careers," 184–85, 199; Loughran and Zissimopoulos, "Are There Gains to Delaying Marriage?" 21; Gerstel and Gross, "Commuter Marriage: A Review," 78–90; and Rapoport and Rapoport, *Dual Career Families Re-examined*, 317–19.

9. Kohen, "The Military Career Is a Family Affair," 402–6; and Segal "The Military and Family as Greedy Institutions," 83.

10. Janofsky, "The Dual-Career Couple: Challenges and Satisfaction," 108–10.

11. Roffey, Woods, Price, and Kallett, "Work and Family Issues," 25–40.

12. Moini, Zellman, and Gates, "Providing Child Care to Military Families," 60.

13. Reeves, "Dual-Military and Single Parents," 34.

14. Fricker, *The Effects of Perstempo*, 46.

15. Evertson and Nesbitt, "The Glass Ceiling Effect," 115.

16. Ibid., 117.

17. Ibid.

18. AFI 36-2110, *Assignments*, 29.

19. Ibid., 301.

20. Air Force News Agency, "Career Fields Database."

21. Air Force News Service, "Air Force Officials Discontinue Permissive PCS Policy."

22. AFI 36-2110, *Assignments*, 49.

23. Ibid., 305.

24. Ibid., 298.

25. Ibid., 351.

26. AFI 36-2908, *Family Care Plans*, 2.

27. Ibid., 7–8.

28. Ibid., 2.

29. Ibid., 4.

30. Long, "Retention and the Dual-Military Couple," 46–69.

31. Moini, Zellman, and Gates, "Providing Child Care to Military Families," 60.

32. Roffey, Woods, Price, and Kallett, "Work and Family Issues," 26.

33. Evertson and Nesbitt, "The Glass Ceiling Effect," 95.

34. Segal, "Military and Family as Greedy Institutions," 84.

35. AFI 36-2110, *Assignments*, 301.

36. Roffey, Woods, Price, and Kallett, "Work and Family Issues," 27; and Evertson and Nesbitt, "The Glass Ceiling Effect," 95.

37. Roffey, Woods, Price, and Kallett, "Work and Family Issues," 27.

38. Segal, "The Military and Family as Greedy Institutions," 96.

39. Williams, *Unbending Gender*, 91–94.

40. Devilbliss, *Women and Military Service*, 43.

41. Frank, "Family Location Constraints," 26.

42. Moini, Zellman, and Gates, "Providing Child Care to Military Families," 60.

43. Evertson and Nesbitt, "The Glass Ceiling Effect," 126.

44. Thie, Harrell, and Thibault, "Officer Sabbaticals," 37–45.

45. Williams, *Unbending Gender*, 88–94.

Bibliography

Air Force Instruction 36-2110. *Assignments*, 20 April 2005. http://www.e-publish ing.af.mil/shared/media/epubs/AFI36-2110.pdf (accessed 21 January 2008).

Air Force Instruction 36-2908. *Family Care Plans*, 1 October 2000. http://www .e-publishing.af.mil/shared/media/epubs/AFI36-2908.pdf (accessed 21 January 2008).

Air Force News Agency. "Career Fields Database." *Airman* 50, no. 1 (Winter 2006). http://www.af.mil/news/airman/0106/career.shtml (accessed 21 January 2008).

Air Force News Service. "Air Force Officials Discontinue Permissive PCS Policy." *Air Force Print News Today*, 25 September 2009. http://www.af.mil/news/ story_print.asp?id=123169622.

Altucher, Kristine A., and Lindy B. Williams. "Family Clocks: Timing Parent-hood." In *It's about Time: Couples and Careers*, edited by Phyllis Moen, 49–59. Ithaca, NY: Cornell University Press, 2003.

Beck, Ulrich. *Risk Society: Towards a New Modernity*. London: Sage Publications, 1992.

Becker, Penny Edgell, and Phyllis Moen. "Scaling Back: Dual-Earner Couples' Work-Family Strategies." *Journal of Marriage and the Family* 61 (November 1999): 995–1007.

Devilbliss, M. C. *Women and Military Service: A History, Analysis, and Overview of Key Issues*. Maxwell AFB, AL: Air University Press, 1990.

Evertson, Adrienne F., and Amy M. Nesbitt. "The Glass Ceiling Effect and Its Impact on Mid-Level Female Military Officer Career Progression in the United States Marine Corps and Air Force." Master's thesis, Naval Post-graduate School, 2004. http://www.js.pentagon.mil/dacowits/research/ glassceiling.pdf (accessed 21 January 2008).

Frank, Robert H. "Family Location Constraints and the Geographic Distribu-tion of Female Professionals." *The Journal of Political Economy* 86, no. 1 (February 1978): 117–30.

Fricker, Ronald D. *The Effects of Perstempo on Officer Retention in the U.S. Mili-tary*. Santa Monica, CA: RAND, 2002.

Gerstel, Naomi, and Harriet Engel Gross. "Commuter Marriage: A Review." *Marriage & Family Review* 5, no. 2 (1982): 71–93.

Janofsky, Barbara J. "The Dual-Career Couple: Challenges and Satisfactions." In *The Organization Family: Work and Family Linkages in the U.S. Military*, edited by Gary L. Bowen and Dennis K. Orthner, 97–115. New York: Praeger, 1989.

Kohen, Janet A. "The Military Career Is a Family Affair." *Journal of Family Is-sues* 5, no. 3 (September 1984): 401–18.

Long, Valarie A. "Retention and the Dual-Military Couple: Implications for Military Readiness." Master's thesis, Virginia Polytechnic Institute and State University, 2008. http://scholar.lib.vt.edu/theses/available/etd-0123

2008-104353/unrestricted/Retention_and_the_Dual-Military_Couple_MA_Thesis_Valarie_Long.pdf (accessed 19 August 2009).

Loughran, David S., and Julie M. Zissimopoulos. "Are There Gains to Delaying Marriage? The Effect of Age at First Marriage on Career Development and Wages." Working Paper WR-207. Santa Monica, CA: RAND, 2004.

Moen, Phyllis, ed. *It's about Time: Couples and Careers.* Ithaca, NY: Cornell University Press, 2003.

Moini, Joy S., Gail L. Zellman, and Susan M. Gates. "Providing Child Care to Military Families: The Role of the Demand Formula in Defining Need and Informing Policy." Santa Monica, CA: RAND, 2006.

Oppenheimer, Valerie Kincade. "A Theory of Marriage Timing." *The American Journal of Sociology* 94 (November 1988): 563–91.

———. "Women's Employment and the Gain to Marriage: The Specialization and Trading Model." *Annual Review of Sociology* 23 (November 1997): 431–53.

Pixley, Joy E., and Phyllis Moen. "Prioritizing Careers." In *It's about Time: Couples and Careers,* edited by Phyllis Moen, 183–200. Ithaca, NY: Cornell University Press, 2003.

Rapoport, Rhona, and Robert N. Rapoport. *Dual-Career Families Re-examined: New Integrations of Work and Family.* London: Martin Robertson and Co, 1976.

———. "Three Generations of Dual-Career Family Research." In *Dual Career Couples,* edited by Fran Pepitone-Rockwell, 23–48. Beverly Hills, CA: Sage Publications, 1980.

Reeves, Connie L. "Dual-Military and Single Parents: What about the Kids?" *Minerva* 3 (December 1995): 25–68.

Roffey, Arthur E., Frank R. Woods, Bridgett R. Price, and Melissa R. Kallett. "Work and Family Issues Affecting Early Career Decisions of the First Women Graduates of the U.S. Air Force Academy." *Minerva* 7 (December 1989): 25–40.

Segal, Mady Wechsler. "The Military and Family as Greedy Institutions." In *The Military: More than Just a Job?* edited by Charles C. Moskos and Frank R. Wood, 79–97. Washington, DC: Pergamon-Brassey's, 1988.

Thie, Harry J., Margaret C. Harrell, and Marc Thibault. "Officer Sabbaticals: Analysis of Extended Leave Options." Santa Monica, CA: RAND, 2003.

Williams, Joan. *Unbending Gender: Why Family and Work Conflict and What to Do about It.* New York: Oxford University Press, 2000.

About the Author

Maj Valarie Long is currently a student assigned to Air Command and Staff College at Maxwell Air Force Base. She is a 1997 graduate of the US Air Force Academy and earned a master's degree from Virginia Tech. As a career intelligence officer, she has deployed in support of Operations Joint Forge, Joint Guardian, and Essential Harvest as well as Operations Enduring Freedom and Iraqi Freedom.

CHAPTER 17

THE FEMININE
AS A FORCE MULTIPLIER

Edith A. Disler

Executive Summary

For the past century, one focus of study in warfare has been operational art—that level of understanding above tactics which brings doctrine, strategy, and operations to bear in a conflict. And it is safe to say that doctrine, strategy, and operations all favor masculine approaches to conflict. However, cognitive science, linguistic study, organizational science, and anthropology all point to the contrasting but complementary characteristics of the masculine and the feminine. The hypermasculinity of the military, while obviously appropriate, necessary, and indeed critical in tactical situations, has hampered commanders' broad vision of past military actions and stands to hinder a favorable outcome in the current and future conflicts. While military authorities are noting the growing importance of qualities like empathy and intuition in soldiering— qualities inherent in the feminine—the military maintains policies which restrict the presence of women, as the feminine, in both military operations and strategy. Historical and contemporary examples clearly show the value of the feminine in woman-centered cognition, interaction, and strategies—value which could easily translate into more effective ways of approaching Department of Defense (DOD) missions and greater success in military actions. Clearly, military implementation of inherently masculine systemic approaches to war-fighting would be much more likely to meet with success, even in the realm of military operations, if fully complemented by feminine empathetic and communicative skills.

During an episode of the cable TV series *Mad Men*—set in the 1950s—a grandfather and World War I veteran reaches into a box of memorabilia to find a battle souvenir. "This was the helmet of a Prussian soldier; finest soldiers in the world," he comments to his grandson, who is about seven years old. While talking, the grandfather pokes his finger through a bullet hole in the helmet.

"Did you kill him, grandpa?"

"Probably," he answers, "we killed a lot of them."

"War is bad," says the little boy.

"Maybe," says grandpa, "but it'll make a man out of you."

Few will argue about the admiration Americans share for those who step forward to serve in the armed forces, under oath—those who are willing to step into the fight and train hard to do so skillfully. In addition, few will argue that the same willingness to fight and war-fighting skills are aspects of the construction of American masculine identity: "they'll make a man out of you." Yet military operations have always required much more than just technical skill. Whether the battles of a century ago or today's battles in Iraq and Afghanistan, winning the battle *and* the war requires both the hard science of technical skill and the softer sciences which help us understand motivation, perspective, culture, and determination. Such all-encompassing considerations fall within the realm of military operational art.

In his January 2009 *Joint Forces Quarterly* article, "Systems versus Classical Approach to Warfare," Prof. Milan Vego of the Naval War College approaches the question of operational art by pointing out that planners and practitioners of warfare "clearly confuse the distinctions between the nature of war and character of war," where the "nature of war refers to constant, universal, and inherent qualities that ultimately define war throughout the ages, such as violence, chance, luck, friction, and uncertainty," and where the "character of war refers to those transitory, circumstantial, and adaptive features that account for the different periods of warfare. They are primarily determined by sociopolitical and historical conditions in a certain era as well as technological advances."[1] This author proposes that as a nation we have long overlooked a fundamental issue that bridges the nature and character of war—a characteristic that is so unflinchingly and unquestioningly taken for granted, we have yet to critically examine its applicability in both the nature of war and the character of war as we now know them. It is this author's opinion that the self-same masculine institutional identity that brings to the battle the willingness to fight and the combatant fighting skill—the qualities that will "make a man out of you"—has resulted in entrenched thinking that has limited our ability to prevail against a low-tech, insurgent enemy.

The connection among males, masculinity, and the military is as American as the "Star-Spangled Banner"—which is to say, Americans have operated on the assumption that it is women's feminine role to sew the star-spangled banner and men's masculine role to defend it. What is remarkable is the fact that this notion persists, despite the well over two million women veterans in the United States. One semantic difficulty here is that males and masculinity, and women and femininity, have been so conflated as to be inextricable. If we are able to consider masculinity and femininity apart from their associations with male and female, we can begin to tease out the qualities which connote the masculine and the feminine, thereby delineating qualities and perspectives which can be considered in a discussion of the character and the nature of war. Both men and women, after all, carry both masculine and feminine traits. What we must ask ourselves is how these traits can be realized and distinguished so as to improve results in the theater of war.

The last two decades have seen quite a boom in the study of masculinity. In his book *Manhood in America*, Michael Kimmel points to the importance men place upon proving themselves to other men (homosocial enactment). "How has American history been shaped by the efforts to test and prove manhood," he asks, "the wars we Americans have waged, the frontier we have tamed, the work we have done, the leaders we admire?"[2] Despite the fact that masculinity scholars do not necessarily point to the military as the definitively masculine model for American males, the qualities of the military certainly mesh point-by-point with the individual characteristics scholars use to collectively define masculinity: willingness or even a desire to fight, homosocial enactment together with an acculturated sense of power and hierarchy, and subordination of the feminine.[3] Most importantly, as Kimmel points out in his 2003 essay, "Whatever the variations by race, class, age, ethnicity, or sexual orientation, being a man means 'not being like women.' This notion of anti-femininity lies at the heart of contemporary and historical conceptions of manhood, so that masculinity is defined more by what one is not rather than who one is."[4] Femininity has been less often and less clearly defined but has consistently been associated with the tasks of the home, with connection, and with community. Even these associations between the masculine and feminine, however, have evolved and continue to evolve as "sociopolitical and historical"[5] conditions change.

The military clearly relies upon the contemporary notions of masculinity and its requisite "anti-femininity"[6] in its organizational structure and specifically in the limits placed upon women in the military. As Frels points out in her Army War College analysis of the current DOD policy, "All of the reasons that support the current policy [restricting women's roles] have one common thread: they are all based on supposition and beliefs rather than facts."[7] Such limitations on women occur at the nexus of masculinity as an aspect of both the nature of war and the character of war. Specifically, as an element of the nature of war, masculine notions of contest have ubiquitously driven both military offensives and defensives; as an element of the character of war, societal attitudes toward women in the military are "transitory, circumstantial, and adaptive features that account for the different periods of warfare."[8] During World War II, for example, women in countries the world over were incorporated into various missions—combatant and noncombatant—as necessary for that period of warfare, but were expected to return to their feminine domestic sphere upon cessation of hostilities. Even in the space of the last three decades, acceptable roles for women (as the embodiment of the feminine) in the US military have been "determined by sociopolitical and historical conditions . . . as well as technological advances."[9] The 1993 removal of some combat restrictions, particularly as regards high-tech aircraft and ships, for example, was a product of both increasing sociopolitical pressure and technological advances which rendered physiological justifications irrelevant.

The Sociopolitical and Historical Condition

In considering the relevance of the masculine and the feminine as elements of today's American military, it is important to remember that some attitudes stem from a post–World War II and Cold War anomaly: the view that the United States needed a standing army. Kimmel points out that, as of the turn of the twentieth century, many believed "decades of peace had made American men effeminate and effete; only by being constantly at war could frontier masculinity be retrieved."[10] That masculinity was retrieved by the end of World War II, and a standing army would provide that state of being constantly at war that would drive, consciously or not, American masculine ideals. As we saw in the 1990s, the United States military fairly floundered in its sole superpower status because it was no longer at war, wasn't content with the high tempo of peace operations such as peacekeeping and nation building, and didn't have a named enemy against which it could fight, or at least compare itself, for military or ideological superiority. This is one reason the military incessantly prepares for the last war—it needs to have a yardstick against which to feel superior in firepower; or, since many of the last wars are conflicts we didn't win, perhaps fighting the last war is a "do over," in a sense, in order to prove belated superiority. As we have seen in the wars in Iraq and Afghanistan and as we saw in the war in Vietnam, the nature of war is such that superiority in firepower and technology, which are elements of the character of war, have little to do with prevailing in conflict.

Interestingly, during WWII all of America's assets, including its workforce of women, were brought to bear. In fact, as many women served in uniform during World War II as there are people—men *and* women—in the entire US Air Force of 2010. And, of course, that total doesn't include women in the Office of Strategic Services (whose ranks included famed chef Julia Child, who was too tall for the Women's Army Corps); women working with the Manhattan Project; women building tanks, ships, and aircraft; and thousands if not hundreds of thousands of other women who contributed directly or indirectly to that war effort. In other words, the wartime enterprise was so massive and its mission so overarching that the presence of the feminine as embodied in this gargantuan female workforce, though not without its problems, was largely seen as a grand form of teamwork—women's and men's talents complemented one another when applied to a common overarching goal. In addition, since the nation had not formerly known a standing army, all—male and female—assumed that they would fight the good fight and then return to the same jobs, homes, and roles they had before the war. The inherently masculine nature of the military was not in question as "our boys" battled against the Axis powers; and the more feminine talents, as embodied in Rosie the Riveter (who, interestingly, was a symbol of slightly masculinized feminine strength), were considered both necessary for the war effort and transient.

Today, however, women are joining a standing army which is home to a masculine identity that has become deeply embedded, both in the military's institutional ethos and in the American public's perception of the institution of the military. Since World War II, women's presence in the military has been tolerated at best, and gay men—apparently perceived as harboring elements of the feminine—are today completely unwelcome. In his book *From Chivalry to*

Terrorism, Braudy observes, "Like previous efforts in the United States to keep out women, blacks, and, elsewhere, Jews, Gypsies, or other minority groups, the assumption is that [homosexual men] are 'feminine.' Because they lack the virile qualities necessary to engage the enemy, their mere presence will undermine camaraderie, loyalty, and the fighting will of the heterosexuals who stand in the trenches with them."[11]

Throughout the 1950s and 1960s, women's presence as a percentage of the services was limited, and women could not marry or have children and remain in the service. The military's limited tolerance for the presence of the female, as feminine, was especially apparent in the 1970s, when the social progress of that era's women's movement provided a stark contrast to the severe restrictions placed upon the few women in the military. As a case in point, it was during the 1970s that military leadership openly resisted women's presence at the military service academies. Having crossed the threshold of the twenty-first century, the well-documented problems of sexual assault and harassment at the military service academies and in our current theaters of war demonstrate continued resistance to the presence of women as the feminine. Were women perceived as full partners with men in military service, the dictates of all leadership training and even unofficial rules of camaraderie would ensure their safety and inclusion. Yet, as masculinity scholars point out, masculine identity virtually requires subordination of the feminine. Therefore, though the institutional identity of the military should value the more constructive task cohesion required to complete a mission, higher emphasis is placed on a destructive level of social cohesion which, probably unconsciously, places higher priority on maintenance of masculine identity than it does on protection of the mission and all who contribute to it.

Even today women are still forbidden to serve in many military specialties, including most of special operations, battle tanks, and infantry—the most masculine of specialties. Since it is individuals from such combat specialties who are promoted into the highest positions of leadership, such as the leadership of major commands and joint component commands, both women and the more feminine perspectives involving community building, empathy, and cooperation have been carefully filtered from the positions most involved in strategic thinking, operational planning, and force structuring. The heart of this issue is this: women's participation in the military, hence the presence of the feminine, has been limited because of women's polarity from the masculine. What the military has yet to realize is that it has limited its own ability to prevail by filtering out women as a category of persons together with the most useful propensities of the feminine. Specifically, the military has limited its capacity to employ qualities unique to women and the feminine which would be particularly useful in conflicts like those in Iraq and Afghanistan: interpersonal communication skills that could build support in-country and make intelligence and information gathering much more fruitful, empathetic skills that could help the military better understand and act against its opponents and bolster its allies, community building skills which could go far in the effort to help the communities of Iraq and Afghanistan become self-sufficient, and a capacity for understanding the importance of those subjective and unpredictable aspects of war fighting and conflict that influence operations and can become integral to victory in the long run.

The following anecdote illustrates the latter point. Early into the post–Cold War era a young officer was approached at an officers' club by a man who had overheard her talking with friends. "Did I hear you say you were a missileer?" he asked. "Yes—you heard right." He then told her about nuclear inspection opportunities at what was then the On Site Inspection Agency (OSIA), now the Defense Threat Reduction Agency. "We've found out that women do really well in nuclear arms control work in Russia in ways that men just can't." The difference stemmed from a contrasting set of cultural assumptions about women, as well as women's ability to judge the efficacy and veracity of information they had gathered. Women, the OSIA had discovered, bring unique qualities to the job. This is, by the way, a realization which corporations have discovered via their bottom line, rather than by sociological theory.

To raise a slightly more abstract, but more current, example, in his criticism of Field Manual (FM) 3-0 Gian Gentile notes that "the recently released current version of FM 3-0 states that, for the commander, operational art involves 'knowing when and if simultaneous combinations [of offense, defense, and stability operations] are appropriate and feasible.'"[12] Gentile is openly hostile to doctrine which elevates any consideration above what he sees as the Army's priority: "Fighting and winning the nation's wars." "By placing nationbuilding as its core competency over fighting," he writes, "our Army is beginning to lose its way, and we court strategic peril as a result."

At the heart of Gentile's concern, a concern shared by many others, is the blunting of essential tactical skills, such as infantry, artillery fire, and tank combat—skills that are, without question, essential to a standing army. Men are particularly well suited for both those tactical skills and the hierarchical structure of the military. This assertion is borne out by findings within social, psychological, and linguistic investigations, which have consistently observed male predisposition to contest and hierarchy, as well as the operation of mechanical systems and even systems of thinking such as military strategy. Yet no one has asked a simple question: if men are predisposed to contest and are therefore uniquely suited to military combat and fighting, to what parallel quality are women predisposed and what role can it play in winning the nation's wars? Gentile indulges in the logical fallacy of false dichotomy: tactical skills and nation building skills within an institution as large as the military—or across government and nongovernment agencies—are not, as Gentile implies, mutually exclusive, unless of course your institutional predisposition resents, and therefore resists, their inclusivity. In other words, the masculine prides itself on suitability for tactical skills, has carried the enshrinement of the masculine nature of those tactical-level skills into the operational and strategic spheres, and has therefore limited itself in its ability to perform functions which it now realizes it needs and has written into doctrine "such things as establish local governance, conduct information operations, build economies and service infrastructure, and provide security, all of which are elements of building a nation."[13]

Following the American penchant for technology, then, fighting skills are critical and the United States has superior tactical skills and equipment, yet in nine years it has been unable to secure Afghanistan, and Iraq is but limping toward democracy and self-sufficiency. Gentile and others would argue that is because the military has neglected its true function: fighting. Yet others have noted that the technical and technological skills are, in and of themselves, simply insufficient in conflicts such as those in Iraq and Afghanistan. In his essay "Clausewitz and World War IV," retired Maj Gen Robert Scales addresses this directly, pointing out that "victory will be defined more in terms of capturing the psycho-cultural rather than the geographical high ground. Understanding and empathy will be important weapons of war. Soldier conduct will be as important as skill at arms. Culture awareness and the ability to build ties of trust will offer protection to our troops more effectively than body armor."[14]

This author would argue that the military's self-imposed limitations on the presence of women, as well as its lack of strategic appreciation for the wisdom of the feminine in the countries we invade and occupy, has placed concomitantly self-imposed limitations on its ability to break away from entrenched methods of thinking. Scales, for example, in discussing the contributions of social science to victory in conflict enumerates nine areas in which soldiers must improve their social science skills. One of the critical skills he names is the value of tactical intelligence: "The value of tactical intelligence—knowledge of the enemy's actions or intentions sufficiently precise and timely to kill him—has been demonstrated in Iraq and Afghanistan. Killing power is of no use unless a soldier on patrol knows who to kill," he notes. In Afghanistan in particular, fully half the population can pinpoint the enemy with tremendous accuracy, yet they are not valued as allies. In her book *Veiled Threat: The Hidden Power of the Women of Afghanistan*, Sally Armstrong notes that even today little has changed since the rule of the Taliban for the women of Afghanistan who were once quite free to be educated, dress as they liked, work where they liked, and move about as they liked. "They told me they're still poor, they haven't seen any of the UN money everyone is talking about, and Al-Qaeda members still roam the streets and scowl at the women when they walk by," Armstrong notes.[15] Were the United States to provide security and safety for those women and their families, value their contributions, and not write off their inhumane treatment as a cultural norm, security and victory would be close at hand. The military must, as Scales expounds, think and operate in new ways and with new perspectives.

Systems and Empathy

One way to think in new ways and with new perspectives is to change the pool of people to whom you turn while devising your strategies and implementing them. As noted earlier, industry has done this quite well. In industry, diversity is not a compliance issue or a public relations issue; it is a matter of corporate and fiscal success. The last thing that the most creative of industries want is the kind of homogeneity of training and thought that one finds in the military. It is well proven that this homogeneity has its usefulness in a highly disciplined organization like the military, but the military would be prudent to note that such sameness has its down side as well. One inherent difference of perspectives the military should exploit resides in the difference between male and female thought processes and ways of knowing.

In his book *The Essential Difference: The Truth about the Male and Female Brain*, Simon Baron-Cohen boils the difference between the two down to this: "The female brain is predominantly hard-wired for empathy. The male brain is predominantly hard-wired for understanding and building systems."[16] For too many years, women's predisposition for empathy—some men complain women are "too emotional" to be in the stressful situations of war—has been interpreted as weakness. Together with the fact that empathy, or emotion, is a feminine trait, weakness and empathy are, in the masculine military environment, unthinkable. Yet military thinkers like General Scales are pinpointing the value of empathy in current and future conflict, as though we do not already have access to it via the thousands of women on active duty and the hundreds of thousands of women in the countries we have invaded and occupied.

In a military environment populated by males and imbued with masculine identity, naturally the thinking will be as Baron-Cohen stipulates: system oriented. Women's talk is very relationship oriented, argues Baron-Cohen, and men's very systems oriented, with an emphasis on topics like technology, traffic and routes, power tools, and computer systems. "Systemizing," he says, "is the drive to understand a system and to build one. By a system, I do not just mean a machine. . . . Nor do I even just mean things that you can build (like a house, a town, or a legal code). I mean by a system anything which is governed by rules specifying input-operation–output relationships," to include military strategy.[17] This assertion fits perfectly into Vego's skepticism of the military penchant for theories and strategies like effects-based operations and systemic operational design, even well-respected analyses postulating that the enemy can be stopped if a strategy attacks the right nodes in a "system of systems."[18] While some of these systemic treatments of military operations acknowledge the importance of human response, they do not, and cannot, accurately account for the unpredictability of human response.

Greater receptiveness to the properties of the feminine, however, may help with this aspect of conflict. Baron-Cohen notes that women's empathetic skills are cognitive and affective. As he points out, "the cognitive component entails setting aside your own current perspective, attributing a mental state (sometimes called an 'attitude') to the other person, and then inferring the likely content of their mental state, given their experience. The cognitive component also allows you to *predict* the other person's behavior or mental state" (emphasis in original).[19] The affective component involves the emotional response to the cognitive component—sympathy, for example.

But the presence of women does not equate to the presence of the strengths of the feminine. This is partly due to the fact that systemic (masculine) thinking tends to discount alternative thinking. As Baron-Cohen points out, "If the other person makes a suggestion, boys are more likely to reject it out of hand by saying, 'Rubbish,' or 'No, it's not,' or more rudely, 'That's stupid.' It is as if the more male style is to assume that there is an objective picture of reality, which happens to be *their* version of the truth. The more female approach seems to be to assume from the outset that there might be subjectivity in the world."[20] This masculine "objective picture of reality," combined with the reality of military hierarchy, which silences subordinates, severely limits receptiveness to alternative perspectives, especially feminine perspectives, at a time when new thinking is sorely needed.[21]

With this in mind, let's look again, as but one example, at that list of functions in FM 3-0: "establish local governance, conduct information operations, build economies and service infrastructure, and provide security." Interestingly, these functions are named in a manner consistent with Baron-Cohen's assertion that men are "hard-wired for understanding and building systems"—all those elements are, after all, systems. Clearly nation building is a priority for national security, and clearly national security and nation building are both matters of building systems, specifically government, information operations, economic systems, systems of infrastructure, and presumably physical security. However, scratch the surface of each of these systems and you quickly discover the need to understand people's experiences, knowledge base, ideologies, concerns, history, needs, and priorities—all of which require the ability to interact with and accurately "read" the people involved. Those systems have just entered the feminine realm: empathy.

Such fuzzy factors, as some might call them, have been increasingly acknowledged by those who have been able to reflect on their personal experiences in Iraq and Afghanistan. Like Scales, both Niel Smith and John Patch, military members and veterans of military operations in Iraq, acknowledge the importance of more human problems and disparage the lack of preparation their military training gave them for such human understanding—for empathy. Smith, having returned to Germany from Iraq in 2004, bemoans the fact that "a year of operations in Baghdad and three months fighting the first Sadr rebellion made it clear to me that our strategies and methods were inadequate to meet the demands of the environment."[22] As he explored the literature of counterinsurgency, specifically the experience of Vietnam, he was dismayed to learn that the Army "failed to realize the fight was for the loyalty of the population, which we had placed secondary to engaging the enemy in battle."[23] In other words, the military possessed a model—a system—for engaging in battle but did not have a model, or system, for engaging the people. Similarly, Patch learned the value of understanding "fundamental regional human problems"[24] in the Balkans, not from his military training, but by reading David Kaplan's *Balkan Ghosts*, which gave Patch an invaluable sense of cultural awareness that could otherwise have come only from engagement with the people of the Balkans, on the ground in the Balkans—what the Army now euphemistically calls the "human terrain." "The great gift of *Balkan Ghosts* is its insights into the simple, powerful lesson that it is all about the people: their history, passions (good and bad), collective guilt, rulers, gods, food, drink, festivals, and, of course, their fears. Neither expansive technology nor unlimited funds (or boots on the ground) can trump the basic truism that it is about the people," writes Patch.[25]

To what generalization do these realities point? In this author's studied opinion, military implementation of inherently masculine systemic skills is much more likely to meet with success, even in the realm of military operations, if fully complemented by feminine empathetic and communicative skills. As Robert Gates pointed out in a 2008 speech at National Defense University, "Never neglect the psychological, cultural, political, and human dimensions of warfare, which is inevitably tragic, inefficient, and uncertain. Be skeptical of systems analysis, computer models, game theories, or doctrines that suggest otherwise. Look askance at idealized, triumphalist, or ethnocentric notions of future conflict." It is the masculine propensity for creating and maintaining the military system, together with a more feminine empathetic analysis and involvement, that can move closer to performing the daunting challenge posed by Secretary of Defense Gates.

So, what is hampering the military's ability to employ the best of what its own feminine presence has to offer? Quite simply, the Department of Defense is hampering itself. Scales observes that

> strategic success will come not from grand sweeping maneuvers but rather from a stacking of local successes, the sum of which will be a shift in the perceptual advantage—the tactical *schwerpunkt*, the point of decision, will be very difficult to see and especially to predict. As seems to be happening in Iraq, for a time the enemy may well own the psycho-cultural high ground and hold it effectively against American technological dominance. Perceptions and trust are built among people, and people live on the ground. Thus, future wars will be decided principally by ground forces, specifically the Army, Marine Corps, Special Forces and the various reserve formations that support them.[26]

The place where General Scales argues empathic and intuitive skills are most needed is precisely the place where military women are not permitted: in ground infantry and special forces units.

Were women and the strengths of the feminine appreciated and valued, women of the American military and civilian women in the theaters of war would be invaluable resources in both accruing "local successes" and building trust.

Women's ability to build trust, gain local successes, and even glean intelligence derives from Louann Brizendine's notion of "emotional congruence." Brizendine, psychiatrist and author of *The Female Brain*, notes that women are naturally suited to establishing emotional congruence[27]—the ability to mirror and understand "the hand gestures, body postures, breathing rates, gazes, and facial expressions of other people as a way of intuiting what they are feeling.... This is the secret of intuition, the bottom line of a woman's ability to mind-read."[28] Emotional congruence, however, requires close involvement with others over time, both of which are anathema to a masculine and technologically focused military that prefers to exert force for a quick win, rather than invest in face-to-face interaction over time. In addition, while skeptics might be inclined to argue that such a skill does not translate across cultures, as in a wartime environment, Baron-Cohen and his research colleagues found otherwise. In the United States many law enforcement officers for example, male and female, develop such skills over the course of experience. Such skills are not, however, "issued" to infantry soldiers.

Another way to think about these fuzzy problems was put forward by Center for Strategic and Budgetary Assessment researcher Barry Watts. Watts notes in his report *US Combat Training, Operational Art, and Strategic Competence: Problems and Opportunities* that "tactical problems are 'tame' in that they generally have definite solutions in an engineering sense. So-called 'wicked' problems are fundamentally social ones. They are ill-structured, open-ended, and not amenable to closed, engineering solutions. Operational and strategic problems appear to lie within the realm of wicked or messy problems." Comparing Watts's findings with those of Baron-Cohen, one might come to the conclusion that women, with empathetic thought processes geared to social issues, would offer valuable insights in complement to the "closed, engineering" or, as Baron-Cohen would posit, "systemic" solutions to operational and strategic problems. Watts goes on to point out that "because human brains exhibit only two fundamental cognitive modes—intuition based on pattern recognition, and the deliberate reasoning associated most closely with the cerebral cortex—the logical place to locate a cognitive boundary between the intuitive and reasoned responses in terms of the traditional levels of war—tactics, operational art, strategy—is between tactics and operational art."[29] Perhaps, then, full cognitive understanding of all levels of war would be better served with both masculine and feminine thought processes on the job. This notion is substantiated when, with his model for a cognitive divide between intuition and reasoning on the table, Watts states, "the cognitive skills underlying tactical expertise differ fundamentally from those demanded of operational artists and competent strategists."[30] Again, weeding women and the feminine from the level of strategy and operational art via tactical exclusions is fundamentally limiting the military's ability to develop well-considered strategy and operational art.

Given the American military's Western predisposition, it naturally defers to Clausewitzian views of the nature and character of war and operational art. Students of military strategy would do well to also consider the precepts of Sun Tzu's *Art of War*. Sun Tzu addresses the more masculine logical and systemic requirements of armies and warfare. He also, however, shows respect for what may be regarded as the more feminine, empathetic elements. Consider the applicability of empathy, emotional congruence, and subjectivity to the following:

- "If you know the enemy and know yourself, you need not fear the result of a hundred battles. If you know yourself but not the enemy, for every victory gained you will also suffer a defeat. If you know neither the enemy nor yourself, you will succumb in every battle."

- "All warfare is based in deception."
- "We cannot enter into alliances until we are acquainted with the designs of our neighbors."
- "There are not more than five primary colors (blue, yellow, red, white, and black), yet in combination they produce more hues than can ever be seen."

The more masculine inclination would be to "know the enemy and know yourself," for example, according to demographics: numbers of troops, pieces of equipment, firepower, and logistical limitations. The more feminine inclination would be to "know the enemy and know yourself" according to will, motivation, and support on the home front. As a non-Western thinker, Sun Tzu saw the importance of empathy and emotional congruence, though by other names, in his principles of war.

Possibility and Precedent

So how might some of these attributes of the feminine play out in less theoretical, more practical ways? The list of tasks derived from FM 3-0, *Information Operations*, for instance, subsumes skills in the field of influence operations, and influence operations are based on communication, facial expression, perception, and interpretation, to name a few characteristics. Economic systems writ large start with micro elements of the system, such as incentives to Afghan farmers to replace poppy crops with something less harmful, and communicating the notion that opiates are "harmful" requires those farmers to experience a level of empathy. Economic systems also thrive on small loans to individual entrepreneurs, and programs involving micro-loans to women have been remarkably successful in India, Africa, and the Middle East. Systems of infrastructure are important, but how do you prioritize which infrastructure project should come first? Logical analyses of population density, existing repairable infrastructure, and availability of new equipment are irrelevant in an area which is still populated by thieves, vandals, and insurgents. How do you know the thieves, vandals, and insurgents are present? I assure you the sorely neglected women of those communities, who are trying desperately to care for their children and the elderly, will know. And they would be more likely to confide it to another woman than to a man who appears much more threatening. One might also discover the existing threats to physical security in the course of the right kind of conversation with the same women.

As many in the military have finally realized, the DOD role in peacekeeping and nation building is a reality that the military must deal with, is trying to deal with, and yet is obviously not comfortable with. However, long marginalized because of the "women in combat" question, women's feminine inclinations toward cooperative strategies and community focus, properly viewed, may play a large role in the talents needed for nation building and peacekeeping. Consider the following examples.

In 1986, when she finished basic training, the Army sent Eli PaintedCrow, nicknamed "Taco" because of her ability to speak Spanish, to Honduras as an interpreter. The United States was building bases and airstrips in the country to help the Hondurans fight the Sandinistas. "She would mingle with the Hondurans when she could, curious to get to know them and uneasy about whether her government was in the right," writes Helen Benedict, author of *The Lonely Soldier*. The US military knows by now that operational success is limited if the people among whom we operate are not friendly to our intentions. Or as Sun Tzu put it, "We cannot enter into alliances until we are acquainted with the designs of our neighbors." PaintedCrow could see such problems brewing, but the "intelligence" she had gathered was shrugged off.

Benedict also writes about women serving in Iraq. Separately they complained about being told they were going to Iraq to help the Iraqis, to liberate them; yet they had little or no training regarding Iraqi culture or way of life. From a masculine frame of reference, if you're going over to "help" people, then you're either (1) going over to kill the bad guys, so how much do you need to know? or (2) already superior and simply have to tell them what to do and how to do it. From the feminine perspective, empathy and understanding are important elements if one intends to "help" someone.

To take a more historical example, such skills employed by Sacagawea rescued the Lewis and Clark expedition from oblivion on a number of occasions. In her book *Ladies of Liberty*, Cokie Roberts notes several entries in William Clark's journals which pay tribute to Sacagawea's many skills, including interpersonal ones. Her knowledge of the edible roots, berries, and vegetables across the West saved the group from disease, if not starvation, on a number of occasions. On several more occasions Clark points out that "the wife of Charbonneau our interpreter we find reconciles all the Indians as to our friendly intentions," and that Sacagawea forged friendships in the various tribes, thereby discovering shortcuts for the journey.[31] Sacagawea, Roberts writes, served as guide, interpreter, and protector.

In more recent decades, several models of engagement of women and the feminine have rebuilt communities and nations. PBS commentator Maria Hinojosa interviewed female legislators and cabinet ministers regarding the recovery of Rwanda in the wake of the genocidal killings of 1994. Rwandan Pres. Paul Kagame, Hinojosa points out, made a concerted effort to bring women into the political system. Nearly half the members of the lower house of Parliament in that country are women—a greater percentage than anywhere in the world. "Many Rwandans," notes Hinojosa, "believe that women are better at reconciliation and maintaining peace and are less susceptible to corruption." While connections between cause and effect can be debated, Rwanda has rebounded quite well from its dark experience at the end of last century. Its economy has recovered partly due to businesses opened by Rwandan women who wanted to help in that nation's recovery. Actions like those taken by a former government minister who oversaw a program which placed all but 4,000 of the country's 500,000 orphans in Rwandan homes by encouraging

Rwandan women to take them in have helped the culture recover as well. Were the United States to promote similar large-scale recruitment of women and their strengths in Iraq and Afghanistan, those countries' recoveries would be well under way. The United States and various nongovernmental organizations have employed small-scale programs targeting women, but such token responses are not likely to take root without support on a much larger scale. Skeptics would argue that such is not the job of the US military, but without an assurance of safety and security, particularly where women are prey to the brutal and illiterate members and mullahs of the Taliban and Al Qaeda, some countries—Afghanistan, in particular—are certain to slip back into chaos with the barbaric treatment of women and assured poverty.

Which large-scale programs focused on women have been successful? On a very large scale, the Hunger Project has seen remarkably positive results with a long period of success in Africa, South Asia, and Latin America according to a model which can be duplicated in nation building and peacekeeping environments. The Hunger Project uses proven strategies to bring villages out of poverty and hunger and make them self-sufficient—typically within five years. Core to the Hunger Project's philosophy, though, is empowerment of women and girls in order to achieve lasting change—a philosophy which has also found some success in Afghanistan. Had a similar strategy been implemented in Iraq and Afghanistan from the outset, though on a much grander scale, those countries might have been well on their way to self-sufficiency by this point. The Hunger Project's theory of change relies upon three pillars of thought: (1) mobilize grassroots people for self-reliant action, (2) empower women as key change agents, and (3) forge effective partnerships between people and local government.

The Hunger Project's remarkably successful theory of change is clearly in the domain of the feminine. Interestingly, mobilizing the grassroots population and forging partnerships with local government are also principles which are subsumed in FM 3-0 and the guidance for current military operations in Iraq and Afghanistan. Would the latter have met with greater success in Iraq and Afghanistan if it had also employed the second element and empowered women as key change agents, both as military participants and community participants? This author believes the answer to that question is clearly in the affirmative. The Hunger Project's model for success further incorporates integrated community development, complete with established indicators and millennium development goals meant to achieve community development—indicators and goals which could be replicated in US action in Afghanistan, in particular. The Hunger Project's success in building communities will seem antithetical to the masculine military aim of "fighting and winning"; however, the reality is that the Hunger Project has had great success doing what the military is being called upon to do—building secure communities—even though the Hunger Project hasn't had to do it in a definitively masculine way.

Summary

The United States' masculine military model has produced the most powerful military in the world with the most well-trained personnel and the world's most powerful and accurate weapons. But unlike its ability to deter Soviet aggression, this Cold War model of strength and preparation has not deterred genocide instigated by tyrant leaders, it has not removed threats from the flow of drugs across borders, nor has it deterred terrorist attacks by ideological despots. We must then ask ourselves what is missing—why can't the highly logical, technological, and democratically ideological US military prevail against tyrants, drug runners, and terrorists? The reason is obviously manyfold, but this author contends that it is because the US military is not using the full measure of its potential. In capitalizing upon the qualities of the masculine to create and perpetuate its appropriately combatant institutional identity, the US military has created a culture which maintains masculine strictures in its thought processes, its force structure, its tactics, and its strategy. Despite the many strengths of the military which have resulted from the masculine mindset, the requisite subordination of the feminine that masculine identity demands has limited the military's own ability to employ all available human wisdom, experience, instinct, and talent. The military has devoted decades of effort to defending its cultural assumptions regarding what women in the military should not be allowed to do. But, if the United States wants to maintain the world's best military, it should not focus on the feminine as weakness—it should instead focus on the possibilities of the feminine as a force multiplier.

Notes

(All notes appear in the shortened form. For full details, see the appropriate entry in the bibliography.)

1. Vego, "Systems versus Classical Approach to Warfare."
2. Kimmel, *Manhood in America*, 2.
3. Braudy, *From Chivalry to Terrorism*; Kimmel, "Masculinity as Homophobia"; and Connell, *Masculinities*.
4. Kimmel, "Masculinity as Homophobia," 58.
5. Vego, "Systems versus Classical Approach to Warfare."
6. Kimmel, "Masculinity as Homophobia," 58.
7. Frels, "Women Warriors," 27.
8. Vego, "Systems versus Classical Approach to Warfare."
9. Ibid.
10. Kimmel, *Manhood in America*, 111.
11. Braudy, *From Chivalry to Terrorism*, xiv.
12. Gentile, "Let's Build an Army," 27.
13. FM 3-0, *Operations*, quoted in Gentile, "Let's Build an Army," 27.
14. Scales, "Clausewitz and World War IV."
15. Armstrong, *Veiled Threat*, 198.
16. Baron-Cohen, *Essential Difference*, 1.
17. Ibid., 61.
18. Vego, "Systems versus Classical Approach to Warfare."
19. Baron-Cohen, *Essential Difference*, 26.

20. Ibid., 48.

21. See Samuels and Samuels, "Incorporating the Concept of Privilege into Policy and Practice," this volume.

22. Smith, "Lost Lessons of Counterinsurgency,"

23. Ibid.

24. Patch, "Ground Truth," 33.

25. Ibid., 44.

26. Scales, "Clausewitz, and World War IV."

27. Brizendine, *Female Brain*, 121.

28. Ibid., 122.

29. Watts, *US Combat Training*.

30. Ibid., 11.

31. Roberts, *Ladies of Liberty*, 131.

Bibliography

Armstrong, Sally. *Veiled Threat: The Hidden Power of the Women of Afghanistan.* New York: Four Walls Eight Windows, 2002.

Baron-Cohen, Simon. *The Essential Difference: The Truth about the Male and Female Brain.* New York: Basic Books, 2003.

Benedict, Helen. *The Lonely Soldier: The Private War of Women Serving in Iraq.* Boston: Beacon Press, 2009.

Bly, Robert, and Marion Woodman. *The Maiden King: The Reunion of Masculine and Feminine.* New York: Henry Holt and Co., 1998.

Braudy, Leo. *From Chivalry to Terrorism: War and the Changing Nature of Masculinity.* New York: Alfred A. Knopf, 2003.

Brizendine, Louann. *The Female Brain.* New York: Morgan Road Books, 2006.

Connell, R. W. *Masculinities.* Berkeley: University of California Press, 1995.

Drexler, Peggy. *Raising Boys without Men: How Maverick Moms Are Creating the Next Generation of Exceptional Men.* Emmaus, PA: Rodale, 2005.

Frels, Mary C. "Women Warriors: Oxymoron or Reality." US Army War College Strategy Research Project. Carlisle Barracks, PA: US Army War College, 1999.

Gentile, Gian. "Let's Build an Army to Win All Wars." *Joint Forces Quarterly* 52 (2009): 27–33.

Hinojosa, Maria. "Transcript: Women, Power, and Politics." Aired 19 September 2008. http://www.pbs.org/now/shows/437/transcript.html.

The Hunger Project. http://www.thp.org/home.

Hunt, Krista, and Kim Rygiel. *(En)Gendering the War on Terror: War Stories and Camouflaged Politics.* Hampshire, UK: Ashgates, 2006.

Kimmel, Michael S. *Manhood in America: A Cultural History.* New York: The Free Press, 1996.

———. "Masculinity as Homophobia: Fear, Shame, and Silence in the Construction of Gender Identity." In *Privilege: A Reader*, edited by Michael S. Kimmel and Abby L. Ferber, 51–74. Boulder, CO: Westview Press, 2003.

Newberger, Eli H. *The Men They Will Become: The Nature and Nurture of Male Character.* Reading, MA: Perseus Books, 1999.

Patch, John. "Ground Truth and Human Terrain." *Armed Forces Journal*, November 2008, 33.

Roberts, Cokie. *Ladies of Liberty: The Women Who Shaped Our Nation*. New York: William Morrow, 2008.

Scales, Robert. "Clausewitz and World War IV." *Armed Forces Journal*, July 2006. http://www.armedforcesjournal.com/2006/07/1866019.

Smith, Niel. "Lost Lessons of Counterinsurgency." *Armed Forces Journal*, November 2008, 32.

Vego, Milan N. "Systems versus Classical Approach to Warfare." *Joint Forces Quarterly* 52 (2009): 40–47.

Watts, Barry. *US Combat Training, Operational Art, and Strategic Competence: Problems and Opportunities*. Center for Strategic and Budgetary Assessment. 2008. http://www.csbaonline.org/4Publications/PubLibrary/R.20080821 .US_Combat_Training/R.20080821.US_Combat_Training.pdf.

About the Author

Dr. Edie Disler is a 25-year veteran of the Air Force who has served as an intercontinental ballistic missile (ICBM) crew member, executive support officer to the secretary and deputy secretary of defense, conventional arms control inspector, speechwriter, and faculty professor. She earned a bachelor's in English language and literature from the University of Michigan, a master's in technical and expository writing from the University of Arkansas at Little Rock, a master's in national security and strategic studies from the Naval War College, and a PhD in linguistics from Georgetown University. Dr. Disler is the author of *Language and Gender in the Military: Honorifics, Narrative, and Ideology in Air Force Talk* and has regularly published and presented works regarding women in the military, construction of sexual identity in the military, and discourse in a military environment. She is a founding partner of the consulting firm Interactional Strategies LLC.

CHAPTER 18

ADDRESS BY THE CHAIRMAN TO THE MILITARY LEADERSHIP DIVERSITY COMMISSION

ADM Mike Mullen

Arlington, Virginia, 17 September 2009

Good morning. I really appreciate the opportunity to be able to spend some time with you today. I'll try to give you some of my perspectives on diversity in the military.

This is a really tough subject. You know, I grew up coming out of the '60s in the Navy. Again, I grew up in a nice white middle-class neighborhood in Southern California. And I remember going home after my second—my plebe year, actually. I was home August of '65, and I was watching my black-and-white television, 15 miles from Watts. Watts was burning, and I didn't have a clue where it was. Except somewhere down by the Coliseum, where I would go to watch, as a kid—you know, watch the Rams play, or go down and watch the Lakers play. But it was sort of to-and-from. And it—I mean, it was a searing experience for me, because I didn't know, and yet I was so close.

And that stays with me today, in terms of what I know and what I don't know—and what I can know, having grown up where I did. And I tell the story—and some of you have heard the story. You know, my dear friend and classmate Charlie Golden—you know, we both went into the Naval Academy at the same time. And we came from different places—believe me. And Charlie taught me that. And he taught me in such a graceful, dignified way. That, again, is something else that has stuck with me, as he was blazing trails I didn't even understand. And many of you have done the same thing.

So key to this, as far as I'm concerned, is, what do leaders who are not minorities understand about what it takes to get here? And if we don't understand that—or we don't have some ideas about that—it's pretty difficult to lead in an area that's as challenging as this. And particularly for us in our culture. And we've come a long way—and, certainly, when I was young, I had no expectations to be in the Navy a long time, and certainly no expectations, despite what Les said, that I'd ever be the CNO [chief of naval operations].

Some prefatory remarks and the question-answer session are omitted from this transcript. The complete transcript is available at http://www.jcs.mil/speech.aspx?id=1249.

But when I was CNO, I actually thought I was in charge—acted like I was in charge. And you can do something there, and I made it a top priority. And I went down and addressed the NNOA [National Naval Officers Association] conference in New Orleans, I think a week or two before Katrina—mid-July. And I walked in there first: This is a priority. Mike Hagee and I did this, and Mike's another classmate, and you get someone else very focused and dedicated to diversity. And we almost did a Mutt-and-Jeff kind of thing. And I walked in with an all-white male staff and tried to tell—and there were many young junior officers there, as well. And one of the pieces of feedback I got from that visit was: You know, nice try. You know, what about your staff? A big message.

And so, two years later, when I—and I think Les has heard this story. But two years later I was having a farewell party for my personal staff—for four or five of them. And we had this party at the quarters. And there were probably, I don't know, 20 of us or so in the quarters. And I looked around. And as a—going back to that visit to New Orleans, that somebody called me on. And, literally, you know, from that moment forward, my staff diversified greatly, in terms of women and minorities—because of, obviously, just the message itself that that sent in terms of priority.

And I made it a priority, and I found I could do a lot. And then Gary has sustained that. I see Mark Ferguson here, and I know that. I just told Les Jeff Fowler at the Naval Academy has made it a priority, and 33 percent of the class that has just entered the Naval Academy is minority—33 percent. And that is, actually, the only way we're going to solve this long-term—and I'll speak to that—because of who we are. I mean, you can't lateral in at the O6 level, you know, and try to—so whatever decisions we make right now, that's where we are for 30 years. Or you pick—that's how you generate flag officers—general officers. And you do that when you're recruiting them at 16.

But back to that party. When I sat—as I sat and looked at my staff, and what an unbelievable talent pool—because I didn't—you know, as CNO—and, at this level, you don't have a lot of time to suffer the individuals who cannot deliver. So, actually, the best staff I ever had—and I looked, and there—I can't remember if there was a white male on that staff. And what was sad to me about that, as I looked at that picture in my own home, is: Look at what I had missed. It only took me until I got to be CNO.

So that's what we're missing, and that's what—and we don't know that. We don't know what's—you don't know it until you figure out you're missing it. So, as you do your work, how you penetrate leaders—and as I was thinking back to that New Orleans conference. And one of the things—while they gave me feedback later on, one of the things I challenge them is—because they were mostly minorities: You know, where are your Caucasian, you know, brothers and sisters here? How come they're not in the room? And as I look around this panel, you know, I worry about—I mean, a lot of you I don't know.

But how do we engage the leadership across the board on this? Not just lay it on minority organizations to generate requirements that we don't understand and we can't execute. Which has been a model for a long time, by the way. And

I'm not saying it's a bad model, but I think there's a better way. So how do we pull leaders in to understand where we are, and what are the possible—and, to that story, what are we missing?

So I go to the Naval Academy—the 33 percent right now. The only reason that happened is because Jeff has made it a priority, and Admiral Fowler has made it one of his top two, I think. I think his top two—it may be number one. Do you know, Leo? It's in the—so it's one or two. And because the leader makes that decision, you end up with 33 percent—and people go out and work on it. And there's nothing—those of us in the military, we think there's nothing we can't do if we put our mind to it. And that is so critical.

I'll use the Naval Academy, just because I know that number. But unless we get it right at the service academies and at our accession points, across the board, we're going to live with whatever we are—whatever our entrance requirements—however we're meeting our entrance requirements. And I think we need to be aggressive, and I think we need to, from a leadership standpoint, continue to do that.

And it becomes a very difficult issue. I go back to the quality of my staff when I was CNO and I looked around the room that night. Absolutely the best I'd ever seen, and look what I'd missed.

And then the other thing that I learned as CNO, as I engaged on this, is—and it really goes to General Becton (sp.) and others, who engaged in the education side: We're not going to get there without education. And, in fact, I can remember meeting with the—(inaudible)—college presidents and chancellors to discuss this. And one of them said: If you don't get them by the time they're six, they're gone. Six years old! I mean, we're focused on high-school juniors and seniors, across-the-board, to recruit to the Naval Academy; recruit into OCS [Officer Candidate School]. Six years old.

Now, fortunately or unfortunately, I'm old enough where, you know, a 10—so that's from six to 18, that's 12 years. Well, 12 years isn't that much anymore to me, even though it is way back then. But if we don't get programs in effect that, in fact, start kids off on an education that will support those of us in the military, literally at six years old from that, then we're going to fall short—we're going to fall short. And we must do that.

And that doesn't mean—I mean, that kind of investment is a great investment, no matter where that young boy or girl ends up. And it's an investment, quite frankly, that I'm happy to make on the part of the Department of Defense in those kinds of programs. And there are those kinds of people—there are those kinds of programs out there. Look—how do we know where they are, and how are we connected to them? And when I say "we," I'm talking about leaders like me. And then how are we then producing—how are we connected to programs all over this country that are focused on diversity?

The other thing—and it goes back to sort of emphasis throughout my life. Because I went through this in the '70s—when Zumwalt was the CNO and said, Boom, we're changing—as a junior officer. And it was extraordinary. And I was open to this, and it was jammed. And it was at a time some of you would remember—Les and others—but many would not. And we had riots on ships. This was—and all of that cried for change.

We had—it was a very dangerous situation, and it was jammed and extremely painful. And I would argue we're better than that in terms of making this a priority and execution of change. But it's still got to be pretty aggressive—and it's got to be, in my view, top-down leadership. And if we don't understand it, we're not—we can't lead. Even if we make it a priority, if we really don't understand it. And it's got to be near-term.

So what do I do at the leadership level now, with the classes that have all been commissioned? And then what do I do to make sure that those numbers—? And it's 6 percent for I think—as I look at all the services, 6 percent flag officers, for instance, for women. And that's about what it was when these young women were commissioned in all our services. And it's a little high—it's about the same in terms of flag and general officers across-the-board, because that was the commissioning class.

And then there is a—and I think you just need to look at this, and ask the services to talk to you about this. And get the service leadership to come in and talk to you about it. It'll get their attention. And I don't mean the vice chiefs—I mean the chiefs. Have them come and see you, and talk about it. And then it's: How do I put a position—who do I put in position to move along here? And when does that happen? And, actually, it starts happening about the O5 level. And that's a pretty deep look.

I didn't understand that one when I was an O5—I didn't have many people doing that. But, actually, as I got more senior, I started to understand that a lot better. And there are key jobs, and everybody knows that—and I'm a big believer of putting somebody that's qualified in it and giving them an opportunity. And they either sink or swim, quite frankly. And it's the opportunity issue. And how do we measure that in our services? How do we understand that we're being—you know, we're assigning people to the assignments that generally garner success? And that is through opportunities, and then it's sink or swim—as it is for anybody.

And, speaking to that, and how services look at that—and how we measure it. You know, how do we know? And, obviously, very tightly wound inside the selection processes that are obviously legally binding. The screening processes that we have, which are, by and large, almost as rigid, legally, as the selection processes for promotion are, as well.

And those kinds of—I mean, so there's a near-term issue; there's a midterm issue; then there's a long-term issue—that I think we, as a department, have got to do a whole lot better on. And we are better, but we still have—I believe we still have a long way to go.

And my fundamental belief is that we, as a military, must represent our country. We must represent the demographics of it. It is the greatest strength of our country. And if it's going to take us—you know, the clock's ticking here. If it's a race between that commissioning group and 30 years, where is my country in 30 years? Because I know, at least a couple years ago—the statistic was, in 2050, you know, the majority becomes the minority in our country. And leadership has to be represented, and we're not. And we're not a track to do that. And that's a really—given that we are hierarchical, that's a really tough problem to solve.

But that's fundamentally what I believe. And it is dangerous for the military to not be representative—because I think, in the long run, the criticality of the military to our national security—and if it doesn't represent our country at the leader—we just drift away. We drift away over time. And that would be a really bad outcome for our country.

So I am—in fact, Ted Chiles, who some of you may know—who's not a quiet, shy guy—came to see me recently—you know, to just remind me of what I wasn't doing with respect to diversity. And, as I was telling Les, when you're the head of a service, and you're a Title 10 person, you actually do own a lot of things. And in my current position I either don't own anything or I own everything. (Laughter) But it was a good wake-up call from Ted, who I have great respect for, and is someone that can—is—he gives me great insight and is very free of that—you know, he is not shy about giving me that insight.

So I appreciate—I guess my counsel would be to encourage you to be as aggressive as you can be here. Don't be shy, because we're in much better shape than we used to be, but we still have a long way to go. And what I found in the Navy—again, sort of growing up—going back to that, when Zumwalt was the CNO. And he literally changed our—started to change our culture overnight. But what I also found was that because we had prioritized on African-Americans, we were nowhere with Hispanics—nowhere.

And look at the Hispanic population that we have in our country. And the underpinnings of lack of education—propensity to even go to high school, much less college—which the statistics are mind-boggling. And so we're way behind there. And then extend that to the rest of our diverse population that we have as a country, and I just think we all ought to wake up.

So there's a great opportunity here. And I'm delighted that the wisdom of the country saw fit to both sign this legislation and create this commission. And I just hope you can ring bells at the highest possible levels, because I think it's a strategic imperative for the security of our country. So thanks.

About the Author

ADM Mike Mullen was sworn in as the 17th chairman of the Joint Chiefs of Staff on 1 October 2007. He serves as the principal military advisor to the president, the secretary of defense, the National Security Council, and the Homeland Security Council. A native of Los Angeles, he graduated from the US Naval Academy in 1968. He commanded three ships, Cruiser-Destroyer Group 2, the George Washington Battle Group, and the US 2nd Fleet/NATO Striking Fleet Atlantic. Ashore he has served in leadership positions at the Naval Academy, in the Navy's Bureau of Personnel, in the Office of the Secretary of Defense, and on the Navy Staff. Mullen is a graduate of the Advanced Management Program at the Harvard Business School and earned a master of science degree in operations research from the Naval Postgraduate School. Prior to becoming chairman, Mullen served as the 28th chief of naval operations.

CHAPTER 19

WHY DIVERSITY EFFORTS IN THE DEPARTMENT OF DEFENSE AND INTELLIGENCE COMMUNITY HAVE COME UP SHORT

Charles V. Bush   Alfredo A. Sandoval
Joseph P. Calderon   Juan H. Amaral

Diversity efforts within the US government, particularly the Department of Defense (DOD) and the intelligence community (IC), have proven to be inadequate. Their failure is largely due to organizations approaching diversity more as a personnel program than a critical mission element imperative to national security. Leaders often discuss and study the importance of diversity, but little evidence has emerged over recent years to indicate they fully embrace it. Hence, military organizations (broadly defined to include the larger intelligence communities outside of the armed services) fall woefully short in establishing diversity within their senior executive and officer ranks. Moreover, DOD and IC leaders continue to establish and communicate incongruent department diversity mission statements, objectives, and goals that lack prescribed, mandatory performance standards. Because there are no prescribed performance standards, there exists no leadership accountability and thus no leadership responsibility for monitoring diversity. Therefore, organizations deliver poor and unacceptable outcomes on diversity objectives, which leaders regrettably accept.

Despite the rhetoric to the contrary, promoting diversity issues and programs has not been a leadership priority within the military, nor has implementing diversity historically been attractive to career-minded senior executives or senior military officers.

A lack of diverse leadership has the potential to produce severe negative consequences. In an amicus curiae brief,[1] the friends of the court stated:

> In the 1960s and 1970s, while integration increased the percentage of African-Americans in the enlisted ranks, the percentage of minority officers remained extremely low, and perceptions of discrimination were pervasive. This deficiency in the officer corps and the discrimination perceived to be its cause led to low morale and heightened racial tension. The danger this created was not theoretical, as the Vietnam Era demonstrates. As that war continued, the armed forces suffered increased racial polarization, pervasive disciplinary problems, and racially motivated incidents in Vietnam and on posts around the world. In Vietnam, racial tensions reached a point where there was an inability to fight.[2]

Diversity cannot be viewed as another personnel program, but rather must be considered a national security imperative. The evolving interagency global mission requires an integrated effort in leading a diverse enlisted force in defense of this nation. Former CIA Director George Tenet warns:

> The IC is a global enterprise and requires its most important resource—its people— to be as diverse as the global environment in which it operates. Critical thinking in the IC depends on the inclusion of a diverse workforce at all levels with a wide range of expertise and deep knowledge of other societies, religions, cultures, and languages. To combat the national security threats our country faces, the IC needs collectors and analysts from diverse cultural and ethnic backgrounds who can think and communicate like our targets and penetrate their human and technical networks. The IC also needs employees from a broad spectrum of the society who, based on their upbringing and experiences, can provide a view of the world from different and unique perspectives.[3]

The result: diversity efforts within the military domain remain ineffective. Military organizations are often predisposed to engage in "paralysis by analysis" by either initiating or reviewing diversity studies or diversity commissions to study the issue. The rapidly growing literature warns of the shortcomings of ineffective diversity policies while extolling the criticality of much-needed diversity within the military ranks. Evolving threats and a worldwide mission require that US forces maintain an officer corps and analyst cadre possessing a global perspective, a cultural education, and an innate ability to accommodate and communicate with allies and adversaries with differing perspectives and agendas in a world where "sticks and carrots" are defined quite differently.

For example, there is an absence of diverse senior leader role models, both male and female, in the Senior Executive Service (SES), in the intelligence community's analyst cadre, and in senior military officer grades.[4] According to Tenet, "The need to increase work force diversity in the Intelligence Community (IC) has long been seen as a strategic imperative."[5] Leadership must be comprised of individuals from widely diverse backgrounds and experiences with the ability to lead a diverse global military force of people from every culture and viewpoint. Absent this, the service components and IC agencies have an uphill battle in adequately diversifying their ranks and, hence, in defending this country. Addressing this diversity issue is not simply a matter of balancing demographics. A lack of diversity in DOD and IC leadership positions, coupled with the underrepresentation of minorities among intelligence analysts and military officers, has a direct negative impact on how US forces operate. It threatens the morale, welfare, and effectiveness of our military and intelligence services and therefore compromises the safety of our nation.[6] Without DOD and IC senior leadership held accountable and responsible for the success of diversity, our nation is at risk.

Today's DOD executive corps is a homogeneous group possessing limited diversity. A DOD executive is defined as a one- to four-star general or admiral, grades O-7 through O-10, and the equivalent civilian SES grades. Recent findings indicate that Caucasians, regardless of gender, possess superior promotion rates to the executive corps as compared to diversity members. Of males in the DOD executive ranks, 93 percent are white. Of females in the DOD executive ranks, 89.5 percent are white. Yet of males in the total DOD forces, 75.6 percent are white, while 64.76 percent of females in the total forces are white. Whites comprise 75.1 percent of the US population.[7] In 2007, African-Americans comprised 12 percent of the US population and 16.2 percent of DOD forces, yet only 4.4 percent of the executive corps. Similarly, Hispanics comprised 12.5 percent of the US population and 9.1 percent of DOD forces, but a mere 1.8 percent of the executive corps. In looking at this data, it is hard to argue that diversity initiatives have been organizational priorities for DOD organizations.

These poor diversity statistics are partly a result of the past and current low numbers of qualified diversity candidates and graduates from our nation's service academies and other officer-commissioning sources. This underrepresentation directly and adversely affects the pool of qualified diversity candidates available for senior promotions 25 years into the future.[8]

The minority political caucuses have not sufficiently utilized the full measure of their clout in influencing the DOD to remedy its poor track record of diversity in the senior civilian and military executive ranks. In fiscal year 2007, the DOD's executive corps was comprised of 2,781 executives, who provide the leadership and guidance for an overall force of 2,860,896 Airmen, Soldiers, Sailors, and civilians.

In this population, Caucasians possess statistically superior promotion rates to the executive corps when compared to diversity members. The data indicates that Caucasians rise to the DOD executive ranks at a rate three times greater than Native Americans and Asian Pacific Islanders, four times greater than African-Americans, and six times greater than Hispanics.[9]

Furthermore, the minority political caucuses have neither significantly participated in nor closely monitored military officer commissioning programs, particularly in the service academy nomination processes. This inattention has allowed the service academy cadet/midshipman populations and Reserve Officer Training Corps (ROTC) scholarship award recipients to skew heavily to Caucasian majorities.[10] The aforementioned amicus curiae brief stated, "The primary sources for the nation's officer corps are the service academies and the ROTC." It further stated, "At present, the military cannot achieve an officer corps that is both highly qualified and racially diverse unless the service academies and the ROTC use limited race-conscious recruiting and admissions policies."[11] This diversity underrepresentation in service school appointments is disappointing, and the opportunity cost is staggering. The opportunity cost to the African-American community results in a potential annual loss of approximately 257 appointments, valued at $107 million, and to the Hispanic community, it results in a loss of 247 service academy appointments, valued at $102 million.[12]

The *amici* signatories included former secretaries of defense William Cohen and William J. Perry; President Reagan's former national security advisor, Robert "Bud" McFarlane; notable military leaders such as Gen Ronald

Fogleman, former USAF chief of staff; Gen H. Norman Schwarzkopf, former commander, USCENTCOM; and Adm William J. Crowe, former chairman, Joint Chiefs of Staff. In their brief, they argued, "Based on decades of experience, *amici* have concluded that a highly qualified, racially diverse officer corps educated and trained to command our nation's racially diverse enlisted ranks is essential to the military's ability to fulfill its principal mission to provide national security."[13]

The friends of the court also concluded, "It is obvious and unarguable that no governmental interest is more compelling than the security of the Nation." Furthermore, "the absence of minority officers seriously threatened the military's ability to function effectively and fulfill its mission to defend the nation."[14]

The military has come up short in accomplishing any real measurable or sustainable progress in the diversity of their respective senior ranks.[15] Additionally, within these ranks, there is confusion, ambiguity, and in some quarters, a lack of vigorous leadership committed to diversity efforts and programs. Senior leaders have failed to make the case for diversity, to clearly differentiate it from affirmative action, to debunk the myth that it lowers standards, and to successfully implement diversity efforts and programs that materially affect the composition of their departments. It is mission critical that the DOD and IC have leaders, officers, and analysts possessing an innate ability to discern and decipher various cultures, languages, and perspectives in defense of our nation. This global reality requires our military to foster and expedite effective DOD and IC strategic language thinkers and speakers. Currently, the DOD and IC are falling far short in accomplishing this objective.

US military forces now operate with multinational organizations more diverse than NATO.[16] Twenty-first century commanders will face unfamiliar tasks in unfamiliar places—such as dining with local tribal leaders in Afghanistan, conducting coordinated operations with Chinese warships against Somali pirates, providing leadership in joint exercises in East Asia, and building schools with women's groups in Africa. "The Diversity Senior Advisory Panel for the Intelligence Community (DSAPIC) urges the director of central intelligence (DCI) to refocus efforts to increase among the IC population the diversity of skills, languages, talents, expertise, and people that are critical to the success of the IC's mission."[17]

DOD leadership proclaims that officers must be able to communicate, relate, and recognize cultural attributes to be effective in the area of operation.[18] They say that a lack of diversity in the officer corps challenges the credibility of the military hierarchical system and the sense of good order and discipline in both the enlisted force and officer corps.[19] Yet in both the DOD and IC, there is no measurable progress in the diversification of our country's officer commissioning sources or in the composition of our intelligence community's analyst cadres.

If senior DOD and IC leadership continue to ignore diversity, we will lose the war for strategic talent, which will undoubtedly result in a loss of strategic advantage for our military commanders—an advantage which may well be the deciding factor in future conflicts. It is the responsibility of senior civilian and military leadership to ensure US forces effectively and efficiently operate with—and in—diverse organizations, cultures, languages, and environments. Leadership must recognize that traditional American cultural values are not easily transferable to other parts of the world, nor are they always welcomed. US military forces and intelligence assets must be diverse to successfully communicate, perform outreach, and fight in multiple mediums and languages simultaneously.

Diversity is key to unit cohesiveness, which in turn is critical to mission effectiveness. DOD and IC leadership must be accountable and responsible for the success of diversity in their commands.[20] According to chief of naval operations (CNO) Adm Gary Roughead:

> If you look at the Navy in its entirety, it's a representative mix of America's society. But if you look at the leadership, it tends to be very white male. And that is not the direction where the country is going and nor is it from that non-diverse group that you get, in my mind, the best solutions to problems. . . . It's from diversity that I think you get many different perspectives, different ideas, you get different experiences and that gives you a richness of solutions that you otherwise wouldn't have. . . .

> We are beginning to see things that are moving in the right direction. For example, at our Naval Academy, we are bringing in the most diverse group in its history. And that is a function of not offering any type of special programs or compromising on any standards that have existed at the Naval Academy, but rather on a commitment to getting out and talking to people and making young people aware of the opportunities that exist in the Navy.[21]

It is *sine qua non* (1) for senior DOD and IC leaders to aggressively make the case for diversity, to clearly differentiate it from affirmative action, to debunk the myth that it lowers standards, and to successfully implement diversity efforts and programs that materially affect the composition of their departments; (2) to request that senior leadership establish diversity goals as key performance metrics to effect real change and that accountability and responsibility for their implementation and success be directly related to promotions; (3) to impart to leadership the understanding that a worldwide mission requirement requires diverse personnel so IC forces can effectively support the US officer corps in leading a diverse enlisted force in defense of this nation; and (4) to have leadership understand that an absence of diversity in our nation's DOD and IC forces has a direct negative impact on operational effectiveness and threatens their good order and discipline.

Finally, diversity is a critical mission element and a national security imperative that must be established in the DOD's and IC's SES ranks, the military's officer corps, and the intelligence community's analyst cadre to effectively protect our nation and to have leadership begin to reflect the forces they command and the population they defend.

Recommendations

Recruitment

To realize progress on diversity, senior DOD and IC leaders must begin to treat this issue as a critical mission element imperative to national security. They should establish attainable, measurable, and coordinated diversity efforts and programs within their departments and be held accountable for achieving them. These agencies should recruit diverse applicants for mid- and senior-level executive positions from the private and public sectors. The secretary of defense and the director of central intelligence should establish coordinated DOD and IC diversity recruitment policies across their departments and agencies to maximize resources and improve results.

DOD service components and the IC agencies must coordinate and align diversity recruitment efforts among their respective commissioning sources and personnel recruitment efforts and also be held accountable for their successful implementation.

Development

The DOD and IC should review promotion and selection processes for key assignments to ensure that all selection criteria are essential for successful job performance and that no systemic barriers to diversity exist.

Civilian and military leadership must ensure that all employees receive relevant training, mentorship, and access to education, management, and leadership experiences necessary for professional advancement.

Congressional members need to examine, prioritize, and appropriately staff their districts' nomination efforts to provide the service schools with a full slate of qualified candidates who can successfully compete for academy appointments. Moreover, diversity students from these districts must be mentored, supported, and tutored much earlier to prepare them for the rigors of the federal service academy nomination process.

It is essential that a strategic diversity program fostering diversity recruitment goals be adopted by all officer commissioning sources and IC agencies. Metrics should be established to monitor and measure those responsible for the performance of these diversity programs.

Notes

1. Lt Gen Julius W. Becton, Jr., et al., "Amicus Curiae Brief of Diversity among the Armed Forces Officer Corps," *Supreme Court of the United States, No. 02-241, 02-516.*

2. Ibid., 6.

3. George Tenet, *A National Security Imperative for the Intelligence Community*, 2005, 3.

4. Charles V. Bush, Alfredo A. Sandoval, and Joseph P. Calderon, "DOD Executive Diversity Study" (unpublished manuscript, 2008).

5. Tenet, *National Security Imperative*, 7.

6. Lt Gen Richard Newton, "An Air Force Perspective" (speech, Liaison Officer Director Conference, US Air Force Academy, Colorado Springs, CO, 15 January 2009).

7. Bush, Sandoval, and Calderon, "DOD Executive Diversity Study."

8. Ibid.

9. Ibid.

10. Jason Boman, Trier-Lynn Bryant, and Christopher Goshorn, "Diversity Recruiting Office, USAFA DFM Operations Research Capstone Study" (unpublished manuscript, 2006).

11. Becton et al., "Amicus Curiae Brief," 5.

12. Lt Col Alfredo A. Sandoval and Lt Col Joseph P. Calderon, "USAFA Diversity Study, Classes 2006–2009" (unpublished manuscript, 2007).

13. *Supreme Court of the United States, No. 02-241, 02-516*, 5.

14. *Haig vs. Agee*, 453 U.S. 280, 307, 1981, in *Supreme Court of the United States, No. 02-241, 02-516*, 7.

15. Bush, Sandoval, and Calderon, "DOD Executive Diversity Study."

16. Newton, *Air Force Perspective*, 2009.

17. Tenet, *National Security Imperative*, 8.

18. Newton, *Air Force Perspective*, 2009.

19. *Supreme Court of the United States, No. 02-241, 02-516*, 13-4.

20. Tenet, *National Security Imperative*, 4.

21. Lindsay Wise, "Sunday Conversation: Navy Chief Says It's Time for a Change, Roughead Sees Need for More Diversity in Leadership Roles," *Houston Chronicle*, 25 July 2009.

About the Authors

Charles V. Bush is a retired executive, having served as a senior corporate officer of several multinational corporations. He finished his career as president and chief executive officer of a competitive telecom corporation of which he was a founder. In 1954, Mr. Bush was selected by Chief Justice Earl Warren for appointment as the first African-American page on Capitol Hill. He earned a commission from the US Air Force Academy in 1963 as the institution's first African-American graduate. Mr. Bush served in Vietnam as an intelligence officer, where he was awarded the Bronze Star. He earned a master's degree from Georgetown University and an MBA from Harvard University.

Lt Col Alfredo A. Sandoval is an Air Force Academy liaison officer and a founding member of the Air Force Academy's Diversity Advisory Panel. In 2007, he was recognized as the Air Force Academy's Outstanding Diversity Officer of the Nation. He serves as chairman of congressional district CA-45's Academy Selection Committee and has coauthored several studies pertaining to diversity in the military. Colonel Sandoval graduated from the US Air Force Academy, earned an MBA from Wright State University, and completed Executive Development Training at the Wharton School of Business. He is a military graduate of both Air Command and Staff College and Air War College. Colonel Sandoval is currently a partner in the asset management firm The Private Investment Group, Inc. and a founding partner of the real estate firm Creosote Partners LLC.

Lt Col Joseph P. Calderon is an Air Force Academy liaison officer and a founding member of the Academy's Diversity Advisory Panel. He is currently a commercial airline pilot and has served as a military aviator, aviation security consultant, and martial arts instructor. He is also a founding partner in the Talon Group, specializing in developing, analyzing, and implementing Native American development projects. Colonel Calderon is the US Air Force Academy's 2009 Outstanding Diversity Officer of the Nation. He serves as the chairman of CA-44's congressional Academy Selection Committee. He has coauthored several studies reviewing diversity and continues to advise senior leadership and members of Congress in ways to improve diversity representation within the Department of Defense. Colonel Calderon is a 1984 graduate of the US Air Force Academy and a distinguished graduate of Squadron Officer School and Air Command and Staff College.

Juan Amaral is a 1984 graduate of the US Air Force Academy and a former cost analyst for the Air Force. He also holds a master's degree from the Air Force Institute of Technology. He worked for Raytheon before starting his own consulting firm specializing in earned value. Mr. Amaral retired in 2007 as a major from the US Air Force Reserve, having served 12 years as an Air Force admissions liaison officer.

CHAPTER 20

THE TANNING OF THE MILITARY

AaBram G. Marsh

G iven that the US population is expected to become "majority-minority" over the next few decades, the evolving "tanning of America" highlights the urgent need for more diversity among its military officer corps.[1] This chapter calls attention to a trend that is remarkably similar to that which brought awareness of the bias and discrimination feeding discontent in the ranks during the late 1960s and early 1970s. At that time, the armed services' leaders were largely unaware of the issues of their growing black enlisted populations but were eventually shocked into action when disgruntlement turned into protest.

Despite projections that new recruits over the next decade will become majority-minority, as was the case over 40 years ago, there is growing concern that the Department of Defense (DOD), and the US Air Force specifically, may not be developing its officer corps to be representative of wider society. Moreover, many within the higher ranks do not appear to fully appreciate the necessity for mentorship, which is critical in fostering professional relationships. Consequently, if not proactively addressed, this trend of an increasingly culturally isolated officer corps exposes the Air Force to two profound risks within the next two decades: (1) officer recruiting and retention risks, and (2) growing tension between majority white senior officers and majority-minority junior officers and enlisted members.

This is an analysis of the historical career progression of active-duty US Air Force African-American commissioned officers to assess potential discrepancies concerning their recruitment, promotion, and retention. Composing 17 percent of the entire active-duty military force and 8.5 percent of the officer corps, blacks remain the largest minority group in the US military.[2] Therefore, addressing the concerns and inequities faced by them can have implications for understanding similar underrepresentation and differentiations in career development for other minority groups and offers a prescription for guidance and action as US society adjusts to the reality of a more racially and ethnically diverse society resulting from the tanning of America over the next decade.

This chapter suggests that inconsistent and delayed mentorship, which many minority officers receive, is a major limiting factor to healthy diversity of the US Air Force officer corps. Despite similar capabilities, motivation, and aptitude, minority officers often experience fewer professional development opportunities because of their limited access to critical informal mentor relationships early in their military careers. As a result, since the 1990s, many of the service's gains in improving diversity among its officers are now reversing.

Without adequate and continual mentorship, large proportions of minority officers suffer difficulties in receiving career-enhancing assignments, resulting in less competitive records and disproportionately lower promotion rates. More alarming, the glaring underrepresentation of African-American, Hispanic, and Asian officers in the service's senior ranks signals to all in uniform that only a small proportion will rise to senior positions as the nation's population becomes considerably more multiethnic and multiracial over the next few decades. Therefore, it is imperative that the US Air Force take action to ensure its officer corps improves its diversity, actively recruiting and retaining the "best and brightest" from all groups.

Brief History of Black Officers in the US Air Force from 1940 to 1970

Americans of African heritage have served faithfully and honorably in defense of the United States in all the nation's wars since the American Revolution. However, the opportunity for blacks to lead and command troops under the banner of their nation was denied until the Civil War. During that war, approximately 75 to 100 blacks served as officers in the US Army, although they were restricted to units commanded by white officers.[3] After the Civil War, it was not until World War I that significant numbers of blacks were again commissioned as officers in the US Army.

During and after World War I, the US Army Air Service and Air Corps and the US Marine Corps refused to accept any African-Americans.[4] By 1940, the entire US military's black officer strength stood at five on active duty: three chaplains, Brig Gen Benjamin Davis, Sr., and his son Capt Benjamin Davis, Jr. (a drop from 1,408, 0.7 percent of the US Army's officers in World War I).[5]

Despite vehement objections from Army Air Corps leaders, Pres. Franklin D. Roosevelt directed them to accept African-American personnel into their ranks. This led to the creation of the Tuskegee Airmen, the 99th Fighter Squadron, the 332nd Fighter Group, and the 477th Bomber Group. Subsequently, black participation in the Army Air Forces mushroomed from zero in 1940 to 138,903 in 1945 (of whom 1,559 were officers).[6] During the war, the 332nd Fighter Group downed over 400 enemy aircraft, earned numerous combat citations, and gained the respect of the bomber crews they valiantly escorted on missions.[7] Despite accolades, black officer mobility was stunted due to Army Air Forces policies mandating segregated units and forbidding African-American officers from commanding white officers.

When the US Air Force formed in 1947, blacks made up 6.1 percent of its personnel strength, but only 0.6 percent of its officer corps.[8] The proportion of black officers was not due to lack of quality or quantity of available black aviators. Rather, it resulted from the Air Force's arbitrary limitation of their numbers to approximately 500. Air Force support of strict segregation and quotas remained in place despite post-war recommendations that commissions be granted to qualified personnel regardless of race.[9]

In 1948, Pres. Harry S. Truman, acting on the recommendations of his President's Committee on Civil Rights, issued Executive Order 9981, which declared "equality of treatment and opportunity for all persons in the armed services without regard to race, color, religion, or national origin."[10] President Truman then created the President's Committee on Equality of Treatment and Opportunity in the Armed Forces, which abolished quota systems and started integrating training programs.[11] Primarily due to concerns about the human capital wasted by preserving two segregated air forces, Secretary of the Air Force Stuart Symington deactivated the 332nd Fighter Group and integrated its personnel into previously all-white organizations.

Personnel demands of the Korean War and the exemplary performance of blacks in integrated units debunked many misperceptions, so that by 1956 the remnants of the overt "Jim Crow" segregation had effectively disappeared from the US military.[12] Yet, although African-Americans could serve throughout the Air Force, the phenomenon later defined as "institutional racism" remained pervasive and resulted in persistent, unofficial racial prejudice and discrimination. Very few black officers were recruited or promoted during this period and consequently, by 1964, despite the proportion of blacks growing to more than 10 percent of the enlisted population, the population of black officers rose to a mere 1.5 percent.[13]

In 1962, Pres. John F. Kennedy reestablished the President's Committee on Equal Opportunity in the Armed Forces, under the leadership of Judge Gerhard Gesell, to help remedy the on- and off-base racial discrimination facing black service members.[14] The committee made a number of recommendations, which were ignored by the individual services.[15] Commanders relied on voluntary compliance with antidiscrimination laws and regulations, ignored court-martial and promotion rate disparities, and refused to confront local businesses and law enforcement officials when service members complained of off-base prejudice.[16]

The racial divide increased during this period due largely to ineffective leadership by junior officers and noncommissioned officers.[17] The few black officers in uniform had neither the numbers nor the influence within their services to alter the spiral of racial unrest spreading worldwide in the late 1960s. Additionally, many of their white peers were either unaware of, or uninterested in, complaints of prejudice raised by their black subordinates. Consequently, racially charged incidents flared at dozens of installations in all military branches between 1968 and 1973, with ever-greater magnitude, damage, and casualties. The US Air Force experienced the largest race riot in its history at Travis Air Force Base beginning on 22 May 1971, during which violence between several hundred Airmen raged for two days.[18]

A Period of Progress

In the early 1970s, Secretary of Defense Melvin Laird promised to eliminate "every vestige of discrimination" from the military. Air Force Chief of Staff Gen John Ryan required all commanders to support the "Equal Opportunity and Race Relations Education program with the same vigor and enthusiasm as that given the flying mission."[19] With forceful support from leaders at higher levels, implementation of these initiatives ended the violent and costly racial conflict that had risked erosion of the discipline and trust essential to the armed forces. By the late 1970s, the Air Force had all but eliminated overt racial bias.

Twenty-two years after President Truman issued Executive Order 9981, Lt Gen Benjamin Davis, Jr. was still the sole black to have attained the grade of general officer in the US Air Force when he retired in 1970. During the 1970s, however, the US Air Force made a concerted effort to promote outstanding senior black leaders who were qualified for the general officer ranks but would not have been selected due to earlier institutional discrimination. From 1970 to 1990, 24 African-American Air Force officers attained the rank of general. Of these officers, two were promoted to the four-star rank, Gen Daniel "Chappie" James and Gen Bernard Randolph, and the first black female general officer was selected, Maj Gen Marcelite Jordan-Harris.[20] These officers all entered the Air Force prior to the all-volunteer force.

Despite the rise of prominent blacks to the general officer ranks, the Air Force still struggled with equity in the promotion system as it had from the earliest days. During the first several decades following the creation of the Air Force, policy decisions influencing assignments, command opportunities, and subsequent promotions favored pilots. Pilots were promoted disproportionately to their numbers in the officer corps. In 1970, while only 30 percent of all officers were pilots, 88 percent of all generals and 65 percent of all colonels were pilots.[21] Although personnel policies slowly changed to allow more nonaviation officers to rise to the higher grades, the situation remained grim for mission support officers in the late 1980s. As late as 1989, pilots were promoted at a significantly higher percentage, both in the promotion zone (IPZ) and below the promotion zone (BPZ), as compared to mission support officers:

To colonel:	IPZ (49.0 percent vs. 42.3 percent)
	BPZ (4.7 percent vs. 1.9 percent)
To lieutenant colonel:	IPZ (67.3 percent vs. 65.1 percent)
	BPZ (5.1 percent vs. 2.4 percent)
To major:	IPZ (93.1 percent vs. 81.3 percent)
	BPZ (3.3 percent vs. 2.2 percent)[22]

Such disproportionate pilot promotion rates significantly affected the situation of black officers because they were least concentrated in tactical operations career fields (i.e., pilot, navigator). In fact, the proportion of blacks in this area never rose above 4.0 percent and was concentrated primarily in engineering/ maintenance, supply, and administrative careers.[23] Consequently, with promotion opportunities to the highest grades in the Air Force favoring pilots, the underrepresentation of black officers in tactical operations fields limited their promotion opportunities. A report conducted by the Office of the Under Secretary of Defense for Personnel and Readiness further revealed that the root cause of the limited numbers of minorities in aviation career fields was a general lack of recruitment.[24]

Blacks who were selected for tactical operations career tracks tended to suffer higher attrition rates than their white peers despite similar qualifications. In 1984, a study to determine possible factors influencing disproportionately high attrition rates of black officers during undergraduate pilot training (UPT) from 1977 to 1983 reported the attrition rate for blacks was as much as 27 percent higher than that of white males.[25] Additionally, unlike their white counterparts, there was limited correlation between black officer washout rates and Air Force Officer Qualifying Test scores. This report concluded that the primary reasons for black underrepresentation in UPT were limited role models/recruiters and limited mentorship in preparation for and throughout UPT. Despite the progress made evident by the increasing number of senior African-American officers and the growing proportion of Air Force officers of African descent, various indications toward the end of the Cold War revealed that some progress made by black officers over the previous two decades started to stagnate and even reverse.

Recruitment

In 1971, seeking to attain equal opportunity, the USAF developed an initiative to recruit qualified blacks into its officer corps.[26] For the first time since being forced to admit blacks in 1940, the service "made it possible for [African-Americans] to enter upon a career of military service with assurance that [their] acceptance will be in no way impeded by reason of [their] color."[27] As a result, within the first 10 years of the implementation of the all-volunteer military, African-American representation in the Air Force officer corps tripled (see fig. 20-1). During the 1970s, the numbers of black officers increased in part due to Vietnam-era efforts to expand the AFROTC presence at historically black colleges and universities, as well as targeted recruitment by the US Air Force Academy (USAFA). The emphasis placed on the recruitment of black officers partially resulted from the antiwar backlash at predominately white schools that reduced the numbers of white male volunteers.[28]

From the late 1970s into the early 1980s, the US Air Force leveraged the nation's precarious economic situation, which included skyrocketing inflation and persistent unemployment, to recruit new black college graduates.[29] For most of the 10-year period from 1973 to 1983, the Air Force generally outperformed most of its sister services, recruiting the highest proportion of blacks.[30] The military proved attractive to blacks as it came to be widely recognized for its efforts to eliminate vestiges of overt racism.

The class of 1963 was the first at the US Air Force Academy to commission blacks, with three black graduates earning commissions.[31] Over the next decade, only 33 more black cadets would graduate. As a result, Air Force leaders sought to recruit more black cadets and augmented the Air Force preparatory school curriculum to provide remedial academic support to "assist otherwise

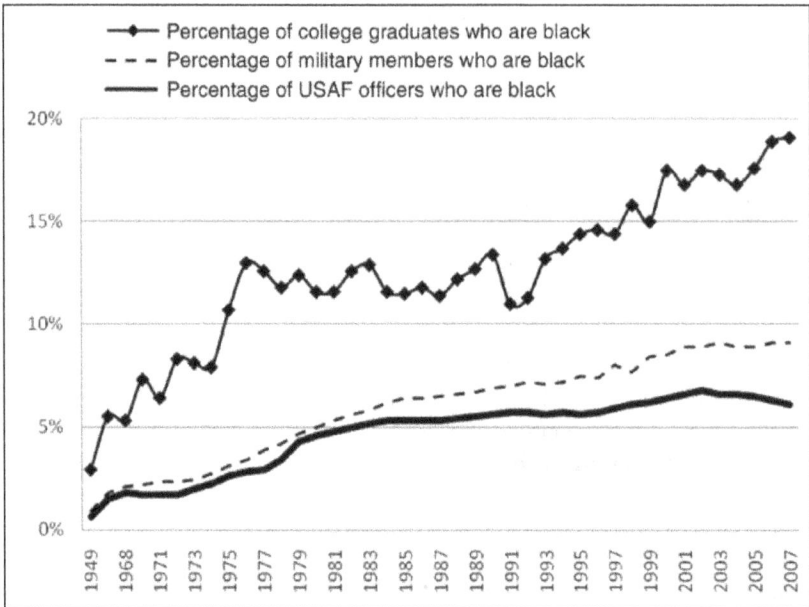

Figure 20-1. Percentage of college graduates, military members, and USAF officers who are black.[32]

qualified black applicants to overcome minor education deficiencies."[33] This was necessary given the inferior secondary educations that many African-Americans received due to residual segregation effects in the South and in urban areas elsewhere throughout the country. Consequently, it was not until 1973 that the Academy consistently graduated blacks with each successive class.[34] So successful were the Academy's early efforts to recruit, develop, and provide equal treatment of black students that in 1977, Cadet Edward A. Rice became the first black designated as a cadet wing commander. (Cadet Rice eventually became a decorated bomber pilot and three-star general.) USAFA progressively graduated more black officers between 1971 and 1986, with their percentages growing from 1.3 percent to a high of 7.3 percent.[35]

Due to the targeted recruitment of black students, African-American populations within AFROTC detachments also exploded. In 10 years, the percentage of black AFROTC graduates climbed from 2.6 percent in 1972 to 13.8 percent in 1982.[36] Over time, however, the focus on recruiting qualified black officer candidates diminished. After the mid-1980s, the proportion of blacks commissioned from either AFROTC or USAFA shrank steadily. By 1991, the proportion of black officers commissioned fell to 6.2 percent from USAFA and 3.6 percent from AFROTC programs, returning to levels not seen since 1983 and 1973 respectively.[37]

Once aware of these negative trends, in 1993 Congress directed the General Accounting Office (GAO), now called the Government Accountability Office, to compile a report on gender and racial disparities at USAFA. The GAO found that minorities had higher attrition rates, that they were subjected to proportionally more academic and honor reviews, and that proportionally fewer were represented in the top 50th or 15th percentiles of their graduating classes. Additionally, it was discovered that minority and white students possessed opposing views of equal treatment, with higher percentages of whites viewing minorities as being treated better and equally large proportions of minorities perceiving worse treatment. Finally, the GAO found that although the USAFA leadership was aware of many of the issues raised in the report, it did not analyze discrepancies in student performance, establish criteria to determine performance differences, or document actions taken or plans to implement in the future to improve equal opportunity.[38]

The Department of Defense found that key factors inhibiting the early development and advancement of black officers included educational/precommissioning preparation, "slow starts" in initial assignments, and limited access to peer and mentor networks.[39] Weak or nonexistent mentorship and peer networking proved significant because information important to career success is made available to junior officers only through networks comprised of more senior officers. More than anything else, the need for mentorship of junior officers emerged.

Career Progression from 1991 Onward

In 1991, the Bush administration and Congress sought a 25 percent reduction-in-force of active-duty force levels (from the 1987 baseline of 2,174,000 active-duty personnel) by fiscal year 1997. In 1991, the drawdown target was amended to 1,630,500 troops by 1995. The Clinton administration later revised the reduction-in-force to include an additional 200,000 active-duty members by the end of fiscal year 1997.[40] Many of those recruited in the late 1970s and early 1980s were most vulnerable to the military's force reduction programs. Consequently, there was significant concern that force drawdown efforts would remove a disproportionate number of black officers, thus reversing hard-fought gains in career progression during the 20 years since the beginning of the all-volunteer era.

Surprisingly, the proportion of black officers in the US military increased overall by a full percent (6.5 percent to 7.5 percent) during this decade. Yet of all service branches, the Air Force made the smallest gains, with an increase from 5.4 percent to 5.9 percent.[41] Stagnant growth in the representation of black officers since the 1990s across the services is attributable to the end of targeted recruiting of eligible black college students after 1983. Air Force leadership observed the decline in African-American accessions in the 1990s and incorporated the Gold Bar Program.[42] This program selected newly commissioned minority second lieutenants to serve as AFROTC and USAFA recruiters with the dual benefit of providing youthful, energetic officers who could more easily relate with high school and college-age young adults, encouraging them to consider serving as an officer in the Air Force. Unlike the earlier recruiting programs that took military officers out of standard career tracks, Gold Bar selected newly commissioned officers to serve for no more than one year, therefore not affecting their overall career progression. This resurgence in targeted recruiting returned the black accessions percentage to over 7 percent in 1996.[43] Since 1997 most of the distinctions between each commissioning source and perceived future promotion eligibility have disappeared.[44] In 1997, the Air Force began aerospace studies courses at historically black colleges and universities to increase the recruitment of blacks desiring careers in aviation and to impart skills necessary to improve graduation rates in pilot training. As a result, the proportion of black Air Force pilots increased to 5.7 percent, and attrition rates declined for both pilot and navigator training by the late 1990s.[45]

Retention

Blacks are more likely to remain in the service than are their white counterparts, but less likely to be promoted within the field grade ranks.[46] Even though blacks continue to be concentrated in mission support career fields, this did not translate into improved promotion rates when mission support officers began to achieve parity in promotion percentages with pilot cohorts after 2000. In seven of eight O-5 (lieutenant colonel) promotion boards from 2001 to 2007 and five of six O-6 (colonel) promotion boards from 2003 to 2007, mission support officers outperformed even pilots, yet blacks underperformed promotion averages in 11 of 14 boards.[47]

It is useful to compare the promotion rates of blacks and women for three reasons: (1) women are also predominately concentrated in support career fields; (2) their representation is also considered under equal opportunity programs; and (3) their population numbers most closely compare with those of blacks. There is a stark difference, however, in the promotion rates of these two groups through the field grade ranks. Unlike black officers, who on average have been significantly underrepresented in promotion boards since 1989, women have outperformed male promotion rates—by 2.2 percent to O-4, 4.1 percent to O-5, and 6.4 percent to O-6—during this same period.[48]

The Importance of Mentorship

Since the drawdown, the proportion of black Air Force officers continues to fall behind the proportion of the US population of African heritage with college degrees (8.4 percent in 2007).[49] Most alarming is that since 2002, the proportion of black Air Force officers dropped nearly a full percent, to 6.1 percent in 2007.[50] This trend erases the gains that the service made through the early 1980s in recruiting and retaining promising black college graduates.

African-Americans continue to graduate from the same accession sources and train in the same career fields, yet they suffer reduced opportunities for key developmental assignments, selection for in-residence PME, and thus, ultimately future promotions. This suggests that despite the fundamental nature of mentorship in "determining an individual's success within an organization," black officers continue to face difficulty in cultivating peer and mentor relationships.[51] One likely reason for this is that despite the end of overt racial prejudice within the services, a certain degree of social segregation remains.[52] Although officers, regardless of race, agree that the promotion board process is fair, many blacks comment that consistent disparities in promotion results are attributable to limited mentorship and difficulties receiving career-enhancing assignments, which all result in less competitive records.[53]

Mentorship is essential to the development and retention of all officers. There are three types of mentoring relationships: situational—information/advice provided to help protégés make near-term decisions; informal—flexible, unstructured relationship in which the mentor provides insight as the need exists; and formal—a systematic, structured forum for mentors and protégés (usually organizationally sponsored).[54] Informal mentorship is the most common form and because associations are generally based on an individual's sense of familiarity, there is a tendency for individual bias in selecting potential protégés. Yet in various studies conducted since the 1970s, minority and female officers argued invisible and undisclosed cultural biases prevented them from receiving the same guidance and feedback as is readily provided to their white male peers.

According to the DOD report *Career Progression of Minority and Women Officers*, minority officers may face invisible barriers that can be impediments to receiving mentorship. One such obstacle takes the form of actions to test an individual. For example, during focus groups white officers routinely expressed the belief that minorities generally possessed weak academic and military educational backgrounds, which led them to subject minority and female officers to intense scrutiny, generally at the beginning of an assignment.[55] Another challenge is discomfort with individuals of different backgrounds. Consequently, white senior officers may not include junior ranking minorities in informal work or social activities.[56] Failing to mentor others restricts diversity in professional relationships and stunts individual career progression. Therefore, it is important that mentorship actively plays an integral role throughout every aspect of officer progression.

In 2008, an Associated Press article, "After 60 Years, Black Officers Rare," highlighted the dearth of black officers in high-ranking positions and suggested that the cause was mostly lack of mentorship.[57] Addressing the issue of dwindling black officer proportions and disparities in their promotion rates requires a long-term focus, and the Air Force should consider taking a page from the Army in addressing this concern. In 2005, it created the Army Diversity Office to analyze steps to improve diversity among its entire military/civilian workforce. To work towards its aim of workforce diversity, the Army outlined five success factors: leadership commitment, strategic planning, accountability/assessment/evaluation, employment involvement, and mentoring.[58] Focused mentorship and development is the centerpiece of this program. Unsurprisingly, the proportion of black officers increased from 11.4 percent in 2000 to 13.1 percent in 2007.[59]

The US Army can serve as a model because it took action after discovering stagnation in the progression of black Army officers during the 1990s, when the proportion of black officers bumped slightly from 11.2 percent in 1990 to 11.4 percent in 2000.[60] In 1997, then–Lt Col Remo Butler studied this trend and wrote that for the Army to improve the representation and career viability of black officers, it must make mentorship a central element at all levels of career development and educate its personnel on real cultural awareness (instead of political correctness), thereby minimizing the impact of "the good old boy network."[61] It appears that the Army observed the trend, reoriented its efforts, decided it needed to make a change, and took action to ensure it remains a first-choice profession for all of America's young adults into the twenty-first century.

Despite the obvious tanning of America and contrary to the trends in its sister services, the US Air Force's officer corps was less diverse in 2007 than it was in 1999. It appears that the US Air Force has acknowledged its need to reverse the declining minority representation with the creation of the Air Force Personnel Center Diversity Council in 2009.[62] This was a positive first step, yet based on available evidence and inconsistent strategic communications on diversity over the last decade, the Air Force still needs to enact immediate and far-reaching changes to correct negative patterns.

Recommendations for the Air Force's Future: Focused Recruitment and Twenty-First Century Mentorship

As the composition of US society becomes more pluralistic, it behooves the service to simultaneously target focused recruitment programs to draw more minority officers into the Air Force and also educate its officers to actively (and comfortably) mentor all subordinates to ensure they enjoy similar opportunities to attain the highest grades of leadership. This urgency is marked by the decline in the representation of African-Americans in the most junior officer grade—second lieutenant—which evaporated from 7.8 percent in 2001 to 3.8 percent in 2007.[63] This precipitous drop, if not rectified immediately, bodes poorly for the service both as the nation transitions from its historically binary racial approach (majority/minority) to a more pluralistic society and simply because such a miniscule proportion of the officer corps is unsustainable for future career progression.

To address improving minority officer recruitment, the Air Force should: (1) establish and publish any targeted accession goals for minority officers; (2) re-institute a diversity-focused Gold Bar program; and (3) reconsider its historical approach and transition from focusing primarily on recruiting students pursuing degrees in hard sciences to those involved as campus student leaders.

As is often quoted by many senior officers, "what gets measured gets done." Likewise, if the Air Force is serious about getting the best and brightest, it should establish goals for recruiting into accession sources and commissioning increasing numbers of minority officers. The aim should be to eventually draw proportions of each population group similar to their respective proportions among the college educated in the United States. To attain these increasing numbers of minority officers, the Air Force should reinstitute the Gold Bar program, as it did in the 1990s. However, it should be broadened to include primarily underrepresented population groups (i.e., female, black, Asian, and Hispanic). These newly commissioned officers should be assigned to one-year stints to serve as AFROTC and USAFA recruiters, encouraging the growing numbers of young Americans who look like them to consider serving as an officer in the Air Force.

Finally, the Air Force should shift its focus from recruiting college students pursuing degree programs in hard sciences to recruiting those who are recognized student leaders for two reasons. The first is most obvious to those already in uniform. Except for a few line officer career fields like communications, engineers, and scientists, there is little correlation between students' baccalaureate degree program and their capabilities to perform their duties as an officer and a leader. As such, focused recruitment of students with technical degrees tends to dissuade others who may have more leadership experience in other capacities like student government, fraternities/sororities, and internships with large corporations and would serve as impressive future Air Force officers. The other reason for considering an approach to recruit more campus leaders is that significantly smaller proportions of women and minorities tend to pursue technical degrees. However, many more are often leaders of very large and influential campus, national, and international organizations. Therefore, the US Air Force's focus on recruiting individuals pursuing hard science degrees may cause it to fail to attract young Americans who have already established significant credentials leading large organizations, programming/budgeting large amounts of money, and mentoring future leaders.

Once recruited, these officers must be provided sufficient mentorship so as to arm these individuals with a fair opportunity to succeed in their respective military careers. Such mentorship should take on a modern approach. This twenty-first century mentorship should focus on two areas: (1) educate the Air Force's more senior officers to actively (and comfortably) mentor all subordinates and (2) take advantage of new social networking capabilities to bridge generational and cultural gaps between older and younger officers.

One way to educate officers on the primacy of mentorship for career development, which requires limited effort and provides long-term benefit, would be to incorporate techniques on mentoring and cross-cultural communications into all levels of officer PME. This should not be labeled another "equal opportunity" training event, but as a commander/officer development activity. In addition, the Air Force should expand the mentorship utility of its professional social networking communities of practice like Commander's Connection and Lieutenant's Bar and require participation as mentors by officers attending in-residence PME. Given their propensity to use social networking Web sites, officers of the current and future generations are most prone to employ these "universal mentoring" forums for communication, networking, and advice. Consequently, leaders' participation will keep them in contact with junior officers and further exploit cyber technology as a means to transcend normal "stovepipe" organizational boundaries.[64] This also enhances the likelihood that more senior officers will mentor others with whom they might not necessarily interact.

The tanning of America is perhaps the most important reason for the US Air Force's implementation of a comprehensive diversity effort. Blacks and the other two large minority groups, Asians and Hispanics, constitute smaller proportions of Air Force officers in comparison to respective civilian population groups.[65] The US Air Force should take immediate efforts to reverse these trends to ensure its senior leaders of tomorrow (today's junior officers) are neither culturally isolated nor uncomfortable with relating to and mentoring the more diverse Airmen of the future.

For those who believe that this issue can be dealt with in the future, it must be mentioned that this tanning of America has already started. In metropolitan areas throughout the nation, children under the age of five are already majority-minority.[66] Consequently, when these children start joining the armed forces en masse by 2023, more than half of the US population under age 18 is projected to be majority-minority.[67] If these trends continue unabated, this will occur when today's majority-white-male second lieutenants are majors and lieutenant colonels commanding tomorrow's squadrons of racially and ethnically diverse Airmen.

The single most lethal "weapon system" of the US military is its personnel. Therefore, it is imperative that each service continue to make every effort to attract and retain the most talented and capable individuals embodying the nation's more diverse citizenry. A less representative officer corps runs the risk of the armed services losing its edge on recruiting the nation's best and brightest because many of these young Americans may choose other, more diversely representative military branches or professions. Accordingly, as its senior officers become culturally isolated from the youth of America's more pluralistic society, they may fail to address the disparities in career progression between racial and ethnic groups. Perhaps a more ominous outcome of a future unrepresentative officer corps is the potential for intercultural conflict within the service, as was the case in the late 1960s and early 1970s. Such forecasts of crisis are increasingly likely if the Air Force does not act to improve diversity now, starting with increased formal and informal officer mentorship programs.

Notes

1. "A McKinsey of Pop Culture? Steve Stoute Is Making Hot Sellers Out of Cold Brands by Turning Execs on to 'the Tanning of America,'" *BusinessWeek*, 26 March 2007, http://www .businessweek.com/magazine/content/07_13/b4027062.htm?chan=top+news_top+news+index _companies (accessed 9 February 2009). The phrase "tanning of America" is drawn from Steve Stoute's marketing plan to make brands attractive to a growing multicultural US society.

2. Defense Equal Opportunity Management Institute (DEOMI), *Annual Demographic Profile of the Department of Defense and U.S. Coast Guard FY 2007*, September 2007, 4.

3. Henry E. Dabbs, *Black Brass: Black Generals and Admirals in the Armed Forces of the United States* (Freehold, NJ: Afro-American Heritage House Publishers, 1984), 35.

4. L. D. Reddick, "The Negro Policy of the United States Army, 1775–1945," *Journal of Negro History*, January 1949, 22.

5. Reddick, "Negro Policy," 22; and Bernard C. Nalty, *Strength for the Fight: A History of Black Americans in the Military* (New York: The Free Press, 1986), 136.

6. Alan M. Osur, "Black-White Relations in the U.S. Military, 1940-1972," *Air University Review* 33, no. 1 (November-December 1981), http://www.airpower.maxwell.af.mil/airchronicles/aureview/1981/nov-dec/osur.htm; and Department of Defense (DOD), *Black Americans in Defense of Our Nation* (Washington, DC: Office of the Deputy Assistant Secretary of Defense for Civilian Personnel Policy/Equal Opportunity, 1991), 119.

7. Vance O. Mitchell, *Air Force Officers Personnel Policy Development, 1944-1974* (Washington, DC: Air Force History and Museums Program, 1996), 326; and DOD, *Black Americans in Defense*, 71 and 92.

8. William Bowman, Roger Little, and G. Thomas Sicilia, *The All Volunteer Force after a Decade* (McLean, VA: Pergamon-Brassey's International Defense Publishers, 1986), 75.

9. Ibid., 328.

10. Osur, "Black-White Relations."

11. Ibid.; and Charles C. Moskos, Jr., "Racial Integration in the Armed Forces," *The American Journal of Sociology*, September 1966, 134–35.

12. Moskos, "Racial Integration," 147–48.

13. Bowman, Little, and Sicilia, *The All Volunteer Force*, 75.

14. Nalty, *Strength for the Fight*, 282–84.

15. Alan M. Osur and Charles Moskos, Jr., *Public Opinion and the Military Establishment* (Beverly Hills, CA: Sage Publications, 1971), 149–79.

16. Nalty, *Strength for the Fight*, 290–302.

17. Ronald H. Spector, *After Tet: The Bloodiest Year in Vietnam* (New York: Vintage Books, 1993), 247–49.

18. "Disorders Erupt at Coast Base," *Facts on File*, Thursday, 20 May–Wednesday, 26 May 1971, 388.

19. Osur, "Black-White Relations."

20. DOD, *Black Americans in Defense*, 187–90.

21. Mitchell, *Air Force Officers*, 356 and 362.

22. Air Force Personnel Center (AFPC), "Active Duty Officer Promotions Line of the Air Force (LAF) Historical," Air Force Personnel Center Statistics, http://wwa.afpc.randoph.af.mil/demographics/ReportsSearch.asp (accessed 24 January 2009).

23. LT Jon E. Lux, "The Effects of the Military Drawdown on Recruiting Minority Officers" (master's thesis, Naval Postgraduate School, Monterey, CA, 1995), 97–103.

24. DOD, *Career Progression of Minority and Women Officers* (Washington, DC: Office of the Under Secretary of Defense for Personnel and Readiness, 2002), viii.

25. Maj James G. Powell, "Determining Factors Which Influence Black Attrition Rates in Undergraduate Pilot Training," research paper (Maxwell AFB, AL: Air Command and Staff College, 1984), vii, 41, 49–50.

26. Mitchell, *Air Force Officers*, 336.

27. President's Committee on Equal Opportunity in the Armed Forces, *Equality of Treatment and Opportunity for Negro Military Personnel Stationed within the United States* (initial report) (Washington, DC: US Government Printing Office, June 1963), 11.

28. James E. Westheider, *The African-American Experience in Vietnam: Brothers in Arms* (Lanham, MD: Rowman & Littlefield Publishers, Inc., 2008), 111.

29. Westheider, 111.

30. DOD, *Population Representation in Fiscal Year 2000*, table D-22.

31. General Accountability Office (GAO), *Air Force Academy: Gender and Racial Disparities*, Publication No. GAO/NSAID-93-244 (Washington, DC: US Government Printing Office, September 1993), 2.

32. Data from 1947, 1964, and 1968 provided independently because information on racial composition of USAF officer personnel prior to 1970 could be ascertained from available resources from these earlier dates. DEOMI, *Semi-Annual Race/Ethnic/Gender Profile by Service/Rank of the Department of Defense & U.S. Coast Guard, 2000* (Patrick AFB, FL: Defense Equal Opportunity Management Institute Research Directorate, September 2000); DEOMI, *Semi-Annual Race/ Ethnic/Gender Profile by Service/Rank of the Department of Defense & U.S. Coast Guard, 2001* (September 2001); DEOMI, *Semi-Annual Demographic Profile of the Department of Defense and U.S. Coast Guard, 2002* (September 2002), 8; DEOMI, *Semi-Annual Demographic Profile of the Department of Defense and U.S. Coast Guard, 2003* (September 2003), 8; DEOMI, *Semi-Annual Demographic Profile of the Department of Defense and U.S. Coast Guard, 2004* (September 2004), 8; DEOMI, *Annual Demographic Profile of the Department of Defense and U.S. Coast Guard FY 2005* (September 2005), 8; DEOMI, *Annual Demographic Profile of the Department of Defense and U.S. Coast Guard FY 2006* (September 2006), 8; DEOMI, *Annual Demographic Profile of the Department of Defense and U.S. Coast Guard FY 2007* (September 2007), 8; DOD, *Population Representation in the Military Services, Fiscal Year 2000* (Washington, DC: Office of the Under Secretary of Defense for Personnel and Readiness, February 2002), table D-2, http://www.defenselink.mil/prhome/poprep2000/html/chapter8/chapter8.htm (accessed 18 January 2009); and Bowman, Little, and Sicilia, *The All Volunteer Force*, 75.

33. Mitchell, *Air Force Officers*, 336.

34. DOD, *Black Americans in Defense*, 267.

35. Donald Giglio, *AF Black Officer Accession Numbers 1971-1994: AFROTC, OTC, & USAFA* (Maxwell AFB, AL: Air University Holm Center, January 2009).

36. Ibid.

37. Ibid.

38. GAO, *Air Force Academy*, 2–3.

39. DOD, *Career Progression*, viii.

40. LT Roy L. Nixon, "Defense Downsizing and Blacks in the Military" (master's thesis, Naval Postgraduate School, Monterey, CA, 1993), 27.

41. DOD, *Population Representation in Fiscal Year 2000*, table D-22.

42. DOD, *Career Progression*, 42.

43. DOD, *Population Representation in Fiscal Year 2000*, table D-22.

44. DOD, *Career Progression*, 25.

45. Ibid., 50–54.

46. Susan D. Hosek, Peter Tiemeyer, M. Rebecca Kilburn, Debra A. Strong, Selika Ducksworth, and Reginald Ray, *Minority and Gender Differences in Officer Career Progression*, RAND Monograph M–1184 (Santa Monica, CA: RAND, 2001), xv.

47. AFPC, "Active Duty Officer Promotions Line of the Air Force (LAF) Historical."

48. Ibid.

49. US Census Bureau, "Education Attainment in the United States: 2007," http://www.census .gov/population/www/socdemo/education/cps2007.html (accessed 1 February 2009).

50. DEOMI, *Annual Demographic Profile, FY 2007*, 28.

51. DOD, *Career Progression*, 84.

52. Hosek et al., *Minority and Gender Differences*, 71–73.

53. DOD, *Career Progression*, 85.

54. Lt Col Emerson A. Bascomb, "Mentoring Minorities and Women in the United States Military" (Maxwell AFB, AL: Air War College, 1998), 4.

55. DOD, *Career Progression*, 82–85.

56. Ibid.

57. Lolita Baldor, "After 60 Years, Black Officers Rare," *USA Today*, 23 June 2008, http://www .usatoday.com/news/topstories/2008-07-23-4210460986_x.htm (accessed 10 January 2009).

58. Lt Col Anthony D. Reyes, *Strategic Options for Managing Diversity in the U.S. Army* (Washington, DC: Joint Center for Political and Economic Studies, June 2006), xi–xii.

59. DEOMI, *Annual Demographic Profile, FY 2007*, 2.

60. DOD, *Population Representation in Fiscal Year 2000*, table D-27.

61. Lt Col Remo Butler, "Why Black Officers Fail in the US Army," *Parameters* 29, no. 3 (Autumn 1999): 23–25.

62. AFPC, "AFPC Establishes Diversity Council," *Air Force Print News Today*, http://www .afpc.randolph.af.mil/news/story_print.asp?id=123139871 (accessed 22 March 2009).

63. AFPC, "Regular Officer History FY94–FY02," Air Force Personnel Center Statistics, http://wwa.afpc.randoph.af.mil/demographics/ReportsSearch.asp (accessed 24 January 2009); and AFPC, "Regular Officer History FY03–FY07," Air Force Personnel Center Statistics, http:// wwa.afpc.randoph.af.mil/demographics/ReportsSearch.asp (accessed 24 January 2009).

64. Maj Darrell E. Adams, "Mentoring Women and Minority Officers in the US Military," research paper (Maxwell AFB, AL: Air Command and Staff College, 1997), 32.

65. DOD, *Population Representation in Fiscal Year 2005*, 2.

66. N. C. Aizenman, "In the Under-5 Set, Minority Becoming the Majority," *Washington Post*, 7 August 2008, http://www.washingtonpost.com/wp-dyn/content/article/2008/08/06/AR200808 0603683_pf.html (accessed 4 January 2009).

67. N. C. Aizenman, "U.S. to Grow Grayer, More Diverse: Minorities Will Be Majority by 2042," *Washington Post*, 14 August 2008, http://www.washingtonpost.com/wp-dyn/content/article/ 2008/08/13/AR2008081303524.html (accessed 4 January 2009).

About the Author

Maj AaBram G. Marsh is currently the commander of the 49th Materiel Maintenance Support Squadron at Holloman Air Force Base in New Mexico. He was previously a student at Air Command and Staff College and the operations officer for the 8th Logistics Readiness Squadron, Kunsan Air Base, Republic of Korea. Marsh holds both a bachelor's and master's degree in history from the University of Florida and a master's degree in military operational art and science from Air Command and Staff College.

CHAPTER 21

REMARKS AT THE DEPARTMENT OF JUSTICE AFRICAN-AMERICAN HISTORY MONTH PROGRAM

Eric Holder

Washington, DC, 18 February 2009

Every year, in February, we attempt to recognize and to appreciate black history. It is a worthwhile endeavor for the contributions of African-Americans to this great nation are numerous and significant. Even as we fight a war against terrorism, deal with the reality of electing an African-American as our President for the first time and deal with the other significant issues of the day, the need to confront our racial past, and our racial present, and to understand the history of African people in this country, endures. One cannot truly understand America without understanding the historical experience of black people in this nation. Simply put, to get to the heart of this country one must examine its racial soul.

Though this nation has proudly thought of itself as an ethnic melting pot, in things racial we have always been and continue to be, in too many ways, essentially a nation of cowards. Though race related issues continue to occupy a significant portion of our political discussion, and though there remain many unresolved racial issues in this nation, we, average Americans, simply do not talk enough with each other about race. It is an issue we have never been at ease with and given our nation's history this is in some ways understandable. And yet, if we are to make progress in this area we must feel comfortable enough with one another, and tolerant enough of each other, to have frank conversations about the racial matters that continue to divide us. But we must do more—and we in this room bear a special responsibility. Through its work and through its example this Department of Justice, as long as I am here, must—and will—lead the nation to the "new birth of freedom" so long ago promised by our greatest President. This is our duty and our solemn obligation.

We commemorated five years ago, the 50th anniversary of the landmark *Brown v. Board of Education* decision. And though the world in which we now live is fundamentally different than that which existed then, this nation has still

---

These remarks are reproduced from the US Department of Justice, http://www.justice.gov/ag/speeches/2009/ag-speech-090218.html.

not come to grips with its racial past nor has it been willing to contemplate, in a truly meaningful way, the diverse future it is fated to have. To our detriment, this is typical of the way in which this nation deals with issues of race. And so I would suggest that we use February of every year to not only commemorate black history but also to foster a period of dialogue among the races. This is admittedly an artificial device to generate discussion that should come more naturally, but our history is such that we must find ways to force ourselves to confront that which we have become expert at avoiding.

As a nation we have done a pretty good job in melding the races in the workplace. We work with one another, lunch together and, when the event is at the workplace during work hours or shortly thereafter, we socialize with one another fairly well, irrespective of race. And yet even this interaction operates within certain limitations. We know, by "American instinct" and by learned behavior, that certain subjects are off limits and that to explore them risks, at best embarrassment, and, at worst, the questioning of one's character. And outside the workplace the situation is even more bleak in that there is almost no significant interaction between us. On Saturdays and Sundays America in the year 2009 does not, in some ways, differ significantly from the country that existed some fifty years ago. This is truly sad. Given all that we as a nation went through during the civil rights struggle it is hard for me to accept that the result of those efforts was to create an America that is more prosperous, more positively race conscious and yet is voluntarily socially segregated.

As a nation we should use Black History month as a means to deal with this continuing problem. By creating what will admittedly be, at first, artificial opportunities to engage one another we can hasten the day when the dream of individual, character based, acceptance can actually be realized. To respect one another we must have a basic understanding of one another. And so we should use events such as this to not only learn more about the facts of black history but also to learn more about each other. This will be, at first, a process that is both awkward and painful but the rewards are potentially great. The alternative is to allow to continue the polite, restrained mixing that now passes as meaningful interaction but that accomplishes little. Imagine if you will situations where people—regardless of their skin color—could confront racial issues freely and without fear. The potential of this country, that is becoming increasingly diverse, would be greatly enhanced. I fear however, that we are taking steps that, rather than advancing us as a nation are actually dividing us even further. We still speak too much of "them" and not "us." There can, for instance, be very legitimate debate about the question of affirmative action. This debate can, and should, be nuanced, principled and spirited. But the conversation that we now engage in as a nation on this and other racial subjects is too often simplistic and left to those on the extremes who are not hesitant to use these issues to advance nothing more than their own, narrow self interest. Our history has demonstrated that the vast majority of Americans are uncomfortable with, and would like to not have to deal with, racial matters and that is why those, black or white, elected or self-appointed, who promise relief in easy, quick solutions, no matter how divisive, are embraced. We are then free to retreat to our race protected cocoons where much is comfortable and where progress is not really made. If we allow this attitude to persist in the face of the most significant demographic changes that this nation has ever confronted—and remember, there will be no majority race in America in about fifty years—the coming diversity that could be such a powerful, positive force will, instead, become a reason for stagnation and polarization. We cannot allow this to happen and one way to prevent such an unwelcome outcome is to engage one another more routinely—and to do so now.

As I indicated before, the artificial device that is Black History month is a perfect vehicle for the beginnings of such a dialogue. And so I urge all of you to use the opportunity of this month to talk with your friends and co-workers on the other side of the divide about racial matters. In this way we can hasten the day when we truly become one America.

It is also clear that if we are to better understand one another the study of black history is essential because the history of black America and the history of this nation are inextricably tied to each other. It is for this reason that the study of black history is important to everyone—black or white. For example, the history of the United States in the nineteenth century revolves around a resolution of the question of how America was going to deal with its black inhabitants. The great debates of that era and the war that was ultimately fought are all centered around the issue of, initially, slavery and then the reconstruction of the vanquished region. A dominant domestic issue throughout the twentieth century was, again, America's treatment of its black citizens. The civil rights movement of the 1950's and 1960's changed America in truly fundamental ways. Americans of all colors were forced to examine basic beliefs and long held views. Even so, most people, who are not conversant with history, still do not really comprehend the way in which that movement transformed America. In racial terms the country that existed before the civil rights struggle is almost unrecognizable to us today. Separate public facilities, separate entrances, poll taxes, legal discrimination, forced labor, in essence an American apartheid, all were part of an America that the movement destroyed. To attend her state's taxpayer supported college in 1963 my late sister in law had to be escorted to class by United States Marshals and past the state's governor, George Wallace. That frightening reality seems almost unthinkable to us now. The civil rights movement made America, if not perfect, better.

In addition, the other major social movements of the latter half of the twentieth century—feminism, the nation's treatment of other minority groups, even the anti-war effort—were all tied in some way to the spirit that was set free by the quest for African-American equality. Those other movements may have occurred in the absence of the civil rights struggle but the fight for black equality came first and helped to shape the way in which other groups of people came to think of themselves and to raise their desire for equal treatment. Further, many of the tactics that were used by these other groups were developed in the civil rights movement.

And today the link between the black experience and this country is still evident. While the problems that continue to afflict the black community may be more severe, they are an indication of where the rest of the nation may be if corrective measures are not taken. Our inner cities are still too conversant with crime but the level of fear generated by that crime, now found in once quiet, and now electronically padlocked suburbs is alarming and further demonstrates that our past, present and future are linked. It is not safe for this nation to assume that the unaddressed social problems in the poorest parts of our country can be isolated and will not ultimately affect the larger society.

Black history is extremely important because it is American history. Given this, it is in some ways sad that there is a need for a black history month. Though we are all enlarged by our study and knowledge of the roles played by blacks in American history, and though there is a crying need for all of us to know and acknowledge the contributions of black America, a black history month is a testament to the problem that has afflicted blacks throughout our stay in this country. Black history is given a separate, and clearly not equal, treatment by our society in general and by our educational institutions in particular. As a former American history major I am struck by the fact that such a major part of our national story has been divorced from the whole. In law, culture, science, athletics, industry and other fields, knowledge of the roles played by blacks is critical to an understanding of the American experiment. For too long we have been too willing to segregate the study of black history. There is clearly a need at present for a device that focuses the attention of the country on the study of the history of its black citizens. But we must endeavor to integrate black history into our culture and into our curriculums in ways in which it has never occurred before so that the study of black history, and a recognition of the contributions of black Americans, become commonplace. Until that time, Black History Month must remain an important, vital concept. But we have to recognize that until black history is included in the standard curriculum in our schools and becomes a regular part of all our lives, it will be viewed as a novelty, relatively unimportant and not as weighty as so called "real" American history.

I, like many in my generation, have been fortunate in my life and have had a great number of wonderful opportunities. Some may consider me to be a part of black history. But we do a great disservice to the concept of black history recognition if we fail to understand that any success that I have had, cannot be viewed in isolation. I stood, and stand, on the shoulders of many other black Americans. Admittedly, the identities of some of these people, through the passage of time, have become lost to us—the men, and women, who labored long in fields, who were later legally and systemically discriminated against, who were lynched by the hundreds in the century just past and those others who have been too long denied the fruits of our great American culture. The names of too many of these people, these heroes and heroines, are lost to us. But the names of others of these people should strike a resonant chord in the historical ear of all in our nation: Frederick Douglass, W.E.B. DuBois, Walter

White, Langston Hughes, Marcus Garvey, Martin Luther King, Malcolm X, Joe Louis, Jackie Robinson, Charles Drew, Paul Robeson, Ralph Ellison, James Baldwin, Toni Morrison, Vivian Malone, Rosa Parks, Marion Anderson, Emmit Till. These are just some of the people who should be generally recognized and are just some of the people to whom all of us, black and white, owe such a debt of gratitude. It is on their broad shoulders that I stand as I hope that others will some day stand on my more narrow ones.

Black history is a subject worthy of study by all our nation's people. Blacks have played a unique, productive role in the development of America. Perhaps the greatest strength of the United States is the diversity of its people and to truly understand this country one must have knowledge of its constituent parts. But an unstudied, not discussed and ultimately misunderstood diversity can become a divisive force. An appreciation of the unique black past, acquired through the study of black history, will help lead to understanding and true compassion in the present, where it is still so sorely needed, and to a future where all of our people are truly valued.

About the Author

Eric H. Holder is the 82nd attorney general of the United States. He received his undergraduate and law degrees from Columbia University. Prior to assuming his current position, he served in a variety of federal offices. He was appointed by Pres. Ronald Reagan in 1988 as a superior court judge in the District of Columbia and in 1993 was appointed by President Clinton as the US attorney for the District of Columbia. He was later appointed as the deputy attorney general in 1997 and then acting attorney general briefly under President Bush in 2001. Attorney General Holder was the first African-American to hold each of these federal offices.

ATTITUDES AREN'T FREE

THINKING DEEPLY ABOUT DIVERSITY
IN THE US ARMED FORCES

SOCIAL POLICY

PERSPECTIVES 2010

You cannot escape the responsibility of tomorrow by evading it today.
—Abraham Lincoln

Fortunately for the military, its ability to close the door on the torture chamber is simple—it must cease torturing any human being that ever winds up in its custody.
—Matthew Harwood

If DOD chooses to use nonmilitary personnel to conduct activities outside the scope of traditional supply and logistics, then it must accept that these individuals are no longer entitled to civilian status.
—Maj Lynn Sylmar, JD, USAF

Healthy civil–military relations necessitates that the Executive perceive that all military advice is borne out of high levels of expertise and not ideological beliefs.
—Dr. Rachel Sondheimer et al.

I liken the existing honor condition to a "dishonorable pit" where the cadets try to avoid falling into the pit but don't mind getting as close to the edge of the pit as possible with their sometimes "questionable" behavior.
—Brig Gen Ruben Cubero, USAF, retired

We need all our soldiers and leaders to approach mental health like we do physical health. No one would ever question or ever even hesitate in seeking a physician to take care of their broken limb or gunshot wound, or shrapnel or something of that order. You know, we need to take the same approach towards mental health.
—BG Gary S. Patton, US Army

SOCIAL POLICY PERSPECTIVES 2010

Any discussion about policy reform in 2010 should take into account the profoundly difficult conditions of the US economy at present. At the end of the 1990s, the economy grew at a blistering pace in the wake of emerging markets driven by globalization, the expansion of the Internet, and the fall of Communism. As the only remaining superpower, the United States took center stage as the dominant player in the global free market. Then came Y2K. Many feared this computer glitch would wreak havoc on the world. It seems silly looking back. Yet despite the "all flash, no boom" of Y2K, the United States was unknowingly about to enter one of its most challenging eras.

First came the tragedy of September 11, 2001, soon followed by the famed "Internet bubble," which burst in March 2002, sending the American stock market into a tailspin. Soaring oil demand between 2005 and 2008 caused the price of gasoline to double. For the first time in decades, the threat of inflation became real. However, the threat vaporized in 2007 with the onset of the global economic downturn. Faced with the subprime mortgage crisis, investment bank failures, falling home prices, and tight credit markets, the United States found itself in a recession by mid-2008. To help stabilize financial markets, the US Congress established an emergency $700 billion Troubled Asset Relief Program (TARP) in October 2008. The government later used these funds to purchase equity stakes in some of the largest American banks and automotive manufacturers. In January 2009 the US Congress passed and Pres. Barack Obama signed a bill providing a near-trillion-dollar fiscal stimulus. And in January 2010, the country is embarking on one of the most ambitious social programs in US history—national healthcare reform. Economically speaking, the country looks very different than it did a decade earlier, and the future challenges we will face appear arduous.

From a military perspective, the challenges have been no less profound. The Department of Defense (DOD) took center stage in 2002 as it began a full-scale prosecution of the Global War on Terror, pursuing campaigns in Iraq and Afghanistan. Although Saddam Hussein was overthrown and executed, rebuilding the nation of Iraq into a self-governing country turned out to be far more challenging than former leaders and policymakers believed. Al-Qaeda's leader, Osama bin Laden, remains at large, while US forces continue combat operations against the Taliban in Afghanistan. At the periphery, the governments of Iran and North Korea remain belligerent to US interests in the region as they both continue to develop nuclear programs despite opposition by the world community. It isn't bad enough for military leaders who face the prospect of a failed war in an era of increasingly constrained resources with the threat of nuclear conflict, but they must also account for the domestic allegations of abuse and torture. The mention of Guantanamo Bay or Abu Ghraib creates spirited debate about what constitutes appropriate treatment of those captured by US forces and what rights they should be given. In a nutshell, military leaders of 2010 face a multitude of unparalleled challenges.

In the final section of this volume, the authors address a wide array of unanswered questions regarding what the "right" policies should be to deal with the issues that profoundly affect the human beings for which the DOD finds itself responsible.

The section begins with an introspective look into ideological perceptions reported by West Point cadets. A team of experts led by **Dr. Rachel Sondheimer** from the Social Sciences Department at the United States Military Academy finds that cadets believe military and civilian populations occupy drastically different ideological spaces. The authors argue that this could lead to problems for the military in meeting the intent of DOD directives to avoid inferences of partisan political approval and endorsements. They provide a staunch warning about the moral hazard in the gap between perceived ideological leanings in the civilian sphere and those in the military sphere and how that gap may affect the way military advice is perceived in the policy arena.

One of the most noteworthy aspects of twenty-first century warfare is the increased presence of warrior-civilians on the battlefield. **Maj Lynn Sylmar**, a USAF staff judge advocate, develops an original taxonomy of an emerging phenomenon to illustrate how past policies of using government contractors, DOD civilians, and other civilian organizations are jeopardizing conventional notions of a combatant as defined under international law. She evaluates the increased risks personnel may face if new policies and legislation aren't mandated soon.

Dr. Carla Sizer, a civilian DOD contractor and retired Air Force officer, and **Dr. Claude Toland**, a scholar from DeVry University, capture the latest understanding of the phenomenon that has become known as post traumatic stress disorder (PTSD). In addition to articulating a historical perspective of PTSD by other names, they set forth specific recommendations for senior leaders and policy makers to consider as the debate continues over how best to deal with the fastest growing combat-related issue facing American war veterans.

In "Enjoining an American Nightmare," **Matt Harwood** from *Security Management* magazine presents a historical perspective of America's long-standing opposition to torture. Reflecting on the recent debates about questionable interrogation methods employed in Iraq, he argues that anyone who tortures another human being should be prosecuted to the fullest extent of the law.

In the first of two chapters dealing with ethics, **Dr. Tom Gibbons** from the Naval War College presents a study which demonstrates the importance of a college honor-code experience to reinforce military core values. He argues for further reinforcement of honor codes in military schools and service academies and ongoing ethics training within the curricula of war colleges and senior enlisted schools.

Having spent a majority of his career studying honor codes and ethics at the United States Air Force Academy (USAFA), **Dr. Chuck Yoos**, a retired Air Force colonel and USAFA professor emeritus, develops a framework he believes necessary to bring "honor" back into the honor system at the US Air Force Academy. After providing the context in which the current honor code and system have emerged, Dr. Yoos provides a step-by-step approach that he contends is essential to sufficiently reform the honor system to achieve the fundamental objectives it had been created to achieve decades earlier. Former USAFA dean **Brig Gen Ruben Cubero, retired**, offers a foreword.

The final two chapters address economic issues. In the first, **Drs. Bill Gates and Peter Coughlan** from the Naval Postgraduate School review the economics literature regarding one of the latest and most innovative methodologies used to efficiently solve complex problems among a "market" of individuals. They consider the use of auctions as an alternative to the recent force-shaping boards to reduce officer end-strength, and they make several recommendations for senior leaders to consider before adopting such a concept wholesale. They argue that, despite the allure of auctions, the inherent nature of the military structure makes their use questionable for large-scale personnel issues.

Finally, **Dr. Steve Fraser**, a retired Air Force officer and current finance professor at Florida Gulf Coast University, closes out the discussion of social policy perspectives for both the section and volume by discussing one of the most closely guarded benefits of career service members: the military retirement system. Dr. Fraser provides a review of changes to the DOD defined-benefit retirement systems and raises questions about the validity of "cliff-vesting." He suggests that future service members likely will not view military service in the same way as previous generations and suggests the time has come for senior leaders and policy makers to try something new.

CHAPTER 22

IDEOLOGICAL PERCEPTIONS AND CIVIL-MILITARY RELATIONS

Rachel Milstein Sondheimer
Isaiah Wilson III *Thomas Greco*
Kevin Toner *Cameron West*

Introduction

During the height of the 2008 presidential primary season, the Department of Defense (DOD) issued Directive 1344.10 updating its policy on political activities in relation to members of the armed forces. Active duty members of the armed forces may, among other things, register and vote, express partisan political opinions, join partisan organizations, contribute time and money to political campaigns, and display a bumper sticker on a private automobile as long as these activities are conducted as a private citizen, not as a representative of the armed forces. Active duty members may not, however, participate in partisan fundraising, speak at a partisan political gathering, march or ride in a partisan political parade, or display a large political sign on a private automobile. As Section 4 of the directive indicates, the DOD charts a fine line in encouraging its members to "carry out the obligations of citizenship" while simultaneously "keeping with the traditional concept that members on active duty should not engage in partisan political activity, and that members not on active duty should avoid inferences that their political activities imply or appear to imply official sponsorship, approval, or endorsement."[1]

Civilian control of the military is rooted in America's traditional distrust of standing armies and is lawfully provided for in the US Constitution. By tradition and, to a large degree, by statutory regulations, the military identifies itself as an "apolitical" body. In many respects, this is readily apparent as representatives of the armed forces and senior military leaders play a strictly advisory role to their appointed civilian counterparts. However, by nature and intent, the military cannot be apolitical; the military, whether it be in its expert role of advising civilians on defense issues or interacting with other departments concerning policy issues and appropriations, is an active player in the political process. After all, Clausewitz defines war as a continuation of politics by other means.[2] Rather than deny the political nature of one of the key players in the executive branch, the military embraces this role but must work to protect it by carefully considering the deleterious consequences of appearing partisan or of a particular ideological bent. Public perception of the nature of military involvement in the political arena is integral to healthy civil-military relations.

As DOD Directive 1344.10 indicates, there are prohibitions on certain behaviors in the political arena, but there is no clear line between what can be done and what should be done by members of the military in achieving a healthy civil-military relationship. Nor is there much, if any guidance, provided to members of the armed forces on the intent and spirit of these guidelines. The question we consider here is whether this directive and its predecessors are sufficient to "avoid inferences . . . to imply official sponsorship, approval, or endorsement"[3] of political and/or partisan political activities.

Given the increased prominence of the active military in this time of war, there is surprisingly little recent research on the role of perceptions of the ideological and partisan beliefs of the military within a republic generally and within the United States specifically. In this work, we seek to fill this gap by discussing perceptions of ideology and the consequences of these perceptions on civil-military relations. We begin by presenting an overview of the evolving nature of the military's involvement in political affairs. We use the principal-agent framework to argue that ideological perceptions of the military matter. We then present data on military perceptions of its own ideological bent based on a survey of cadets at the United States Military Academy. We conclude with a discussion of the ramifications of the results including consideration of why Directive 1344.10 may prove insufficient, on its own, in avoiding the appearance of implicit partisan political endorsements.

It is important to note that the intent of this work is quite conservative in nature. We make limited recommendations on specific policies and particular courses of action. Our modest goal is to inform policy makers and scholars of the potential ramifications of the perceptions of military ideology on the military's ability to function as an admittedly political yet nonpartisan and nonideological expert on defense issues so that they may incorporate this knowledge into training and fostering a proper command climate in the armed forces.

Ideology and Civil-Military Relations

Politics and the Military

While the Constitution establishes civilian supremacy over the military, debates have long endured over the role of the military in political matters. "Political matters" as a descriptive term has taken on many meanings and conceptions over the years. In the post–Civil War era, leading military reformers like Gens William T. Sherman, Emory Upton, and Rear Adm Stephen B. Luce adamantly opposed military intervention in any endeavor not amounting to their "real endeavor—war."[4] In other words, the military should collectively avoid tinkering in, or worse being dragged into, the business of government decision making. On the individual level, these men, Sherman in particular, believed that members of the military (officers in particular) should not "form or express an opinion on party politics."[5] Much of this thinking was not attributed so much to maintaining civilian control of the military, but to professionalizing a fighting force in peacetime to prepare for future conflict. The greatest obstacle to realizing such a goal was the political establishment generally and politicians specifically.[6]

Within this great push to form a uniquely "apolitical" body, pioneers of modern Army professionalism fostered a culture that sought to avoid politics altogether. Huntington explains:

> In sharp contrast to the opinions of the officer corps in the 1830s, after the Civil War officers unanimously believed that politics and officership do not mix. Not one officer in a hundred, it was estimated, ever cast a ballot. In part this was a result of shifting stations and state restrictions. But to a much larger extent the abstention of the officer corps, stemmed, in the words of an Army major, "from settled convictions, from an instinctive sense of its peculiar relation as an organization to the Republic." . . . The concept of an impartial, nonpartisan, objective career service, loyally serving whatever administration or party was in power, became the ideal for the military profession.[7]

A convenient byproduct of this was a "virtually nil" contribution by civilians to this movement, which would prove highly effective in maintaining professional autonomy in matters of a purely military nature. Indeed, by self-design, the military kept its distance from the policy process, entering only when absolutely necessary. The work of these men, especially their views on politics, became institutionally ingrained in the minds of future military leaders.

In the years following the Second World War, the military's role in politics became increasingly more frequent due to a variety of factors. On the policy-making level, the complex nature of international affairs, technology, and military strategy necessarily thrust the military into the policy process—an inherently political process. This is not to imply that the military's influence on policy was preeminent. In fact many members of the military still remained ambivalent about getting involved in politics, and numerous government reforms aimed at the national military establishment arguably enhanced civilian control of the military.[8] Largely in reaction to this gradual shift into the policy-making process, the study of professionalism took center stage in civil-military relations theory, where scholars like Samuel Huntington and Morris Janowitz sparked a dialogue about the proper role of the military in society that continues to this day.

Huntington saw the military as almost completely distinct from society at large. Through his discussions of objective and subjective control, he sought to clearly delineate the boundaries of each sphere (civilian and military). Under the objective control framework, matters of policy would remain squarely in the hands of civilian policy makers while the military was free (devoid of civilian meddling) to carry out such policy in a way it deemed proper. Under this construct, the military would be given a free hand in executing stated policy as there would be no overlap of responsibility between the two spheres. Subjective control, the less desirable construct, asserted a degree of overlap between the civilian and military spheres of responsibility. Here, the interests of the policy makers and the military would be conjoined to a greater or lesser degree during both the planning and execution stages of military policy. For example, the military would provide, and the civilians would accept, military advice in making policy. The drawback to this construct in the eyes of Huntington, however, was that the relationship would be reciprocal—the civilians would have the opportunity to delve into purely military matters, that is, the execution of military policy.[9]

Morris Janowitz provided a decidedly different point of view on military professionalism. Although he believed the military member should avoid unnecessary influence in the political *process*, he cannot be considered, by tradition or design, completely detached from politics—nor should he be:

> The professional soldier is "above politics." Under democratic theory, the "above politics" formula requires that, in domestic politics, generals and admirals do not attach themselves to political parties or overtly display partisanship. Furthermore, military men are civil servants so that elected leaders are assured of the military's partisan neutrality. . . . But partisan neutrality does not mean being "above politics" to the point of being unpolitical.[10]

Janowitz's views on the political nature of military officers are well-supported today. Senior ranking members of the military are indeed very much involved in the policy-making process because of their many responsibilities outlined under Title X of the US Code. Of course, the extent to which officers should influence policy outcomes is still very much debated.

It would seem that the primary role of the military in the policy-making process is "to do no harm" to American democracy.[11] We do not seek to wade into the debate over what constitutes the scope of possible harms, instead focusing on a single instance: the military providing policy advocacy rather than policy advice to its civilian counterparts. In a lecture to cadets at the United States Military Academy, Don Snider argued that the military's role in the policy-making process is to use its *expert knowledge* to *advise* the civilian leadership and then to *execute* the civilian authority's decision.[12] During the advisory process, the military must refrain from *advocating* for personal or professional policy preferences. Advice is separate and distinct from advocacy, and the military is obligated to provide the former while abstaining from the latter. The consequences of appearing to advocate rather than to advise are best understood through the lens of the principal-agent model.

The Principal-Agent Framework

When thinking about the expert and advisory role of the military, we argue that perception of ideology matters. This is apparent when applying the principal-agent framework, as outlined by Feaver and Kohn,[13] to the realm of civil-military relations. Using this scheme, the military acts as the agent for multiple principals to include the executive, the Congress, and the American people.[14] This organizational framework provides services that are useful for the principal but inefficient to perform on one's own. As such, control over these services is delegated to the agent, which has the expertise to carry out the assignment. As such, while the executive acts as commander in chief, the president has neither the time nor the resources to tackle defense issues, instead delegating much of this role to the Department of Defense.

The problem inherent in this framework is that this delegation of power and development of expertise and ensuing presence of asymmetric information flow leave open the possibility that the agent will behave in ways contrary to the principal's desires—a concept known as the moral hazard. Opportunities for moral hazard develop when the principal has neither the expertise nor the physical resources to monitor the agent's actions. Most concerns over moral hazard involve the possibility of shirking in the implementation of a given policy. While this is a potential problem, we are interested in a moral hazard that occurs much earlier in the public policy process—policy formulation and adoption. This is where we see the role of the military as providing expert advice to civilian policy makers in the executive and legislative branches.

The cultivation of the military and high-level military leaders as experts charged with providing advice to their civilian counterparts and superiors necessitates situations with asymmetric information. The underlying assumption of this military role is that advice is borne out of the military's role as an expert in defense affairs. This assumption is undermined and the moral hazard develops if there is reason to believe that such advice, and perhaps advocacy, is the result, not of defense expertise, but rather of ideological and/or partisanship preferences. Overt partisanship and/or political ideology, defined as a consistent and coherent set of beliefs about what goals government ought to pursue, provide alternative explanations for policy advice. In considering advice provided by a member of the military, does a civilian policy maker believe it was given due to the counsel's military and defense expertise or because the policy maker believes that the counsel has an inherent view of the world that informs his or her opinion on such affairs? Moreover, does the civilian policy maker feel that this ideology is specific to the individual providing counsel, or is it due to an overarching perception of the larger organization of which the counsel is a member?

The perception of ideological differences matters just as much as real ideological differences. In a policy climate with incomplete information on the beliefs of actors in the system, perception often has more sway than reality. In other words, in the case of advising on policy matters, what the agent thinks often matters less than why the principal believes the agent thinks this. Healthy civil-military relations necessitate that the executive perceive that all military advice is borne out of high levels of expertise and not ideological beliefs.

One might assume that concerns over this moral hazard are most acute when the principal and agent are perceived as having different political ideologies. We argue that while this instance might allow for more salient examples of the perceived undermining of military expertise (e.g., tension between the Clinton administration and the Department of Defense concerning gays in the military), any situation of ideology hampering the ability to provide expert advice is troublesome to the military's role within the republic. Even if the military is thought to be of a similar ideological bent to its principal, players in the policy process may never know whether advice is based on expertise or ideological beliefs.

We argue simply that any such perception can lead to a lack of trust and that lack of trust can lead to suboptimal policy decisions and undermine the role of the military in the policy-making and implementing process. This is the intent behind the permissions and prohibitions outlined in DOD Directive 1344.10—to avoid inferences or implicit appearance of partisanship or endorsement of particular candidates, policies, and ideas by members of the armed forces and thus by the armed forces themselves. By trying to divorce itself from partisanship, the DOD is attempting to present itself and its members as lacking an inherent view of the role of government (i.e., an ideology) so that it can dispense expert military advice when called upon to do so. Here we seek to discern whether the current policy concerning political activities goes far enough to create the perception of a nonideological (and nonpartisan) military force.

Previous Research

The Triangle Institute for Securities Studies undertook an extensive investigation of the party identification and ideological leanings of members of the military in the late 1990s.[15] It found that 61.7 percent of the officers surveyed self-identified themselves as Republicans, 9.9 percent as Democrats, and 18 percent as Independents, whereas nonmilitary respondents answered 28.8 percent, 35.4 percent, and 30.8 percent respectively. In terms of political ideology, 64.5 percent of the officers consider themselves conservative, 27.5 percent as moderate, and 6.8 percent as liberal compared with 38.7 percent, 26.6 percent, and 27.5 percent of nonmilitary respondents, respectively. The data thus indicate what Feaver and Kohn and other scholars refer to as the "gap" in civil-military relations. In this case (a sample from the late 1990s), the officer population differs significantly from the civilian population on matters of party and ideology (except "moderate").

The Triangle Institute data, however, is dated and does not ask about perceptions. How does the military perceive itself as a body and the civilian population as a whole? We seek to begin to measure the *perceptions* of the ideological persuasions, if any, of military and civilian populations. Rather than begin by uncovering how civilians perceive military populations, we turn our gaze into the looking glass to see how a subsample of the military perceives itself and other subpopulations. We discuss our methodology and results in the following sections.

Data Collection and Methodology

The intent of our study is to ascertain the military's perceptions of its own ideological persuasion, of the ideological persuasions of the civilian population, and of the civilian population's assessment of the military's ideological persuasion. By examining how a subpopulation of the military views itself as a whole and the civilian population as a whole, we hope to discern whether or not members of the military perceive themselves as a distinct ideological group within society. Does the military appear to harbor a different view towards the role of government in comparison to the broader civilian population, implying that decisions made and advice offered by members of this group could be seen as the result of a particular ideology and not the result of defense expertise? Because we are interested in the role of perceptions in policy climate with asymmetrical information, we are not interested in whether or not the military actually has a different ideology but whether or not we think that it does.

Data was collected from a cohort of cadets at the United States Military Academy enrolled in American Politics during the spring semester of 2009. The course is required of all cadets and is generally taken during a student's yearling (sophomore) year although some plebes (freshmen) enroll in the course. Instructors taught a total of 508 students across 29 sections of the course. 470 students took part in the survey.[16]

The convenience of surveying the Corps of Cadets aside, this sample population provides us with insight into the beliefs and perceptions of the future officer corps. While West Point produces only about 20 percent of the junior officer corps, its graduates constitute a disproportionate number of more senior-level and general officers. These senior officers are the ones placed in key advisory roles to the civilian leadership.

Data collection was integrated as part of a lesson on political ideology and attempted to ascertain the perceptions of cadet and civilian political ideologies using a four-quadrant grid. At the beginning of the lesson, instructors displayed the four-quadrant grid (fig. 22-1) in their classrooms. The cadets were expected to have read a chapter on political ideology prior to the day's class

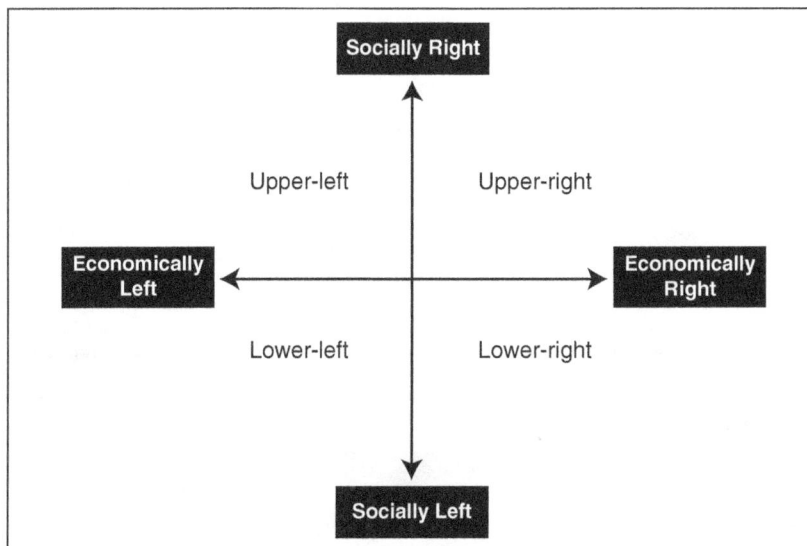

Figure 22-1. The four-quadrant ideological grid.

meeting, and instructors were told not to answer any questions seeking to explain the meaning of the diagram.

Students were first asked a series of questions regarding perceptions of their own political ideology, the political ideology of the Corps of Cadets and the military, and the political ideology of portions of the civilian population (see appendix for complete survey instrument). Specifically, cadets were asked to place each of these populations within one of the above quadrants. Students were also asked to reflect on how they believe the military is viewed by their peers in civilian colleges as well as by the civilian population at large.

Before delving into the results, it is important to briefly review our interpretation of the political ideology connoted by each quadrant. The upper-left quadrant connotes a populist ideology in popular parlance, the upper-right quadrant a conservative political ideology, the lower-left quadrant a liberal ideology, and finally, the lower-right quadrant a libertarian political ideology. Our expectation is that results scattered across all of these quadrants would indicate an inability to define or lack of an ideological perception of a given group while any sort of dominance of one quadrant over the others would indicate a defined perception of an ideological leaning.

Results

Table 22-1 provides an overview of the pertinent results. Overall, the results are quite stark, with cadets perceiving the Corps and the military as conservative and their civilian college peers and the civilian population at large as more liberal. Below, we walk through the results in more detail.

The trend toward perceived conservative political ideology among the Corps is evident with 69 percent of respondents describing the Corps of Cadets as conservative. Eleven percent described the Corps as falling in the populist quadrant, 5 percent as liberal, and 11 percent as libertarian. Approximately 4 percent chose the "other" category, with each of the 18 cadets who responded "other" describing the Corps of Cadets as having no dominant ideological

Table 22-1. Cadet placement of each group on the four-quadrant ideological grid

Quadrant	Cadet Placement of Corps of Cadets	Cadet Placement of Military	Cadet Placement of Civilian College Students	Cadet Placement of Civilian Population	Cadet Placement of Civilian Perception about Military
Upper Left	11%	19%	8%	21%	8%
Upper Right	69%	60%	4%	8%	78%
Lower Left	5%	6%	73%	37%	6%
Lower Right	11%	10%	12%	22%	7%
Other	4%	4%	2%	12%	1%

quadrant. A clear majority of cadets surveyed perceive themselves as an organization to be politically conservative.

When asked to place the civilian student population in a quadrant, cadets overwhelmingly responded in almost the exact opposite manner as they did when placing the Corps. In placing the ideological leanings of the civilian student population, 344 of the cadets, or 73 percent, responded by placing the civilian student population in the liberal quadrant of the diagram. As evident in figure 22-2, this is close to the mirror opposite of the 69 percent of respondents who placed the Corps as conservative. Likewise, only 4 percent of cadets placed civilian college students as conservative, similar to the small percentage, 5 percent, of cadets who placed the Corps as liberal. When comparing themselves as a Corps to the broader population of college students, a majority of cadets see the Corps of Cadets as conservative yet view the civilian college student body as liberal.

We expanded our research beyond a comparison of college students to see how cadets viewed the larger society. The survey asked cadets to place the military's political ideology as well as the civilian population's ideology, with the results contrasted in figure 22-3. In placing the military, 60 percent of cadets responded that the military falls in the conservative quadrant of the diagram, 19 percent placed the military in the populist quadrant, 6 percent in the liberal quadrant, 10 percent in the libertarian quadrant, and 4 percent fell in the "other" category. Similar to the rest of the survey questions, responses in the "other" category described the military as not falling predominantly in one particular quadrant.

In placing the civilian population as a whole, cadet responses did not present nearly the disparity between civilians and the military as between the Corps of

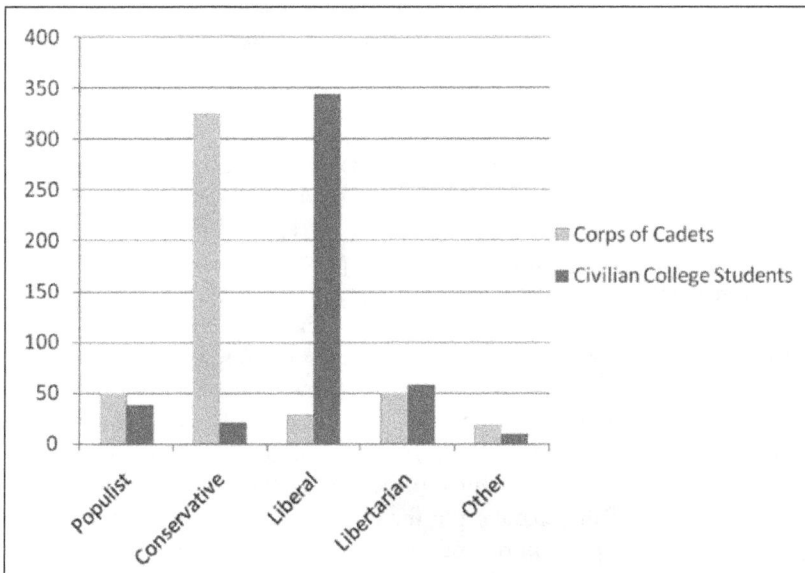

Figure 22-2. Cadet placement of Corps ideology and civilian college student ideology.

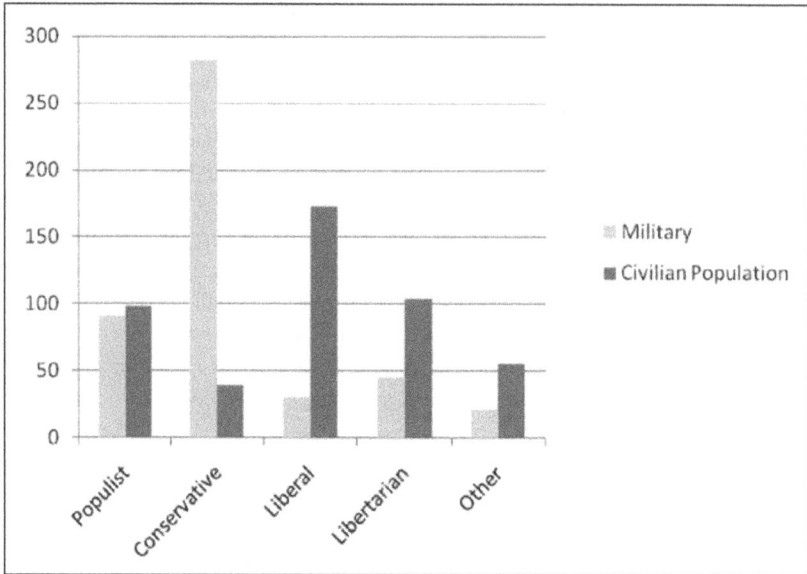

Figure 22-3. Cadet placement of military ideology and civilian population ideology.

Cadets and civilian college students. However, there is still a stark contrast in the way cadets see the military and the civilian population in terms of political ideology. A plurality of responses, 37 percent, placed the civilian population in the liberal quadrant, while 22 percent responded that the population falls in the libertarian quadrant, 21 percent in the populist quadrant, and 12 percent in the "other" category. Interestingly, however, only 8 percent placed the civilian population in the conservative quadrant, the lowest percentage in any quadrant. While there was no dominant quadrant in this question indicating that the sample does have a specific view of the ideology of the society writ large, the conservative quadrant has a distinctly lower number of responses than all other quadrants and the "other" category. Thus, we can say that a majority of the respondents view the military as conservative and the civilian population as not conservative.

The last question asked cadets about their perceptions of the civilian population's evaluation of the military in terms of political ideology. In other words, if we were to ask members of the civilian population to place the military in one of these quadrants, what would they say? Returning to table 22-1, we see that responses to this question presented the strongest trend towards one quadrant, with 78 percent responding that the civilian population perceives the military as falling in the conservative quadrant. The next highest response proportion was 8 percent placing civilian perception of the military in the populist quadrant, while 6 percent placed civilian perception in the liberal quadrant and 7 percent placed civilian perception in the libertarian quadrant. Only 1 percent placed civilian perception in the "other" category. The data here indicate that a vast majority of survey respondents believe civilians perceive the military as an ideologically conservative organization.

Conclusion

The results of our study of cadets are quite stark. Cadets believe that military and civilian populations, whether in college or in the larger society, occupy drastically different ideological spaces. The military is perceived to be ideologically conservative while the civilian sector is perceived to be liberal. Moreover, an overwhelming majority of subjects in our sample believe that civilians perceive the military to be conservative. The data is particularly troubling in light of the Huntington-Janowitz divide. While the military has become involved in the political process, which Janowitz would argue is a healthy necessity, its apparent perception of itself as occupying an ideological space separate and distinct from the civilian sphere is worrisome. The military has subjective control with seemingly little perception of neutrality.

The results of this study and other work in this vein are valuable across a range of subjects of importance to the Department of Defense including, but not limited to, professionalization, behavior during elections, military participation in politics generally, and the role of the Federal Voting Assistance Office specifically. A subsample of the military population perceives itself to be ideologically distinct from its civilian counterparts, raising clear concerns over the deleterious consequences for the ability of the military to meet the intent of DOD Directive 1344.10 in avoiding inferences of partisan political approval and/or endorsements. While we avoided asking about partisanship, we can interpret these results to mean that cadets perceive their organization and the military as having a particular view of the role of government that is distinctly different from that of the civilian population. The moral hazard that DOD Directive 1344.10 attempts to stave off seems glaringly apparent to our sample of cadets.

While the intent of DOD Directive 1344.10 in delineating between permissible and prohibited political activities is justified, it ought to be reviewed and further expounded upon to highlight the necessity of these distinctions for the proper functionality of the military within our republic. In light of our findings, it seems that perhaps the directive does not go far enough in explaining the necessity of this and similar regulations. Specifically, we advise policy makers to reconsider the specific delineations drawn in DOD Directive 1344.10 between permissible and prohibited political activities and how they contribute to the perceptions of the military as a partisan and/or ideological body. Increased education on professionalism, civil-military relations, and the dual role of active military personnel as guardians of the state and citizens of the state will foster a deeper understanding of the need for Directive 1344.10 and why there may be activities that are *permissible* but not *advisable* for healthy civil-military relations. Further study and contemplation may also lead policy makers to conclude that even more partisan political activities may need to be prohibited due to the image conveyed to the public at large. The privacy afforded at the ballot box does not extend to the world of campaign contributions, where campaign finance laws mandate that individuals list their employers when donating over a certain amount to a partisan political candidate.

Further, the DOD ought to consider engaging the civilian population in this discussion of civil-military relations. To our knowledge, few, if any, civilian colleges address the proper role of the military in the policy-making process as part of their introduction to American politics courses. The notion of a nonpartisan military designed to provide expert knowledge on defense affairs is a virtually unknown concept in policy courses. The public should be educated to understand that, while active duty military members may fulfill their obligations as citizens, these private opinions do not reflect the armed forces as an expert and professional body.

Overall, DOD policy makers must be aware of the potential moral hazard in the gap of perceived ideological leanings within civilian and military spheres and how it may affect the perception of military advice in the policy arena. It is also vital that the Department of Defense be proactive in this debate and continue to put forth scholarship and sponsor data collection in both military and civilian spheres to understand and counter the consequences of this potential moral hazard, which threatens to undermine the role of the military as key provider of defense expertise on policy matters.

Any errors of fact or judgment are the responsibility of the authors. The views expressed in this paper are solely those of the authors and do not represent the views of the United States Military Academy, the Department of the Army, and/or the Department of Defense.

Notes

(All notes appear in shortened form. For full details, see the appropriate entry in the bibliography.)

1. DOD Directive 1344.10, *Political Activities*.
2. Clausewitz, *On War*.
3. DOD Directive 1344.10, *Political Activities*.
4. Huntington, *The Soldier and the State*.
5. Ibid.
6. Ibid, 259. A common quote from Huntington's work is "if any convictions ... were acquired by the cadet, noted one officer, they were generally of contempt for mere politicians and their dishonest principles of action."
7. Ibid.
8. The organizational reforms enacted by Congress in the decades following World War II (starting with the National Security Act of 1947 and ending most recently with Goldwater-Nichols in 1986) were aimed not at excluding the military from the political process, but at regulating their influence in it.
9. Of course, one could argue that there is no such thing as a "purely military matter" in contemporary affairs of state.
10. Janowitz, *The Professional Soldier*.
11. Ulrich, "Infusing Normative Civil-Military Relations."
12. Snider, "The Army Profession," 2008.
13. Feaver and Kohn, *Soldiers and Civilians*.
14. We limit our discussion to the relationship between the military and the executive because this relationship is most visible to the American people and is the area of most concern.
15. Feaver and Kohn, *Soldiers and Civilians*.
16. It is important to note that the sample population is not a representative sample of the military, the Army, or even the Army junior officer corps. However, we argue that accounting for cadet perceptions is a useful subsample in gauging future Army leadership.

Appendix
Survey Questions

1. In which quadrant do you place yourself? (upper-left, upper-right, lower-left, lower right)

2. Which quadrant best characterizes the Corps of Cadets? (upper-left, upper-right, lower-left, lower-right, other [please describe and be specific])

3. Which quadrant best characterizes the military? (upper-left, upper-right, lower-left, lower-right, other [please describe and be specific])

4. In which quadrant do you place your mother/mother surrogate? (upper-left, upper-right, lower-left, lower-right, n/a)

5. In which quadrant do you place your father/father surrogate? (upper-left, upper-right, lower-left, lower-right, n/a)

6. Which quadrant best characterizes civilian college students? (upper-left, upper-right, lower-left, lower-right, other [please describe and be specific])

7. Which quadrant best characterizes the civilian population? (upper-left, upper-right, lower-left, lower-right, other [please describe and be specific])

   Please go to politicalcompass.com and take the ideology quiz.

8. What are the results from your ideology quiz? (actual coordinates)

Bibliography

Clausewitz, Carl von. *On War.* Translated by J. J. Graham. London: Kegan Paul, Trench, Trübner and Co., 1908.

Department of Defense Directive 1344.10. *Political Activities by Members of the Armed Forces*, 21 February 2008.

Feaver, Peter, and Richard H. Kohn. *Soldiers and Civilians: The Civil-Military Gap and American National Security.* Cambridge, MA: MIT Press, 2001.

Huntington, Samuel P. *The Soldier and the State: The Theory and Politics of Civil-Military Relations.* Cambridge, MA: Belknap Press of Harvard University Press, 1957.

Janowitz, Morris. *The Professional Soldier: A Social and Political Portrait.* Glencoe, IL: Free Press, 1960.

Snider, D. M. "The Army Profession." Lecture. United States Military Academy, West Point, NY, 2008.

Ulrich, M. P. "Infusing Normative Civil-Military Relations Principles in the Officer Corps." In *Future of the Army Profession*, edited by D. M. Snider and L. J. Matthews, 655–682. 2nd ed. New York: McGraw-Hill, 2005.

About the Authors

Dr. Rachel M. Sondheimer is an assistant professor in the American Politics, Policy, and Strategy stem of the Department of Social Sciences at the United States Military Academy at West Point. She teaches courses on political analysis, the public policy-making process, and campaigns and elections. Sondheimer's research interests include voting behavior, experimental methods, and civil-military relations. One strain of her research uses natural and randomized experiments to isolate the influences of education and family background on political and civic participation. Her recent research includes analysis of the political behaviors of active members of the military and military communities with a focus on the impact of these behaviors on civil-military relations. Sondheimer received her PhD in political science at Yale University in 2006 and her undergraduate degree in government from Dartmouth College in 2001.

LTC Isaiah "Ike" Wilson III is an associate professor with the Department of Social Sciences at the United States Military Academy. He holds a bachelor's degree from the United States Military Academy as well as graduate degrees from Cornell University to include a PhD. He also received graduate degrees from the US Army's Command and General Staff College and the School of Advanced Military Studies. Wilson commanded in Germany and the Balkans and is a combat veteran of Operation Iraqi Freedom and Operation Enduring Freedom. He served as the chief of war plans and military strategist for the 101st Airborne Division in Northern Iraq, and later as the division's chief architect for the 101st Airborne's reorganization. Wilson has also served as a special advisor on civilian-military planning for US Forces-Afghanistan, NATO-International Security Assistance Force, and the US Embassy-Kabul, assisting in the development and authoring of the US Government Integrated Civilian-Military Campaign Plan for Afghanistan and Pakistan.

MAJ Thomas Greco is an active duty officer currently assigned as an Iraq Transition Team deputy chief in the 2nd Brigade Combat Team of the 10th Mountain Division. His most recent assignments include the Command and General Staff College at Fort Leavenworth, Kansas, and as an assistant professor of American politics, policy, and strategy with the Department of Social Sciences at the US Military Academy. He also serves as an Army engineer, having commanded twice. He holds a bachelor's degree from West Point, a master's degree from the University of Missouri at Rolla, and a master's degree from Georgetown University.

MAJ Kevin Toner is currently an instructor of American politics, policy, and strategy in the Department of Social Sciences at the United States Military Academy. He has served as a tank platoon leader in Bosnia with First Brigade, First Cavalry Division, and a brigade planner and cavalry troop commander in Iraq with First Brigade, First Infantry Division. Toner earned his bachelor's degree from the United States Military Academy in 1997 and his master's degree from Columbia University. Following his duties at West Point, he will become a public affairs officer.

2LT Cameron B. West is a 2009 graduate of the United States Military Academy where he earned his bachelor's degree. Lieutenant West is an armor officer stationed at Fort Bragg, North Carolina. He worked on this project as part of his senior thesis.

CHAPTER 23

WARRIOR–CIVILIANS
NONMILITARY PERSONNEL ON THE BATTLEFIELD

Lynn R. Sylmar

Introduction

Three military vehicles make their way down a dirt road outside of Al Basrah, Iraq. Air Force Office of Special Investigations (AFOSI) Special Agent Jill Thomas,[1] a Department of Defense (DOD) civilian employee, rides along with another agent and several military personnel. Outfitted in desert uniforms and military protective gear, the team is on its way to pick up a suspected al-Qaeda collaborator at his home. AFOSI is responsible for collecting intelligence in the area and intends to collect any computers, documents, or information located in the suspect's home after the military special operators apprehend him. As they approach the home, they begin to take on small arms fire. They expected resistance, but not to this extent.

An explosion overturns the lead vehicle. Personnel from the vehicle are quickly recovered and the convoy attempts to retreat. Agent Thomas is wounded in the exchange, but she understood the dangers of being in a combat zone. As a federal agent, she expected that she would be shot at, but in a combat zone, was she a combatant? She was briefed by a wing judge advocate general (JAG) that she was a civilian and couldn't lawfully be targeted by the enemy—unless she took part in hostilities. She wondered, for the first time, what that meant and how the enemy was expected to distinguish her from the combatants in the vehicle. In the event of her capture, to what protections was she entitled?

Jill's DOD identification card indicates she is a civilian. Under international law, civilian status protects her from direct attack by the enemy. However, she looks just like the other members of the team. She is wearing a military uniform, military protective gear, and carrying a weapon. Additionally, as an agent for AFOSI, she interrogates suspected al-Qaeda affiliates, conducts human intelligence (HUMINT) activities, and acts as a security escort—functions that have traditionally been performed by members of the military. Based on her conduct, would her captors still consider her a civilian, or had she somehow become an illegal combatant?

437

Problem Background

The presence of nonmilitary personnel on the battlefield is not new; they have supported the military in every major war in US history. During the Revolutionary War, they were used extensively in supply functions,[2] and later amidst the War of 1812, they completed the majority of the labor in the field under the complete command and control of the military.[3] By 1908, the military had sufficient personnel and expertise in armed service to support itself.[4] Yet during World War I and II, inadequate numbers of personnel once again necessitated the use of persons outside the military to support and sustain combat forces.[5]

By 1973, DOD adopted a policy of total force integration. The policy directed the armed services to fully integrate nonmilitary employees into the national defense effort.[6] It wasn't until the end of the Cold War, however, that resource and budgetary constraints forced dramatic reductions in the active force.[7] In response to fewer available dollars, DOD began utilizing persons outside the military to maintain operational readiness with a smaller number of active-duty service members.[8]

DOD's increased dependence on advanced technologies and weapons is another apparent factor driving its growing reliance on nonmilitary personnel. The technical expertise for many of the United States' sophisticated systems already existed within the civilian sector that developed them.[9] Therefore, it seemed to make sense to place contractors—already trained and with system expertise—into positions supporting and maintaining this high-tech equipment. By doing so, the need to train military members to operate or support the systems was eliminated, freeing them up for combat-related duties. Nonmilitary personnel also relocated and deployed less often, providing greater continuity and institutional memory to the support of these systems.[10] As a result, nonmilitary personnel were viewed as a way of achieving greater operational efficiencies at a reduced cost.

Since the early 1990s, individuals outside the military have become increasingly vital to conducting the mission of the armed forces. In some areas, they significantly outnumber uniformed service members and are conducting a broader spectrum of activities than ever before. The use of these nonmilitary personnel to carry out certain functions reduced the number of military troops and therefore the amount of service member entitlements, making the employment of individuals outside of the military force increasingly attractive. In addition, functions performed by contract employees can be purchased as needed. This allows the military to buy expertise without having to maintain the skill on a long-term basis. The use of nonmilitary personnel also provides DOD the flexibility to determine the most effective and efficient composition of the force. Despite all of the benefits of using nonmilitary personnel, there are also risks. Many of these individuals have become indistinguishable from combatants—in both appearance and function—creating uncertainty regarding their status as civilians.

In 1995, Maj Brian Brady, a US Army judge advocate, identified the fact that few deployed commanders and contractors understood the status of non-military personnel in the field.[11] While some military analysts concluded that in a combat zone these individuals had become "legitimate targets,"[12] confusion remained "about their status under the Law of War."[13]

The debate over the status of individuals "accompanying the force" continued in 2001, when Maj Lisa Turner and Maj Lynn Norton, two Air Force judge advocates again identified challenges associated with having nonmilitary personnel on the battlefield. They identified three categories of nonmilitary persons: DOD civilians, contractors, and nonaffiliated civilians—all having "varying statuses, rights and responsibilities under international and domestic law, and under DOD and service regulations."[14]

The resulting development of domestic service doctrine reflected the confusion and uncertainty of the status of nonmilitary persons "accompanying the force." Army Pamphlet 715-16, *Contractor Deployment Guide*, instructed that individuals who accompany the force[15] "can only be used to perform selected combat support and combat service support (CSS) activities."[16] Joint Publication (JP) 4-0, *Joint Logistics*, added that "in all instances, contractor employees cannot lawfully perform military functions and should not be working in scenarios that involve military combat operations where they might be conceived as combatants."[17]

Using nonmilitary personnel to perform "selected combat support and combat service support activities" lacks defined parameters and has not been limited to "traditional" support activities. While JP 4-0 initially limited contractor functions to three support arenas—systems support, external theater support, and theater support[18]—the scope of the contract duties has continued to grow. Systems support contracts designed to use nonmilitary personnel to repair and sustain existing systems have expanded to include system operation. During combat, weapon systems such as unmanned aerial vehicles (UAV) are increasingly being operated by nonmilitary personnel.[19] Additionally, theater support contracts that used to provide goods, services, and minor construction[20] now include security details, facilities protection, and prisoner interrogation.[21]

The varying definitions of support led to differing conclusions by the armed services about the status of nonmilitary individuals executing these functions on the battlefield. The Air Force, for example, concluded that individuals performing "duties directly supporting military operations" were combatants "subject to direct, intentional attack."[22] The Navy, however, contended that these individuals were not combatants and "not subject to direct attack although they assume the risk of [becoming] collateral damage because of their proximity to valid military targets."[23]

Although attempts have been made to create clarity and consistency, doctrine and guidance remain unclear. Today contractors, who had once been restricted to using force only in self defense, can now use force when performing security functions and to protect assets and persons.[24] Bearing in mind this expanded authority, it is unclear how federal law can rationalize their status as civilians.[25]

Consider the following: (1) By engaging in hostilities, individuals not in the military lose civilian status. (2) To the extent this is true, why would the DOD contract for services that place personnel at risk of becoming "illegal combatants"? (3) Moreover, with large numbers of nonmilitary personnel on the front lines wearing military uniforms,[26] how can they be protected from attack?

The current practices, at best, create a real risk that nonmilitary personnel will be intentionally targeted and, worse, if captured, subject to trial by the enemy for hostile acts.[27] From now on, DOD leaders and policy makers should eliminate the use of the term "civilian" except as defined under international law. Furthermore, the service secretaries should take the steps necessary to clearly distinguish personnel who qualify for civilian status from those individuals who do not. Finally, policy makers must consider incorporating nonmilitary individuals who perform activities other than battlefield logistics and supply *into* the armed forces. Making these individuals members of the force is necessary to eliminate the risk that they could be considered unlawful combatants.

Definitions: Who's Really Who?

1. **Combatants**: members of the armed forces; a unique set of individuals authorized to engage in hostilities.[28] Examples: infantry soldier, submariner, and F-15 pilot.

2. **Noncombatants**: a subset of the armed forces who have been prohibited by their nation-state, not international law, from engaging in hostilities.[29] Noncombatants and civilians are mutually exclusive. As members of the force, this group receives no greater protections under the law than combatants.[30] Examples: military chaplains.[31]

3. **Civilians**: persons who are *not* members of the armed forces.[32] These individuals include the indigenous population, nonaffiliated persons, and persons who accompany the armed forces.[33] This group is entitled to civilian status because they are not permitted to "take a direct part in hostilities." Examples: the Cleaver family, Doctors Without Borders, and the Red Cross.

    A. **Nonaffiliated persons**: a subcategory of civilians. Persons not affiliated with an armed force include the media, nongovernmental organizations, private voluntary organizations, intergovernmental organizations, refugees, stateless persons, and internally displaced persons.[34] Examples: the Afghanistan population and Doctors Without Borders.

    B. **Persons accompanying the force**: a subcategory of civilians. This group includes individuals who accompany an armed force but are not members of it.[35] Examples: Blackwater Worldwide Security.

4. **Illegal combatants:** Individuals who engage in combat without the authority of their nation-state.

As illustrated in Figure 23-1, individuals on the battlefield are broadly classified as either "in the military" or "not in the military." Within each of these
two broad categories, there are two subcategories. For individuals in the military, the two subcategories are combatants and noncombatants. For those not
in the military, the subcategories are nonaffiliated individuals and individuals
accompanying the force.

The roles and statuses of both subcategories of "in the military" are fairly
well understood. "Combatants" are those authorized to engage in hostilities
against the enemy. They are obligated to conduct their war fighting in accordance with international law principles and to distinguish themselves from
civilians. They may be directly targeted by the enemy, are entitled to prisoner of
war (POW) status upon capture, and are immune from prosecution for their
law of war (LOW) compliant actions.

"Noncombatants" are the nonfighting personnel of an armed force.[36] These
individuals are not authorized to engage in hostilities because their nation-
state has prohibited them from fighting. However, because they are members
of the armed force, under international law they represent a legitimate target
for attack by the enemy.

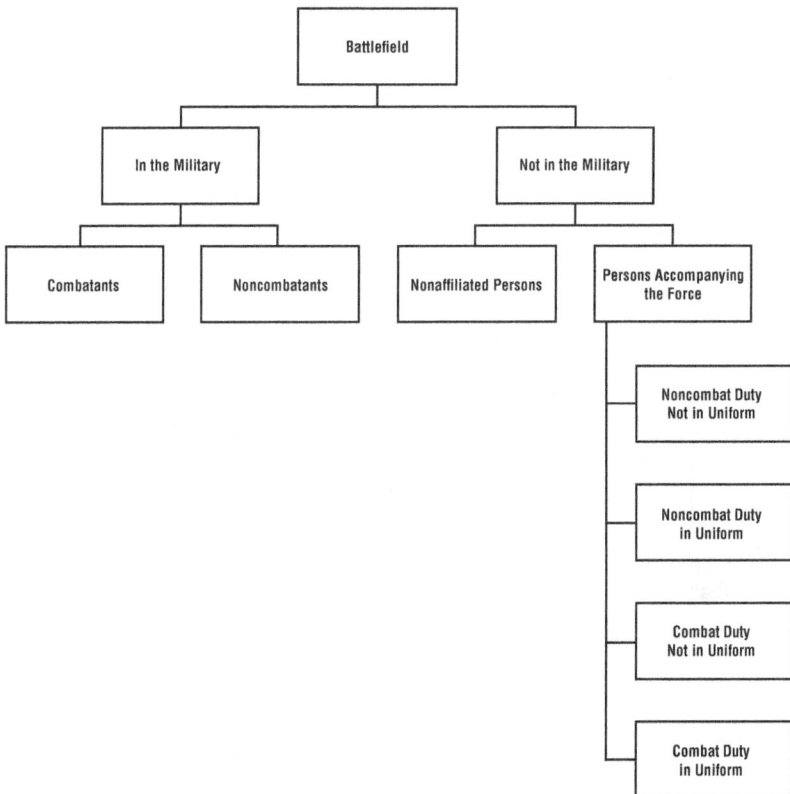

Figure 23-1. Classifications of individuals on the battlefield.

The categories of individuals "not in the military" are more problematic. Within the category of "not in the military," nonaffiliated persons are the most clearly defined. These individuals are not associated with either of the warring parties and are not authorized to engage in combat. Nonaffiliated individuals are entitled to civilian status and thus are entitled to be respected and protected at all times. During hostilities the status and roles of this group of individuals present few legal concerns and are generally well understood.

Of those individuals who are not in the military, the category of "persons accompanying the force" is more complex, creating a great deal of confusion regarding appropriate legal statuses and roles. Within this subcategory of persons not in the military, there are four groups of individuals. The first group consists of individuals who accompany the force but remain distinct from it— this is the traditional definition of persons "accompanying the force."

These individuals do not wear the military uniform, perform support—not combat—functions, and are therefore considered civilians under international law. Some examples of individuals in this group are contractors who provide billeting facilities, provide messing service, or operate the Army and Air Force Exchange Service. Although members of this group risk injury because of their proximity to military operations, they are not a legitimate target for the enemy because they are distinct from combatants in both appearance and function.

The remaining subcategories of persons "accompanying the force" are either not entitled to or are in danger of losing civilian status. These subcategories consist of individuals who are not in the military but perform either (1) noncombat duty, in military uniform; (2) combat duty, not in military uniform; or (3) combat duty, in military uniform.

Among these subcategories of nonmilitary personnel, the first group at risk consists of those persons who, although they do not perform combat duties, wear a military uniform. Persons in this category are at risk of losing their civilian status because they have become indistinguishable from combatants. The second group is made up of individuals who perform combat activities but do not wear a military uniform. Persons in this category violate the international LOW by engaging in combat illegally. Third are those individuals who engage in combat and wear the military uniform. They, like group two, engage in combat illegally. Although they distinguish themselves from civilians, they violate the LOW because they do not have combatant status.

Under international law, only members of the armed force are able to qualify for combatant status. By taking a direct part in hostilities without being members of the armed force, individuals become "illegal combatants." Illegal combatants are not entitled to POW status. Additionally, they may be prosecuted by a detaining nation for any hostile acts they have taken.

Some Practical Examples

Goodwill Gail—Noncombat duty, in military uniform.[37] Under international law, persons "accompanying the force" are not members of the military.

These individuals do not qualify for "combatant" status. They support the force and typically include members of "labour units," or they are "responsible for the welfare of the soldier,"[38]—like Gail. Gail is an Army morale, welfare, and recreation specialist and a DOD civilian. When she deployed to Iraq, she was issued a military uniform, which she wears daily. She travels around to different units to provide soldiers with game stations, videos, and magazines—anything to help them feel like someone cares. The problem for Gail is that by wearing the military uniform,[39] she has become indistinguishable from the armed force she supports.

Covert Chris—Combat duty, not in military uniform. Chris is an intelligence analyst and a DOD contract employee. In Iraq, he wears his jeans and a company shirt while accompanying the Army reconnaissance team. He wants to ensure he remains distinct from the military. He has been instructed by his contract manager that he is a civilian and cannot lawfully engage in activities that may be considered combat. The problem for Chris is that no one can tell him exactly what constitutes "activities that may be considered combat." While some may not consider Chris to be a combatant, international law experts and a recent Israel Supreme Court decision define intelligence gathering against an enemy army as direct participation in combat.[40]

G. I. Jill—Combat duty, in military uniform. As discussed, Jill Thomas, our heroine from the opening scenario, is an AFOSI agent and DOD civilian employee. She both wears the military uniform and performs a combat activity. Her job often requires the use of force, a key characteristic of a combatant.[41] Further, she was hired to conduct prisoner interrogations and security activities formerly executed by uniformed service members.[42] Jill has become a replacement for or an augmentee of the military force. However, she is not a member of it. Thus, although she distinguishes herself from those entitled to civilian status, she is engaging in apparent hostilities without authority. Her activities create the risk that she, like Chris, will be considered an "illegal combatant."

Gail and the Need for Distinction

International law requires warring parties to distinguish their combatants through a distinctive uniform or symbol which makes them discernable from civilians. Over time, nation-states developed the practice of having combatants wear a military uniform.[43] This requirement is the result of the desire to restrict warfare to acts of violence against combatants and military targets. It is believed that forces unable to distinguish enemy combatants from civilians would likely resort to targeting all individuals in an area.

Article 48 of Additional Protocol I dictates that "to ensure respect for and protection of the civilian population and civilian objects, the Parties to a conflict are required at all times to distinguish between the civilian population and combatants and between civilian objects and military objectives and accordingly must conduct their operations only against military objectives."[44] DOD's conduct during current combat operations, however, fails to adequately differentiate its combatants from nonmilitary personnel. In fact, a recent policy memorandum grants geographic combatant commanders the authority to direct uniform wear for deployed nonmilitary personnel, undermining the uniform's use as a traditional method of distinction.

DOD contends that, despite the international law requirement of distinction, uniform wear by nonmilitary personnel is *not* inconsistent with international law.[45] Directing individuals who are otherwise entitled to civilian status to wear a military uniform, however, makes distinguishing them from combatants impossible. This action by DOD, therefore, increases the likelihood civilians will be intentionally targeted by the enemy. While international law does not require combatants to wear a "military uniform," this practice has evolved over years of combat as the fundamental method of identifying combatants. Even so, DOD has ignored this tradition,[46] citing safety concerns. While it may be true that nonmilitary personnel in uniform are more easily identified at a distance by friendly forces, they are also easily misidentified as a combatant by the adversary.

This existing misuse of the uniform only adds confusion to the battlefield. DOD has prescribed some methods to distinguish combatants from civilians, but they are ineffective. One method, attaching the word "civilian" in place of the service name over the uniform pocket, is impractical. The identification tags are written in English and are often difficult, if not impossible, to see at a distance or under protective gear. Ultimately, military uniforms, even with the distinct name tape, are for all intents and purposes combatant uniforms. Arguing that uniform wear in a hostile environment increases the security of nonmilitary personnel contradicts years of tradition.

Chris and Jill and the Need for Combatant Status

The term "civilian" as defined by DOD is a US citizen or foreign national hired to work for the DOD.[47] The term identifies persons who are affiliated with the armed forces but are not service members. Individuals who are not in the military, however, are not necessarily entitled to civilian status on the battlefield. Under international law, civilian is a status afforded only to those persons who do not engage in hostilities.

In the scenario of Chris and Jill, both have directly participated in hostilities. As a result, neither of them would qualify for civilian status. Additionally, because they are not members of the armed forces—that is they are not combatants—international law would not recognize their authority to engage in hostilities. Absent appropriate authority, both of them could be considered criminals facing potential prosecution for their actions under the law of the detaining state. If either of them killed an enemy combatant, he or she could be tried for murder. Furthermore, because neither of them is entitled to status as a POW,[48] he or she could not expect repatriation at the cessation of hostilities.

Direct Participation in Hostilities

The complicated legal framework regarding "direct participation in hostilities" creates ambiguity about the types of activities that can be performed by nonmilitary personnel. While international law does not prohibit nonmilitary personnel from engaging in combat, they may lose civilian status and are not protected as authorized combatants. Combat is defined by some experts as "kill[ing] or take[ing] prisoners, destroy[ing] military equipment, or gather[ing] information in the area of operations."[49] Others argue for an expanded definition based on the changing nature of warfare that includes persons who "operate a weapons system, supervise such operation, or service such equipment."[50] These ambiguities make it difficult to determine when an individual may be engaging in combat.

Too Much Legalese

Scholars of international armed conflict such as W. Hays Parks and Geoffrey Corn have attempted to clarify the activities that constitute "direct participation in hostilities." Parks emphasizes that direct participation in hostilities is only an action which "cause[s] actual harm to the personnel and equipment of the enemy armed forces."[51] Corn, on the other hand, advocates a "functional discretion" test.[52] Under Corn's test, if an individual's decision-making authority could result in a violation of the LOW, that activity should be considered a direct part in hostilities.[53] The problem with this type of delineation necessitates an assessment of every activity being conducted to determine if a prohibited level of discretion exists.

Parks' definition is equally problematic. According to this definition, it is difficult to determine what constitutes "actual harm." For example, it is unclear if an intelligence analyst in the area of hostilities would qualify as a combatant. It may be argued the intelligence analyst is not causing actual harm to an enemy because the analyst is not killing anyone. According to the Israeli Supreme Court, however, "direct participation in hostilities" does not require the use of arms.[54] Harm can be done without the use of arms at all. In this case, although the analyst is not shooting a bullet at the enemy, he is causing direct harm by providing targeting information that may be used by a B-1 bomber aircraft to drop bombs on the enemy.

Under Corn's functional discretion test the same analyst's activities would have to be assessed under the four LOW principles—distinction, necessity, proportionality, and minimization of unnecessary suffering—to determine the level of discretion the analyst possesses. Generally, for intelligence analysts, the principle of necessity is an essential consideration. An analyst is the primary individual responsible for identifying valid military objectives. The principle of necessity requires that a target be an object which by its nature, purpose, location, or use effectively contributes to the war-fighting, war-sustaining capabilities of the enemy and whose partial or total destruction will result in a distinct military advantage for friendly forces.[55] Because the identification of targets is a fundamental combat operation, the misapplication of the principle of necessity could create a LOW violation. Thus in the Corn analysis, although the analyst may not have discretion with regard to other principles of the LOW, he or she may still be considered a combatant.

A major problem with the functional discretion test is that mental discretion is difficult to measure and can change with seniority, rank, and level of responsibility. It is possible, then, to have personnel with the same duty title but different legal statuses based on the level of discretion they exercised during a particular event. A junior analyst deployed to the field, for instance, may not have the authority to designate targets while she is working at the air operations center. When she goes forward with the brigade combat team, however, her target designation authority may change. Attempting to ascertain her legal status based on her daily or perhaps hourly discretion is of little value.

Clearly identifying the status of persons on the field is critical in ensuring adequate protections for civilians and necessary entitlements for combatants. However, neither of these legal constructs provides much clarity for commanders or affected nonmilitary personnel. Personnel in combat need clear, simple guidelines and procedures that reduce the potential for diverse legal conclusions that may have devastating consequences.

The Risks

Gail, the goodwill specialist mentioned earlier, is a mother. She has a daughter and a son. She remembers when they headed off to college—the calls home and the care packages she sent. It was these memories that motivated her to bring compassion in the form of Sony PlayStations® and cookies to the troops—her troops. She never imagined that she would be considered a combatant. Today, however, she is in the crosshairs of Abdulla Sayeed,[56] a 17-year-old member of al-Qaeda. She would not be the first American Abdulla has killed. He has been fighting since he was nine. No time for school, but he doesn't need to read. He knows the uniform of the Americans. He aims and squeezes the trigger.

Meanwhile, in a small concrete room across town, Jill waits. She is alone in the room. She has been alone for about three hours now. The adrenaline from the earlier firefight has worn off. Surprisingly, she isn't worried. She understands that under international law, she is a POW and will be treated humanely. Suddenly, outside the door she hears yelling. She hears the English words "terrorist" and "criminal," and a man is thrown into the room. It's Chris. She doesn't know him, but she recognizes his face. What did her captors mean by "terrorist" and "criminal"? Were they talking about Chris? He isn't a terrorist or a criminal. He's an intel guy. He wears jeans and carries only the 9MM he is authorized for self defense.

"MY GOD," she thinks. Maybe they were talking about her. They couldn't be. Admittedly she is in a uniform, but DOD wouldn't direct her to wear it if it weren't appropriate. And certainly they would not use her to carry out activities that were not lawful. But she looks like a combatant, and she is the one who was carrying an M4 assault rifle. The adrenaline is back.

What Now?

Under the current regime, nonmilitary personnel on the battlefield are at significant risk. They are wearing uniforms and protective gear that make them indistinguishable from their military counterparts. Additionally, the activities they conduct have expanded, closing the gap between support activities and actions which may be considered "direct participation in hostilities." Both of these factors put the civilian status of these individuals in jeopardy. It is imperative that policy makers act to eliminate this risk. The following represent four simple, yet necessary actions to ensure adequate protections for nonmilitary personnel accompanying the force:

1. **Stop using the term "civilian" except as defined under international law.** Policy makers need to stop deceiving themselves. Not all nonmilitary personnel are civilians under international law. Using the term "civilian" to define all nonmilitary personnel leads to the misunderstanding that they all qualify for civilian status. They do not. Within national policy and guidance, DOD must limit the use of the term "civilian" to qualifying personnel.

2. **Clearly distinguish those who do qualify as civilians.** Individuals who are entitled to protection from attack must look like they are protected, not like a target. Directing nonmilitary persons to wear a military uniform undermines their protections and is inconsistent with the traditional practice of nation-states. A name tape with the word "civilian" is not easily seen or understood by an enemy. Nowhere under international law is anyone required to speak or read English. To ensure civilians are protected, they cannot continue to wear the uniform of the US military forces. If the purpose is to ensure quick, clear, and easy identification of civilians by *both* friendly and enemy forces, a reflective orange safety vest would be more effective.

3. **Limit the activities performed by nonmilitary personnel.** One of the primary purposes of international humanitarian law is to regulate the conduct of combat. Those not involved in the fighting must remain distinct—not only in appearance, as discussed above, but in function—from those who *are* involved in the fighting. The question over how much involvement in combat results in the loss of civilian status must be answered more clearly. Direct participation in hostilities, actual harm, and functional discretion tests are not easy to understand or apply. The traditional functions historically performed by civilians, however, provide a simple basis for characterizing noncombat activities. Examples from history include support activities[57] and logistics.[58]

To ensure that nonmilitary personnel are entitled to civilian status, the functions they perform must remain limited to logistics and supply. Logistics is defined as "moving and supplying armies,"[59] while supply is the act of "providing"[60] items such as parts, food, and ammunition. It is inappropriate for civilians to supply services that involve the use of force[61] against, provision of battlefield intelligence about, or the damaging of the opposing military's force or property. These three activities are inextricably combat and cannot be carried out by an individual in civilian status. Prohibiting nonmilitary personnel from executing these three activities creates greater clarity regarding the status of these individuals without limiting their ability to perform the type of logistics and supply roles needed and envisioned under international humanitarian law.

4. **Incorporate nonmilitary personnel who perform functions other than logistics or supply into the armed forces.** If DOD chooses to use nonmilitary personnel to conduct activities outside the scope of traditional supply and logistics, then it must accept that these individuals are no longer entitled to civilian status. Using civilians in this manner is inconsistent with the intent of international law and places true civilians on the battlefield at risk. Enemy combatants, witnessing the hostile acts by individuals who are not in the military, cannot readily identify which persons present a danger. As a result, all individuals in a contested area may be considered a threat, creating the risk that those associated with the military as well as those who are not will be killed.

To prevent the risk of attack against civilians, it is necessary to craft legislation to incorporate nonmilitary persons who perform functions other than those historically carried out by nonmilitary personnel into the armed forces. These individuals should be considered an "auxiliary" military force, identified by military uniform and capable of engaging in hostilities.[62] As an auxiliary US force, these individuals would be entitled to combatant status and all the relevant protections. Additionally, incorporation of these individuals would create clearer lines of distinction and reduce the risk of any unintentional targeting of legitimate civilians.

The Merchant Marine Act of 1936 provides a solid foundation for crafting the necessary legislation. The act is a mechanism for nonmilitary mariners to become an auxiliary force during times of war.[63] In a similar fashion, selected nonmilitary personnel could become an auxiliary to the armed forces during deployments to areas of combat. Their membership in the force would additionally provide clear command and control for commanders while further enabling the nonmilitary personnel to carry out all activities, without the danger of being considered illegal combatants.

Legislating the incorporation of nonmilitary personnel into the force can be simple and does not necessarily have to entitle them to full service-member benefits. This issue, however, requires further consideration to determine what is appropriate. Currently, thousands of individuals are operating in hostile areas without an expectation of service-member benefits. However, because closely affiliated entities such as the Women's Air Force Service Pilots and some Merchant Marines have received some entitlements,[64] further research in this area is warranted.

Conclusion

The current policies and generalities leave personnel accompanying the force in uncertain and dangerous conditions. On the battlefield, every day they are taking chances. The risk that any individual supporting the military is inappropriately attacked or prosecuted for illegal combatant activities is a risk that US leadership should not continue to take.

Several layers of unclear or contradictory domestic policy and guidance currently exist, much of which is confusing even to legal experts. However, because the use of nonmilitary personnel during combat is likely to continue, policy makers must act to protect them. The recommendations outlined in this paper are simple, yet they provide clear parameters that will more effectively protect those present on the battlefield.

Notes

(All notes appear in shortened form. For full details, see the appropriate entry in the bibliography.)

1. The fictional scenario and character of Special Agent Jill Thomas are based on an interview of Special Agent Julie Lecea, Air Force Office of Special Investigations (AFOSI) detachment commander, Luke AFB, Arizona, regarding actual AFOSI activities in Iraq.

2. Harvell, *Department of Army (DA) Civilians in Support of Military Operations*, 1–2.

3. Ibid.

4. Ibid.

5. Ibid., 2.

6. General Accounting Office (GAO), *DOD Force Mix Issues*, chap. 0:2.

7. Heaton, "Civilians at War," 3.

8. GAO, *DOD Force Mix Issues*, chap. 0:1.

9. Ibid., chap. 2:3.

10. Ibid.

11. Brady, "Notice Provisions," 3.

12. Ibid.

13. Ibid.

14. Turner and Norton, "Civilians at the Tip of the Spear," 2.

15. Department of Army Pamphlet 715-16, *Contractor Deployment Guide*. Although the Army Handbook specifically addresses the limits of contractor support, similar legal arguments exist for civilian employees.

16. Army Pamphlet 715-16, 138.

17. Joint Publication 4-0, *Joint Logistics*, chap. V.

18. Ibid., V-I.

19. Dunn, "Contractors Supporting Military Operations," 5.

20. Ibid.

21. Ibid.

22. Ibid., 12.

23. Ibid.

24. 48 Code of Federal Regulations, Part 252.225-7040, b(3)(ii).

25. Ibid., b(3).

26. Army Pamphlet 715-16, App. B-1, para. 5-1.

27. Guillory, "Civilianizing the Force," 4.

28. Additional Protocol I, Article 48.

29. Fleck, *Handbook of Humanitarian Law*, 84. (Medical service and religious personnel, although often categorized as "noncombatants" are granted the "benefit of neutrality," and unlike other "noncombatants" are to be "respected and protected under all circumstances." See *Handbook*, 88–89).

30. Ibid., 84.

31. Medical service and religious personnel, although often categorized as "noncombatants" are granted the "benefit of neutrality" and unlike other "noncombatants" are to be "respected and protected under all circumstances." See Fleck, *Handbook*, 88–89.

32. Ibid., 210–11.

33. Ibid., 95.

34. Turner, "Civilians at the Tip of the Spear," 1.

35. Fleck, *Handbook*, 95.

36. Ibid., 84.

37. Based on the real-world example in Bosnia cited by Katherine Peters in "Civilians at War."

38. Fleck, *Handbook*, 95.

39. Haynes, "Combat Civilians."

40. Israel Ministry of Foreign Affairs, "Israel Supreme Court Decision."

41. Bailes, Schnecker, and Wulf, *Revisiting the State Monopoly*, 1.

42. Maginnis, "Security Contractors in War."

43. Fleck, *Handbook*, 75.

44. Additional Protocol I, Article 48.

45. Department of Defense Directive (DODD) 1404.10, *DOD Civilian Expeditionary Workforce*, para 6.9.8.

46. Ibid.

47. DODD 1400.31, *DOD Civilian Work Force Contingency*, para 3.1.

48. Additional Protocol I, Article 47, 1977.

49. Fleck, *Handbook*, 232.

50. Ibid.

51. Maxwell, "Law of War and Civilians on the Battlefield," 18.

52. Corn, "Unarmed but How Dangerous?" 261.

53. Ibid., 261.

54. Israel Ministry of Foreign Affairs, "Israel Supreme Court Decision."

55. Department of the Navy, *Commander's Handbook on the Law of Naval Operations*, para 5.3.1.

56. Abdulla Sayeed is a fictional character. Any resemblance to any actual person, living or dead, is purely coincidental.

57. Green, *The Contemporary Law of Armed Conflict*, 106.

58. Ibid.

59. Dictionary.net, "What Does Logistics Mean?"

60. Dictionary.net, "What Does Supplying Mean?"

61. Nothing in international law prohibits anyone's use of force in self defense; therefore nothing in this article should be interpreted to add such a restriction.

62. Merchant Marine Act, 1936, Title I, Section 101 (b).

63. Ibid.

64. US Maritime Service, "Mariners' Struggle for Veteran Status."

Bibliography

Additional Protocol I to the Geneva Conventions of 12 August 1949. Articles 48, 50, and 51.3, 1977.

Additional Protocol II to the Geneva Conventions of 12 August 1949. Article 1, 1979.

Albrisketa, Joana. *Blackwater: Mercenaries and International Law.* Madrid, Spain: Fundación para las Relaciones Internacionales y el Diálogo Exterior (FRIDE), October 2007.

Bailes, Alyson, Ulrich Schnecker, and Herbert Wulf. *Revisiting the State Monopoly on the Legitimate Use of Force.* Policy Paper 24. Geneva, Switzerland: Geneva Center for the Democratic Control of Armed Force, 2007.

"Blackwater Worldwide." *New York Times Online.* 29 January 2009. http://topics .nytimes.com/top/news/business/companies/blackwater_usa/index.html (accessed 10 March 2009).

Brady, Brian H. "Notice Provision for United States Citizen Contractor Employees Serving the Armed Forces of the United States in the Field: Time to Reflect their Assimilated Status in Government Contracts?" *Air Force Law Review* 147 (1995): 1–83.

Commentary on the Additional Protocols of 8 June 1977, Article 73.

Corn, Geoffrey. "Unarmed but How Dangerous? Civilian Augmentees, the Law of Armed Conflict, and the Search for a More Effective Test for Permissible Civilian Battlefield Functions." *Journal of National Security Law and Policy* 2 (2008): 257.

Department of Army Field Manual 100-5. *Operations,* 14 June 1993.

Department of Army Pamphlet 715-16. *Contractor Deployment Guide,* 27 February 1998.

Department of Defense Directive (DODD) 1400.31. *DOD Civilian Work Force Contingency and Emergency Planning and Execution,* 1 December 2003.

DODD 1404.10. *Emergency-Essential DOD U.S. Citizen Civilian Employees,* 1 December 2003.

Department of Defense Instruction (DODI) 2311.01E. *DOD Law of War Program,* 9 May 2006.

DODI 3020.41. *Contractor Personnel Authorized to Accompany the U.S. Armed Forces,* 3 October 2005.

Dunn, Richard L. "Contractors Supporting Military Operations." In *Excerpts from the Third Annual Acquisition Research Symposium,* May 17–18, 2006.

England, Gordon, deputy secretary of defense. Memorandum, 25 September 2007.

Fleck, Dieter. *Handbook of Humanitarian Law in Armed Conflicts.* New York, NY: Oxford University Press, 1995.

General Accounting Office. *DOD Force Mix Issues: Greater Reliance on Civilians in Support Roles Could Provide Significant Benefits,* 19 October 1994.

Geneva Convention. *Relative to the Treatment of Prisoners of War (Article 4).* Geneva, Switzerland, 1949.

Green, Leslie. *The Contemporary Law of Armed Conflict*. Manchester, England: Manchester University Press, 1993.

Guillory, Michael E. "Civilianizing the Force: Is the United States Crossing the Rubicon?" *Air Force Law Review* 51 (2001).

Harvell, Thea, III. *Department of Army (DA) Civilians in Support of Military Operations: How Should Current Policies Change to Better Support Them on Today's Battlefield?* Carlisle Barracks, PA: Army War College, 18 March 2005.

Haynes, Mareshah. "Combat Civilians: Managing the Airfield." *Air Force News*, 27 March 2008. http://www.offutt.af.mil/news/story.asp?id=123092773 (accessed 14 August 2008).

Heaton, J. Ricou. "Civilians at War: Reexamining the Status of Civilians Accompanying the Armed Forces." *Air Force Law Review*, Winter 2005.

International Committee of the Red Cross (ICRC). "Commentary on the Additional Protocols of 8 June 1977 to the Geneva Conventions of 12 August 1949," 1987.

———. "Direct Participation in Hostilities: Questions and Answers," 6 February 2009. http://www.icrc.org/web/eng/siteeng0.nsf/htmlall/direct-participation-ihl-faq-020609 (accessed 13 June 2009).

———. *Report on Direct Participation in Hostilities under International Humanitarian Law*. Geneva, Switzerland: ICRC, September 2003.

ICRC and TMC Asser Institute. "Report on the Third Expert Meeting on the Notion of Direct Participation in Hostilities." Geneva, Switzerland, 23–25 October 2005.

Israel Ministry of Foreign Affairs. "Israel Supreme Court Decision on Targeting Terrorist Operatives," 20 December 2006. http://www.mfa.gov.il/MFA/Government/Law/Legal+Issues+and+Rulings/Israel%20Supreme%20Court%20decision%20on%20targeting%20terrorist%20operatives%2020-Dec-2006 (accessed 31 March 2009).

Joint Publication (JP) 1-02. *Department of Defense Dictionary of Military and Associated Terms*, 12 April 2001 (as amended through 17 October 2008).

JP 4-0. *Joint Logistics*, 18 July 2008.

Maginnis, Robert. "Security Contractors in War." *Human Events.com*, 21 September 2007. http://www.humanevents.com/article.php?id=22499 (accessed 17 March 2009).

Maxwell, David. "Law of War and Civilians on the Battlefield: Are We Undermining Civilian Protections?" *Military Law Review*, September–October 2004.

Merchant Marine Act of 1936. US Code. Title 46A, chap. 27.

Navy Warfare Publication 1-14M. *Commander's Handbook on the Law of Naval Operations*, July 2007.

Peters, Katherine. "Civilians at War." *Government Executive*, 1 July 1996. http://www.govexec.com/reinvent/downsize/0796s2.htm (accessed 14 August 2008).

Quadrennial Defense Review Report. Washington, DC: DOD, 6 February 2006.

Rawcliff, John, and Jeannie Smith. *Operational Law Handbook.* Army Judge
  Advocate General's Legal Center and School, 2006.
Schmitt, Michael N. "Humanitarian Law and Direct Participation in Hostili-
  ties by Private Contractors or Civilian Employees." *Chicago Journal of Inter-
  national Law,* Winter 2004.
Turner, Lisa, and Lynn Norton. "Civilians at the Tip of the Spear–Department
  of Defense Total Force Team." *Air Force Law Review,* Spring 2001.
Uniform Code of Military Justice. US Code. Title 10.
US Merchant Marine. "Mariners' Struggle for Veteran Status." USMM.org.
  http://www.usmm.org/strugglevetstatus.html (accessed 18 February 2009).

About the Author

Maj Lynn Sylmar is currently the staff judge advocate at the 65th Air Base Wing, Lajes
Field, Azores, Portugal. From July 2008 to June 2009, she was a student in residence at
Air Command and Staff College, Maxwell AFB, Alabama. During her career she has
served in various legal positions, including operational attorney at the National Security
Agency and chief of operational law at the Joint Functional Component Command for
Network Warfare. Major Sylmar earned both her undergraduate and law degrees from
Florida State University.

CHAPTER 24

POST-TRAUMATIC STRESS DISORDER
FIGHTING THE BATTLE WITHIN

Carla Sizer
Claude Toland

There is no instance of a nation benefitting from prolonged warfare.

—Sun Tzu
The Art of War

On a night in July 2007, in Sparta, Minnesota, Noah Pierce raised a gun to his head and pulled the trigger. Noah was a 23-year-old Army veteran who spent two tours in Iraq. Noah had been diagnosed with post-traumatic stress disorder (PTSD) as a result of his multiple tours in Iraq. While there, Noah was directly involved in the killing of several enemy combatants in house-to-house raids. Regrettably, Noah also ran over a child by accident as she dashed into the road in front of his Bradley. He witnessed the deaths of friends killed in combat as well as roadside bomb attacks. On a morning in July 2007, Noah's friend found him slumped over his steering wheel with a suicide note indicating the horrors with which Noah lived.[1] Unfortunately, Noah's story is neither unique nor uncommon in 2010. The suicide rate at the time of publication had surpassed the record high for the US Army set in 2008, and it continues to rise. There is little disagreement that many, if not most, who chose to take their own lives suffered from PTSD. The evidence suggests that the time has come to formally address PTSD from a comprehensive policy perspective.

Background

The United States military has troops deployed in nearly 130 countries around the world performing a vast range of missions. Some US deployments are a result Cold War–era commitments that have existed for more than 50 years, while others are the direct result of US involvement in Iraq and Afghanistan. Sources suggest more than 1.7 million military personnel have been deployed in support of Operation Iraqi Freedom (OIF) and Operation Enduring Freedom (OEF) since 2001—of which 500,000 have been deployed more than once and 330,000 have sustained severe combat injuries. The result: hundreds of thousands of service members are at risk for PTSD, depression, and traumatic brain injury (TBI).[2] According to the Department of Veterans Affairs, approximately 1,800 US troops have been maimed by penetrating head wounds, and hundreds of thousands more may have suffered a mild TBI as a result of improvised explosive device (IED) blast waves.[3] No one questions the impact physical wounds such as amputations, brain injuries, burns, and shrapnel can have on a service member. However, what has been less understood until now is how enduring long-term pain can disable individuals who suffer from the invisible wounds of war—and there is none more profound facing the military today than PTSD.

What Is PTSD?

PTSD is an emotional trauma that can have debilitating long-term negative effects, which are not always visible. PTSD typically develops after exposure to a traumatic event in the face of grave physical harm or threats manifesting in severe depression or generalized anxiety. This combat-related affliction is not new to the military. In fact, the psychological experiences of war have likely been a problem of earlier wars for as long as warfare itself has existed.

Although PTSD has existed for centuries, it has been often overlooked and misunderstood. During the 1800s, soldiers were regularly diagnosed with "exhaustion" following battle. War-weary soldiers were commonly sent to the rear for a short time only to be returned to the front lines a short time later. In 1876, the common diagnosis for Civil War soldiers was "soldier's heart." Symptoms included startle responses, hyper-vigilance, and heart arrhythmias. During World War I, mental fatigue became known as the "effort syndrome" or "shell shock," and later, during World War II, as "combat fatigue." All of these terms describe military veterans who were exhibiting symptoms of stress and anxiety as the result of combat trauma. Despite the changing terminology, the effects were the same. In 1980, PTSD became a formalized diagnosis after experts determined anxiety disorders were commonly triggered by exposure to traumatic events.[4]

A Tale of Two Pattons

Consistent with the attitudes documented in the 1800s, the controversial World War II American general, George S. Patton Jr., embraced a "get over it" type of attitude and was known to have slapped soldiers for what he believed was malingering in US field hospitals. Ironically, the current director of manpower and personnel on the Joint Staff at the Pentagon is US Army Brig Gen Gary S. Patton (no relationship to the World War II hero). He spoke out publicly in 2009 admitting that he suffered from PTSD as a result of his combat experience in Iraq and has since sought counseling for emotional trauma. Although the names of these two generals are the same, their perspectives on what we now know to be PTSD could not be more different. The latter General Patton has emerged as an exemplar for those suffering from PTSD by talking publicly about his own battles with stress and how counseling has

helped him deal with PTSD. By removing some of the stigma associated with
seeking professional help, he hopes to influence contemporary perceptions of
PTSD as a battle fought within long after the cessation of combat hostilities.
Given the traditional stigma associated with military personnel seeking mental
health assistance, service members tend to be very reluctant in pursuing treatment. A 2005 study by the National Center for PTSD reported approximately
40 percent of service members experiencing PTSD indicated an interest in
receiving treatment.[5] Nevertheless, many believe coming forward could put
their careers at risk. Among the proportion who do seek treatment, many do so
at their own expense to maintain privacy. However, the greatest concern are for
those who forego treatment altogether.

A Modern Perspective of PTSD

The large number of injuries produced as a result of the current conflicts has
brought PTSD to the forefront of the debate by military experts. A study by
the RAND Corporation found that a huge gap exists between understanding
mental health needs of deployed veterans and the need to focus on PTSD as a
major issue.[6] Nevertheless, the full extent to which mental health problems are
being detected and appropriately treated remains unclear. More detailed and
increased mental health screening is required for all battlefield-injured veterans.[7]
Unfortunately, PTSD is emerging as the signature injury for many US military
service members deployed to Iraq and Afghanistan since 2001. Preliminary
findings suggest that PTSD will be present in at least 18 percent of those serving in Iraq and 11 percent of those serving in Afghanistan. The notable increase
in suicide among military personnel can be directly linked to a stressed, strained,
and exhausted military. The unprecedented numbers of redeployments to Iraq
and Afghanistan only exacerbate the stress placed on soldiers. The killing of five
fellow soldiers by one of their own in May 2009 serves to illustrate the mental
toll that the current wars are taking on our troops.[8] Research indicates an estimated 550 to 650 veterans committing suicide each month as a direct result of
PTSD. Sadly, their names aren't considered part of the more than 5, 100 military war deaths, which rise with each passing month.[9] Nor will the psychological disorders caused by the wars in Iraq and Afghanistan be included as part
of the 50,000 severe combat wounds inflicted thus far.[10]

Recently, the Department of Defense (DOD) and the Veterans Administration (VA) have come under Congressional and public scrutiny regarding
their capacity to address PTSD. Under current policy, veterans must prove a
service connection where they "engaged in combat with the enemy" to get the
VA to cover care related to PTSD and provide benefits. VA rules generally require a combat action decoration, unit records, or other documentation to prove
a veteran engaged in combat with the enemy. On a claim for PTSD, veterans
must show credible third-party evidence that they suffered combat-related
stress, such as eyewitness verification. Unfortunately, the VA's standard of proof
for what constitutes "engaging in combat with the enemy" is often too high for

many veterans suffering from PTSD to prove their case since this standard, devised in 1993, fails to recognize what we now know about this debilitating illness and the nature of today's counterinsurgencies.

The Current Situation

While all soldiers deployed to a war zone will feel some degree of stress, Pentagon surveys suggest that most will manage to readjust to normal. However, as many as 30 percent of troops with three or more deployments are likely to suffer from serious mental-health problems.[11] The current 12 months between deployments seems inadequate for soldiers to recover from the stress of a combat deployment before heading back to war. Thus, the number of soldiers requiring long-term mental-health services will continue to increase with both the increased frequency and duration of combat deployments. Service members and their families need the immediate attention of the nation to ensure successful reintegration, transition, and recovery pre- and post-deployment.

Conclusions

From the words of retired Gen Colin Powell, former chairman of the Joint Chiefs of Staff:

> This country has a profound obligation to honor its commitments to our veterans—including the lifetime medical care they were promised. Moreover, healthcare is an important incentive in attracting quality recruits to today's all-volunteer armed forces, on which our very national security depends.[12]

Since the 1930s, the VA has provided primary care, specialized care, and related medical and social support services for veterans of the United States military. Our military service members earned the best care this country can afford. Unfortunately, there is a shortfall in our knowledge and understanding about the mental health needs of our combat veterans, as well as gaps in the access and quality of care that must be addressed. With PTSD symptoms on the rise, it is up to the DOD and the VA to improve their ability to assist our veterans. The cost of mental health care is not and should never be an excuse to ignore the needs of these service members.

Recommendations

Ensure DOD and VA Commitment

The first step in the journey to resolving issues surrounding PTSD is to make everyone aware that PTSD is real. Second, the DOD and VA must make a comprehensive commitment to provide mental health services for our combat veterans. Service members must be aggressively encouraged to seek care without being concerned with the stigma often associated with it.

Understand Veteran Needs

While there is wide policy interest and concern by the VA and the DOD, there are still significant gaps in understanding the needs of our veterans. The DOD and VA need to find ways to work effectively with civilian and military practitioners in an effort to discover the best treatments for combat-related PTSD and to improve both the efficiency and transparency of the system. Otherwise, we risk facing another generation of combat veterans similar to that of Vietnam. Military leaders and mental health physicians should consider screening combat veterans for PTSD immediately upon their return to the United States. Preemptive treatments should also be considered, including PTSD recognition for family members, service members, and battle buddies.

Reduce Stigma

Online tools should be developed to assist service members who wish to find out privately if they are experiencing symptoms of PTSD and guide them to appropriate resources to get help. Although most service members will initially be treated in military treatment facilities, the mental health clinics can then contact the service member privately so that they may receive a referral for off-base counseling. Perhaps if service members had access to confidential treatment, there might be an increase in veterans seeking out the help they need. To help eliminate the stigma associated with seeking psychological health treatment, it is imperative to change military culture and encourage service members to seek treatment. Furthermore, it is equally important to provide commanders, families, and friends the information they need to effectively guide those suffering with PTSD.

Along with physical, mental, and social screening, there should also be mandatory screening that includes brain scans for traumatic brain injury. These screenings should be as common as receiving anthrax vaccinations before deployments. Long-term assessments should also be made, not just with the service member, but with the immediate family as well. If there are underserved areas, then civilian mental health centers should be incentivized to care for these combat vets.

Improve Access to Care

The VA faces challenges in providing access to care for OEF/OIF veterans, many of whom have difficulty securing appointments, particularly in facilities that have been resourced primarily to meet the demands of older vets. This new group of veterans needs special attention and a high priority. The DOD and the VA need to consider that there is a substantial unmet need for treatment of PTSD and major depression among military service members following deployment. With more than 300,000 new cases of mental health conditions among OEF/OIF veterans, a commensurate increase in treatment capacity and qualified providers is desperately needed. Incentivizing qualified professionals is one way to attract and retain well-trained mental health professionals.

The bottom line is that veterans should follow Brig Gen Gary S. Patton's example and be empowered to seek appropriate care. Likewise, commanders, supervisors, and family members must encourage individuals to seek health care before problems become critical. A comprehensive system to monitor the follow-up care of PTSD victims also needs to be established. It is important to keep in mind that it is not just the traumatic experience of war, but it is also the constant reminders and meaning of those events that actually create the trauma our service members experience. Unfortunately, the prevalence of traumatic mental injuries among veterans is high and is destined to continue growing in the future.

Notes

(All notes appear in shortened form. For full details, see the appropriate entry in the bibliography.)

1. Gilbertson, "The Life and Lonely Death."
2. Ibid. See also Murdough, "Pain, Depression, PTSD."
3. Glasser, "A Shock Wave of Brain Injuries"; Hoge et al., "Association of Posttraumatic Stress Disorder," 150–53; Hoge et al., "Mild Traumatic Brain Injury," 453–63; Tanielian et al., *Invisible Wounds of War.*
4. MacGregor et al., "Psychological Correlates of Battle and Nonbattle Injury," 224.
5. National Center for PTSD, *Returning from the War Zone.*
6. Shen, Arkes, and Pilgrim, "The Effects of Deployment Intensity," 217.
7. MacGregor et al., "Psychological Correlates of Battle and Nonbattle Injury."
8. Ibid.
9. Murdough, "Pain, Depression, PTSD."
10. Gilbertson, "The Life and Lonely Death."
11. Polusny et al., "Impact of Prior Operation Enduring Freedom/Operation Iraqi Freedom Combat Duty," 35.
12. National Association for Uniformed Services, http://www.naus.org/news/news_veterans.html.

Bibliography

Bumiller, Elisabeth. "Army Addresses Soldiers' Suicide Rate." *New York Times,* 18 November 2009, A19.

Davis, John. "Parents of Marine Found Dead Seek PTSD Awareness." *Sarasota-Herald Tribune,* 12 March 2008. http://www2.tbo.com/content/2008/mar/12/parents-marine-found-dead-seek-ptsd-awareness/ (accessed 16 July 2009).

Department of Veterans Affairs (VA). "Treatment of PTSD." http://ncptsd .va.gov/ncmain/ncdocs/fact_shts/fs_treatmentforptsd.html.

Garamone, Jim. "Army Brig. Gen. Gary S. Patton Urges Servicemembers to Seek Help for Stress Disorder." *Militaryinfo.com,* 6 April 2009. http://www.militaryinfo.com/news_story.cfm?textnewsid=2951.

Gilbertson, Ashley. "The Life and Lonely Death of Noah Pierce." *Virginia Quarterly Review* 84, no. 4 (2008).

Glasser, R. A. "A Shock Wave of Brain Injuries." *Washington Post*, 8 April 2007. http://www.washingtonpost.com/wp-dyn/content/article/2007/04/06/AR2007040601821.html (accessed on 24 May 2009).

Government Accountability Office (GAO). *VA Health Care: Spending for Mental Health Strategic Plan Initiatives Was Substantially Less Than Planned.* Washington, DC: GAO, November 2006. www.gao.gov/cgi-bin/getrpt?GAO -07-66 (accessed 24 May 2009).

Hoge, Charles W., Artin Terhakopian, Carl A. Castro, Stephen C. Messer, and Charles C. Engel. "Association of Posttraumatic Stress Disorder with Somatic Symptoms, Health Care Visits, and Absenteeism among Iraq War Veterans." *American Journal of Psychiatry* 164, no. 1. (January 2007): 150–53.

Hoge, Charles W., Dennis McGurk, Jeffrey L. Thomas, Anthony L. Cox, Charles C. Engel, and Carl A. Castro. "Mild Traumatic Brain Injury in U. S. Soldiers Returning from Iraq." *New England Journal of Medicine* 358, no. 5 (31 January 2008): 453–63.

House Committee on Veterans' Affairs. *Legislative Hearing on H.R. 952, the "Compensation Owed for Mental Health Based on Activities in Theater Posttraumatic Stress Disorder Act."* http://veterans.house.gov/legislation/111 legislation.shtml.

MacGregor, Andrew J., Richard A. Shaffer, Amber L. Dougherty, Michael R. Galarneau, Rema Raman, Dewleen G. Baker, Suzanne P. Lindsay, Beatrice A. Golomb, and Karen S. Corson. "Psychological Correlates of Battle and Nonbattle Injury among Operation Iraqi Freedom Veterans." *Military Medicine* 174, no. 4 (March 2009): 224.

Murdough, Brenda. "Pain, Depression, PTSD, and the Silent Wounds of War." *EP Magazine*, I May 2009.

National Association for Uniformed Services. http://www.naus.org/news/news_veterans.html.

National Center for PTSD. "Returning from the War Zone: A Guide for Families and Military Members." http://ncptsd.va.gov/ncmain/ncdocs/manuals/GuideforFamilies.pdf (accessed 20 Aug 2009).

National Council on Disability. "Invisible Wounds: Serving Service Members and Veterans with PTSD and TBI." http://www.ncd.gov/newsroom/publications/2009/veterans.doc.

Newman, Richard A. "Combat Fatigue: A Review of the Korean Conflict." *Military Medicine* 129 (1964): 921–28.

Polusny, Melissa A., Christopher R. Erbes, Paul A. Arbisi, Paul Thuras, Shannon M. Kehle, Michael Rath, Cora Courage, Madhavi K. Reddy, and Courtney Duffy. "Impact of Prior Operation Enduring Freedom/Operation Iraqi Freedom Combat Duty on Mental Health in a Predeployment Cohort of National Guard Soldiers." *Military Medicine* 174, no. 4 (April 2009): 353.

Seal, Karen H., Thomas J. Metzler, Kristian S. Gima, Daniel Bertenthal, Shira Maguen, and Charles R. Marmar. "Trends and Risk Factors for Mental Health Diagnoses among Iraq and Afghanistan Veterans Using Department of Veterans Affairs Health Care, 2002–2008." *American Journal of*

Public Health, 16 July 2009. http://www.ajph.org/cgi/reprint/AJPH.2008 .150284v1 (accessed 16 July 2009).

Shen, Y., J. Arkes, and J. Pilgrim. "The Effects of Deployment Intensity on Post-traumatic Stress Disorder: 2002–2006." *Military Medicine* 174, no. 3 (2009): 217.

Simon, Cecilia Capuzzi. "Bringing the War Home." *Psychotherapy Networker* 31, no. 1 (January/February 2007): 28–37.

Tanielian, Terri, Lisa H. Jaycox, Terry L. Schell, Grant N. Marshall, M. Audrey Burnam, Christine Eibner, Benjamin R. Karney, Lisa S. Meredith, Jeanne S. Ringel, Mary E. Vaiana, and the Invisible Wounds Study Team. *Invisible Wounds of War: Summary and Recommendations for Addressing Psychological and Cognitive Injuries.* Santa Monica, CA: RAND Corporation, 2008.

Thompson, Mark. "America's Medicated Army." *Time* in Partnership with *CNN*, 5 June 2008. http://psychrights.org/Articles/080608TimeMagAmericas MedicatedArmy.htm (accessed 24 Aug 2009).

Towrey, Chris. "Healing the Wounds of War." *Massage Magazine*, March 2009.

Tull, Matthew. "Rates of PTDS in Veterans." *About.com*, 22 July 2009. http:// ptsd.about.com/od/prevalence/a/MilitaryPTDS.htm (accessed 20 Aug 2009).

Zoroya, Gregg. "A Fifth of Soldiers at PTSD Risk." *USA Today*, 6 March 2008. http://www.usatoday.com/news/world/iraq/2008-03-06-soldier-stress _N.htm (accessed 29 August 2009).

About the Authors

Dr. Carla Sizer is recently retired from the United States Air Force after 21 years of honorable service. Her last military duty station was as an assistant professor at the United States Air Force Academy. She currently works as a program manager for the Department of Defense at Peterson Air Force Base, where she manages space control sensor programs. Dr. Sizer's interests include researching post-traumatic stress, mentoring others, and public speaking. Dr. Sizer's eldest son was killed in action during Operation Iraqi Freedom on 5 September 2007. After her son's death she founded the Pikes Peak Gold Star Mother's Support Group, which is for mothers who have lost their sons or daughters in war. She also founded the Specialist Dane R. Balcon Junior Reserve Officer Training Corps (JROTC) Scholarship Fund, dedicated for JROTC cadets in high school who plan to pursue higher education. Dr. Sizer completed her undergraduate work at Southern Illinois University at Carbondale and her graduate work at Georgia College and State University in public administration. She earned her doctorate from the University of Phoenix.

Dr. Claude Toland is the president of DeVry University—Houston Metro. He joined DeVry University in June 2007 as the dean of academic affairs, where he provided leadership in developing and growing DeVry's academic programs. Prior to joining DeVry University, Dr. Toland served as campus director for the University of Phoenix. Dr. Toland earned a bachelor's degree from Colorado Christian University in addition to two master's degrees and a doctorate from the University of Phoenix.

CHAPTER 25

Enjoining an American Nightmare

Matthew Harwood

On a sweltering Washington, DC, day in late August 2009, the Central Intelligence Agency (CIA) released a special review from its inspector general entitled "Counterterrorism Detention and Interrogation Activities," dated 7 May 2004.[1] The 162-page report, more than five years old by its release, investigated the agency's use of "enhanced interrogation techniques" (EIT) on high-value al-Qaeda detainees caught by the United States after the terrorist attacks of 11 September 2001. The 10 specific EITs deemed legal by the Justice Department's Office of Legal Counsel for CIA use included stress positions, sleep deprivation, and waterboarding—all commonly considered torture by international law, the human rights community, and the United States prior to 2001.[2] Unfortunately, those techniques would finally make their way into military interrogation rooms in Afghanistan and Iraq, most notoriously Abu Ghraib, in the United States' Global War on Terror.

Despite assurances from the Bush administration that such EITs were legal, the CIA's inspector general launched the investigation after receiving information that "some employees were concerned that certain covert agency activities at an overseas detention and interrogation site might involve violations of human rights."[3] The employees worried that CIA interrogators tortured detainees in contravention of United States and international law. Participants in the Counterterrorist Center (CTC) program, responsible for terrorist interrogation, damningly told investigators that they feared prosecution for torturing detainees.[4] Worse still, the report could not determine whether the enhanced techniques worked, while acknowledging that the techniques' practitioners knew they could harm the prisoner. According to the report, "the fact that precautions have been taken to provide on-site medical oversight in the use of all EITs is evidence that their use poses risks."[5]

Despite this guilt and fear among interrogators, defenders of these "enhanced interrogation techniques" argue that torturing detainees in violation of US and international law was and remains necessary to safeguard the American population against a heartless enemy that could be stopped no other way. Lead among these was former vice president Dick Cheney, who played an intimate role in pushing for the EITs.[6] In the late hours of 24 August 2009, the former vice president issued a statement defending EITs after the CIA documents' release. "The documents released . . . clearly demonstrate that the individuals subjected to Enhanced Interrogation Techniques provided the bulk of intelligence we gained about al-Qaeda," he said. "This intelligence saved lives and prevented terrorist attacks."[7]

While the statement ignored whether or not these detainees gave the "bulk of intelligence" before or after they were tortured, the underlying message was undoubtedly clear: Torture works. It saves American lives. Argument over. The problem, however, is that these arguments were deeply flawed. Torture rarely, if ever, works. And the lives it may save in the immediate present will not equal the multitude of servicemen and civilian lives lost because of torture's power to stoke anti-American violence worldwide. But there are better, moral arguments than these practical concerns for why American officials and servicemen should never torture. Torture violates everything the United States is supposed to stand for: the sanctity of the individual, human rights, and the rule of law.

Drawing on the United States' historic opposition to torture in its darkest days, the practice's prohibition internationally and domestically, its grotesque and counterproductive uselessness, and the irreparable harm it does to both tortured and torturer alike, I hope to convince servicemen that torture is always wrong and harms US national security and prestige. I will stress that any official or serviceman's actions that lead to the torture of another human being should be prosecuted to the fullest extent of the law. Only then can the US military reclaim the moral high ground it has lost in this new century of counter-terrorist conflict.

A Founding Aversion

Proponents of torturing detainees often resort to the "ticking time bomb." In this hypothetical situation, the United States has caught a terrorist with knowledge of an imminent and catastrophic attack against an American city, and torture is the only way to find out where the bomb is located. Putting aside the slim probability that such a scenario could happen outside a television show,[8] there was a time in American history where even the idea of the United States seemed destined for demise. During the darkest days of the American Revolutionary War, Gen George Washington prohibited his rag-tag army of colonists, seething with vengeance, from torturing British prisoners of war (POW). This order came when American POWs, described as traitors and insurgents by the British military, were routinely tortured. After the Battle of Bunker Hill, all 31 colonial captives died in British custody. The circumstances were not pretty.[9]

Despite this knowledge, Washington warned his Northern Expeditionary Force on 14 September 1775 that:

> Should any American soldier be so base and infamous as to injure any [prisoner] ... I do most earnestly enjoin you to bring him to such severe and exemplary punishment as the enormity of the crime may require. Should it extend to death itself, it will not be disproportional to its guilt at such a time and in such a cause ... for by such conduct they bring shame, disgrace and ruin to themselves and their country.[10]

After the Continental Army's victory at the Battle of Trenton, Washington guaranteed the humane treatment of all POWs in colonial custody. "Treat them with humanity, and let them have no reason to complain of our copying the brutal example of the British Army in their treatment of our unfortunate brethren who have fallen into their hands," he wrote. Even when the country could have been nothing more than an aborted dream, Washington chose to outlaw torture rather than desecrate the Enlightenment principles that the Continental Army fought for. Scott Horton, an international lawyer and harsh critic of the Bush administration's enhanced interrogation techniques, writes:

[Washington] made it a point of fundamental honor (and that was his word) that the Americans would not only hold dearly to the laws of war, they would define a new law of war that reflected the humanitarian principles for which the new Republic had risen. These principles required respect for the dignity and worth of every human being engaged in the conduct of the war, whether in the American cause or that of the nation's oppressor.[11]

The decision not to torture derived not only from General Washington's fealty to liberal[12] principles, but from a strategist's cunning. Continental POWs were treated so well that many British soldiers and their Hessian mercenaries defected to the Continental Army, many of whom became citizens when the colonies achieved independence.[13]

Even Pres. Abraham Lincoln, faced with the country's disintegration and ruin, banned torture during the Civil War. Endorsing the Lieber Code for Union soldiers, Lincoln outlawed the use of "torture to extort confessions."[14] The code, named after Francis Lieber, a professor of Columbia College in New York, would become the foundation for international laws of armed conflict. "The governments of Prussia, France, and Great Britain copied it. The Hague and Geneva Conventions were indebted to it," writes historian Richard Shelley Hartigan. "Though buried in voluminous United States government publications, 'the General Orders, no. 100,' remains a benchmark for the conduct of an army toward an enemy army and population."[15]

Despite Washington's historic precedent, reaffirmed by Lincoln, the US military finds itself stained with torture's disgrace for adopting the Bush administration's enhanced interrogation techniques.

A Universal Abomination

Beyond its own military prohibitions not to torture in the Revolutionary War and the Civil War, the United States has agreed multiple times not to torture anyone that falls into its custody since the end of World War II. Underneath the Third Geneva Convention of 1949, the United States pledged not to do "violence to life and person, in particular murder of all kinds, cruel treatment and torture" as well as "outrages upon personal dignity, in particular, humiliating and degrading treatment."[16]

While the previous presidential administration argued in internal memos that the Geneva Conventions did not apply to al-Qaeda detainees because the terrorist organization was not a "High Contracting Party to Geneva," it still reaffirmed its belief that detainees should be treated humanely.[17] Despite the administration's unilateral decision that Geneva did not apply, the United States was also a signatory to another international treaty that banned all forms of torture absolutely. The Convention against Torture and Other Cruel, Inhuman or Degrading Treatment or Punishment bans torture based on the "inherent dignity of the human person." According to the treaty, torture is defined as:

> any act by which severe pain or suffering, whether physical or mental, is inten-tionally inflicted on a person for such purposes as obtaining from him or a third person information or a confession, punishing him for an act he or a third person has committed or is suspected of having committed, or intimidating or coercing him or a third person, or for any reason based on discrimination of any kind, when such pain or suffering is inflicted by or at the instigation of or with the consent or acquiescence of a public official or other person acting in an official capacity.[18]

The Convention finds torture so abominable that "no exceptionable circum-stances whatsoever, whether a state of war or a threat of war, internal political instability or any other public emergency" could justify the practice.

Pres. Ronald Reagan signed the Convention in 1988,[19] and in 1994, the US Congress ratified it, making it the supreme law of the land.[20] In 1996, Congress also passed the War Crimes Act strengthening the rule of law against torture. Much like George Washington more than two centuries before, the United States declared that any citizen, "whether inside or outside the United States," involved in torture would face serious punishment. According to the law, a US citizen convicted of torture "shall be fined under this title or imprisoned for life or any term of years, or both, and if death results to the victim, shall also be subject to the penalty of death."[21]

Taken together, all three laws not only ban torture under all circumstances, but exclude nobody, no matter their position or rank, from prosecution, even execution. Is it any wonder that both CIA and military interrogators adminis-tering EITs feared not only for their reputations, but their very freedom if their actions were exposed? In one memorable passage from the CIA inspector gen-eral report, a CIA officer feared he and his colleagues would find themselves on a wanted list before the World Court for war crimes.[22]

To understand why CIA and military interrogators, as well as other officials and lawyers inside the US government, feared that EITs constituted torture, it is necessary to describe the most notorious practice: waterboarding. In the CIA's own words:

> The application of the waterboard technique involves binding the detainee to a bench with his feet elevated above his head. The detainee's head is immobilized and an interrogator places a cloth over the detainee's mouth and nose while pour-ing water onto the cloth in a controlled manner. Airflow is restricted for 20 to 40 seconds and the technique produces the sensation of drowning and suffocation.[23]

Known as the "water treatment" during World War II, the Japanese com-monly waterboarded their POWs throughout the Pacific theater.[24] After hos-tilities ceased, Gen George McArthur convened the International Military

Tribunal for the Far East (IMTFE), composed of judges from the nations previously at war with Japan, to prosecute Japanese officers for their torture and other inhumane treatment of Allied POWs. Among the techniques listed as torture was waterboarding.[25]

The IMTFE's description of the practice is eerily similar to the CIA's:

> The so-called "water treatment" was commonly applied. The victim was bound or otherwise secured in a prone position; and water was forced through his mouth and nostrils into his lungs and stomach until he lost consciousness. Pressure was then applied, sometimes by jumping upon his abdomen to force the water out. The usual practice was to revive the victim and successively repeat the process.[26]

Some of the Japanese defendants who were found responsible for "ordering, authorizing, and permitting commission of war crimes including, *inter alia*, torture," were sentenced by the IMTFE to death by hanging.[27]

Perhaps more inconvenient to defenders of EITs, especially those inside the Department of Justice (DOJ) that approved the techniques during the Bush administration, is that US courts had ruled on waterboarding before. In 1983, the DOJ successfully prosecuted a Texas sheriff and three of his deputies for waterboarding suspects in violation of their civil rights in 1983. Count one of the indictment alleged the defendants conspired to:

> subject prisoners to a suffocating "water torture" ordeal in order to coerce confessions. This generally included the placement of a towel over the nose and mouth of the prisoner and the pouring of water in the towel until the prisoner began to move, jerk, or otherwise indicate that he was suffocating and/or drowning.[28]

All four were convicted. Sheriff James Parker received 10 years in prison and a $12,000 fine. During sentencing, District Judge James DeAnda called Parker and his deputies "a bunch of thugs," adding, "the operation down there would embarrass the dictator of a country."[29]

And yet practices once reserved for the twentieth century's worst dictators and secret police forces were embraced by the same administration that vowed to destroy tyranny wherever it reared its terrible head. The Bush administration approved the CTC program's using the EITs on detainees, including waterboarding, without setting limits.[30] And waterboard they did. The CIA inspector general's report states al-Qaeda operative Abu Zubaydah was waterboarded 83 times[31] and stuffed inside a cage he referred to as a "a tiny coffin."[32] Zubaydah was subjected to these extreme techniques after interrogators determined he was holding out on them. He wasn't.[33] Khalid Sheik Mohammad, the mastermind of 9/11, was waterboarded 183 times[34] and told his children would be murdered if he did not talk.[35] The effects of such techniques shattered Zubaydah's psyche. He masturbated "like a monkey," a former CIA officer told journalist Jane Mayer, adding, "[Zubaydah] didn't care that they were watching him. I guess he was bored, and mad."[36]

Many of the same EITs that were used against Zubaydah and Mohammad migrated to the detention facilities at Guantanamo Bay, Cuba, (GTMO) when Secretary of Defense Donald Rumsfeld approved 15 special "counter-resistance techniques"[37] for use against Mohammed Mani' Ahmad Sha' Lan al-Qahtani, otherwise known as "detainee number 063," by American officials, on 2 December 2002.[38] Qahtani endured a 54-day span of harsh interrogation techniques: 20-hour interrogations; standing sessions that swelled his feet and hands; sexual humiliation, including a forced enema; and denial of bathroom breaks.[39] Qahtani's health began to fade; his heart rate plunged. A psychiatrist who viewed Qahtani's medical history over the interrogation span questioned whether it put him "in danger of dying."[40] Qahtani begged his interrogators to let him commit suicide.

And just as the CTC program's harsh interrogation practices spread to GTMO, Rumsfeld's "counter-resistance techniques" also spread to detention facilities in Afghanistan and Iraq, most notoriously Abu Ghraib prison—the same facility in which deposed dictator Saddam Hussein tortured his political prisoners. There detainees were subjected to horrifying abuses, according to an internal military report authored by Maj Gen Antonio M. Taguba:

> Breaking chemical lights and pouring the phosphoric liquid on detainees; pouring cold water on naked detainees; beating detainees with a broom handle and a chair; threatening male detainees with rape; allowing a military police guard to stitch the wound of a detainee who was injured after being slammed against the wall in his cell; sodomizing a detainee with a chemical light and perhaps a broom stick, and using military working dogs to frighten and intimidate detainees with threats of attack, and in one instance actually biting a detainee.[41]

At least 100 detainees died during American interrogation sessions.[42] In one autopsy report obtained by the American Civil Liberties Union (ACLU), a military medical examiner deemed the death of a 52-year-old man at a detention facility in Nasiriyah, Iraq, a homicide. The cause of death: strangulation.[43] The ACLU has compiled many more such autopsy reports.[44]

Even for the most ardent serviceman who believes in the necessity of breaking a few eggs sometimes, there's the military's own binding legal regime, the Uniform Code of Military Justice (UCMJ). According to internal memos from military lawyers entered into the Congressional record in July 2005, many of the EITs were deemed illegal underneath the UCMJ. "Several of the more extreme interrogation techniques, on their face, amount to violations of domestic criminal law and the UCMJ (e.g., assault)," wrote then-Maj Gen Jack L. Rives, deputy judge advocate general for the US Air Force. "Applying the more extreme techniques during the interrogation of detainees places the interrogators and the chain of command at risk of criminal accusations domestically."[45]

To stand before such legal and historical precedence and defy it because someday, somewhere in the future, a terrorist could attack the United States isn't patriotism: it is reckless vigilantism. "Cruelty disfigures our national character," former general counsel of the Navy, Alberto J. Mora, a heroic critic of EITs, told Mayer. "It is incompatible with our constitutional order, with our laws, and with our most prized values. . . . Where cruelty exists, law does not."[46]

Torture is a self-defeating proposition for any military, especially one committed to protecting the US Constitution against all enemies, foreign and domestic.

Torture's Blowback

There seems no reason to doubt that the United States' use of torture occurred out of a commendable duty to protect American soil from another terrorist attack by squeezing actionable intelligence out of detainees with ties to al-Qaeda. Yet the best of intentions cannot turn bad policies with even worse consequences into legal policies with salutary consequences.

Putting aside the paramount moral and legal concerns that torture raises, it is important to focus on the various reasons why the US military's decision to torture is already considered a strategic failure in its fight against jihadism.[47] Torture creates more enemies, produces bad intelligence, and leaves US servicemen vulnerable to the same treatment when the enemy captures them, whether that be another state or a substate actor, like al-Qaeda. In the strongest sense, it is contrary to the national security of the United States.

Maj Matthew Alexander, a pseudonym, is a military interrogator who followed the rules in Iraq while conducting 300 interrogations and supervising over a 1,000. According to him, torture has the second-order effect of increasing the level of insurgents and terrorists in the fight against US forces overseas.[48] "I listened time and time again to foreign fighters, and Sunni Iraqis, state the number one reason they had decided to pick up arms and join al-Qaeda was the abuses at Abu Ghraib and the authorized torture and abuse at Guantanamo Bay," said Alexander.[49] The Navy's former top lawyer agreed. During his testimony before the Senate Armed Services Committee in June 2008, Mora, general counsel of the Navy under then-secretary of defense Donald Rumsfeld, said, "US flag-rank officers maintain that the first and second identifiable causes of US combat deaths in Iraq—as judged by their effectiveness in recruiting insurgent fighters into combat—are, respectively the symbols of Abu Ghraib and Guantanamo."[50]

Torture's ability to radicalize its victims and those who identify with the victims shouldn't be surprising. In fact, two of the military's biggest targets in its war against al-Qaeda were produced by torture: Dr. Ayman al-Zawahiri and Abu Musab al-Zarqawi. Al-Qaeda's second-in-command and its chief intellectual, al-Zawahiri was active in underground Islamist activity dedicated to bringing down the Egyptian government of Anwar Sadat. After Sadat's assassination in 1981, Egypt's current president, Hosni Mubarak, swept up thousands of Islamists and threw them into prison, including al-Zawahiri. During torture sessions, al-Zawahiri broke down and gave up his comrades. According to the *New Yorker*'s Lawrence Wright, al-Zawahiri "was humiliated by this betrayal. Prison hardened him; torture sharpened his appetite for revenge."[51] Al-Zarqawi, on the other hand, was a Jordanian street thug and sex offender imprisoned in the country's notoriously harsh prison system. The leader of the most bloodthirsty segment of the Iraqi insurgency, al-Qaeda in Iraq, he was killed in Iraq by an American airstrike in June 2006. Like al-Zawahiri, he is believed to have been systematically tortured while he embraced Islam during his prison term, learning to memorize the Koran.[52] Both regimes received substantial security assistance from the United States, which wasn't lost on either of these men. According to Chris Zambelis of the Jamestown Foundation's *Terrorism Monitor*:

> For radical Islamists and their sympathizers, US economic, military, and diplomatic support for regimes that engage in this kind of activity against their own citizens vindicates al-Qaeda's claims of the existence of a US-led plot to attack Muslims and undermine Islam. In al-Qaeda's view, these circumstances require that Muslims organize and take up arms in self-defense against the United States and its allies in the region.[53]

Torture, as Zambelis notes, is a frequent topic of discussion for al-Zawahiri. In a May 2007 statement, he savaged US relations with Egypt. "American hypocrisy, which calls for democracy even as it considers Hosni Mubarak to be one of its closest friends, and which sends detainees to be tortured in Egypt, exports tools of torture to Egypt and spends millions to support the security organs and their executioners," he said, "even as the American State Department, in its annual report on human rights, criticizes the Egyptian government because it tortures detainees!"[54] So if indirect support of regimes that torture can produce such enemies, imagine the unknown number of enemies the United States will face in the future because US servicemen and intelligence agents personally battered and psychologically harmed detainees.

Moreover, torture doesn't only produce more enemies to detain or kill; it produces extremely unreliable intelligence. According to Army Field Manual (FM) 34-52, *Human Intelligence Collector Operations*, which outlines the military's acceptable interrogation standards:

> Experience indicates that the use of force is not necessary to gain the cooperation of sources for interrogation. Therefore, the use of force is a poor technique, as it yields unreliable results, may damage subsequent collection efforts, and can induce the source to say whatever he thinks the interrogator wants to hear.[55]

The United States has known this at least from the beginning of the Cold War when the government produced the survival, evasion, resistance, and escape program after 36 US Airmen were tortured into giving "stunningly false confessions during the Korean War." The program taught US servicemen captured by the enemy how to resist torture techniques by subjecting them to those same techniques, including waterboarding, under highly controlled circumstances. After 9/11, the program was tragically "reverse-engineered" into an instrument of torture by the US government.[56]

Not surprisingly, "stunningly false confessions" followed from detainees during harsh interrogations. One of the most unjust cases was that of Maher Arar, an innocent Canadian telecommunications engineer. American officials snatched Arar during his trip home to Canada from Tunisia while he was trying to board his connecting flight at John F. Kennedy Airport in New York City. Implicated by confessions extracted by torture in Syria, Arar was extraordinarily rendered to the same country where he was also tortured. During these torture sessions, Arar confessed to training with al-Qaeda in Afghanistan. He had never been to the country. "I was ready to do anything to get out of that place, at any cost," he told reporter Mayer.[57]

Another case was low-level al-Qaeda member Zubaydah, who "reportedly confessed to dozens of half-hatched or entirely imaginary plots to blow up American banks, supermarkets, malls, the Statue of Liberty, the Golden Gate Bridge, the Brooklyn Bridge, and nuclear power plants."[58] The government dispatched federal law enforcement to follow up on these leads, wasting time and resources.[59] The CIA inspector general's report seems to confirm the fantastical nature of Zubaydah and other detainees' confessed plots, stating "this Review did not uncover any evidence that these plots were imminent."[60]

In 2006, the Intelligence Science Board investigated what was scientifically known about interrogation and intelligence gathering for the US intelligence community in the wake of the torture scandals. Its answer: not much. In a chapter reviewing the *KUBARK Counterintelligence Interrogation Manual*, the CIA manual notorious for discussing coercive methods, Col Steven Kleinman, an Air Force reservist and experienced intelligence officer, wrote that there is absolutely no empirical evidence that torture works:

> The scientific community has never established that coercive interrogation methods are an effective means of obtaining reliable intelligence information.

> In essence, there seems to be an unsubstantiated assumption that "compliance" carries the same connotation as "meaningful cooperation" (i.e., a source induced to provide accurate, relevant information of potential intelligence value).[61]

But there is scientific evidence mounting that torture biologically impairs a victim's ability to recall information from long-term memory and thus is an ineffective interrogation technique. Writing in the journal *Trends in Cognitive Science*, Prof. Shane O'Mara of the Trinity College Institute of Neuroscience argues that the EITs approved by the Justice Department legal memos would not elicit truthful information.[62] Rather, O'Mara contends that extreme stress on captives would degrade their memory and could even produce false memories, or confabulations. In these situations, interrogators would be hard-pressed to distinguish accurately between what was truth and what was stress-induced fiction.[63] The KUBARK manual, Kleinman writes, essentially agrees with O'Mara's findings. Even if captives had intelligence information, its authors state, torture is so psychologically damaging that it could degrade their ability to communicate it accurately.[64]

O'Mara dismisses the belief that torture works as a "folk psychology that is demonstrably incorrect."[65] Everything neurobiologists know about the brain, he says, proves EITs will not likely help detainees remember critical intelligence information. "On the contrary, these techniques cause severe, repeated and prolonged stress, which compromises brain tissue supporting memory and executive function," O'Mara writes. "The fact that the detrimental effects of these techniques on the brain are not visible to the naked eye makes them no less real."[66]

The kicker in the fight over whether torture works or not is that there is another way to get good, solid, actionable intelligence from terrorist detainees: be nice to them. This is a style of interrogation known as rapport-building, outlined by Army FM 34-52 and used by the FBI and police departments. It was used to kill Iraq's most vicious terrorist, who, ironically, torture helped produce. Major Alexander recounts how in only six hours time, his rapport-building technique convinced a man to give up the location of al-Zarqawi. "The old methods of interrogation had failed for 20 days to convince this man to cooperate," Alexander said in an interview. "The American public has a right to know that they do not have to choose between torture and terror."[67] Ironically, Kleinman writes in his review of the KUBARK manual that its authors spent considerable time discussing how important rapport-building skills are to any interrogator.[68]

The final practical reason why the United States, especially its military, should never torture is simple self-interest. There is no way to tell what repercussions will follow from the United States' embrace of EITs. The Judge Advocate General School's (TJAGS) dissenting memos understood this. As MGEN Thomas J. Romig, judge advocate general for the Army, observed in his memo weighing the legality of EITs, "the implementation of questionable techniques will very likely establish a new baseline for acceptable practice in this area, putting our service personnel at far greater risk and vitiating many of the POW/detainee safeguards the US has worked hard to establish over the past five decades."[69] Or as Lt Gen Jack Rives, the Air Force's judge advocate general, put it, "Treating [Operation Enduring Freedom] detainees inconsistently with the Conventions arguably 'lowers the bar' for treatment of US POWs in future conflicts."[70] In other words, the US flight from the international legal paradigm it helped create would open captured servicemen to torture.

There was something else the US government opened its service personnel to by condoning torture: prosecution. The TJAGS understood this as well. Romig argued that the administration's legal argument that the commander-in-chief could do anything to protect national security in wartime would not likely prevail in either US courts or internationally. "If such a defense is not available," he wrote, "soldiers ordered to use otherwise illegal techniques run a substantial risk of criminal prosecution or personal liability arising from a civil lawsuit."[71] Rives' analysis also agreed, and much like the guilt-riddled CIA interrogators that first used EITs on high-value detainees, he believed implementing the proposed interrogation techniques "places interrogators and the chain of command at risk of criminal accusations abroad, either in foreign domestic courts or in international fora, to include the [International Criminal Court]."[72] Sure that one technique amounted to torture, Rear Adm Michael F. Lohr, judge advocate general for the US Navy, argued that servicemen could not serve as interrogators when the technique was administered because "they are subject to UCMJ jurisdiction at all times."[73]

In addition to US history, the law, and the national security of the nation as well as its servicemen, there's one more intimate reason why torture is wrong: it destroys the humanity of all who come into contact with it.

Crossing Over

Torture is a wrenching experience for torturer and tortured alike. The descriptive notion of "breaking someone" should be explanation enough. To break something means to damage it irreparably. Once broken, something can never again be the same. Yet torture is nevertheless described this way, with little regard that the object is a human being.[74] Even less regard is given to the person commanded to strip another human being of his or her integrity—a ghastly responsibility that ultimately cracks the torturer as well as the tortured. According to Mayer:

> Experts on torture . . . often write of the corrosive and corrupting effect that such animalistic behavior has on discipline, professionalism, and morale. [One] former officer said that during "enhanced" interrogations, officers worked in teams, watching each other behind two-way mirrors. Even with this group support, he said, a friend of his who had helped to waterboard Khalid Sheikh Mohammed "has horrible nightmares." He went on, "When you cross over that line of darkness, it's hard to come back. You lose your soul. You can do your best to justify it, but it's well outside the norm. You can't go to that dark place without it changing you." He said of his friend, "He's a good guy. It really haunts him. You are inflicting something really evil and horrible on somebody."[75]

Indeed, Professor O'Mara notes that there is overwhelmingly evidence in "the historical literature" that former torturers fall into alcohol and drug abuse.[76]

No service member should be asked while defending his country to sacrifice his humanity, whether willingly or unwillingly. But this is exactly what happened as the US military, at the direction of civilian leadership, condoned torture in an ill-advised gamble to protect the country from further terrorist attacks. Torture, as the military's recent history shows, cannot be contained. Rather, as author and journalist Andrew Sullivan argues, it is a virus infecting its practitioners. Once it is unleashed, it has a way of spreading uncontrollably. "Remember that torture was originally sanctioned in administration memos only for use against illegal combatants in rare cases," Sullivan writes. "Within months of that decision, abuse and torture had become endemic throughout Iraq, a theater of war in which, even Bush officials agree, the Geneva Conventions apply."[77]

The US military's widespread use of torture once again shows good people are capable of very bad things when the right pressures are selected. This was dramatically illustrated during a classic social psychology experiment in the early 1970s by Philip Zimbardo at Stanford University. Using college students, Zimbardo randomly divided his test subjects into two groups of volunteers: guards and prisoners. The guards were given the authority to do whatever was needed within limits to maintain law and order inside the prison. When a rebellion broke out, the guards reacted fiercely, doling out arbitrary punishments and humiliating prisoners by stripping them and calling them names. According to Zimbardo, "In only a few days, our guards became sadistic and our prisoners became depressed and showed signs of extreme stress." The experiment was supposed to last for two weeks; it made it only six days. The college students, suddenly thrust into an unfamiliar environment, enacted their roles in a profound and unexpected fashion. The experiment was so intoxicating, Zimbardo wrote that:

> Even the "good" guards felt helpless to intervene, and none of the guards quit while the study was in progress. Indeed, it should be noted that no guard ever came late for his shift, called in sick, left early, or demanded extra pay for over-time work.[78]

Sparked by a paper trail of memos from the Bush administration, the United States has replicated a Stanford prison experiment of global proportions. In the process, it has jeopardized the humanity of every service member associated with the harsh interrogation regime it created. Along the way, the US military has seemingly forgotten its grandest historical mission: to protect the Enlightenment values enshrined in the Constitution every service member pledges to protect.

What Is to Be Done?

Fortunately for the military, its ability to close the door on the torture chamber is simple—it must cease torturing any human being that ever winds up in its custody. That's the easy part, and, unfortunately, the military will not get any credit for undoing abominable practices that should have never been done in the first place. Indeed, it will take a long slough for the US military to regain its prestige and honor domestically and internationally.

Regaining the high ground will mean difficult and unpopular decisions to investigate, try, and prosecute American servicemen who tortured detainees in their custody along with their superior officers. The military, as an honorable institution, must ignore the fact that the orders came from the secretary of defense and civilian lawyers. Military prosecutors must ignore any Nuremberg-style defense that relies on following orders, owing to the Torture Convention's blanket prohibition on such treatment regardless of the circumstances. The fact that a subordinate carried out a superior officer's orders should, however, be taken into account during sentencing. And like the International Military Tribunal (IMT) that prosecuted Axis officers, superior US officers who conspired in torture or knew of the abuse and did nothing to stop it must also be prosecuted under the doctrine of command responsibility. As just war theorist Michael Walzer argues, command responsibility means "military commanders, in organizing their forces, must take positive steps to enforce the war convention and hold the men under their command to its standards."[79] During the IMT, as previously noted, the Allied powers executed Axis officers under command responsibility, even those that arguably had no control over their subordinates.[80] The US military, by adhering to the same standards it applied to Axis officers, would show that the rule of law does indeed guide the US military, however long overdue its application is. Otherwise, the United States will retain its tarnished image as "a law unto itself," as General Romig observed in his 2003 memo criticizing EITs. Accountability is the only way to conquer impunity, the hallmark of tyranny.

Second, the US military must continue to adhere to the interrogation guidelines established by Army FM 34-52, which stresses the rapport-building approach and forbids the use of force and any inhumane treatment of prisoners. In this effort, the military recently received a push. On 24 August 2009, the same day the CIA released its report on EITs used against high-level detainees, the Special Task Force on Interrogations and Transfer Policies concluded that Army FM 34-52 should not only govern military interrogations, but any interrogation undertaken by any federal agency.[81] The task force, created by Pres. Barack Obama, also recommended forming specialized interrogation teams that recruit the government's best interrogators to question high-level terrorism suspects. According to a Justice Department press release, the High-Value Detainee Interrogation Group "would bring together officials from law enforcement, the US Intelligence Community and the Department of Defense to conduct interrogations in a manner that will strengthen national security consistent with the rule of law."[82] That last part, of course, is the most critical. MAJ Matthew Alexander, the rapport-building interrogator who got the intelligence that led to al-Zarqawi's demise, observes that the:

> success of an elite interrogation team will be dependent upon the leadership of the team ... and leadership of the interrogation team will be as important as the actual interrogations. It involves prioritizing detainees and information requirements, matching interrogators to detainees, and advising on interrogation strategies. The bureaucratic hurdles that are sure to arise given the inevitable power struggles will make the leadership challenge difficult.[83]

According to Major Alexander, the US government should focus on training elite interrogation leaders to ensure the United States never tortures anyone in its custody again. The US military should find interrogators like Alexander and recommend them for such distinction. Men of Alexander's caliber should also be tapped to teach servicemen the various historical, legal, practical, and personal reasons why torture conflicts with the best of the nation's ideals.

Despite the US military's entanglement with the dark side, it is important to remember that dissent coursed throughout the entire hierarchy. None, however, were more eloquent than CAPT Ian Fishback, who fought unsuccessfully to get his commanders to end the systematic abuse he witnessed, erect clear interrogation guidelines, and abide by command responsibility. After 17 months of consulting the military's chain of command for clear guidance, Captain Fishback finally broke down and wrote Senator John McCain, a former Vietnam POW tortured by the North Vietnamese, begging him for clear guidelines on "the lawful and humane treatment of detainees."[84] Fishback understood the enormous millstone torture strapped around all servicemen's necks and didn't want to see the American military and its honorable traditions end up in the abyss. He wrote:

> I am certain that this confusion contributed to a wide range of abuses including death threats, beatings, broken bones, murder, exposure to elements, extreme

forced physical exertion, hostage-taking, stripping, sleep deprivation, and degrading treatment. I and troops under my command witnessed some of these abuses in both Afghanistan and Iraq.

This is a tragedy. I can remember, as a cadet at West Point, resolving to ensure that my men would never commit a dishonorable act; that I would protect them from that type of burden. It absolutely breaks my heart that I have failed some of them in this regard.

That is in the past and there is nothing we can do about it now. But, we can learn from our mistakes and ensure that this does not happen again. Take a major step in that direction; eliminate the confusion. My approach for clarification provides clear evidence that confusion over standards was a major contributor to the prisoner abuse. We owe our soldiers better than this. Give them a clear standard that is in accordance with the bedrock principles of our nation.[85]

But Captain Fishback wasn't done. He, like other patriots inside the US military, understood that the military must be bound by the rule of law, as an example of its oath to preserve individual freedom against those like al-Qaeda, whose indiscriminate slaughter decapitates it. In a stirring crescendo of idealism and duty, Fishback asks, "Will we confront danger and adversity in order to preserve our ideals, or will our courage and commitment to individual rights wither at the prospect of sacrifice?"[86]

His answer exemplifies the citizen-soldier American service members pledge to be. "My response is simple. If we abandon our ideals in the face of adversity and aggression, then those ideals were never really in our possession. I would rather die fighting than give up even the smallest part of the idea that is 'America.'"[87] His letter is a testament that inside the US military lies redemption.

Notes

1. Central Intelligence Agency (CIA) Inspector General, *Counterterrorism Detention and Interrogation Activities* (*September 2001–October 2003*) *Special Review*, 2003-7123-IG (Washington, DC: CIA, 7 May 2004).

2. Ibid., 19–21.

3. Ibid., 1–2.

4. Ibid., 94.

5. Ibid., 89.

6. Jane Mayer, *The Dark Side: The Inside Story of How the War on Terror Turned into a War on American Ideals* (New York, NY: Anchor Books, 2008), chap. 4.

7. Stephen Haynes, "Cheney Statement on CIA," *The Weekly Standard,* The Blog, 24 August 2009, http://www.weeklystandard.com/weblogs/TWSFP/2009/08/cheney_statement_on_cia_docume.asp.

8. Critics of the "ticking time bomb" scenario, such as the Association for the Prevention of Torture, lampoon its proponents for "fetishizing" the television show *24*, where counterterrorist agent Jack Bauer routinely tortures suspects to get information to avert a terrorist act at the last minute. One notable military critic is US Army Brig Gen Patrick Finnegan, the dean of the United States Military Academy at West Point, as reported by Jane Mayer, "Whatever It Takes," *The New Yorker,* 19 February 2009.

9. Scott Horton, "A Tale of Two Georges," *Huffington Post*, 19 February 2007, http://www.huffingtonpost.com/scott-horton/a-tale-of-two-georges_1_b_41091.html.

10. Ibid.

11. Ibid.

12. *Liberal* in this instance refers to the philosophy of classical liberalism, which stressed the protection and sanctity of the individual person from all forms of tyranny.

13. Ibid.

14. Human Rights Watch, "Accountability for Torture: Questions and Answers," 13 May 2009, http://www.hrw.org/en/news/2009/05/13/accountability-torture.

15. Francis Lieber and Richard Shelley Hartigan, *Lieber's Code and the Law of War* (Piscataway, NJ: Transaction Publishers, 1983).

16. International Committee of the Red Cross (ICRC), "Convention (III) Relative to the Treatment of Prisoners of War, Geneva, 12 August 1949," http://www.icrc.org/IHL.nsf/FULL/375?OpenDocument.

17. Pres. George Bush, "Humane Treatment of Taliban and al-Qaeda Detainees," memorandum, 7 February 2002, http://www.pegc.us/archive/White_House/bush_memo_20020207_ed.pdf.

18. Office of the United Nations High Commissioner for Human Rights, "Convention against Torture and Other Cruel, Inhuman or Degrading Treatment or Punishment," 10 December 1984, http://www2.ohchr.org/english/law/cat.htm.

19. Department of State, "U.S. Signs UN Convention against Torture," *US Department of State Bulletin*, August 1988, http://findarticles.com/p/articles/mi_m1079/is_n2137_v88/ai_6742034/.

20. Michael John Garcia, "The U.N. Convention against Torture: Overview of U.S. Implementation Policy Concerning the Removal of Aliens," Congressional Research Service (CRS) Report for Congress, 11 March 2004, http://www.law.umaryland.edu/marshall/crsreports/crsdocuments/RL32276_03112004.pdf.

21. *US Code Collection*, Title 18, Section 2441, "War Crimes," http://www.law.cornell.edu/uscode/18/usc_sec_18_00002441----000-.html.

22. CIA Inspector General, *Counterterrorism Detention and Interrogation Activities*, 94.

23. Ibid., 15.

24. Howard S. Levie, *Terrorism in War—The Law of War Crimes* (Dobbs Ferry, NY: Oceana Publications, Inc., 1993), 357.

25. Evan Wallach, "Drop by Drop: Forgetting the History of Water Torture in U.S. Courts," *Columbia Journal of Transnational Law* 45, no. 42 (April 2007): 478.

26. International Military Tribunal for the Far East (IMTFE), *IMTFE Judgment (English Translation)*, Chapter VIII: Conventional War Crimes, http://www.ibiblio.net/hyperwar/PTO/IMTFE/IMTFE-8.html.

27. Wallach, "Drop by Drop," 493 n. 110.

28. Ibid., 502.

29. Ibid., 504.

30. Philippe Sands, *Torture Team: Rumsfeld's Memo and the Betrayal of American Values* (New York, NY: Palgrave Macmillan, 2008), 5.

31. CIA Inspector General, *Counterterrorism Detention*, 36.

32. Mayer, *The Dark Side*, 165.

33. Ibid., 178. According to Mayer's sources, Zubaydah was of marginal intelligence value. FBI agent Daniel Coleman, who translated the detainee's diaries, described Zubaydah as having "a schizophrenic personality."

34. CIA Inspector General, *Counterterrorism Detention*, 91.

35. Ibid., 43.

36. Mayer, *The Dark Side*, 175.

37. Philippe Sands, "The Complicit General," *New York Review of Books*, 24 September 2009, http://www.nybooks.com/articles/23071.

38. Mayer, *The Dark Side*, 190; Sands, "The Complicit General," 7.

39. Sands, "The Complicit General," 10–13; Mayer, *The Dark Side*, 206–07.

40. Mayer, *The Dark Side*, 208–9.

41. Seymour Hersh, "Torture at Abu Ghraib," *New Yorker*, 10 May 2004, http://www.newyorker.com/archive/2004/05/10/040510fa_fact; and the original Maj Gen Antonio M. Taguba re-

port, "Article 15-6 Investigation of the 800th Military Police Brigade," http://news.findlaw.com/hdocs/docs/iraq/tagubarpt.html.

42. John Sifton, "The Bush Administration Homicides," *Daily Beast*, 5 May 2009, http://www.thedailybeast.com/blogs-and-stories/2009-05-05/how-many-were-tortured-to-death/.

43. Office of the Armed Forces Regional Medical Examiner, "Final Autopsy Report," obtained by the American Civil Liberties Union (ACLU), http://action.aclu.org/torturefoia/released/102405/3164.pdf.

44. ACLU, "Autopsy Reports Reveal Homicides of Detainees in U.S. Custody," ACLU, http://action.aclu.org/torturefoia/released/102405/.

45. Maj Gen Jack L. Rives, "Memo from Major General Rives, Deputy Judge Advocate General, U.S. Air Force, dated 5 February 2003."

46. Mayer, *The Dark Side*, 236.

47. Since 9/11, US military actions have been commonly called the War on Terrorism. This is nonsensical. You do not fight a tactic—terrorism; you fight an enemy—al-Qaeda and its jihadist fellow travelers.

48. Patrick Cockburn, "Torture? It Probably Killed More Americans than 9/11," *The Independent*, 26 April 2009, http://www.independent.co.uk/news/world/middle-east/torture-it-probably-killed-more-americans-than-911-1674396.html.

49. Scott Horton, "'The American Public Has a Right to Know That They Do not Have to Choose between Torture and Terror': Six Questions for Matthew Alexander, author of *How to Break a Terrorist*," interview, *Harper's Magazine* online, 18 December 2008, http://www.harpers.org/archive/2008/12/hbc-90004036.

50. Alberto Mora, "Statement of Alberto Mora: Senate Committee on Armed Services Hearing on the Treatment of Detainees in U.S. Custody," 17 June 2008, http://armed-services.senate.gov/statemnt/2008/June/Mora%2006-17-08.pdf.

51. Lawrence Wright, "The Rebellion Within," *The New Yorker*, 2 June 2008, http://www.newyorker.com/reporting/2008/06/02/080602fa_fact_wright?currentPage=all.

52. Chris Zambelis, "Is There a Nexus Between Torture and Radicalization?" *Terrorism Monitor*, 26 June 2008, http://www.jamestown.org/single/?no_cache=1&tx_ttnews[tt_news]=5015.

53. Ibid.

54. Ibid.

55. Army Field Manual 34-52, *Human Intelligence Collector Operations*, September 2006, Chap. 1, http://www.globalsecurity.org/intell/library/policy/army/fm/fm34-52/chapter1.htm.

56. Mayer, *The Dark Side*, 157–61.

57. Ibid., 129–34.

58. Ibid., 178–79.

59. Ibid.

60. CIA Inspector General, *Counterterrorism Detention*, 88.

61. Steven M. Kleinman, "KUBARK Counterintelligence Review: Observations of an Interrogator," in *Educing Information: Interrogation: Science and Art* (Washington, DC: Intelligence Science Board, National Defense Intelligence College, 2006), 130.

62. Shane O'Mara, "Torturing the Brain," *Trends in Cognitive Sciences*, 24 September 2009, http://www.cell.com/trends/cognitive-sciences/fulltext/S1364-6613(09)00199-5.

63. Ibid.

64. Kleinman, "KUBARK Counterintelligence Review," 132–33.

65. O'Mara, "Torturing the Brain."

66. Ibid.

67. Horton, interview with Major Alexander, "The American Public Has a Right to Know."

68. Kleinman, "KUBARK Counterintelligence Review," 102.

69. Maj Gen Thomas J. Romig, "MG Thomas J. Romig, U.S. Army, the Judge Advocate General, dated 3 March 2003," memorandum.

70. Rives, memo.

71. Romig, memo.

72. Rives, memo.

73. Rear Adm Michael F. Lohr, "Memo from Rear Admiral Lohr, dated 13 March 2002."

74. Andrew Sullivan, "The Abolition of Torture," *New Republic*, 19 December 2005, http://www.tnr.com/article/politics/the-abolition-torture.

75. Mayer, *The Dark Side*, 175.

76. O'Mara, "Torturing the Brain."

77. Sullivan, "The Abolition of Torture."

78. Philip G. Zimbardo, "Stanford Prison Experiment: A Simulation Study of the Psychology of Imprisonment Conducted at Stanford University," http://www.prisonexp.org/.

79. Michael Walzer, *Just and Unjust Wars: A Moral Argument with Historical Illustrations*, 3rd ed. (New York, NY: Basic Books, 2000), 317.

80. Ibid., 319. The standard case of command responsibility is that of Japanese General Yamashita, who was executed in 1946 for the crimes his troops committed in the Philippine campaign, even though it was credibly established that he had no control over them.

81. Department of Justice, "Special Task Force on Interrogations and Transfer Policies Issues Its Recommendations to the President," press release, 24 August 2009, http://www.usdoj.gov/opa/pr/2009/August/09-ag-835.html.

82. Ibid.

83. Matthew Alexander, "Interrogation Elite," *Huffington Post*, 24 August 2009, http://www.huffingtonpost.com/matthew-alexander.

84. Capt Ian Fishback, "A Matter of Honor," *Washington Post*, 28 September 2005, http://www.washingtonpost.com/wp-dyn/content/article/2005/09/27/AR2005092701527_pf.html.

85. Ibid.

86. Ibid.

87. Ibid.

About the Author

Matthew Harwood is an associate editor of *Security Management* magazine. He holds a master of letters degree in international security studies from the University of St. Andrews in Scotland. He is a frequent contributor to *The Guardian* and has written for *The Washington Monthly*, *Columbia Journalism Review*, and the *Huffington Post*. Harwood has also worked for the television show *NOW* with Bill Moyers and was part of the team which produced the Peabody-award-winning documentary, "A Question of Fairness."

CHAPTER 26

HONOR CODES AND MILITARY CORE VALUES

SOCIAL POLICIES THAT WORK

Thomas J. Gibbons

Ethical transgressions within the ivy-covered walls of academia often make front-page news. In 2001, the University of Virginia charged 130 students with honor code violations.[1] In 2006, Ohio University announced 44 possible plagiarism cases within the School of Engineering.[2] And in 2007, the superintendent at the US Air Force Academy confined all 4,300 cadets to the campus in response to cheating, while the Army reported widespread cheating among Soldiers completing correspondence courses online.[3] Such large-scale violations of acceptable practices in military academic programs are particularly troubling and raise questions about the persistent belief that honor codes help to set high ethical standards of conduct throughout the ranks.

Of course, ethical lapses are not confined to colleges and universities. The US economy is still recovering from the Enron debacle. The public learned how Bernard Madoff swindled thousands of investors out of billions of dollars in 2009. In 2004, the principal deputy assistant secretary of the Air Force for acquisition and management was convicted of conspiring with Boeing executives to help the company obtain a lucrative multimillion-dollar defense contract.[4] Scandals like these create problems and raise doubts in all walks of life.

Social policies, such as honor codes, aimed at the development and maintenance of core values for young adults, are increasingly important to leaders in the military and higher education. Issues related to the use of honor codes have been the subject of many studies. Experts have demonstrated the relationship between college honor codes and business codes of ethics and their effect on reducing unethical behavior in the workplace.[5] However, a limitation of the research was that they only studied graduates from two universities, one with an honor code and another without.

Overview

This chapter reviews responses from military personnel who graduated from a wide range of colleges and universities, comparing college honor code experiences, military core values, and self-reported unethical behavior.[6] The sample of military personnel surveyed were students attending classes at the Naval Station in Newport, Rhode Island, in one of the four schools: US Naval Academy Preparatory School (NAPS), Surface Warfare Officer School (SWOS), College of Naval Command and Staff (CNC&S), and the College of Naval Warfare (CNW). The sample included members from all services and educational backgrounds at different stages in their careers. The 688 students who completed the questionnaire were divided into subgroups by schools: 171 NAPS students at the beginning of their careers; 127 SWOS students in the military for an average of 10 years; 204 CNC&S students in the military for an average of 14 years; and 186 CNW students in the military for an average of 20 years.

The questionnaire contained questions concerning self-reported unethical behavior both in college and in the military workplace and demographic information. To establish content validity of the questionnaire items, a group of higher education experts, senior military officers, and high school teachers were asked to rate the severity of the unethical behaviors. The scores from each group were averaged to yield a severity rating for each behavior. The respondents' score was the sum of the products of the severity rating times the frequency for each of the admitted behaviors.[7] Following the data analysis, select one-on-one interviews were conducted to develop and explore common themes from the questionnaire data.

College Honor Code Experience

Overall, the relationship between the respondent's military education level and his or her self-reported unethical behavior in college was statistically significant. At each of the levels, except NAPS, students with honor code experience had lower scores for self-reported unethical behavior in college than those without college honor code experience. Students at NAPS had only been at the school and exposed to its honor code for five months when the questionnaire was administered, and this may have impacted the results. College honor codes have been found to be significantly related to lower levels of cheating among college students.[8] Honor codes tend to create a sense of community.[9] Students become members of the honor code community, with the code being the bond that keeps them together. In many ways, the military is also a special community that bonds its members. There is a unique military culture with distinct customs, courtesies, and social policies. Members wear the same uniforms to strengthen these ties.

There were no statistically significant relationships for military education level, college honor code experience, and the interaction between the two with respect to self-reported unethical behavior at work. This finding was consistent with previous research, which also found that college honor code experience alone would not reduce work-related unethical behavior.[10]

During the interviews, 70 percent of respondents expressed an opinion that honor code experience had an impact on their ethical behavior at work.

Each respondent with an honor code experience noted that the experience was effective in encouraging and sustaining ethical behavior. The more senior the student's rank, the more likely there was to be a decline in the perception that the college honor code experience was effective in promoting and sustaining ethical behavior at work. A possible explanation may have been that these students had been away from college for a longer time and did not recollect their honor code experience.

Core Values Understanding

There was a statistically significant relationship between core values understanding and self-reported unethical behavior at work. The mean scores for self-reported unethical behavior at work were lower in all subgroups with a strong core values understanding. 90 percent of those interviewed agreed that military core values had an impact on their daily behavior in the workplace, 80 percent verified that college honor codes were reinforced by military core values, and 85 percent indicated that unethical behavior was either not a problem or just a minor problem in the military today. The predominant theme during the interviews was that leadership sets the ethical tone for an organization. These results are consistent with the extant literature.

Also of interest were some comments made by respondents. Many students said that current military training was on track to promote more ethical behavior. Others suggested positive feedback and mentoring programs with one-on-one conversations were needed to improve ethical behavior. Still others recommended more ethics training for both officers and senior enlisted personnel as a part of the curriculum of the military schools. A CNW student said, "We must constantly be aware of unethical behavior in our ranks. No matter how well we address it, it will always be a potential problem. There should be a core course in our war colleges and senior enlisted schools on ethics, not just Power-Point presentations or seminars."

This study examined two different factors that have been shown to influence ethical behavior in a military environment—college honor codes and core values. One of the major findings was that the college honor code experience reinforces military core values to support ethical behavior in the workplace. This is especially important for military leaders, as are the conclusions that can be drawn from the findings of this study.

Finding #1: A firm grounding in military core values coupled with a college honor code experience improves military workplace behavior.

In the military setting, the understanding of core values was significantly related to more ethical behavior in the workplace. In the qualitative portion of the study, honor code experience in college was identified as having an impact on ethical behavior in the workplace and, in fact, provided lifelong values. Many of the respondents internalized their honor code experience, and one officer mentioned that having internalized the honor code was one of his most valuable lessons from college.

Students with college honor code experience and a strong understanding of core values had significantly lower self-reported unethical behavior at work than those with no college honor code experience and a weak understanding of core values. Moreover, most of the interviewed students expressed the opinion that college honor codes were reinforced by military core values. Thus, the combination of honor codes and core values seemingly work well together.

The military core values act as a code of ethics that promotes a sense of community within the military. Officers and recruits are exposed to the core values for their careers from entry level, and this training is continually reinforced. As military personnel progress and are promoted, the core values are internalized and become second nature. Almost all senior CNW students denoted strong understanding of military core values in contrast to entering NAPS students, who had considerably less understanding. Continual reinforcement of core values throughout the military experience appears to be successful in instilling these values in military personnel.

Finding #2: Leaders are important in establishing and maintaining a positive ethical climate.

Leaders set and enforce ethical standards and promote ethical training. A SWOS student said, "The CO, XO, and Master Chief all set the standard on a ship. The top three have a big role in the ethical behavior that goes on. What they do trickles down to the wardroom, the Chief's Mess, and the crew." A CNC&S student commented, "Everyone has input, but the Captain sets the ethical tone on the ship for how things will happen. Everyone sees what he does or does not do." A CNW student responded, "Leadership by example is crucial, especially when it concerns ethical issues." In the military, with its commitment to and tradition of strong leadership, the example set by leaders on ethical matters may carry more weight than in the civilian workplace.

Finding #3: The current military ethical climate is promising.

Despite frequent press reports to the contrary, most of the students interviewed responded that unethical behavior was either no problem or a minor problem in the military today. This response was consistent with the information these same students provided concerning types of unethical behavior they had observed their colleagues commit. It should be noted, however, that students at all levels who were given the opportunity to study in these four military schools were a select group of individuals and may not be representative of the entire military workforce. They did, however, represent current and future military leaders.

Recommendations

The military community should continue to reinforce honor codes in military schools and service academies.

New recruits and officers are exposed to the honor code concepts early in their military training, and these honor codes are supported in subsequent military schools they attend. Honor code violations are dealt with quickly. The study findings support the concept that these are valuable lessons learned because honor codes provide an ethical base and support ethical behavior at work.

The military community should continue to emphasize military core values in military schools and in the military workplace.

Mentoring programs that link senior and junior officers and senior non-commissioned officers with young enlisted service members are an excellent tool to ensure that lessons learned are passed through the ranks and that core values continue to serve as a moral compass for military personnel. If, as the results of this study suggest, core values have an impact on ethical behavior in the workplace, these values should be emphasized throughout the basic training, in military schooling, and at military worksites.

Formal ethical training should be a part of the curriculum in war colleges and senior enlisted schools.

Leaders set the ethical tone for the organization. Discussions with peers and classes on ethical dilemmas prepare leaders for positions of responsibility and enable them to mentor and guide subordinates. Therefore, all curricula at war colleges and senior enlisted schools should stress the importance of ethical education.

Social policies such as college honor codes and military core values are an important part of the military culture. Honor codes are embedded at military academies and colleges and military schools across the country. Each branch of the military has established core values and reinforces these values in required training throughout the course of a military career. A major finding of this study is that the college honor code experience reinforces military core values to support ethical behavior in the workplace. Military leaders should build upon this finding and continue to emphasize the importance of honor codes and core values as rules to internalize and to operationalize in all walks of military life.

Notes

(All notes appear in shortened form. For full details, see the appropriate entry in the bibliography.)

1. Gilgoff, "Click on Honorable College Student."
2. Bartlett, "Ohio U. Investigates Plagiarism Charges."
3. Bender and Baron, "Army Knew of Cheating."
4. Wait, "Defense Returns Some Programs."

5. McCabe, Trevino, and Butterfield, "The Influence of Corporate Codes."
6. A complete version of the study can be obtained by contacting the author.
7. Sims, "The Relationship between Academic Dishonesty and Unethical Business Practices."
8. See Bowers, *Student Dishonesty*; Crown and Spiller, "Learning from the Literature"; May and Loyd, "Academic Dishonesty"; and McCabe and Trevino, "Academic Dishonesty Honor Codes."
9. May and Loyd, "Academic Dishonesty"; McCabe and Trevino, "Academic Dishonesty Honor Codes"; and McCabe, Trevino, and Butterfield, "The Influence of Corporate Codes."
10. McCabe and Trevino, "What We Know about Cheating in College."

Bibliography

Bartlett, Thomas. "Ohio U. Investigates Plagiarism Charges." *The Chronicle of Higher Education* 52, no. 27 (10 March 2006): A9.

Bender, Brian, and Kevin Baron. "Army Knew of Cheating on Tests for Eight Years." *Boston Globe*, 16 December 2007. http://www.boston.com/news/nation/articles/2007/12/16 (accessed 17 Dec 2007).

Bowers, William J. *Student Dishonesty and Its Control in College*. New York, NY: Bureau of Applied Social Research, Columbia University, 1964.

Crown, Deborah F., and M. Shane Spiller. "Learning from the Literature on Collegiate Cheating: A Review of Empirical Research." *Journal of Business Ethics* 17, no. 6 (1998): 683–700.

Gilgoff, Dan. "Click on Honorable College Student." *U.S. News & World Report* 130, no. 20 (21 May 2001), 51. http://web.lexis-nexis.com/universe/document?_m=c10087190334 (accessed 26 Jan 2006).

Lipka, Sara. "Air Force Academy Restricts Students to the Campus for Cheating and Other Misconduct." *Chronicle of Higher Education*, 23 February 2007. http://chronicle.com/article/Air-Force-Academy-Retricts/7256.

May, Kathleen M., and Brenda H. Loyd. "Academic Dishonesty: The Honor System and Students' Attitudes." *Journal of College Student Development* 34, no. 2 (1993): 125–29.

McCabe, Donald L., and Linda Klebe Trevino. "Academic Dishonesty Honor Codes and Other Contextual Influences." *Journal of Higher Education* 64, no. 5 (1993): 522–38.

———. "What We Know about Cheating in College: Longitudinal Trends and Recent Developments." *Change* 28, no. 1 (1996): 28–33.

McCabe, Donald L., Linda Klebe Trevino, and Kenneth D. Butterfield. "The Influence of Collegiate and Corporate Codes of Conduct on Ethics-Related Behavior in the Workplace." *Business Ethics Quarterly* 6, no. 4 (1996): 461–76.

Sims, Randi L. "The Relationship between Academic Dishonesty and Unethical Business Practices." *Journal of Education for Business* 68, no. 4 (March–April 1993): 207–12.

Wait, Patience. "Defense Returns Some Programs to Air Force Control." *Government Computer News* 1, no. 1 (2 February 2006). http://www.gcn.com/vol1_no1/dailyupdates/3806-1.html (accessed 2 Feb 2006).

About the Author

Dr. Thomas J. Gibbons reported to the US Naval War College as the Army advisor in 2003. He retired in 2008 and became an associate professor teaching electives on operational leadership and ethics. He was commissioned in the US Army as a field artillery officer in 1979 and later attended flight training to become an Army aviator. He commanded the 1st Battalion, 10th Aviation Regiment at Fort Drum, New York. He holds a bachelor's degree from the US Military Academy, a master's degree from George Washington University, a master's degree from the US Naval War College, and a doctorate from Johnson & Wales University.

CHAPTER 27

HONOR SYSTEM RENOVATION
DEVELOPMENT AND AUTHENTICATION

Charles J. Yoos II

Foreword

Col Charles J. "Chuck" Yoos and I have been colleagues and friends for close to two decades. When I became the dean of the faculty (in 1991) and chairman of the Character Development Commission (in 1992) at the United States Air Force Academy, Chuck Yoos became my honor liaison officer on the faculty and the chief architect for the Character Development Program for the entire Academy. It was Chuck Yoos who early on became aware of the inherent conflict between an "adjudicated" honor system and a "developmental" character program coexisting at the Academy. For these last two decades, this inherent conflict has taken its toll on the cadets' commitment to honor, and as he predicted, there is little development and even less recognition among cadets for higher and more honorable conduct among them in their progress to becoming officers in the United States Air Force. I liken the existing honor condition to a "dishonorable pit" where the cadets try to avoid falling into the pit but don't mind getting as close to the edge of the pit as possible with their sometimes "questionable" behavior. This is in contrast to the "honorable hill" where the cadets would climb higher and higher on the hill based on their expanded and refined honorable behavior in everything they do.

No one is more qualified to comment on the honor system than Chuck Yoos, who graduated from our Academy in 1968 and has spent 19 of his 30 active duty years at the Academy, involved in the cadet honor system. He has mentored hundreds of cadets on honor. His expertise in organizational theory allows him to understand fully how people grow within any organization. His call in this paper is for a truly "developmental" honor system, one that will put the focus back on honor at the Academy and allow the cadets to engage each other, without fear, on what constitutes the "right" cadet conduct on their way to becoming the type of honorable officers that we expect will graduate from our beloved alma mater. As the 2005 Graduate Leadership Conference held at the Academy confirmed, this is no trivial matter, and there is no greater problem that the current administration needs to resolve than the honor system at the Academy. It is apparent that the majority of cadets are of like mind.

At the same time, Chuck is well aware that some graduates will never agree to anything less than what they experienced at the Academy with regard to the honor system. Their solution is greater "shock and awe!!!" He and I both accept that graduates might recoil at allowing cadets to make mistakes on their way to internalizing the concept of honor. So he has revised his proposal to include the venerable code and system during the last two years, as a period of traditional honor authentication. Please read Chuck Yoos' paper carefully, with ample time for reflection, while grasping the "mentor" relationship that is being proposed rather than the "warden" relationship that exists today. Once read, the concepts in the paper need to be critically reviewed and debated, and through this trial and tribulation will come the solution that we all seek. If this procedure is followed, I am certain the final resolution of the honor system will not be far from what Chuck Yoos has proposed in this paper. Enjoy!

Brig Gen Randy Cubero, retired
USAFA Class of 1961
Former Dean of the Faculty 1991–98

Author's Preface

Il faut cultiver notre jardin d'Académie avec honneur. That is to say, paraphrasing the fabled remark of Voltaire's Candide, *we must cultivate our Academy garden with honor.*

This paper presents a viable solution to the honor system problem. Full background on the problem and details of the solution are omitted for brevity; explanation and examples are provided only to clarify the concepts. For background on the honor system problem, and more evidence of concept validity of the solution, please read my papers, "Blessent mon Coeur," "Habitat for Honor," and "Honor System Transformation: A Time for Debate," on my Web site: http://soba.fortlewis.edu/yoos. For details, please be patient—build the concept, and the details will come.

This honor system renovation is conceived to be *feasible*—that is, to renew the honor system with a valid development process while retaining the revered code and a form of the traditional system, so that the diverse group of Academy stakeholders—cadets, graduates, and Air Force and Academy leaders—can *all* embrace it, own it, and therefore achieve it. It is more than modification to the current dysfunctional system, but less than full transformation.

I propose a renovated honor system *process*. Obviously, content is also vital— what *are* the virtues to be developed and authenticated by the process? As you will read, I have used the imperative virtues of honorable officership as reflected in the Air Force and Air Force Academy core values, the Air Force Academy character development objectives, and the Cadet Honor Code, but the process can accommodate other values and objectives. However, I warn against what elsewhere I have called the *cornucopia of goodness* approach—over time, it naturally is tempting to add more and more virtues and objectives, until the system bogs down indiscriminately.

With rare exception in this paper, I use the words *honor* and *moral* to label the essence of our intent, that graduates do the right thing. Other labels, like character, integrity, ethics, and values, are preferred by some and objected to by others. Let's not get sidetracked by labels.

I gratefully acknowledge the support of the Air Force Academy's Association of Graduates and Falcon Foundation in providing a place and a computer for me to process these words. I assume they thought it might be worthwhile— I hope they find it so.

This paper is dedicated to . . . wait, that's not right. Rather, my process of development, in which I have tried to use my mistakes to become better at proposing a viable solution to the honor system problem, is dedicated to four specially honorable persons: my wife Linda, who is the love of my life; Randy Cubero, who is pathologically honest; Anne Morrissey, who is most perspicacious; and Tyla Sparks, who seeks moral accountability.

Development and Authentication

I propose a renovated honor system of two basic processes—*development* and *authentication*. By *development*, I mean the process of active engagement with the real life, the actual conduct, of the person being developed, by that person and others, where on the dimension being developed, intent and introspection, deed, and consequences and retrospection are continuously connected. A good visual depiction of this process is a spiral, where thinking and doing and then thinking again and then doing again, on and on, spiral ever upward in the direction of higher achievement on the dimension (of honor) being developed. The depiction of this process as a spiral was developed by Col Randy Stiles and me and put into the original strategy for character development in the mid-1990s. I notice that a similar depiction of development as a spiral is the Leadership Growth Model (LGM) of the Officer Development System, in Cadet Wing Manual 36-3501, *The Cadet Sight Picture*, dated March 2004, and associated literature. I see this as very good news, signifying that the Academy apparently has made an explicit commitment to a form of the development process.

It is vital to grasp two imperatives of valid moral development, whether of soldier, scholar, athlete, or other sort of honorable person. First, it must integrate the cognitive and conative mental domains, the domains of knowing and resolve, respectively. To do without knowing what to do is blind; to know what to do without doing it is sterile. They must intertwine.

Second, moral development proceeds on a basis that must admit *error in actual practice* (i.e., not error in discussion or simulation). This is not one man's opinion, nor one school of thought. Rather, it is axiomatic—the very basis of the concept. All authorities on development agree with it, and all development literature supports it. In my paper "Honor System Transformation: A Time for Debate," I wrote:

> Our knowledge of moral development in college-aged youth is conclusive, that mistakes are not only inevitable, but *necessary* for development—practice *is **not*** perfect, it ***makes*** perfect. And in that practice, cadets must be continually challenged greatly to ever higher levels of honor, so *if there weren't any error, there wouldn't be any challenge*! However, for mistakes to be developmentally useful, they must be out in the open—scrutinized and forthrightly acknowledged.[1]

That is, actual mistakes are the "stuff" of development via practice. Another way to think about this, especially for those who find the labels *error* and *mistake* somehow unpalatable, is that development is *diagnostic*—what needs to happen next is based on what just happened, relative to some appropriate expectation about what should have happened. That is sometimes called *deviation*. By whatever name, a process of development is necessarily driven by actual feedback, as is depicted in the Academy's LGM, and is inherent in the honor system development process I propose here.

By *authentication*, I mean the process whereby the honor of an officer candidate is validated against a set of professional standards. The Cadet Honor Code and attendant Wing Honor Board System have traditionally served this purpose—the code states several categorical honor offenses (i.e., lying, cheating, stealing, and tolerating), and the system makes a determination as to whether an allegation is a violation. A cadet who is found by the system to have violated the code is presumed to be unworthy of being commissioned, unless an extraordinary dispensation (probation, suspension) is granted. Cadets who are not found in violation of the code, in most cases because no allegation is ever made against them, are presumed to have demonstrated their worthiness of being commissioned, from an honor perspective. They could be said to have authenticated their honor.

Not to dwell on the problem, but the current system has foundered because it violates the two imperatives of valid moral development I mentioned above. First, the current honor education process is cognitive but not conative—honor lessons are based on "Cadet X letters" and other case studies, exercises, even role-playing, fortified by earnest discussions of honor in academic classes, and so forth, all of which are laudable food for thought, but involve only simulation. Second, since authentication is *by default* (no allegation or finding of violation), with actual moral error *inadmissible*, cadets come to avoid active engagement with their real honor lives—for a full discussion of this point, see parts 0 and I of my paper "Habitat for Honor." In other words, from an honor standpoint, they are "good to go" (graduate)—or, it is more correct to say, "not bad to not go"—as long as they don't trip the honor system wire with evidence of actual dishonorable conduct per the code, so they take steps away from it. From a cadet perspective, it's all well and good to proclaim that it isn't that hard not to lie, cheat, or steal, but it is hard to know when something you might have said or done could be construed that way, especially something you might not be very proud of and otherwise might like to come clean with to be better. So, all things considered, cadets determine that it is best to keep their actual honor conduct to themselves, put it out of their minds, or at least rationalize it, so if it were to be discovered, they could claim lack of intent. Thus, by definition, honor development is stymied, and over time, being apart from their own honor, many cadets are afflicted with an amoral condition that elsewhere I have called "moral anesthesia." Capt Matt Holston, an ex-cadet who recently critiqued my "Habitat for Honor" paper, concurs, calling this a "'searing' of the conscience." To return to Randy Cubero's apt analogy, cadets sometimes wander in a moral fog in the vicinity of the "pit" of dishonor . . . and therefore sometimes fall in. Those cadets who have read my paper and responded to me, or with whom I have spoken, invariably have understood and agreed and expressed a sincere desire for a system in which they could be *honest about their honor*.

Thus, we face an honor system *incongruity*—the need for honor development in cadets, which treats error as valid feedback; and the current honor system, which treats error as allegation of a code violation, to be adjudicated and verdicted, with a finding of violation probably leading to disenrollment, which by definition obviates development.

The crux of the renovated honor system I propose here is to remove this incongruity via an integration of these processes, development and authentication, in a particular way: *honor development, followed by honor authentication enveloped by higher honor development, based on two linchpin criteria.*

First Phase—Honor Development

When new cadets are accepted into the Cadet Wing (i.e., after basic cadet training), rather than taking the Honor Code Oath, they will swear to live by the Honor Creed (a label proposed by Captain Holston most recently, and others previously):

> *We put integrity first.*
> *We will strive to have forthright integrity*
> *By voluntarily deciding the right thing to do*
> *and doing it.*
> *We will not tolerate anything less.*

You may notice that the first line states the principal Air Force Academy core value. The second line states an Air Force Academy character development objective, and you will read below that "forthright" is a pivotal concept in this system. The third and fourth lines convey the essence of true honor—willingness to do the right thing, confirmed by action. The final line adopts the classic honor system tenet of collective self-control. Also, if you read my "Habitat for Honor" paper, you may recall that I proposed this as a restatement of the Honor Code. Here, it would not replace the venerable statement of the code, but would supplement it, for developmental purposes.

The new cadets will be immersed in a true honor *development* process for their first two years, as Fourth and Third Classmen (hereinafter called underclassmen). The nature of that process is described in part II of my paper "Habitat for Honor," but I will summarize here. During this phase of the development process, the underclassmen will pursue fulfillment of the Honor Creed in everything they do. Each cadet will have two honor mentors in that pursuit, an officer or professional staff member and an upperclassman—a three-person relationship, or triad.

The nexus of the process is the developmental spiral of thought and deed, around each cadet's *actual real life* (not case studies or other abstractions). Mistakes will likely be made, constituting the diagnostic feedback which development requires, and also gauging challenge—no mistakes, no challenge. Mistakes will result in consequences, to include punishments and remedies commensurate with the circumstances (the nature of the mistake, the experience of the cadet, the situation, etc.). Even though during this phase of development cadets will not have entered the authentication process—not have taken the Honor Code Oath, not be subject to the Wing Honor Board adjudication process—still, serious mistakes, egregious or repeated, could lead to disenrollment via normal administrative processes. Underclassmen are not absolved of blame for lying, cheating, stealing, or tolerating, plus a host of other dishonorable things not mentioned in the code. Make no mistake about mistakes—this is a no-nonsense developmental process, not a free-pass one. Indeed, it is tougher than the traditional system—more conduct subject to honor consequences, with higher standards. That every mistake does not lead to disenrollment does not make it a "soft" system.

The key to the process is mentoring, whereby thought and deed are scrutinized and forthrightly acknowledged, enabling development. Indeed, the activity of the triad provides a surrogate measure of development—active, regular, robust activity signals success, while dormant, sporadic, meager activity reflects failure.

But to reiterate, the premise of this phase of the honor developmental process is that, unless mistakes demonstrate unsuitability for further development toward commissioning, they are used developmentally.

Cadet Honor Development Panel

Even though mistakes are, in the main, used developmentally, still there must be some fair and consistent way to determine *punishments and remedies commensurate with the circumstances.* Above a threshold level of misconduct, this is done, not by a Wing Honor Board as per the code, but rather by a Cadet Honor Development Panel. The purpose and process of the panel is not to *adjudicate* allegations and render a binary verdict. Rather, the panel seeks to understand in what, if any, ways the conduct brought before it was not *becoming an honorable cadet* and recommend to the commandant of cadets appropriate consequences, punitive and remedial, and if warranted, disenrollment. Due process is not the master control, because in most cases, a constitutional right is not in jeopardy, and, if it is, additional processes come due.

Provisionally, a Cadet Honor Development Panel is made up of the Cadet First Class honor representative from the cadet's squadron, as chair; a member of the cadet's immediate chain of command, varying from the cadet element leader for a Cadet Fourth Class, to the cadet squadron commander for a Cadet First Class; and a cadet peer, chosen at random from the cadet's class, but ascertained to have no close ties to the cadet. These three members listen, question, discuss, deliberate, and decide the recommendations. A Cadet Second Class honor representative, not from the cadet's squadron, serves as recorder, without voice or vote.

The cadet's officer mentor and cadet upperclass mentor (if the cadet is an underclassman) participate in discussion but do not deliberate or decide. An officer in the cadet's immediate chain of command, usually the air officer commanding, and an officer from the Character Development Center observe and answer questions.

Discussions and deliberations are full but not formal—this is a hearing of explanation and the rule of conscience, not a trial of testimony and the rules of evidence. As previously stated, if the Cadet Honor Development Panel recommends disenrollment, the matter proceeds under the jurisdiction of the commandant of cadets, using well-established official procedures for determining if a cadet remains eligible for a commission.

Also, the Cadet Honor Development Panel is *not* governed by the Wing Honor Board System proviso that both act *and intent* must be found. To honor system traditionalists, this may seem like a non sequitur, but in honor *development*, ignorance, confusion, and so forth are not justification for excusal from *honor* consequences, as well as military, academic, or athletic ones. Bluntly, it is dishonorable to be dense, as well as deceitful. This is another illustration of how the developmental system is more rigorous than the traditional one—no excuses, no rationalizations, no equivocations regarding intent.

How is it decided what sort and severity of unbecoming conduct would be brought to a Cadet Honor Development Panel, and what would be handled elsewhere by the mentor or the chains of command—in effect, what is *major* and *minor*? On the one hand, there needs to be reasonable consistency; on the other, to create strict categories and rules risks making honor a "reg book" mentality. To achieve the former and avoid the latter, that decision is made by the chairperson of the Cadet Honor Committee, after consultation with a staff officer in the Character Development Center specifically assigned this duty (for consistency) and the cadet's air officer commanding (for best knowledge of the cadet).

Mentor

Obviously, the honor mentor plays the key role in the honor development process. I present the role more fully in previous papers, plus General Cubero and I have considerable experience with it, which time and space do not allow me to record here. Suffice it to say that from our combined knowledge, experience, and therefore judgment, which we hope constitute some degree of wisdom, being an honor mentor is highly valuable and rewarding. Also, it is *doable*. Not only is it not "rocket science," it is not science at all. Rather, it is an innate capacity of a professional to work with young professionals in progress, not by telling them war stories or the "ropes to skip and the ropes to know" to succeed over their peers, but rather shaping and guiding them. We describe the relationship as intimate, responsible, and consequential. It does not succeed because the mentor is a certified therapist or degreed counselor, but rather because the mentor *holds sacred* the honorable upbringing of the protégé, believes in the values being developed, and is honest and forthright in the developmental relationship.

Organizationally, it is important that the duty of honor mentor be considered a primary one, not a casual additional one. Every Academy professional staff member, officer and civilian, should seek this role and do it well. Indeed, honor mentoring becomes a moral imperative of Academy duty.

Second Phase—Honor Authentication
Enveloped by Higher Honor Development

The beginning of the second class year represents a distinct one-way passage in the cadet experience. Officially, that's when cadets incur an active-duty service obligation. And, according to the LGM, it is when cadets take on important new leadership and developmental responsibilities. Thus, it seems fitting that it also be the point where cadets "step up to the [traditional] line" by taking the Cadet Honor Oath and become subject to the Wing Honor Board system for the three codified offenses and toleration of same and to the presumption of disenrollment for code violations, thus entering the authentication process.

At the same time, these cadets (herein referred to as upperclassmen) do *not* leave the developmental process. That would seem to put the system in the same old predicament—if certain mistakes, that is, those under the code, are now subject to adjudication and dismissal, they cannot at the same time be used as diagnostic feedback in development. It would seem necessary to doff the one to don the other. And if they had left the development process, it would seem that the authentication process would experience the same failure mode— cadets who stow away their honor for their last two years.

As upperclassmen, they will be in both processes—developmental and authentication—and the one (authentication) will be enveloped by the other (development). The key to success, to solve the predicament and eliminate the incongruity, is to satisfy two linchpin criteria: *forthrightness* and a *desire to be morally accountable*. These two criteria are hallmarks of an emerging valid sense of honor, the beginning of true moral maturity. They give credence to what we have always said we wanted to do with honor in cadets—inculcate it in them. Above and beyond a particular dimension of honor that may be involved in a specific situation, only if cadets are forthright and want to be morally accountable can we really verify that the honor development process has taken hold by inculcation.

First Linchpin Criterion—Forthrightness

Forthrightness is not just an instinct or willingness to be open about personal honor conduct, but an *imperative* to be so. Forthright people *must* be openly honest; it is not in them to be otherwise. Elsewhere this has been called "pathological honesty"—not to be taken literally, of course, but to accentuate the notion that forthright people will be honest "*if it kills 'em!*" Of course, this is not about bona fide national security, personal privacy, or innocent social conventions—a forthright person does not blurt classified or sensitive information nor divulge data about persons that is properly private nor be rude and tactless in social situations. On the contrary, forthright people are trustworthy, humane, and polite. But regarding personal conduct, forthright people volunteer exactly, unequivocally what they did and why, without "any mental reservation or purpose of evasion."

Think about forthrightness for a moment, and I believe you will recognize and agree that the developmental value of forthrightness is not as some running public announcement of personal moral conduct. Rather, consider that another way of thinking about forthrightness is reflected (pun intended!) in the venerable saying that "sunshine is the best disinfectant." That is, we are least likely to do wrong when we know that our conduct will be out in the open to be viewed by all. And if we are forthright, then that is, at least ostensibly, always the case for us. So, in many instances, with that as an omnipresent prospect, honorable conduct is confirmed and assured.

Forthrightness Always Works—Examples

Let me illustrate by actual examples. Some years ago at the Academy, a commercial TV cable was installed in part of a cadet dormitory, and while the cable was a service to be subscribed and paid for room by room, it ran in plain sight along the baseboard of all rooms in that area. Believe it or not, a serious discussion broke out among some cadets as to whether or not it would be an honor violation to splice into the cable and enjoy cable TV without subscribing. After all, the pro argument went, the cable company wouldn't really be losing anything, so it wouldn't really be stealing. Yes, those cadets had an occluded sense of honor—they were engaging in honor system sophistry, or quibbling—but that's not the point of the example. Rather, the point is, what brought the argument to a screeching halt was when someone suggested, if you think it's ok to do that, you should be willing to call the cable TV company and tell them (that is, to be forthright). Of course, the immediate reaction was, "Gee, I couldn't do *that*." Just the prospect of being forthright shined light on the situation to illuminate the honorable course. If that example doesn't work for you, try the soft drink machine in the dorm that sometimes doesn't disgorge a soft drink at all and other times gives out several (is it an honor violation to keep the extras?). Sure, these are mundane, but cadets face them every day, and they involve honor.

In other cases, cadets face choices that are more perplexing from an honor standpoint. While there is usually an obviously safe, or at least safer, course from an honor standpoint, being forthright virtually guarantees it—if the honorable course is honestly in doubt, forthrightly discuss it with other persons of honor. If time and circumstance do not allow that, do that thing about which you are most willing to be forthright afterwards.

And forthrightness does not always lead to *not* being able to do the thing you'd otherwise prefer to do. Here's another actual example. Some years ago at the Academy, several cadets with some degrees of color blindness found out they would stand a better chance of passing the standard colorblindness test (distinguishing numbers in colored dot patterns—you know the one) by spending time beforehand staring intently at certain shades of colors in crayons. Never mind whether that works—I don't know. Pilot training physical qualification was at stake, so without saying anything, they did it and took the test. Word got out, and the question arose— had they cheated? I like this example because reasonable and honorable people may disagree. The point is, if they had been forthright and made known how they had prepared, they still might have gotten a favorable medical diagnosis, and certainly they wouldn't have exposed their personal honor to jeopardy. Forthrightness always works.

Second Linchpin Criterion—The Desire to Be Morally Accountable

Similarly, a desire to be morally accountable to someone else confirms that inculcation of honor is happening. Going hand in hand with forthrightness, it signifies not only being openly honest, but taking personal responsibility and ownership of honor. The devil didn't make me do it; it's not my environment or upbringing, not the bad crowd I hung out with, not that everybody is doing it—*I'm* in charge of my honor, and I *want* to be accountable. Again, as with forthrightness, the developmental value is not that cadets will go around thumping their chests about accountability, but rather the prospect that choosing to be morally accountable, especially to someone they respect, is a constant source of moral (en)courage(ment) to do the right thing. The available data, via the Defining Issues Test associated with Kohlberg's and later Rest's model of moral development, suggest that as adults, we transition in our sense of moral accountability from our referent social group to an abstraction (like, in our profession, the Air Force, the Constitution, even the noble ideal of national security). But neophyte professionals, aka cadets, still need to commit themselves to moral accountability personally, to look that respected other person, the mentor, directly in the eye and say, "I *want* to be morally accountable to *you*."

When these two developmental linchpin criteria, *forthrightness* and a *desire to be morally accountable*, are achieved, the second phase of authentication enveloped by continued but higher development can proceed and succeed. Thus, they are overarching—the keystone(s) in the developmental arch. In view of the developmental spiral, I depict them as the pivotal linchpin(s) which affix the developmental spiral to keep it from swaying or toppling.

By any analogy, they are required. So, the question is, are they realistically achievable?

Affirming Achievement of the Linchpin Criteria

The answer is *affirmative*—they *are*, and that *is to be affirmed*. These are not lofty terms to feature in glossy brochures. Rather, they are the honor "sight picture," the fruits of the first (underclass) two-year strictly developmental phase. They are also neither a ritual nor a perfunctory condition to be rubber-stamped. From the beginning, everyone associated with the process is mindful of these criteria—cadets, mentors, and so forth. The developmental process will yield assessment data against these criteria—during the process, *is the cadet becoming*, and nearing the two-year point, *has the cadet become* that forthrightly honorable person who desires moral accountability? Not only will the individual underclassman need to signify that, but the upperclassman and officer or professional staff member who mentored that cadet will need to attest to that. This is not unlike a student pilot being ready to solo, and the instructor pilot attesting to that. In each case, it shouldn't happen unless all parties affirm it.

Moreover, is there any valid reason why any Academy cadets, officer candidates at a prestigious national institution to which they sought admission, would *not* want to satisfy that institution's most precious developmental objectives? The only reason I can imagine is in itself dishonorable—a desire to shield conduct from moral scrutiny and evade responsibility. So, if there is recalcitrance or chronic incapacity, while there might be a remedial deviation in the process, the suitability of that cadet to become an officer *ought* to be carefully reviewed, and it is perhaps best for all that the cadet be disqualified.

But I have no doubt that most cadets will respond with enthusiasm and sincerity. The Academy must do its part by removing any impediments to honor (not challenges, but rather the silly, meaningless obstacles that can trip up well-meaning but unwary cadets). For example, there are sometimes administrative processes that allow a limited number of responses, putting respondents in the dubious position of choosing one that best fits, even though it isn't true. To use a non-Academy yet real example, let's say a municipal traffic ticket for speeding, rather than going to trial, is pled down to agreement by offering the offender a lesser number of points, and then administratively, the offense associated with those points is substituted for the ticketed offense. I know someone who was accused of speeding, carrying a penalty of four points, but was offered two points, for "defective vehicle." To save the trouble, that person accepted but had now agreed to a lie because that person's vehicle was not defective. A small matter, to be sure, but in a system geared to honorable development, removing such small matters habituates honesty and forthrightness.

Honor Authentication (continued)

Back to the honor authentication process. With respect to the categories of the Honor Code (lying, cheating, stealing), the premise has shifted from the previous developmental phase, where mistakes were handled developmentally unless disenrollment was clearly warranted, to authentication via the code, where code violations are presumed to result in disenrollment unless mitigation is clearly warranted. But now, for cadets who have made a sincere commitment to be forthright and accountable for honor, this should not be daunting or even precarious. Far from being morally anesthetized, with their consciences "seared," they will be morally acute, aware that in general, honor is an aspect of everything they do, and in particular, lying, cheating, and stealing are offenses they can easily avoid by being forthright and morally accountable. To put it in a way that cadets might well think about and express it, *"Now let me get this straight. After two years of honor development, and having promised to be forthright and wishing to be morally accountable, you can't even live up to the three basic elements of the code? Get out of here!"*

What about *tolerating*, the final offense recited in the code, yet arguably the most difficult for the wing to uphold, especially because of the strong social taboo against ratting on a mate? My extensive experience with cadets convinces me that there is another enduring taboo in the wing—*don't try my patience and put me in jeopardy, mate.* Previously, cadets drew away from the honor system. In my paper "Habitat for Honor," I described toleration as being moribund—

on the verge of extinction. That is, cadets were *not* tolerating, because they had "learned" not to notice the honor-related behaviors of mates in the first place—*don't see, can't tolerate*. But now cadets will have succeeded in two years of honor development and will be living a forthright life above the code. Along comes a mate who places them in jeopardy by not doing the same thing when it isn't that hard to do? *"Now let me get this straight . . . ."*

Regarding the Wing Honor Board process by which an alleged code violation will be adjudicated, I am aware that modifications to that process have been proposed in the Academy's recent White Paper, including changes to board composition and the evidentiary standard for a finding of violation. This is not the place for me to redesign that process or critique their modifications, but the renovated honor system I have proposed would function best with a Wing Honor Board process which cadets embrace, and that is most likely if the upperclass Cadet Wing at large determines it. No offense to current members of the Cadet Honor Committee, who are typically excellent cadets with deeply ingrained honor, but it is exactly the point that they are also steeped in the status quo system, having operated it to the best of their ability, and it is that system most other cadets have learned to keep at bay.

Higher Honor Development (continued)

Meanwhile, the development process continues for the upperclass cadets as well. But now, with the authentication process operating via the code, honor development moves to a higher plane. No longer are the basic levels of honor being developed, but higher and wider levels of "conduct becoming an officer" (pun intended) are involved. We all know that there are many sorts of such conduct that may not violate the Honor Code *per se*, but clearly reflect a compromise of honor.

Let me use an example that recently became painfully apparent to all who cherish the Academy—sexual misconduct. Let's lay aside sexual assault—dishonorable, but also properly actionable under the Uniform Code of Military Justice. Rather, there is a range of sexual misconduct, from the repugnant to the boorish, which still is conduct unbecoming an honorable person. The work is to continue to use the honor developmental spiral to further sharpen moral acuity, not only to recognize in principle that unwelcome sexual conduct is dishonorable, but to *practice* the awareness, reflect on that practice, and so on, greatly reducing the likelihood of such misconduct.

There is another important change in this phase of honor development that you may have already spotted, or suspected—First and Second Classmen now assume the upperclass role in the mentoring triad—instead of being the protégés, they are the apprentice mentors. In parallel to the LGM, they exhibit honor development and leadership capabilities with respect to the underclassmen. In the mid to late 1990s, under the leadership of General Cubero as chairman of the Character Development Commission, we prototyped this triad arrangement, and I am excited to report that, in addition to being very responsible, the upperclass members of the triads leveraged that responsibility in terms of their own continued development in honor. As you might expect, when someone *below* you is looking *up* to you as a role model of honor, you are obliged to be on your very best behavior. In a functional family, children are reluctant to disappoint their parents, but perhaps even more so, parents are loathe to disappoint their children! This three-pronged relationship (remember, the upperclassmen are still protégés with respect to the officer or professional staff member mentor) is very "*honor(st)able*"—development proceeds in all dimensions.

Outcome—At Graduation and Beyond

The litmus test of this honor system, like any purposeful system, is how it turns out. At and beyond graduation, what will be the *honorability* of our graduates in a future Air Force, where I believe forthright and trustworthy officers will be perhaps more important than ever? Well, let's take stock. Under the renovated honor system that I have proposed, cadets will have been engaged in a valid honor developmental process for four years, with a record of their progress and performance. At the two-year point, they will have exhibited forthrightness and a desire to be accountable for honor, and those criteria will have been attested to by their honor mentor(s). Then, while development continued, they will also have proceeded to authentication, living successfully by the tenets of the code, and based on forthrightness, this authentication will not be by default, but rather by design.

Have I proposed a *panacea*, literally, a cure-all? Of course not! But the design *assures* the desired outcome—if the linchpin criteria are achieved, it follows that honor will be developed. Moreover, the first (underclass) phase of development makes it highly likely that for each cadet, those criteria either will be achieved, or that cadet will be identified as unsuitable for commissioning. Viewed from the negative, it is highly unlikely that a cadet can (im)posture through four years of development via mentoring, falsely appearing to be forthright while deceiving an officer or professional staff honor mentor and an upperclassman via the triad, not to mention classmates and others, plus two years of authentication. Among their many indubitable talents, cadets can spot a phony an air mile away—believe me, it's part of their sight picture!

Beyond graduation, as commissioned officers, having had the mentorship experience, they may wish to seek out more senior officers or other professionals as mentors, ones to whom they choose to be morally accountable. And in accordance with our knowledge of moral development beyond the college years, they will become more morally accountable to organizations like the Air Force, expressions like the Constitution, and noble ideals like national security.

They will have the pride of authenticity in honor—the precious tie that binds graduates—and also the experience of honor development, which becomes lifelong. Honor will be inculcated.

Postscript

Permit me to put the initial quotation of this paper in context. I am told that work is underway at the Academy to consider other environmental and organizational issues under the rubric of "culture." I wish to comment parenthetically on the relationship of culture to honor. In doing so, I draw on my professional knowledge credentials, as an academic, consultant, and practitioner. My doctoral degree is in organization theory and behavior, from whence comes our understanding of culture as a social phenomenon. I have consulted with various organizations on the topic, both in and outside the Air Force, and of course, have been an organizational member and leader.

Culture is a label usually affixed to that deepest identity of an organization that serves to define "who we are." In a sense, culture *is* the organization's reality, things that are taken for granted to be real and true about that organization by those who are acculturated. These things are rarely challenged, reflected on, or even mentioned—they just *are*, and everybody knows them.

The Air Force Academy cadet wing is an example of an organization in which a "strong" culture is to be expected—the cadet experience is relatively homogeneous, and despite their best efforts on weekends and leave, cadets are relatively isolated from the broader, less culturally distinct society around them. There is a "what we all know to be true" in the wing on almost every topic imaginable, but particularly those pertaining to the welfare of cadets. Part and parcel of "what we all know" is "how we all get along to go along," en route to graduation.

Thus, it is not surprising that during several decades in which the honor system faltered, still there was "honor among cadets" apart from it. The classic expression of this reality, blurted out by cadets to each other without even considering the implications, was, *"this has no honor (system) implications—just tell me the truth!"*

I don't condone that disjunction, but it serves to illustrate two important points. First, because matters of honor are at the deep level of the soul, then individually and organizationally, they will have a pivotal effect on the culture. But second, and perhaps most important to be understood, you can't design, program, and install a desired culture. By definition of the concept, that doesn't work, any more than you can order bacilli in a Petri dish to grow as you desire.

That is, culture is an organic phenomenon, not a mechanistic one. You nudge a culture in the desired direction, shaping it by adjusting organizational states and processes. And while that is not blind trial and error, still it can't be predicted, and thus arranged, with accuracy. Combining my two points, I observe that *the cadet wing's culture won't shape honor; importantly, it is the other way around.* I have proposed a renovated system which I reckon is the most realistic option for honor to succeed. I also believe that the linchpin criteria, forthrightness and accountability, will have an important positive impact on culture. True to the lights of my knowledge, I can't say exactly what that will be, but I believe it will be valuable. Over 10 years ago, I titled the first version of my seminal paper "Habitat for Honor within a Community of Character" for exactly this reason—we want the Academy culture to be one of good character, inhabited by honor.

Voltaire's Candide observed, "*Il faut cultiver notre jardin.*" That is, it is necessary that we cultivate our garden. It was Candide's final verdict, and thus I end as I began, with mine.

The Academy is our garden, and we must cultivate it with honor.

Note

1. Chuck Yoos, "Honor System Transformation: A Time for Debate," http://soba.fortlewis .edu/yoos/Debate.doc.

About the Author

Dr. Chuck Yoos, a retired colonel, is a professor emeritus at the United States Air Force Academy, where he served on the faculty for a total of 20 years over his 30-year active duty career. He graduated from the US Air Force Academy in 1968. Dr. Yoos earned an MBA from UCLA and a doctorate from the University of Colorado. While on the Academy faculty, Yoos served in many honor-related capacities; in the latter years, he was the dean of the faculty's senior advisor on honor and the chairman of the commandant's Honor Review Commission Executive Panel. He has published journal articles on honor and character development and is a member of the Center for Academic Integrity. He is currently professor of management at Fort Lewis College in Durango, Colorado, and has also been visiting professor in Germany at Fachhochschule Regensburg and Fachhochschule Deggendorf as well as the École Supérieure de Commerce de La Rochelle in France.

CHAPTER 28

AUCTION MECHANISMS
FOR FORCE MANAGEMENT

Peter J. Coughlan
William R. Gates

Introduction

The Challenge (and Expense) of Military Force Management

In the age of an all-volunteer force, the military services face the constant challenge of devising pay and benefit offerings which appropriately and effectively shape and manage the overall workforce. The military services must continuously calibrate these incentives to precisely attain varying overall end-strength objectives and appropriately balance the force across different ranks and specialties (see fig. 28-1).

Basic military pay and allowances are rigidly tied to service-member characteristics such as rank and years of service; however, these core compensation components are ill-suited to flexibly address the continuous challenge of balancing frequent changes in force-management needs with unobservable and irregular shifts in service-member preferences. Consequently, all military services increasingly rely on more flexible "special and incentive" (S&I) pays to (a) retain additional service members at some times, (b) separate service members at other times, and (c) attract service members to understaffed positions or specialties (i.e., occupations or career fields).

Managing the force with such S&I pays has become an increasingly expensive practice, however. According to the 2008 Quadrennial Review of Military Compensation (QRMC), there are currently more than 60 different S&I pays carrying a cumulative annual price tag which has quintupled in just 11 years, rising from under $1 billion in fiscal year (FY) 1995 to more than $5 billion in FY 2006.[1] Such S&I pays also account for an increasing proportion of total service-member cash compensation, representing 7.3 percent (or about $7,000 annually) for the average officer and 6.6 percent (just over $3,000 annually) for the average enlisted service member.[2] Given both the growing importance and escalating cost of these pays, there has been rising interest in alternative S&I pay methodologies to achieve the same force-management objectives but at a lower overall cost and/or with greater service-member satisfaction.

Elements of Military Force Management	
Force Shaping	**Force Distribution**
Achieving & maintaining appropriate... • **numbers** of service members... • of each enlisted & officer **rank**... • within the **overall force** &... • within each **occupation** or **specialty**.	Matching **service members** of... • the appropriate rank &... • the appropriate experience... With **assignments which maximize** • achievement of military priorities &... • service member job satisfaction.

Figure 28-1. Military force management.

Auction Mechanisms as Potential Force-Management Solutions

Auction mechanisms have been among the most commonly considered alternative approaches both to determine the appropriate dollar amount of certain S&I pays and to identify (in the case of retention or separation pays, for example) which service members should actually receive the bonus (and be retained or separated). In fact, the QRMC itself concluded that "the Services should explore [S&I] pays, such as reenlistment bonuses, which could potentially use an auction mechanism to incorporate member preference into payment rates."[3]

The QRMC endorsement of an auction approach to force management is further buttressed by recent studies of the selective reenlistment bonus (SRB) program, the most expensive of all S&I pays with an annual cost of $1.0 billion in FY 2007.[4] In particular, research on these bonuses indicates that potentially dramatic cost savings—as much as 88 percent in one study detailed below—may be achieved if current methodologies for setting SRB levels were replaced by even the most simple auction mechanism. Such retention auctions offer the potential not only to automatically set the precise bonus amount required to achieve end-strength goals at the lowest cost, but also to endogenously identify and retain those service members who are most dedicated to continuing military service.

Chapter Content and Structure

This chapter will explore the potential promise and pitfalls of employing auction mechanisms as force-management tools. While an extensive auction design literature already exists, only a small portion of this literature applies auctions to labor market environments, and little to none captures the specific attributes and idiosyncrasies of military force management. This chapter begins to develop auction design (and implementation) literature specifically applied to military labor markets.

This chapter analyzes the manner in which force-management S&I pays are currently determined and implemented in the military, proposes criteria for evaluating alternative force-management tools, and applies these criteria to evaluate the traditional approach to these programs. The chapter then describes the basic elements of auction design within the labor market context, illustrates the application of a basic force-management auction mechanism in several force-management applications, and assesses this basic auction mechanism relative to the established performance measures. Towards the end of the chapter, we introduce several more advanced auction designs that offer the potential for both broader application and stronger performance of force-management auctions.

Ultimately, the analysis in this chapter indicates that even a basic auction mechanism applied to various S&I pay programs offers the potential for improved precision in both setting bonus levels and hitting staffing targets, allowing for more responsive and cost-effective force-management incentive programs. In addition, advanced auction designs can be customized to unique force-management needs or environments, offering the potential for even greater responsiveness and value, but perhaps more importantly allowing for an S&I pay system which better recognizes and rewards exceptional service-member achievement.

Overview of the Military Compensation System

One-Size-Fits-All: Base Pay, Allowances, Nonmonetary, and Deferred Compensation

The current military compensation system involves a complex mix of base pay, monetary and nonmonetary allowances, and special pays and bonuses. Base pay is the largest component of military compensation, comprising about 60 percent of the total cash compensation.[5] Base pay is determined by a service member's rank and years of service and is updated annually on 1 January in the military pay tables. Base pay does not depend on the service member's specific specialty or assignment.

There are a number of allowances that supplement base pay, including basic allowance for housing, basic allowance for subsistence, and clothing allowances (enlisted personnel only). These allowances are entitlements that supplement basic pay for all qualifying service members independent of specialty or assignment. The military also offers a number of nonmonetary compensation benefits. These include health care, commissary and exchange privileges, recreation facilities, and so forth. Finally, the military offers several forms of deferred compensation, including accrued retirement pay, accrued health care benefits, and veterans' benefits. Again, these forms of nonmonetary and deferred compensation are available to all qualifying service members independent of specialty or assignment.

Long-Term Force-Shaping Tools: "Regular" Special and Incentive Pays

In contrast to the "one-size-fits-all" nature of the standard military compensation elements described above, each military service also offers a number of S&I pay programs, which are targeted at specific specializations or assignments. Also in contrast to other compensation elements, S&I pays are not tied to a service member's age, rank, years of service, or family status, as these bonuses are intended "to respond to retention or recruitment problems that vary by military occupation, by location or assignment or other circumstances."[6]

Some of these S&I pays are employed to selectively address staffing challenges in particular career fields, in hazardous assignments, or in otherwise less desirable billets. Other S&I pays are intended to attract and retain valuable skills or to increase parity with civilian-sector salaries in certain technical and professional fields.

Conditional Force-Shaping Tools: Retention, Separation, and Transfer Incentives

The one element of the military compensation system which clearly offers the most fertile ground for applying auctions or other bidding mechanisms is the class of "conditional" S&I pays, which are intended to have a short-term or even immediate impact on the size and makeup of the military workforce. These pays include reenlistment incentives (or, more generally, retention incentives), voluntary separation incentives, and even (the quite rare) incentives to transfer across specialties or even across service branches.

We classify these S&I pays as "conditional" incentives because the environmental *conditions* at any particular point in time—incorporating both the momentary manpower needs as well as the contemporaneous and perhaps transitory service-member preferences—determine whether or not any of these S&I pays should be offered on that occasion. Moreover, whenever one of these conditional S&I pays is indeed offered, those same short-term environmental conditions are used (or, at the very least, are supposed to be used) to determine the exact dollar amount of the incentive offered.

Note that these conditional S&I pay programs are distinct from most regular S&I pays—such as submarine pay, aviation pay, sea pay, family separation pay, combat pay, and so on—which are relatively stable incentives long associated with particular duties or assignments. In contrast, the conditional S&I pays are employed to induce voluntary choices among service members that help to quickly "right-size" the overall force or particular military specialties.

Also note that each of these retention, separation, and transfer S&I pays constitutes an incentive that a typical service member might be offered once or perhaps twice (or not at all) during the course of his or her entire military career. From the perspective of each service branch, these S&I pays are either (a) periodic bonus programs subject to adjustment with each iteration (in the case of reenlistment/retention incentives); (b) somewhat infrequent incentives offered only when unanticipated changes in needs or preferences have created an unusually difficult overstaffing situation (in the case of separation incentives); or (c) particularly rare or unique incentives offered only when extreme overstaffing in one specialty or branch is simultaneously matched by an extreme understaffing in another specialty or branch (in the case of transfer incentives).

The dollar amount of these conditional S&I pays has traditionally been predetermined, based on econometric models, rules-of-thumb, or other approaches intended to predict the exact bonus amount necessary to either retain, separate, or transfer the target number of service members. As we will discover, however, these conditional S&I pays are uniquely appropriate for price determination and implementation via auction. This reflects the fact that these conditional S&I pays are specifically intended to appropriately shape and balance the force in times of irregularity and uncertainty, when the other regular incentives have (for some reason) proven inadequate to precisely attain end-strength targets across military specialties and when, therefore, it is especially difficult to forecast the magnitude of financial incentive that will be minimally necessary to restore proper force shape and balance. Hence, this chapter is primarily focused on the use of auction mechanisms as a workforce retention, separation, or transfer tool.

Measuring Performance of Conditional S&I Pay Programs

Considering the important influence that conditional S&I pay programs exert on short-term military force-management objectives, it is essential to develop criteria by which we can both measure the performance of current programs and compare these programs to auction mechanism alternatives.

Key Driver of Performance: Precision

Before discussing specific performance measures for conditional S&I pay programs, it is helpful to recognize that the most critical driver of overall performance for these programs is force-management *precision*. In this context, precision refers to the ability for these S&I pay programs to accurately meet their intended force-management objectives, including overall end-strength targets, balance across career fields, and distribution across specific assignments. A precise pay program is one which creates the appropriate incentive to induce the exact (or near exact) target number of service members to voluntarily retain, separate, or transfer, while an imprecise pay program is one which significantly "overshoots" or "undershoots" this target number.

To illustrate, consider the selective reenlistment bonus (SRB) for zone A reenlistments (service members with 17 months to six years of service) in the US Marine Corps. Using an annual statistical analysis from the Center for Naval Analysis (CNA) that estimates reenlistments by military occupational specialty (MOS) as a function of the SRB level, Paul Bock produced figure 28-2 below, which graphs reenlistments versus SRB level for the USMC MOS 03 (infantry) in 2006.[7]

Using this graph to illustrate the importance of force-management precision, suppose that military objectives dictate the need to reenlist 3,000 Marines within this particular MOS. As indicated in figure 28-2, precisely achieving this reenlistment target would require offering an SRB of $7,000 to this population. If the bonus is set below this level, the USMC will fall short of its reenlistment goal (for example, only 2,500 Marines will reenlist if the SRB is set at

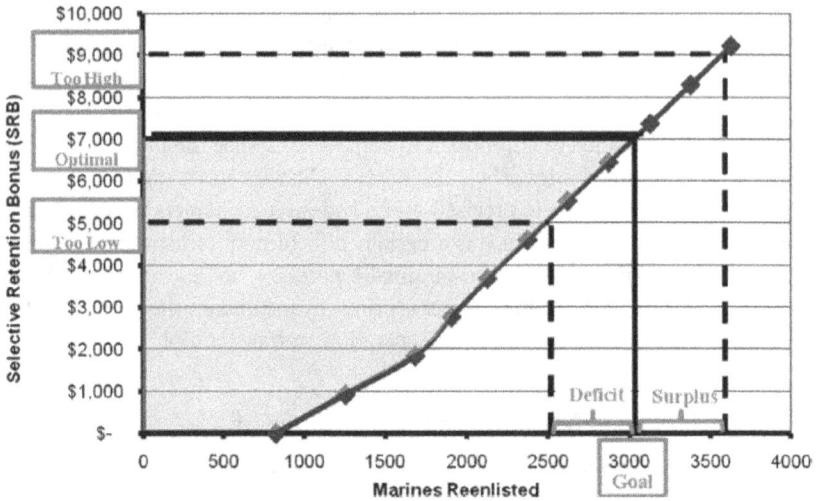

Figure 28-2. The importance of SRB precision. (*Reprinted from* Paul Bock, "The Sequential Self-Selection Auction Mechanism for Selective Reenlistment Bonuses: Potential Cost Savings to the U.S. Marine Corps" [master's thesis, Naval Postgraduate School, Monterey, CA, 2007], 42.)

$5,000), while it will exceed the end-strength target and be forced to suspend reenlistments if the bonus is set too high (for example, 3,500 Marines will want to reenlist if the SRB is set at $9,000).

There is ample evidence that traditional mechanisms relying on statistical analysis or other nonmarket approaches are ill-suited to accurately and consistently approximate the market-clearing SRB level (i.e., the SRB that equates bonus-takers and the targeted end-strength), particularly as underlying economic and national security conditions change. Thus, traditional conditional S&I pay programs, which rely on exogenously predetermined bonus levels, provide low levels of force-management precision.

Unfortunately, imprecision in traditional military force-management programs is not only an empirical regularity but also frequently quite severe. For example, research investigating Zone A reenlistments for the US Marine Corps indicates that the Marines have offered bonuses which were five to 10 times higher on average than the levels that were required to meet reenlistment targets within various specialties. With a USMC Zone A reenlistment bonus budget of $57 million in FY 2007, this single S&I pay program alone could have saved tens of millions of dollars by employing an approach which more accurately determined the appropriate magnitude of force-shaping bonuses.[8]

In sum, while regular S&I pay programs combine with other elements of military compensation to shape and balance the force over the long term, it is the conditional (retention, separation, and transfer) S&I pay programs that perform the "fine-tuning" of the force necessitated by sudden or short-term changes. Unfortunately, however, existing conditional S&I pay programs have proven to be too "blunt" for the precise fine-tuning for which they are intended.

Traditional methodologies for setting pay amounts (via statistical analysis, market research surveys, or simple rules of thumb) are akin to conducting surgery with an axe rather than a scalpel. As will now be illustrated, it is primarily this imprecision that causes existing conditional S&I pay programs to score poorly on key performance measures.

Performance Measures Proposed by the Quadrennial Review of Military Compensation

The QRMC was tasked by the US Department of Defense (DOD) to develop "agile and flexible compensation and benefit tools to optimize force management strategies of the uniformed services."[9] In pursuit of this objective, the 2008 QRMC published four principles for evaluating military compensation programs. Adapting these four principles to the specific application of conditional S&I pays, we generate the following performance measures:[10]

- *Voluntary*: S&I pay programs should be structured such that each service member willingly engages in the associated labor commitment (retention, separation, or transfer) and also perceives that compensation for the assignment is both satisfactory and fair.
- *Flexible and Responsive*: S&I pay programs should be flexible enough to quickly and effectively adjust resources to respond to emerging issues, shifting priorities, and changing market conditions.
- *Best Value*: S&I pay programs should provide cost-effective solutions to address specific service needs while minimizing cost.
- *Support Achievement*: S&I pay programs should successfully compete for talent and reward exceptional performance.

Each of these performance measures will be explored in more detail and applied to existing conditional S&I pay programs in the sections that follow.

QRMC Performance Measure #1: Voluntary

The principle of voluntary service is consistent with the broader theme of an all-volunteer military force and suggests that service-member roles (accession, retention, separation, transfer, and even specific assignments) should be voluntarily chosen to the maximum extent possible. The QRMC actually extends the definition of voluntary, however, in arguing that compensation for any labor commitment (retention, separation, transfer, or assignment) must also be perceived as "both satisfactory and fair."

In the reenlistment context, this principle implies that SRB programs should be precise; bonuses should be set at their market-clearing levels. Clearly, decisions to reenlist given an SRB policy are voluntary for those retained; service members not willing to voluntarily accept the SRB will not be retained. Voluntary retention becomes problematic, however, when DOD sets the SRB above the market-clearing price. In this case, more service members than needed will volunteer to retain, and, if end-strength targets are to be accurately met, many service members eligible and willing to reenlist will be involuntarily separated.

In some cases, service members willing to reenlist after end-strength targets have already been met have, in fact, been allowed to reenlist but without the bonus awarded to others in the same position, an outcome which has clearly not often been perceived as "both satisfactory and fair." Alternatively, some service members will be involuntarily denied the opportunity to accept a separation bonus if DOD terminates the program once it reaches its target number of separations. Similar issues with voluntary decisions extend to S&I pays designed to adjust force levels across career fields or service branches. In either case, an imprecisely set SRB amount leads to a violation of the QRMC principle of voluntary service (or separation).

Assignment incentive pay (AIP) offers an example of one existing S&I pay program in which auctions have already been applied to military force management with the goal of supporting voluntary decisions. The US Navy began offering AIP in June 2003 to alleviate recurrent shortages in hard-fill assignments. AIP is a special pay for active-duty enlisted Sailors. The Navy designates AIP eligible assignments (jobs) and sets a maximum monetary incentive for each job, not to exceed $1,500 per month. Sailors accepting these hard-fill assignments receive monthly incentive pay for the duration of their tour. AIP was expected to make hard-fill assignments attractive to at least one qualified "volunteer."

The AIP program has been implemented using a modified auction format. Sailors interested in AIP-designated billets submit bids, in $50 increments, for their chosen positions. A bid can start at $0 but can't exceed the Navy-determined maximum incentive. After the auction closes, the Navy determines Sailor assignments. The Navy generally selects the lowest-total-cost qualified Sailor, where total cost includes the Sailor's AIP bid as well as any moving and/or training costs necessary for that particular Sailor to fill that particular assignment. AIP has significantly reduced the number of Sailors receiving "involuntary" orders to hard-fill assignments, though such involuntary assignments have not been completely eliminated.

In sum, while existing conditional S&I pay programs are designed to reduce involuntary placements (retention, separation, or assignment) within the all-volunteer force, their effectiveness is limited by their lack of precision. If the amount of these financial incentives is not set at the market-clearing level—where the number of "takers" precisely or approximately equals the specific retention, separation, or transfer target—involuntary labor commitments (or commitments which are voluntary but not perceived as "both satisfactory and fair") will not be eliminated.

QRMC Performance Measure #2: Flexible and Responsive

Conditional S&I pays should be flexible and responsive enough to allow DOD to create financial incentives that quickly reshape the force in response to changing economic and national security conditions. By design, base pay, allowances, and the longer-term "regular" S&I pays are not structured to provide such flexibility; the burden for flexibility and responsiveness falls to the conditional S&I pay programs.

Typically, however, the predetermined levels of conditional S&I pays are publicized through military instructions and administrative messages and are relatively inflexible once announced; once an S&I pay has been announced, the primary opportunity for flexibility involves the decision to suspend the program. While S&I pays can be suspended or extended through future instructions and administrative messages, DOD rarely adjusts the value of the incentive offered.

Considering the precision problems associated with the traditional exogenously determined conditional S&I pays, and the limited opportunity to adjust the incentives once announced, traditional S&I pay programs do not provide the flexibility and responsiveness needed to address the constantly changing economic and national security environments.

QRMC Performance Measure #3: Best Value

The best-value principle essentially states that compensation programs should achieve DOD goals as cost effectively as possible. For the labor market applications described here, this implies minimizing DOD's cost to achieve its force-management objectives. Best value involves at least three considerations: (a) targeting those service members most willing to engage in the desired labor commitment (retention, separation, or transfer), (b) minimizing overpayment for these labor commitments, and (c) identifying the minimum cost incentive packages.

In the retention context previously illustrated in figure 28-2, "best value" first implies retaining those service members who are most willing to serve (assuming military screening mechanisms separate out service members not meeting acceptable quality standards). While only service members willing to retain for the predetermined bonus amount (or less) will reenlist, achieving best value can nonetheless be problematic under traditional SRB programs. If bonuses are set too high, service members are typically selected for retention on a first-come, first-served basis until end-strength targets are met. This favors service members whose service commitments end closer to the time that bonuses are announced; some service members who may be extremely dedicated to continuing military service, but who have poorly timed end-of-service-commitment dates, may be separated rather than retained, reducing efficiency.

To put this another way, service members who may respond late to an SRB offer but who would have been willing to reenlist for a minimal bonus amount might be denied reenlistment in favor of retaining quickly responding service members who are less willing to reenlist in the sense that they required the full amount (or nearly the full amount) of the bonus offered in order to accept reenlistment. Hence, current retention bonus programs may retain service members who are not only more expensive to retain now, but will be more expensive to retain in the future; at the same time, other service members who are much more willing to continue service over both the short and long term may be separated.

"Overpayment" to military service members in SRB programs refers to the fact that many retainees receive a bonus well above what is required to induce their reenlistment, even if the precise market-clearing SRB amount is offered. The magnitude of this overpayment is illustrated in figure 28-2 above, where the shaded area represents the overpayment the retained Marines receive in excess of the minimum reenlistment bonus they require (assuming the USMC accurately identifies the minimum bonus required to meet reenlistment targets). According to the CNA analysis which underlies figure 28-2, approximately 25 percent of the Marines retained would have reenlisted without any bonus, yet they each receive the full SRB offered; each of these Marines receives a $7,000 "overpayment" from the USMC.

At the market-clearing SRB, only the retained Marine "at the margin" (the Marine who is least willing to serve among those retained or the retainee who is most "on the fence" about whether or not to reenlist given the size of the bonus) receives an SRB that is just about equal to the amount required for him to reenlist; the marginal Marine is the only one not being "overpaid" for reenlistment. In figure 28-2, the total SRB cost is $21,000,000 ($7,000 × 3,000 Marines), while the total overpayment is approximately $14,000,000 or two-thirds of the total SRB cost.

At the margin, these overpayments are even more troubling. Consider again the USMC example in figure 28-2 above, and suppose the USMC wanted to increase retention by 100 Marines, from 3,000 to 3,100. To do so, the USMC would need to increase the retention bonus by approximately $300, from $7,000 to $7,300, to attract the 100 additional Marines according to this data. Under a traditional retention bonus program, however, this $300 increase is paid to all 3,100 retained Marines, raising the USMC's total cost to $22,630,000, an increase of $1,630,000 to retain 100 additional Marines, or $16,300 per additional Marine. Moreover, the total overpayment amount would increase by $915,000 to $14,915,000 for the 3,100 retained Marines.

The final factor affecting best value concerns the composition of the incentives offered in conditional S&I pay programs. To date, these programs have relied almost exclusively on monetary incentives. Unfortunately, cash payments will have limited effectiveness with many service members; some people place a higher value on nonmonetary incentives (for example, quality-of-life benefits such as base or port of choice, geographic stability, flexible work arrangements, and so on) than they place on monetary incentives, especially as these monetary incentives are increased to address acute under- or overstaffing concerns. This suggests that best value may be improved if DOD were to augment the monetary incentives of conditional S&I pay programs with nonmonetary incentives, particularly if incentive packages could be individualized to reflect the recipient's personal preferences, as will be discussed later in this chapter.

In sum, existing conditional S&I pay programs (a) often do not select those service members most willing to respond (i.e., willing to retain, separate, or transfer at the least cost), (b) significantly overpay many of the service members who do respond to the incentive offered, and (c) do not allow for a mix of monetary and nonmonetary incentives which could achieve the same objective at a lower overall cost. Consequently, current S&I pay programs do not perform well in terms of the "best value" measure.

QRMC Performance Measure #4: Support Achievement

Supporting achievement involves successfully competing for military talent by encouraging and rewarding performance and recognizing service members' contributions to the mission. Supporting achievement recognizes that service members may differ in their skills and performance; some service members may be more qualified than others (have more or more-appropriate training, have better military bearing, have greater past experience for a job, etc.). Better-achieving service members bring more value to DOD, and the principle that compensation should support achievement suggests that the military services should better compensate more-qualified service members within the same specialty and rank.

As in any labor market, it is important to recognize that preferences in military force management are two-sided. Service members have different preferences for military service (and especially for different assignments), while at the same time the military branches prefer high performers to low performers—in other words, not all infantry sergeants are equally valuable. Despite this obvious reality, existing conditional S&I pay programs consider only one side of the equation; they recognize that willingness to retain or separate varies significantly across service members of the same rank and specialty, but these programs essentially ignore the fact that the military services also have preferences about which particular service members within any given category are retained or separated.

The reason existing conditional S&I pay programs—and virtually all military compensation programs, for that matter—are one-sided is because all service members of the same rank and specialty are offered the same financial incentive. In the case of retention bonuses, for example, this means that the set of service members who are retained are those who are essentially the "cheapest" to retain. To some degree, however, it remains true that "you get what you pay for," and thus the highest-achieving service members within any particular target group are unlikely to also be the least expensive to retain. This is because high achievers within the military will generally also have the best opportunities outside the military and thus will require a greater incentive to remain in service.

To its credit, however, the US Navy has partially incorporated two-sided preferences with its previously discussed AIP program. Recall that the AIP program uses a modified auction approach in which Sailors submit bids indicating the minimum additional pay they would require to serve in each of a set of hard-fill assignments. These Sailor bids obviously reflect preferences on one side of this market, but preferences on the other side of the market are also incorporated to some degree in that any relevant moving and training costs associated with each Sailor-billet combination are added to the Sailor bids to reflect the "true cost" of assigning any particular Sailor to any particular billet. These moving and training costs partially—though not completely—reflect Navy preferences across Sailors, hence incorporating preferences on both sides of the market. As will be discussed later, however, there are problems associated with the manner in which two-sided preferences are currently incorporated in the AIP program, and there are more robust and cost-effective methodologies through which the Navy could do so.

With the notable (yet limited) exception of the Navy's AIP program, conventional S&I pay programs do not consider military preferences across different service members of the same rank and specialty. More importantly, these programs (including AIP) do not directly acknowledge the service members' differential contribution to the mission, instead viewing service members as largely indistinguishable and simply targeting those service members most willing to accept the S&I pay. As a result, the current conditional S&I programs do not perform well in terms of supporting achievement.

The "Hidden" Performance Measure: Practicality

To summarize our performance measurement so far, the conditional S&I pay programs currently used for military force management have largely failed to meet the QRMC's four criteria for military compensation: voluntary, flexible and responsive, best value, and support achievement. In part, these failures reflect the difficulty traditional programs have in precisely determining the market-clearing S&I payment in constantly changing economic and national security environments. These failures also reflect an almost exclusive focus on monetary bonuses and a heavy emphasis on cost-effectiveness at the expense of military achievement. Given these failures, one might ask why these programs have persisted for as long as they have in their current formats. This persistence likely reflects a fifth performance measure not explicitly recognized by the QRMC: practicality.

Practicality in force-shaping and management programs addresses the ease of implementation for the services and ease of participation for service members. The traditional force-shaping and management programs are relatively practical for the services to implement and for service-member participation; the services simply announce a retention, separation, or reassignment bonus policy and service members simply decide whether to accept or reject the offer. Thus, when considering auctions or any other alternative approach to setting S&I pay for military force-management applications, we must address the implicit or "hidden" practicality performance measure as well as the four explicit QRMC principles discussed above.

The Final "Report Card" on Existing Conditional S&I Pay Programs

To summarize, the traditional force-management tools of conditional S&I pay programs can be graded against the performance measures presented in this section as shown in table 28-1:

Table 28-1. Effectiveness of traditional force-management conditional S&I pay programs

Precision	Voluntary	Flexible & Responsive	Best Value	Support Achievement	Practicality
Low	Medium	Low	Low	Low	High

Predetermined S&I bonuses are practical to implement yet face significant risk for under- or over-estimating the bonus level needed to meet end-strength goals, especially in the current environment of global economic and political uncertainty. If the bonus is set too low, the military will not induce enough reenlistments or separations to meet end-strength targets. If the bonus is set too high, the military incurs unnecessary costs and may induce too many reenlistments or separations, unless applications are approved as they arrive and are capped at the desired end-strength target (in which case the services risk retaining/separating a suboptimal group of service members). Furthermore, all eligible service members receive the predetermined incentive, regardless of their individual incentive requirements. This significantly overcompensates some service members. Finally, there is concern that some high-quality service members will not be retained, even when the military might be willing to pay more to retain them.

Concern over the performance of conditional S&I pay for force-management programs has led to consideration of auction mechanisms as an alternative means for setting retention and separation bonuses. The first auction mechanism introduced for military force management on a large-scale basis is the Navy's AIP auction. While initial satisfaction with the AIP auction has generally been very high, before broadly expanding the use of auctions for force management, it is important to fully consider how auction mechanisms could best be designed and applied to achieve military manpower priorities and how such auctions would ultimately compare to traditional force-management programs in terms of the performance measures described above.

The Design of Auctions for Force Management

The Elements of Auction Design

For most people, the concept of an auction is limited to the familiar process, famously used at auction houses such as Sotheby's or on online auction sites such as eBay, in which potential buyers iteratively and openly (perhaps even loudly) attempt to "outbid" each other to purchase an item.[11] The true scope of auction mechanisms, however, is far richer than such simple manifestations. Most generally, an auction is defined as "an economic mechanism whose purpose is the allocation of goods (or services) and the formation of prices for those goods (or services) via a process known as bidding."[12] Hence, an auction is any mechanism in which participants (potential buyers, sellers, or both) submit "bids" which are used to determine (a) what good or service will be sold or provided, (b) by which seller or sellers, (c) to which buyer or buyers, and (d) at what price each transaction will take place.

To understand the role of auctions in the broader field of all economic mechanisms, recognize that the nature of any transaction mechanism depends in part on the number of buyers versus sellers in the particular transaction, as depicted in figure 28-3. Auctions are typically used when there is either a single seller and many buyers or a single buyer and several sellers. More traditional market mechanisms typically set prices when there are many buyers and sellers (although matching auctions or double auctions are sometimes appropriate), while negotiated prices are the norm when there is only a single buyer and seller.

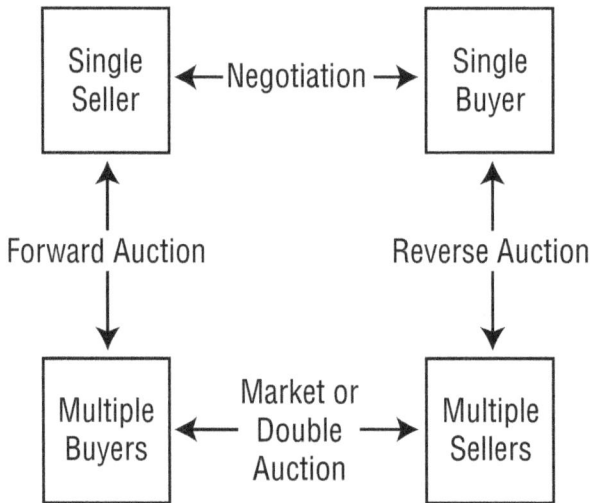

Figure 28-3. Varieties of transaction mechanisms.

Not only are auctions just one of many possible transaction mechanisms, there is an innumerable variety of distinct auction designs within the field of auction mechanisms. For any auction application, therefore, one must carefully select the auction design elements which most appropriately fit the particular context. To further illustrate this point, figure 28-4 depicts a few of the most important auction design considerations, highlighting the design choice that is most appropriate for force-management auction applications.

Forward versus Reverse Auctions

A forward auction is one in which a single seller accepts bids from multiple potential buyers. Forward auctions are the most common form of auction and are observed at auction houses such as Sotheby's or online auction sites such as

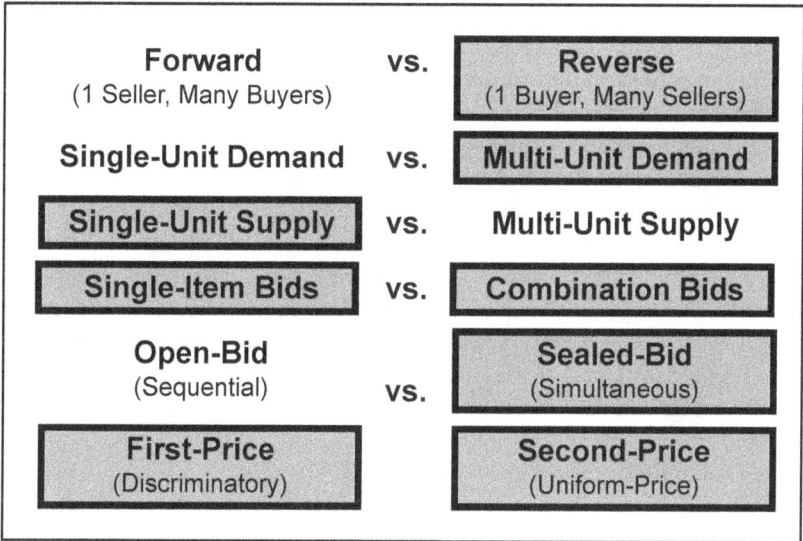

Figure 28-4. Auction design choices in the force-management context—forward versus reverse auctions.

eBay. In these auctions, competition among bidders drives prices higher, and the winning bidder or bidders are those who submit the highest bid or bids.

In contrast, a reverse auction is one in which a single buyer accepts bids from multiple potential sellers. Such reverse auctions commonly occur in the procurement context, in which several contractors bid to sell their products or services to a single buyer. In such auctions, therefore, competition among bidders drives prices *lower*, and the winning bidder or bidders are those who submit the *lowest* bid or bids.

Force-management applications typically involve reverse auctions; the military is the single buyer of military labor commitments (retention, separation, or transfer) while the service members represent multiple sellers of these commitments. Through competitive bidding among service members, auctions have the potential to significantly improve precision and cost effectiveness in setting retention, separation, or transfer bonuses. Force-management auctions can determine the exact market-clearing price which will allow the military to precisely hit its end-strength targets or otherwise accurately manage its force structure at the minimum cost.

Single-Unit Demand or Supply versus Multiple-Unit Demand or Supply

It is also important to classify auction types according to the supply and demand conditions. In particular, each buyer in an auction may be interested in buying only a single unit of the good or service being sold in the auction (single-unit demand) or each may instead be interested in buying multiple units (multiple-unit demand). Similarly, each seller in an auction may have only a single unit of the good or service to sell in the auction (single-unit supply) or each may be selling multiple units (multiple-unit supply). These distinctions apply to both forward and reverse auctions.

In a force-management auction, the military branch (Army, Navy, Air Force, Marine Corps) is the buyer and is interested in buying (retaining, separating, or transferring) multiple units or service members. In contrast, the sellers in a force-management auction are the service members, who each can fill only one assignment or offer one separation. Therefore, the sellers in a force-management auction each can sell only a single unit.

Force-management auctions are therefore characterized by multiple-unit demand and single-unit supply. Moreover, because these auctions are reverse auctions, with sellers bidding against each other, the multiple-unit demand condition means that force-management auctions can also be described as "multiple winner" auctions.

Single-Item Bids versus Combination Bids

Another important element of auction design comes into play whenever the lone seller in a *forward* auction can sell to any single buyer either (a) multiple units of the good or service (especially if those units are not identical) or (b) a single unit with adjustable characteristics. In such cases, a buyer's valuation will often depend on the quantity of units he is buying, the specific combination of distinct units he is buying, and/or the specific characteristics of each unit.

A similar complication arises whenever the lone buyer in a *reverse* auction can buy from any single seller either (a) multiple units of the good or service (again, especially if those units are not identical) or (b) a single unit with adjustable characteristics. In such cases, the price at which a seller is willing to sell will often depend on the quantity of units he is selling, the specific combination of distinct units he is selling, and/or the specific characteristics of each unit.

In either of these scenarios, the auction designer must decide whether to allow combination bids or instead to restrict bidding to single items alone. In particular, will bidders submit bids only for stand-alone items (e.g., $10 for item A, $20 for item B, and so on), or will bidders also be able to (or required to) submit bids for specific quantities (e.g., $10 for one unit of item A, $15 for two units of item A, and so on) and/or specific combinations (e.g., $25 for items A and B together).

Consider, for example, an estate sale in which a dining set, consisting of a table and six chairs, is to be sold via auction. Some potential buyers at the estate sale might be interested in purchasing only the complete dining set as an integrated package, while other potential buyers might be interested in purchasing certain individual pieces (perhaps a few chairs to complement a dining set already owned). In such cases, the auction designer could choose to (a) accept bids only for the complete dining set, (b) accept bids only for individual pieces within the dining set, (c) accept bids for either the complete dining set or for individual pieces, or (d) accept bids for any combination of pieces within the set (e.g., all six chairs without the table, the table plus four chairs, and so on). The auction design decision regarding the structure of bids allowed has critical implications for how buyers should bid, which buyer(s) will ultimately receive which pieces of the dining set, and how much revenue the seller can expect to generate.

For most force-management auctions, it is sufficient to accept bids only for single stand-alone items (retention, separation, or transfer). Thus, most of the auction applications explored in this chapter will assume that bidding service members each submit a single cash bid for a particular labor commitment. It is important to recognize, however, that some of the advanced customized auction applications presented at the end of this chapter do, in fact, call for combination bids. For example, when nonmonetary incentives are incorporated into a reenlistment auction, service members bid not only for reenlistment as a stand-alone labor commitment, but also for combinations such as reenlistment with home-port-of-choice, or reenlistment with geographic stability, and so on.

Open-Bid versus Sealed-Bid Auctions

Another auction design element related to bid submission concerns how bids are to be communicated and whether bids are observable to others while the auction remains open or ongoing. In this design dimension, auctions are generally classified as having either an "open-bid" or "sealed-bid" format.

In open-bid auctions, bidders openly declare or otherwise publicly reveal their bids during the auction (i.e., before a winner is determined). An open-bid forward auction with multiple buyers, for example, starts with a low price, and competitive bidding sequentially raises the price until all but the winning buyer (or buyers, in the case of multiple-unit supply) have dropped out of the bidding. This format is also referred to as an English or sequential-bid auction. In an open-bid reverse auction, note that sellers would sequentially bid the price downward until all but the winning supplier (or suppliers, in the case of multiple-unit demand) have been eliminated.

In sealed-bid auctions, participants submit a single undisclosed bid. All bids are opened simultaneously and the winner(s) declared. In any auction with only a single winner, the bidder who submitted the highest bid would be the winning buyer in a forward auction, while the bidder who submitted the lowest bid would be the winning seller in a reverse auction. Similarly, for auctions with multiple (N) winners, the (N) bidders who submitted the highest bids would be the winning buyers in a forward auction, while the (N) bidders who submitted the lowest bids would be the winning sellers in a reverse auction.

Practicality in force-management applications strongly favors sealed-bid over open-bid auction designs. In any open-bid auction, bidders (or their agents) must periodically or even continuously monitor ongoing price adjustments, as one might observe price movement in person at an auction house or remotely on eBay. Any expectation of active, simultaneous price monitoring by bidders is unreasonable in the military manpower context, however, considering the geographic dispersion, limited connectivity, and overall operating tempo for active-duty service members.

There is no price-monitoring requirement in sealed-bid auctions, however, as bidders need only submit their one-time bids at any point during the prescribed bidding window; all bids are observed simultaneously and the winner(s) determined only after the bidding window has closed. Consequently, we will limit our attention in this chapter to sealed-bid auction formats.

First (or Discriminatory) Price versus Second (or Uniform) Price Auctions

Pricing rules with a single auction winner. The highest bidder wins in virtually all single-winner forward auctions, while the lowest bidder wins in virtually all single-winner reverse auctions.[13] A common variation among auctions, however, is how the price a winning buyer pays or a winning seller receives is determined. Along this dimension, sealed-bid single-winner auctions are most commonly separated into "first-price" auctions and "second-price" auctions.

In a first-price sealed-bid auction, the transaction occurs at the price submitted by the winning bidder, which is—as the name implies—the first price that would be listed if bids were sorted in rank order (from highest to lowest in a forward auction, or from lowest to highest in a reverse auction). In a second-price sealed-bid auction, on the other hand, the transaction occurs at the second-highest price submitted in a forward auction or the second-lowest price submitted in a reverse auction. In other words, the price paid or received in a second-price auction is equal to the bid of the "closest loser"—the losing bidder who came closest to winning.

Generalizing the pricing rules to the multiple-winner context. Auctions in the force-management context can follow either the first-price or second-price approach; however, the price-determination rule must be generalized to the multiple-winner context. The multiple-winner generalization of the first-price auction is the discriminatory price auction (or simply discriminatory auction), in which each winning bidder simply pays (or receives) whatever he or she bid. In a discriminatory forward (reverse) auction with 10 winners, for example, the 10 highest (lowest) bidders would be the winning buyers (sellers) and each winner would pay (receive) the exact amount of his or her own bid.

The multiple-winner generalization of the second-price auction is the *uniform price auction*, in which each winning bidder pays (or receives) the amount of the "first-excluded" bid, which is, as in the single-winner case, the bid submitted by the "closest loser"—the losing bidder who came closest to winning.[14] In a uniform price forward (reverse) auction with 10 winners, for example, the 10 highest (lowest) bidders would be the winning buyers (sellers), and each winner would pay (receive) an amount equal to the 11th highest (lowest) bid submitted. This multiple-winner auction format is called a uniform price auction because all winning bidders pay or receive the same uniform price.

Bidding behavior under alternative pricing rules. Different pricing rules elicit different bidding strategies and potentially different auction outcomes. For the purposes of this chapter, however, it is not necessary to explore and explain these strategic implications in any significant detail.[15] Nonetheless, to understand how auctions with different pricing rules will perform in force-management applications, it is helpful to summarize what is known about the general characteristics of outcomes under the different rules. For simplicity of explanation, we will focus on single-winner forward auctions, as this class of auctions is the most common and familiar. All results easily generalize, however, to the multiple-winner reverse auction context of force management.

To illustrate, consider the situation faced by a person submitting a sealed bid in an auction for a painting that she values at exactly $100. In other words, at any price under $100, she would be willing to buy the painting (and would be happier the lower the price, of course), but she would be unwilling to pay even a penny more than $100 for the painting. Mathematically speaking, the enjoyment or net benefit (or what economists call "surplus") she receives if she purchases the painting at a price of P is equal to 100 - P. What amount, then, should this person submit in her sealed bid for the painting? The answer depends on the pricing rules of the auction.

In a second-price sealed-bid (SPSB) auction, the answer is surprisingly simple: She should bid exactly her true value for the painting, or $100. A thorough proof of this general result can readily (and more appropriately) be found elsewhere.[16] Suffice it to say that if she bids any amount above or below $100 in an SPSB auction, it can only hurt her; doing so will either have absolutely zero effect on the outcome of the auction (i.e., the auction winner and the price paid will both be unchanged), or it will produce an auction outcome that is *worse* for the bidder than if she had instead bid exactly $100. If she bids any amount above $100, for example, the only change that could possibly result is that she could, in fact, win the auction but end up paying a price above $100, which is worse than losing the auction. Moreover, if she bids any amount below $100, the only change that could result is that she could end up *not* winning the auction when she would have won the auction (and paid a price at or below $100) if she had instead bid $100.

Thus, in a second-price sealed-bid auction—and its multi-winner generalization, the uniform price auction—each bidder's optimal strategy is always to submit a bid equal to her true valuation (the absolute maximum she would be willing to pay for the item in a forward auction, or the absolute minimum she would be willing to accept for the item in a reverse auction). In this sense, we say that the SPSB and uniform price auction mechanisms are "truth-revealing," as buyers do best in these auctions by truthfully revealing the maximum amount they are willing to pay and sellers do best by truthfully revealing the minimum amount they are willing to accept. Moreover, experimental simulation of military force-management auctions using enlisted personnel indicates that bidders quickly recognize the optimality of such truthful bidding in these second-price auctions.[17]

Now, what if this same person was bidding for this same painting, but was instead participating in a first-price sealed-bid (FPSB) auction? First recall that in an SPSB auction, the high bidder wins but pays the second-highest bid. Thus, if she optimally bid her true value of $100 and won the painting, she would pay some price P equal to the next highest bid, and therefore she would indeed enjoy a positive net benefit from buying the painting (100 - P ≥ 0). In contrast, however, in an FPSB auction, the winning bidder must pay the amount of her bid; thus it is immediately obvious that she would not want to bid her true value of $100 in an FPSB auction, as this would guarantee zero net benefit (100 - P = 100 - P = 0). Thus, in a forward FPSB auction, the optimal strategy is to bid some amount below your true value for the item being sold.

By how much should she "underbid" her true value of $100? The lower she bids, the higher the net benefit (100 - P) she would receive if she wins the auction, but the lower her probability of winning. In selecting a bid in an FPSB auction, therefore, she faces a tradeoff between her chances to win the auction and her net benefit if she wins. Note that the amount she bids really matters only if she actually wins the auction, so she might as well bid as if she will indeed be the winning bidder, in which case she can also reasonably assume she would be the bidder with the highest *value* for the painting as well.

Identifying her optimal bidding strategy in an FPSB auction thus boils down to answering the following question: If she has the highest value for the painting among all bidders, how low can she bid and still win the painting? The answer is that she can bid as low as the second highest bid, which she can safely assume will be at or below the second highest value for the painting. Thus, her optimal bidding strategy (technically, the "equilibrium" bidding strategy) in an FPSB auction is to bid what she expects the next highest value would be if her value for the object ($100) was indeed the highest value among all bidders.

Revenue or cost equivalence under alternative pricing rules. One interesting and important implication of the above-described optimal bidding strategy is that, on average, the seller of the object can expect to receive the same revenue—whether the object is sold via first-price or second-price auction. To see this, note that the price (or revenue) in a second-price auction will be equal to the second-highest bid. Because the optimal bidding strategy is to bid truthfully, this will be equal to the second-highest value. Under a first-price auction, the price (or revenue) will be equal to the absolute highest bid. According to the optimal bidding strategy above, however, the high bidder in a first-price auction will bid what she expects to be the second-highest value.

Thus, under either the FPSB or SPSB auction format, the expected price in a forward auction is equal to the expected second-highest valuation. In general, the expected revenue for the seller under either auction format is the same. This result is known as "revenue equivalence."

The same holds true for "cost equivalence" in a reverse auction (in which many potential sellers submit bids to a single buyer). Under either the FPSB or SPSB auction format, the expected price (cost incurred by the single buyer) in a reverse auction is equal to the expected second-lowest willingness-to-accept (the absolute minimum price at which a seller will provide a good or service). Thus, in general, the expected cost for a buyer under either auction format is again the same.

Moreover, this revenue/cost equivalence result generalizes to far more complex auction designs. Particularly relevant for the military force-management context, the equivalence result generalizes to auctions with multiple winners. Thus, whether the DOD uses FPSB or SPSB auctions for force management, the total cost to retain, separate, or transfer the target number of service members will be the same under either auction format.[18]

The Basic Force-Management Auction in Practice: Application to the Selective Reenlistment Bonus Program

The Basic Force-Management Auction Design

To understand and evaluate the potential performance of auction mechanisms in the force-management context, it is helpful to consider how such mechanisms might be applied to particular existing or proposed conditional S&I pay programs. It is also most instructive to start with an application of the most basic force-management auction format, involving:

(1) a single auction,

(2) with eligible service members all bidding for the same opportunity,

(3) with all bidders treated equally, such that the set of lowest bidders will win regardless of other bidder characteristics, and

(4) using monetary incentives alone.

With this in mind, this section will explore a conditional S&I pay program in which application of such a basic force-management auction may be both illustrative for this chapter and beneficial to the military: the Selective Reenlistment Bonus Program as executed in the US Marine Corps. Later sections of this chapter will both illustrate the application of this basic force-management auction in other force-management contexts and also introduce more innovative and customized force-management auction mechanisms.

The USMC Selective Reenlistment Bonus Program. The Marine Corps Order on Selective Reenlistment Bonuses (SRB) describes the program in the following way:

> The SRB program was established to assist in attaining and sustaining adequate numbers of career enlisted personnel in designated Military Occupational Specialties (MOSs) and within particular years-of-service groupings. The program provides a monetary incentive for a reenlistment of at least 4 years at three career decision points during the first 14 years of service.[19]

To help set the SRB payments for each MOS, the Center for Naval Analysis provides an annual regression analysis that predicts reenlistments by MOS as a function of the SRB level, although the degree to which this analysis is used to set SRB levels is unclear.[20] All eligible Marines qualify for the SRB, regardless of their willingness to reenlist. As a result, Marines willing to reenlist for less than the SRB offered receive an overpayment from the USMC, which can be substantial. In addition, the USMC appears to set the SRB well in excess of the level predicted by the regression analysis, particularly for critical MOSs. As a result, many SRBs may be unnecessarily high, further increasing the USMC's costs and overpayment to the Marines.[21]

The promise of auctions as selective reenlistment bonus tool. Auctions provide a promising endogenous, market-based approach to setting SRBs in military force-retention programs. To compare the bidding strategies and outcomes for discriminatory and uniform price auctions, consider the following stylized example. Suppose there are 100 militarily qualified officers whose minimum required retention bonuses (opportunity costs of military service) are uniformly distributed between $0 and $60,000 (nonmilitarily qualified officers have been screened out by their fitness reports); DOD wants to retain 75 percent of these officers.

If all officers receive the same bonus, as with the traditional bonus mechanism, the required bonus would be approximately $45,000, the bonus required by the 75th officer; the military would spend $3,375,000 on retention bonuses; 50 percent of this total, or $1,687,500, is overpayment to the retained officers (or costs to DOD in excess of what officers require to keep them in military service). This is money that could be used to buy ships, airplanes, or additional end strength. The difficulty for traditional bonus mechanisms is determining the appropriate retention bonus; past experience indicates that imprecision has been a significant and persistent problem. The implications of over- or underestimating the required bonus were discussed above.

A uniform price reenlistment auction. The optimal strategy in a uniform price auction is for all service members to bid (truthfully reveal) the minimum retention bonus they require to continue in military service. The diagonal line in figure 28-5 depicts the minimum required bonus for the 100 hypothetical officers in this example. In a uniform price auction, this diagonal line depicts the service members' sealed bids. The bonus would be set equal to the 76th lowest (first excluded) bid, and all retained officers would receive this bonus. As above, the expected bonus would be approximately $45,000. The military would spend $3,375,000 on retention bonuses; 50 percent of this total, or $1,687,500, would be overpayment captured by the retained officers.[22]

A discriminatory price reenlistment auction. Suppose instead that DOD set retention bonuses using a discriminatory price auction. In this case, DOD would retain the 75 officers submitting the lowest bids, and each retained officer would receive a retention bonus equal to his or her bid. Recall from the bidding strategies described above, it is not optimal to bid truthfully in a discriminatory price auction. The net benefit (overpayment) maximizing bidding strategy involves a trade-off between expected net benefit and the probability of winning the auction. Officers will bid in excess of their required bonus to increase the overpayment they receive; however, increasing their bid reduces their probability of winning the auction. Thus, the optimal bidding strategy involves a trade-off between risk and return (overpayment).

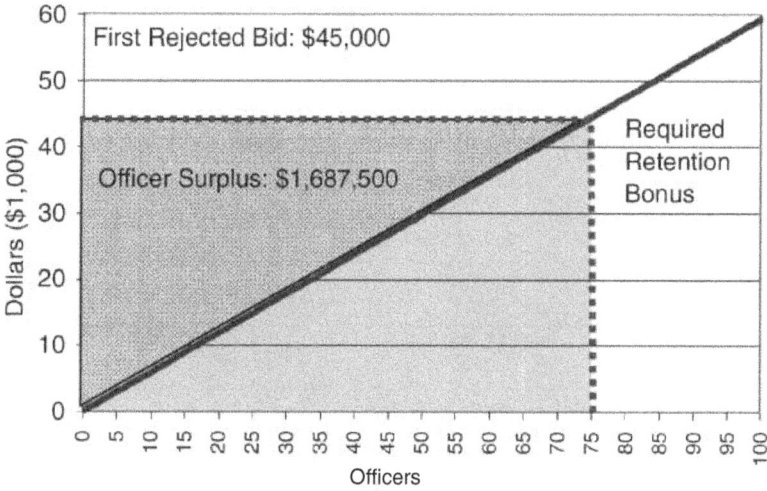

Figure 28-5. Second-price sealed-bid auction result.

In fact, the net benefit-maximizing (risk neutral) equilibrium strategy in a discriminatory price auction is to assume that you are the median winning bidder (e.g., the 37th lowest bid) and estimate how much you can bid above your minimum required bonus and still be one of the 75 officers retained. This situation is depicted in figure 28-6. The optimal bids submitted in a discriminatory price auction are represented by the upper dashed line in figure 28-6.

The first excluded bid is $45,000 in the uniform price auction, which sets the bonus for all retained officers; the 75th lowest bid is $54,000 in the discriminatory price auction. The last retained officer receives this bonus, which is significantly higher than the bonus set in the uniform price auction. In fact, figure 28-6 shows that approximately half the retained officers receive a higher bonus with a discriminatory price auction and half receive a lower bonus (this even split reflects that values are uniformly distributed in this example). More importantly, DOD's total cost is the same in both cases, as predicted by revenue or cost equivalence. DOD's savings from officers receiving a lower bonus with a discriminatory price auction ($165,000 in this example) are exactly offset by increases in DOD's costs for officers receiving higher bonuses.[23] Comparing discriminatory and uniform price auctions, they are equally effective at meeting DOD's precise end-strength targets and equally cost-effective; they differ in the bidding strategy they induce and the distribution of overpayment across retained service members.

Compared to the USMC's traditional process for setting SRBs, a simple discriminatory or uniform price auction provides significantly better performance for the compensation principles described above. An auction allows the USMC to identify the SRB that will precisely meet its end-strength targets within a career field. With market clearing SRB incentives, all service-member

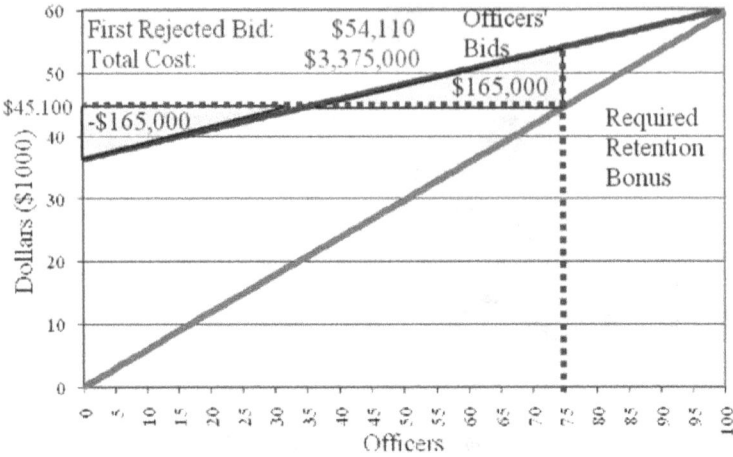

Figure 28-6. Discriminatory price auction results.

decisions will be voluntary; anyone choosing to accept the SRB and reenlist will be able to do so, and no one will be involuntarily separated. An auction also provides the flexibility needed to adjust the SRB to meet current economic and national security conditions. As the USMC collects its auction data, it also has the information and flexibility needed to adjust end-strength targets across career fields, or reenlistment zones, as necessary to best balance the force within the overall SRB budget constraints.

In terms of best value, auctions target those service members most willing to serve, but the simple auctions described so far do not address the overpayment or most cost-effective incentive package issues. Similarly, these simple auctions have the same shortcomings in supporting achievement as the traditional SRB programs (innovative auction solutions to all of these limitations are addressed below). Finally, simple discriminatory and uniform price auctions are slightly less practical to implement; service members would either need some instruction on the optimality of truthful bidding for a uniform price auction or would have an incentive to gather the information required to bid strategically in a discriminatory price auction. Table 28-2 compares a traditional SRB program to an auction-based SRB program.

Table 28-2. Effectiveness of traditional force-shaping and management programs

	Precision	Voluntary	Flexible & Responsive	Best Value	Support Achievement	Practicality
Traditional SRB	Low	Medium	Low	Low	Low	High
Auction-Based SRB	High	High	High	Medium	Low	Medium-High

Further Application of the Basic Force-Management Auction Design

Surface warfare officer continuation pay. The USMC SRB application described above involves enlisted service members. Auction-based retention bonuses can be directly extended to officer retention. As an example, consider Navy junior surface warfare officers (SWO), an officer community with chronic retention shortfalls. Surface warfare officer continuation pay (SWOCP) was established in January 2000 to entice junior officers to remain in the SWO community and fill SWO department head billets.[24] This incentive pay was initially based on a cost-benefit analysis conducted by the SAG Corporation.[25]

The SAG study estimated the expected retention impacts from offering annual bonuses of $5,000, $10,000, or $15,000 for each year of obligated service through two department-head tours. The analysis projected that a $10,000 annual bonus would likely retain the desired end-strength with some margin for error. The US Navy adopted the $10,000 annual payment recommendation. The SWOCP bonus pays officers $50,000 in five installments: the first installment is paid when the SWO retention contract is accepted; the remaining four installments are paid annually on the date that the officer begins his or her department-head assignment or department-head school (whichever is earlier).[26]

Unfortunately, it is difficult to determine the market-clearing bonus that realizes the desired end-strength target using statistical or other nonmarket approaches. Still facing SWO department-head shortfalls, the Navy augmented the SWOCP bonus in 2006 with the junior surface warfare officer critical skills retention bonus (junior SWO CSRB).[27] The junior SWO CSRB pays eligible lieutenants $25,000 to remain in the Navy and the SWO community until they complete two department-head tours. Officers receive $15,000 on the anniversary of their sixth year of service, and $5,000 on the anniversary of their seventh and eighth years of service.[28] Thus, the total retention incentive for junior SWOs completing two department-head tours is $75,000. It is also important to note that all eligible SWOs receive this $75,000 bonus regardless of their willingness to retain. SWOs who are willing to retain for less than $75,000 receive overpayments from the Navy; some are likely significant. Furthermore, SWO retention has remained problematic despite the additional junior SWO CSRB.[29]

As with the USMC enlisted SRB discussed above, an auction-based SWO retention bonus would significantly improve performance across most of the QRMC military compensation principles. The Navy has clearly had trouble identifying the appropriate junior SWO retention bonus. Unlike the USMC enlisted SRBs, which appear excessively high, the SWO bonuses are inadequate to meet the Navy's end-strength targets. An auction-based approach would provide the precision the Navy needs to meet its targets. Alternatively, if the Navy found the market-clearing junior SWO retention cost to be excessive, an auction would give the Navy the information and flexibility needed to adjust its end-strength target and develop an alternative strategy.

As with enlisted SRBs, all decisions to retain or separate would be voluntary and an auction would identify the lowest-cost officers. An auction is practical, with the caveat from above about bidding strategies; a simple auction has the same low performance as traditional retention programs in terms of supporting achievement.

Junior SWO retention concerns help highlight the importance of the best-value metric. One issue related to best value is identifying the most cost-effective incentive package. The junior SWO experience indicates that purely monetary incentives might have limited impacts in some communities and certainly seem to suffer from diminishing returns as the monetary incentive increases (the same appears true for retention in the medical and dental communities). Best-value considerations support a more innovative auction design that might exploit non-monetary as well as monetary incentives, particularly if incentive packages can be tailored to each individual recipient.[30] Such an innovative auction design is discussed in more detail below.

Voluntary separation: defense drawdown 1992–1997. Another example of a past force-management program involves DOD's force drawdown between 1992 and 1997, at the end of the Cold War. During this time, DOD offered three programs to encourage voluntary separations. The programs included the voluntary separation incentive (VSI), the special separation bonus (SSB), and the temporary early retirement annuity (TERA).

VSI was available to service members with at least six years of service and paid an annuity equal to 2.5 percent of the service member's final annual base pay times the service member's years of service; the annuity was paid for twice the service member's years of service. SSB was also available to service members with at least six years of service and paid a lump sum separation bonus equal to 15 percent of the service member's final annual base pay times the service member's years of service. TERA was available to service members with at least 15 years of service and paid a lifetime annuity but at a reduced accrual rate compared to the normal military retirement system at 20 years of service.

These programs did, in fact, increase voluntary separation. However, there are questions about whether they were cost effective to the military. Under these programs, all voluntarily separated personnel received monetary compensation to leave the service. A 1992 RAND study on the drawdown estimated that half of those separated would have left regardless of the program, and others would have left for smaller monetary payments.[31] These personnel received significant overpayments. Furthermore, it is unclear if these incentives accurately hit the desired end-strength targets; the incentives were available to service members in virtually all career fields, ranks, and years of service (six or greater), so the resulting force size and structure was largely unplanned.[32] The separation incentives were set with little idea of the ultimate effect on overall end-strength or the balance of the force across career fields, pay grade (including the balance between enlisted service members and officers), and services.

The discriminatory and uniform price auctions described above could be easily modified to a voluntary separation application. Service members would simply submit bids for the compensation they would require to voluntarily separate from the military (a relatively simple modification could allow service members to submit bids for either lump sum or annuity compensation packages). The services could then set the separation bonus by career field, pay grade, and service to meet the desired total end-strength and force balance. The services could precisely manage their resulting force structure.

As with the SRB, separation pay auctions would significantly improve performance over the traditional voluntary separation programs in terms of precision, voluntary outcomes, flexibility, and best value. There is no difference in supporting achievement; practicality might decrease slightly.

Transfers across services or specialties: Operation Blue to Green. As the United States pursues its Global War on Terror and the conflicts in Iraq and Afghanistan, DOD has found it needs to rebalance end strength across the services. The Army faced a recruiting goal of 80,000 service members in FY 2005, while the Navy and Air Force expected to downsize by 8,000 and 22,000 service members, respectively, in the same year. In response, DOD launched Operation Blue to Green on 19 July 2004. Operation Blue to Green offered bonuses of up to $40,000 for enlisted service members and officers willing to transfer from the Navy and Air Force to the Army. The program targeted the junior enlisted ranks, E-1 through E-5, and junior officers and emphasized specific career fields. The Army hoped to attract up to 8,000 prior-service enlisted service members and officers under this program. The program is still in effect in FY 2009.[33]

Operation Blue to Green provides another example of a force-management program that could be well served by an auction-based incentive. Service members would simply submit bids for the compensation they would require to voluntarily transfer to the Army from the Navy or Air Force. The services could then set the separation bonus by career field and pay grade to meet the desired total end-strength and force balance. The services could precisely manage their resulting force structure, balancing the demands and costs across career fields and pay grade. A similar approach could also be used to adjust end strength across career fields within a service. As with the previous examples, transfer pay auctions would significantly improve performance over the traditional voluntary separation programs in terms of precision, voluntary outcomes, flexibility, and best value. There is no difference in supporting achievement; practicality might decrease slightly.

In this example, supporting achievement might represent a particularly critical principle. In general, the service members most likely to transfer are those in overmanned communities where they face stalled advancement opportunities. If the top performers do not perceive stalled advancement opportunities, the Army may not want to meet its end-strength target by minimizing the cost of those most eager to transfer (i.e., those facing the bleakest promotion opportunities). The Army might be willing to pay a higher bonus for the top performers among those seeking a transfer. This would require a more sophisticated auction that considered both the Army's and the service members' preferences (such auction designs are discussed in the following section).

In general, auction-based S&I pays can significantly improve performance over the traditional S&I programs across the six compensation principles indentified in the QRMC and described above. In some instances, the major performance improvements can be obtained through simple discriminatory and uniform price auction designs. In other cases, where service-member quality is a concern or monetary bonuses are relatively ineffective, DOD might consider a more innovative auction design.

Innovative and Customized Force-Management Auction Designs

The illustrations and evaluations above of the basic force-management auction approach to several different S&I pay programs demonstrate that even the simplest auction format not only offers significant potential benefits but also has a wide variety of potential applications in the military manpower context. The auction approach to force management is more versatile and powerful, however, than even these compelling examples would indicate.

In fact, innovative auction designs—incorporating multiple bidding stages, a broader variety of incentives, or bidding by both sides of the market, for example—can be customized to an even broader array of applications to create even better-performing conditional S&I pay programs. In this section, we will explain several such innovative force-management auctions that are already being designed for some unique force-management challenges.

Reducing Overpayment: Bidding for Retention Contracts of Different Lengths

As noted previously, the basic force-management retention (or reenlistment) auction can significantly improve cost effectiveness (or best value) by assuring that the bonus amount is never set above the market-clearing level. We also observed, however, that the basic retention auction (whether implemented in a discriminatory or uniform-price format) will unfortunately not address the fact that those service members willing to retain for a very small bonus, or even no bonus at all, will necessarily be "overpaid" for their services, as the market-clearing bonus level must be high enough to attract many "takers" who are less easily induced to remain in service. An innovative retention auction approach which does, in fact, address this overpayment issue is a design in which service members bid in two (or more) sequential auctions for reenlistment contracts of different lengths.

To understand the intuition behind why such a sequential auction approach could help reduce the magnitude of overpayment, recognize that the service members who are willing to retain for little or no bonus are those who are most devoted to (or dependent upon) a career in the military. As a consequence, these more-willing-to-retain service members, who would be overpaid in the basic retention auction, are more concerned with assuring the long-term security of their military occupation than those service members who are less willing to retain (i.e., those who require a much larger retention bonus). For this reason, many of the most significantly overpaid retainees would be willing to accept a smaller annual retention bonus (and thus reduce the overpayment) in exchange for a longer guarantee of military employment. This willingness to exchange bonus dollars for job security is especially true in periods of downsizing (of the overall force or of individual specialties), which both the Navy and Air Force have experienced in recent years.

A full explanation and evaluation of this sequential auction approach will be saved for another forum, but it is worth noting here that preliminary work on a mechanism we have titled the sequential self-selecting auction mechanism (S³AM) is quite promising. In this mechanism, service members first bid via auction for a short-term (perhaps one- or two-year) retention contract, and then those successfully retained in this first auction participate in a second auction in which they bid for a longer-term (perhaps four- to six-year) retention contract. Experimental simulation of this S³AM retention approach combined with application to USMC retention data has projected additional cost savings— above and beyond the amount saved via the basic retention auction—between 25 percent and 30 percent.[34]

Incorporating Nonmonetary Incentives Using a Combinatorial Auction

It has also been noted that the traditional approach of focusing exclusively on monetary incentives for retention may not be the most effective or efficient approach. Survey research has clearly shown that many service members may be far more motivated to remain in service if instead offered certain nonmonetary incentives (NMI), such as duty station of choice, geographic stability, sabbaticals, a compressed work week, and so on. Such surveys indicate, moreover, that many service members would be willing to forgo thousands or even tens of thousands of dollars in bonus money in exchange for such NMIs. Unfortunately, however, these same surveys reveal that, for any NMI under consideration, an even larger number of service members (from 30 percent to more than 80 percent) consider the NMI to be essentially worthless and are unwilling to sacrifice even one dollar of bonus money in exchange for the incentive.

So how can NMIs be used as a retention inducement when each NMI is valued very highly by some but not valued at all by many or most? The answer is to use a retention auction allowing combination bids (discussed earlier in this chapter) to create retention bonus packages which not only combine monetary and nonmonetary incentives but also are individualized in the sense that a service member is offered a particular NMI as part of his retention bonus only if he submits a bid for that NMI (in terms of bonus cash he is willing to give up) that is more than it costs the military to provide that incentive to him. Preliminary field and simulation investigations of such a mechanism, which we have titled the combinatorial retention auction mechanism (CRAM), suggest that retention costs could be reduced significantly (ranging from 5 percent to as high as 80 percent depending on NMI costs and DOD retention targets) relative to the use of a basic retention auction using monetary incentives alone.[35]

Incorporating Two-Sided Preferences in Retention Auctions

As noted previously, virtually all existing military pay programs incorporate only one-sided preferences (the preferences of service members), treating all members of the same specialty and rank as identical in terms of their value to the military. To be fair, the same is true for all of the auction mechanisms discussed to this point: The focus has been on retaining, separating, or transferring the target number of service members at the lowest cost, without focusing on whether or not the approach is selecting the "right" service members within a given cohort.

In the retention context, for example, all existing S&I pay programs (as well as the auction alternatives presented so far) will retain those service members who are the least expensive to retain when, in fact, logic would dictate many of the most valuable service members within a given cohort will have the best employment opportunities in the civilian sector and therefore will be among the more expensive to retain. Consequently, the "least cost" approach to retention or reenlistment programs makes it difficult for the military to hold onto those individuals with the general aptitude or skills which make these service members valuable in the civilian sector as well.

Fortunately, however, the auction approach to force management is sufficiently robust to incorporate two-sided preferences—assigning different values to different service members within the same cohort—into *all* of the auction mechanisms presented above. In the basic reenlistment or retention auction mechanisms, for example, two-sided preferences can easily be incorporated (especially using the uniform price auction approach) by essentially assigning "extra credit" to service members who have earned certifications in critical skills, received commendations for their performance, or even simply been identified by senior-level commanders prior to the auction as being key contributors and thus more valuable to retain.

To illustrate how such "extra credit" would work in a uniform-price retention auction, consider a specialty and rank cohort in which a select group of high-performing service members are assessed to be worth $5,000 more to retain than the other members of their group (we leave it to policy makers to decide how such valuations could or should be determined). Each service member in this cohort would bid as before, submitting the minimum bonus amount for which he or she would be willing to remain in service. However, this time the auction mechanism would treat each of the high-performing service members as if his or her cost to retain was actually $5,000 less than the bid amount.

With all bids collected and this "discount" given to the bids submitted by the high performers, the auction mechanism identifies the set of least-cost service members to retain and determines the first-excluded bid or cutoff bid as previously described (treating the high performers' discounted bids the same way regular bids are treated in the basic auction). Any "regular performer" within the cohort would be retained and receive a bonus equal to the cutoff bid amount if and only if he or she submitted a bid below this cutoff. Any of the high performers, in contrast, would be retained if and only if he or she submitted a bid less than $5,000 *above* the cutoff bid, and all retained high performers would receive a bonus equal to the cutoff bid plus $5,000.

The same basic approach could even be used to incorporate two-sided preferences into the customized force-management auctions described above, including both the sequential self-selecting auction mechanism (S³AM) as well as the combinatorial retention auction mechanism (CRAM). Moreover, doing so does not alter the "truth-revealing" nature of the uniform-price approach to any of these auction mechanisms. It remains the optimal strategy for each service member to bid his or her true willingness-to-retain amount, regardless of any extra value that may be assigned (and whether or not this extra value is revealed prior to the auction).

In sum, the versatility of force-management auctions allows for preferences on both sides of the market to be considered, with only minor "tweaking" of the auction mechanisms. Assigning "extra credit" to high performers in these mechanisms will also better align the military's conditional S&I pays with the QRMC principle that compensation should "support achievement."

Incorporating Two-Sided Preferences in the Assignment Process

Having recognized the relative ease with which two-sided preferences can be accounted for in the basic, S³AM, and CRAM retention auction mechanisms, it is important to point out that incorporating two-sided preferences into auction mechanisms applied to the assignment process is a significantly more challenging task. The complication arises from the fact that the process of pairing service members with billets or assignments is a "one-to-one matching" problem. Each service member can be matched with at most one billet and vice versa. This one-to-one matching characteristic makes this force-management problem quite distinct and more complex than the retention, separation, or transfer issues addressed above.

The Navy's Assignment Incentive Pay (AIP) program, which, as discussed previously, attempts to incorporate two-sided preferences in this arena, has itself been subject to the limitations of applying to a one-to-one matching context an auction approach which was not specifically designed for such a scenario. Our own experimental simulations of the current AIP program, for example, have revealed that service members have a strong incentive (which they act on) to manipulate the process by bidding strategically in a manner which increases overall program cost. While the current AIP approach is an improvement in many ways over the previous involuntary manner in which these jobs were filled, an auction mechanism specifically designed for the one-to-one matching context should offer even better performance.

To this end, a mechanism referred to as the truth-revealing assignment and salary calculation (TRASC) mechanism has been adapted from existing matching auction mechanisms with proven success in similar environments. The TRASC mechanism incorporates both the military's valuation of different service members (the maximum bonus the military would be willing to pay to place service members of different experience or quality in each different job assignment) with the service member's bids for all jobs under consideration (the service member's minimum acceptable bonus to serve in each job). Assignments and associated bonus pay are then determined through an iterative process: each service member is initially assigned to whichever job provides him or her the greatest net benefit (bonus offered minus minimum acceptable bonus), but if multiple service members are assigned to the same job, the bonus offered each service member for that job is reduced. The bonuses for oversubscribed jobs are continually reduced until no job is assigned to more than one sailor. This mechanism is not only truth-revealing and hence not subject to the bid manipulation described above, but preliminary research indicates that TRASC offers significant increases in efficiency.[36]

Conclusions and Recommendations

The challenge of military force management has always been a daunting task in the era of the all-volunteer force, requiring a compensation system which both *shapes* the force in terms of carrying and sustaining the required number of service members of each rank in each specialty across the services, and *distributes* the force voluntarily in terms of placing the right service member in the right assignment at the right time. This challenge becomes especially complex in an environment of global economic, political, and security variability and uncertainty.

Most elements of the US military compensation system are too inflexible, however, to serve as proper incentive tools to adapt quickly and effectively to rapidly changing environmental conditions or military needs. The short-term task of force management therefore falls primarily to the set of conditional S&I pay programs, which can be adjusted on a year-to-year or quarter-to-quarter basis to retain, separate, or transfer a targeted number of service members to match immediate defense priorities and economic realities.

How well these conditional S&I pay programs perform this force-management task, and whether an auction-based approach to these programs offers potential improvement, has been the question explored in this chapter. Having investigated and evaluated the traditional approach to S&I pay programs, the basic auction approach to such programs, and finally the more sophisticated customized auction approach, we can present a final "report card" for these alternative approaches to military force management (table 28-3).

Table 28-3. Performance comparison of alternative force-management approaches

	Precision	Voluntary	Flexible & Responsive	Best Value	Support Achievement	Practicality
Traditional S&I Pay Approach	Low	Medium	Low	Low	Low	High
Basic Auction Approach	High	High	High	Medium	Low	Medium-High
Customized Auction Approach	High	High	High	High	Medium	Medium

As indicated in table 28-3, the analysis in this chapter has revealed significant deficiencies in the traditional approach to conditional S&I pay programs. Most fundamentally, the approach of setting the level of retention or separation bonuses based on prior statistical analysis, market research, or basic rules of thumb has resulted in S&I pays which frequently overestimate or underestimate the bonus level necessary to achieve the target number of "takers" among the eligible service members. This lack of precision in setting appropriate bonus amounts ultimately undermines the responsiveness of these programs and significantly raises the associated cost.

A very basic auction approach to these retention, separation, or transfer pay programs offers much better precision in terms of assuring that the bonus amount is set at the exact minimum level necessary to achieve the precise force-shaping outcome desired. The significantly enhanced precision of an auction approach to conditional S&I pays would allow these programs to be much more flexible and responsive to military needs, and to do so at a significantly reduced cost.

More sophisticated auction mechanisms which are customized to meet a specific need or adapted to a specific force-management context offer the potential for even stronger performance improvement. While such advanced auction designs certainly add some complexity—incorporating multiple bidding stages, a broader variety of incentives, or bidding by both sides of the market, for example—they also offer the potential for significantly greater cost savings and the ability to create incentives that efficiently and effectively support achievement and reward service-member performance, a dimension lacking in both traditional S&I pay programs as well as the basic auction approach focused on service-member cost alone.

Recommendations

In sum, the investigation in this chapter supports the following recommendations to the US Department of Defense:

1. The use of basic auction mechanisms for reenlistment, retention, and separation bonuses should be tested via pilot implementations within select military populations or specialties and subsequently refined and expanded to cover additional communities or pay programs as results dictate.

2. The design of more sophisticated and customized auction mechanisms, as well as the testing, calibration, and refinement of these mechanisms in controlled laboratory settings, should be pursued and supported with the intent of ultimately bringing these advanced auction designs to the point of pilot implementation and eventual widespread application.

In closing, our investigation into the use of auction mechanisms as a force-management tool has revealed these mechanisms to be both powerful and complex. While they offer the possibility of a significantly more flexible, cost-effective, and achievement-supporting military incentive system, these auction mechanisms—like any powerful tool—must also be "handled with care" and be designed and implemented in a manner that maximizes their potential.

Notes

1. US Department of Defense (DOD), Undersecretary of Defense for Personnel and Readiness, *10th Quadrennial Review of Military Compensation*, vol. 1, 2008, 39–40.

2. Ibid.

3. Ibid., 13.

4. Ibid., 8.

5. Ibid., 18.

6. Paul F. Hogan, "Overview of the Current Personnel and Compensation System," in *Filling the Ranks: Transforming the U.S. Military Personnel System*, ed. Cindy Williams (Cambridge, MA: MIT Press, 2004).

7. Paul Bock, "The Sequential Self-Selection Auction Mechanism for Selective Reenlistment Bonuses: Potential Cost Savings to the U.S. Marine Corps" (master's thesis, Naval Postgraduate School [NPS], Monterey, CA, 2007), 42.

8. Ibid.; and Peter J. Coughlan, William R. Gates, and Paul Bock, "Innovations in Reenlistment Bonuses" (address, Western Economics Association International Meetings, Seattle, WA, 29 June–3 July 2000).

9. DOD, *10th Quadrennial Review of Military Compensation*, 10.

10. Adapted from DOD, *10th Quadrennial Review of Military Compensation*, 10.

11. For general discussion of auction theory see Paul Klemperer, "Auction Theory: A Guide to the Literature," *Journal of Economic Surveys* 13, no. 3 (1999): 227–86; Preston R. McAfee and John McMillan, "Auctions and Bidding," *Journal of Economic Literature* 25, no. 2 (1987): 702–13; Paul Milgrom, "Auctions and Bidding: A Primer," *Journal of Economic Perspectives* 3, no. 3 (1989): 3–11; and William Vickrey, "Counterspeculation, Auctions, and Competitive Sealed Tenders," *Journal of Finance* 16, no. 1 (1961): 9–20. For further discussion of basic auction formats in force-shaping and force-management applications, see Bock, "Sequential Self-Selection Auction;" William N. Filip, "Improving the Navy's Officer Bonus Program Effectiveness" (master's thesis, NPS, Monterey, CA, 2006); Damian K. Viltz, "Analysis of Separation Pay Options," a Master of Business Administration Professional Report (Monterey, CA: NPS, 2004); and Brooke M. Zimmerman, "Integrating

Monetary and Non-monetary Reenlistment Incentives Utilizing the Combinatorial Retention Auction Mechanism (CRAM)" (master's thesis, NPS, Monterey, CA, 2008).

12. David Henderson, ed., *The Concise Dictionary of Economics*, 2nd ed., s.v. "auctions" (by Leslie Fine) (Indianapolis, IN: Liberty Fund Inc., November 2007).

13. An exception to this general statement is the case of matching or assignment auctions, in which each seller can be matched with only one buyer and/or vice versa. In a force-management matching or assignment auction, for example, it might occur that a single service member is the lowest bidder for two different assignments. Because this service member can "win" only one of these assignments, however, another service member who is not the lowest bidder will necessarily be the "winner" of the other assignment. Such matching and assignment auctions are discussed further later in this chapter.

14. Technically speaking, the broadest multiple-winner generalization of the second-price auction is actually known as the Vickrey auction, which has a more complex price-determination algorithm that essentially requires each winning bidder to pay, for each item he or she wins, the bid of the person who *would have won* that particular item if the actual winning bidder had not bid for that item. Fortunately, however, if each auction winner can buy or sell only one unit, as is the case in force-management auctions in which winning service members can each sell only one labor commitment, the Vickrey auction becomes identical to the uniform price auction described here.

15. For a thorough yet reasonably accessible explanation of the results discussed in this section, see Klemperer, "Auction Theory," 227–86.

16. We recommend Zimmerman, "Integrating Monetary and Non-monetary Reenlistment Incentives."

17. See Benjamin M. Cook, "Using a Second-Price Auction to Set Military Retention Bonus Levels: An Application to the Australian Army" (master's thesis, NPS, Monterey, CA, 2008).

18. The primary qualification to revenue or cost equivalence involves risk aversion. If bidders are risk averse (willing to sacrifice significant expected net benefit to reduce the risk of losing), they would be expected to not bid their actual estimate of the second-highest value in a first-price sealed-bid auction, preferring to raise (lower) their bids slightly to increase their chances of winning a forward (reverse) auction. Risk aversion does not affect the optimal bidding strategy in a second-price sealed-bid auction, however. Thus, a first-price sealed-bid (FPSB) auction (or the discriminatory auction in the case of multiple winners) may be slightly more cost-effective for the military compared to the second-price sealed-bid (or uniform price) auction with risk-averse bidders. As the number of bidders increases, however, note that the impact of risk aversion in this comparison declines significantly (because the estimated second-highest value becomes closer and closer to the highest value anyway).

19. United States Marine Corps, *Marine Corps Order 7220.24M: Selective Reenlistment Program* (Washington, DC: Headquarters United States Marine Corps, 1990), 1–2.

20. Anita U. Hattiangadi, Deena Ackerman, Theresa H. Kimble, and Aline O. Quester, "Cost-Benefit Analysis of Lump Sum Bonuses for Zone A, Zone B, and Zone C Reenlistments: Final Report," Center for Naval Analyses, Alexandria, VA, November 1998.

21. Bock, "Sequential Self-Selection Auction."

22. In this example, there is approximately a $600 difference between the 75th and 76th lowest required retention bonuses. If DOD could accurately forecast the 75th lowest value, it could reduce its retention costs by $600 per officer, or $45,000. The $600 higher bonus payment represents DOD's cost to elicit truthful revelation and ensure precision in force-shaping and management programs.

23. It should be noted that DOD might realize some savings from an FPSB auction if officers are risk-averse. Risk-averse officers would reduce their bids relative to the values depicted in figure 28-6, at least slightly, reducing DOD's total costs.

24. Secretary of the Navy, *SECNAVINST 7220.84: Surface Warfare Officer Continuation Pay (SWOCP)* (Washington, DC: Secretary of the Navy, 14 Jan 2000).

25. Patrick C. Mackin and Kimberly L. Darling, "Economic Analysis of Proposed Surface Warfare Officer Career Incentive Pay," SAG Corporation, 1996.

26. Jennifer L. Lorio, "An Analysis of the Effect of Surface Warfare Officer Continuation Pay (SWOCP) on the Retention of Quality Officers" (master's thesis, NPS, Monterey, CA, 2006).

27. Chief of Naval Operations, *NAVADMIN 012/06: Junior Surface Warfare Critical Skills Retention Bonus*, 10 January 2006.

28. Filip, "Improving the Navy's Officer Bonus Program."

29. Constance M. Denmond, Derek N. Johnson, Chavius G. Lewis, and Christopher R. Zegley, "Combinatorial Auction Theory Applied to the Selection of Surface Warfare Incentives," MBA Professional Report, Naval Postgraduate School, 2007.

30. Ibid.

31. Beth J. Asch and John T. Warner, *An Examination of the Effects of Voluntary Separation Incentive* (Santa Monica, CA: RAND, 2001).

32. Viltz, "Analysis of Separation Pay Options."

33. Courtney Hickson, "Army Launches 'Operation Blue to Green,'" *Army News Service*, http://www.militaryinfo.com/news_story.cfm?textnewsid=1083 (accessed 29 September 2009); "Operation Blue to Green," *Army News Service*, http://usmilitary.about.com/od/armyjoin/a/artransfer.htm accessed 29 September 2009); and "Update: Army Blue to Green Program," NAVADMIN 223/09, 29 July 2009, http://www.corpsman.com/2009/07/update-army-blue-to-green-program-navadmin-22309/ (accessed 29 September 2009).

34. See Bock, "Sequential Self-Selection Auction"; Cook, "Using a Second-Price Auction"; Filip, "Improving the Navy's Officer Bonus Program"; and William J. Norton, "Using an Experimental Approach to Improving the Selective Reenlistment Bonus Program" (master's thesis, NPS, Monterey, CA, 2007).

35. Denmond, et al., "Combinatorial Auction Theory"; Peter J. Coughlan, William R. Gates, and Brooke M. Zimmerman, "The Combinatorial Retention Auction Mechanism (CRAM): Integrating Monetary and Non-Monetary Incentives," NPS Technical Report, Monterey, CA, forthcoming; Zimmerman, "Integrating Monetary and Non-monetary Reenlistment Incentives."

36. Pei-Yin Tan, "Simulating the Effectiveness of an Alternative Salary Auction Mechanism" (master's thesis, NPS, Monterey, CA, 2006).

About the Authors

Dr. Peter Coughlan is an associate professor at the Naval Postgraduate School (NPS) in Monterey, California. At NPS since 2004, he teaches graduate courses in economics and strategic management. Prior to his arrival at NPS, Dr. Coughlan served six years as a professor in the Strategy Unit at the Harvard Business School. He earned both a master's and doctoral degree in economics from the California Institute of Technology, specializing in game theory and behavioral economics, and a bachelor's degree in economics and mathematics from the University of Virginia.

Dr. Bill Gates is the dean of the Graduate School of Business and Public Policy at NPS. A graduate of Yale University (PhD) and UC San Diego, Dr. Gates has been a professor at NPS since 1988. Prior to joining NPS, he was an economist at the Jet Propulsion Laboratory; he has also served as an adjunct professor of economics at Golden Gate University and the Monterey Institute of International Studies. Dean Gates' current research focuses on game theory and mechanism design applied to both military manpower and acquisition.

CHAPTER 29

Is Military Retirement Past Retirement?

Steve P. Fraser

Introduction

Organizations use compensation systems to recruit and retain the best qualified personnel to serve the mission. This is true for private firms, public entities, and the focus of this essay, the military. A recurring question for Department of Defense (DOD) officials is whether or not the current military compensation system is doing the job. The answer of course depends on your criteria for measuring success—and how you measure it. One can argue the current system has been reviewed, studied, or examined to excess. In a Library of Congress research report, Rex A. Hudson reviews no fewer than five major reform proposals conducted since 1975 and a series of smaller efforts from various "committees" and "commissions."[1] The end result is that while much has been discussed, little has changed. Maybe that is okay, maybe not.

Many of the concerns about the existing system that arise from the reviews of the last 35 years are common. John T. Warner provides details on many of the specific concerns.[2] In sum, the existing structure is perceived as too costly, has a vesting problem, and perhaps worst of all, does not serve the intended goal of recruiting and retaining the best qualified people. In the paragraphs that follow, I examine some of the common concerns of the system through the lens of the individual service member. Specifically, I look to highlight some of the factors affecting the individual's decision to stay in the military—and how those factors are influenced or constrained by the military pay and retirement system.

The framework I pursue is one I think most service members follow, though to varying degrees. Each service member essentially performs a marginal cost-benefit analysis at key points in his or her career. The result, of course, is that a service member separates when the opportunity cost associated with staying in the service is too great to ignore. If one looks at the existing compensation and retirement system through this lens, a couple of key insights emerge.

Is Military Compensation and Retirement Expensive?

Obviously. Even trying to calculate the true cost is daunting. Members receive direct cash benefits (pay and allowances), indirect cash benefits (allowances are tax free and some income might be tax exempt), deferred cash benefits (retirement), and those that can be lumped into "other" like health care, education, and so forth. Measuring the real cost is problematic at best. Carla T. Murray and Christian Howlett suggest the estimated annual cost for military compensation for the 1.4 million personnel on active duty in 2005 was more than $193 billion.[3] In fact, trying to reduce the cost of retirement prompted the only significant change to the retirement system in 35 years—and that change did not last. The Military Reform Act of 1986 reduced the amount of the benefit to be received by the service member at the 20-year point from 50 percent to 40 percent of base pay (commonly referred to as REDUX).[4] R. Yilmaz Arguden discussed many unintended consequences of REDUX and estimated personnel losses from the new system would be larger, and occur sooner, than projected.[5] Perhaps Arguden and other critics were clairvoyant. As Hudson notes, the FY 2000 National Defense Authorization Act repealed compulsory REDUX and allowed members to retire under the pre-REDUX system.[6] The estimated long-term cost associated with the retirement system certainly is of direct interest to the government when ascertaining whether the compensation system is working. However, for the service member, the focus is more in the present.

The present for the service member starts with direct pay and benefits received on active duty. Unfortunately, too often the discussion on this topic revolves around simply comparing military pay to that of the pay of comparable positions in the public or private sector. This is too narrow a view. The military pay system is a rather distinct manifestation in the compensation landscape. Its basic elements contain components many would view as being consistent with socialist-like policies. First, individuals are paid based on their rank and tenure, not necessarily on their abilities or how their contributions might be valued by the organization. This approach puts the onus on the service to ensure that job positions are matched with the appropriate rank. Further, two individuals with the same rank and tenure might receive different compensation depending on whether one has dependents. Yes, the married personnel specialist earns more than the single personnel specialist. While surely such policies were introduced with considerable care, not all today's service members see this approach in a similar light. The notion of equal pay for equal added value resonates with many.

The service components must also remember the military is competing with other public and private sectors for attracting qualified personnel. Like it or not, they participate in an active labor market. When the economy struggles, military pay and retirement programs seem attractive. In contrast, during periods of economic growth, the private sector is more likely to attract more talented workers. Given the sheer size of the military and the nature of the vocations, there are times when military pay both lags and exceeds comparable civilian wages. Perhaps surprisingly, there are occurrences where military members are paid more than their civilian contemporaries. We hear little of these cases. Such situations motivate members to support the status quo. Rarely when market forces reduce the civilian side of the pay equation does the service change the manning. Doing so would compound the already tenuous promotion system. How would one re flect being removed from a position on a promotion recommendation because, while qualified, the member was "overpaid"?

Interestingly, where the services note large negative pay differentials between like military and civilian vocations, the first response is to add some special pay or incentive. While Murray and Howlett report that, on average, military compensation exceeds the 75th percentile of civilian compensation, the services offer more than 60 kinds of "special pays."[7] When the Air Force was worried about pilot retention, they first offered leather jackets and then large sums of cash. Due to the specifics of the program, there were cases where junior pilots (both in terms of rank and experience) were being paid more than their supervisors. In addition to the obvious salary inversion brought on by such a program, the crew dynamics were shocked as well. Junior members, who took the money and the associated new service commitments, became the long-term players in the organization. The situation left those more senior members feeling slighted.

Even when one sorts through the idiosyncrasies of the pay system, the question to stay or leave for the service member is not always about the direct cash benefit. A service member's perception of the promotion and assignment system is critical as well. One of the inefficiencies of the existing assignment system is that we move personnel just as they are becoming knowledgeable and qualified to perform their current duty. Because of increasing operations tempos, the services seem to rotate personnel far more than necessary—perhaps with a larger plan to either spread the wealth (perceived "good" assignments) or, probably more realistic, spread the pain (perceived "bad" assignments). There are many cases where very talented people leave the service, even though they continue to want to serve, because they are not allowed to serve in the capacity where they think they can best add value. The all too common "needs of the service" is an apt hammer when every personnel issue looks like a nail.

Equity, promotion, and assignment issues aside, comparing competing civilian salaries is only the starting point for the service member's marginal cost analysis of staying or leaving the military. Another major factor to consider is the *current* value of the *potential* retirement annuity. Fortunately, computing this value is rather straightforward (with one notable exception discussed below). Members who serve for 20 years on active duty are eligible to retire and receive a pension (50 percent of base pay pre-REDUX, 40 percent with REDUX). This pension is in the form of an *immediate* annuity that pays in perpetuity and provides a risk-free, inflation-protected income stream for the service member. Jennings and Reichenstein outline a methodology to compute the present value of the retirement benefit.[8] The authors estimate the present value of the annuity benefit for a 44-year old male retiree with 20 years of service and a $40,000 annual base pay to be approximately $385,000, *at retirement*. This value can be discounted earlier to different decision points (when the service member hits the eight-year or 12-year point, for example). However, such an analysis highlights what is referred to as the *vesting problem*.

Does the Military Retirement System Have a Vesting Problem?

Only if one perceives the best people are leaving the service. The existing military retirement system is both similar and dissimilar to retirement programs offered in other public and private sectors, and the system has been both lauded and criticized for its effects on force shaping. For the benefit of the reader not familiar with the basics of pension plans, I provide a brief overview here.

There are two broad categories of retirement plans, *defined-benefit* (DB) plans and *defined-contribution* (DC) plans. In a DB plan, covered employees serve time with an organization and receive some type of annuitized benefit, the amount of which is based on some formula of tenure and pay within the organization. In a DC plan, the employer (and in many cases the employee) makes a contribution to an employee retirement account, which is invested in assets on behalf of the retiree. These retirement accounts are tax-advantaged instruments. That is, the employee will not pay tax on the earnings in the account until funds are withdrawn.

The primary difference between DB and DC plans concerns the role of the employee. In a DB plan, the employee essentially plays no role. The company (or government entity in this case) is responsible for setting aside funds and investing them so that funds will be available at a later time to pay the specified benefits. In contrast, in a DC plan the choice of investment vehicle, and therefore the responsibility for any wealth created (or lost), is solely up to the employee. Often unwittingly, employees risk the viability of their future retirement income stream by making poor investment choices.

Recently, those investment streams are dependent on an economy currently in crisis. At the time of this publication, the US economy is in the midst of perhaps the worst recession in 75 years. US equity markets fell 40 percent in 2008 and another 10 percent in early 2009. Randell reports the unemployment rate reached 9.7 percent in September 2009, the highest rate since 1983.[9] In an attempt to curtail the economic freefall, the federal government has executed fiscal stimulus programs whose cumulative effect is a projected budget deficit of more than $1.5 *trillion* in the next fiscal year and cumulative deficits of $9 *trillion* over the next decade.[10] In short, we are living in trying economical times.

One casualty of the economy is the DB plan. For those entities with DB plans, investment assets funding the plans have plummeted in value. Even before the recent setback, more and more firms were moving to DC plans and away from DB plans. The Center for Retirement Research at Boston College reports that pension coverage of workers with only DB plans has fallen from 21 percent in 1992 to 8 percent in 2007. In contrast, those workers covered by only DC plans have increased from 19 to 30 percent. A DC plan allows firms to better manage their balance sheets and shift the vast portion of retirement responsibility to the worker.

The existing military retirement system is most similar to a DB plan—with two notable exceptions. First, the military DB plan pays an immediate annuity. Most DB plans do not provide a benefit until a "normal" retirement age. The second notable difference, alluded to above, rests with when the employee is vested. In case of plans subject to the Employee Retirement Income Security Act (ERISA), employees can be vested with as few as three years of service. The military system is not subject to ERISA and is an all-or-nothing system. If a member serves a day short of 20 years, the member earns no retirement benefit. This "cliff vesting," as it is often called, influences behavior not necessarily seen in plans outside the military. Warner suggests the system helps continuation rates for personnel in the 6th to 20th years of service while helping to reduce continuation rates thereafter.[11]

The notion of marginal analysis is not lost on those retiring. It is not uncommon to hear that one is working for 50 cents on the dollar as soon as one reaches the 20-year cliff. Cliff vesting is not necessarily a problem, only a further compounding factor in a complex system. The discussion surrounding the impact of cliff vesting is not a new one for policy makers. As early as 1976 (with the Defense Manpower Commission), there was a call for reducing the 20-year vesting point. Yet the existence of the DB plan (even with a 20-year vesting requirement) can be a strong retention factor for service members, especially for those who see limited opportunity outside the military.

The narrow range of opportunities might exist for two very different reasons. One, the service member might be very well qualified in what he or she does, yet there might be little or no demand for those skills outside the military. The DB plan works for this individual (and perhaps appropriately so). In contrast, the limited employment opportunity set outside the military might exist because the service member has been able to accomplish the minimum, still achieve promotions, and move from position to position without building a resume that might allow him or her to compete well in the labor market outside the military. Warner notes the additional influence of 20-year vesting on supervisors—that is, they might be reluctant to separate those without vested benefits.[12] When talented people see the system rewarding those personnel, and ultimately those beneficiaries become leaders of the system, those with greater marginal benefits outside of the service are more motivated to leave the service.

Is the Existing Military Compensation and Retirement System Working?

It depends. On one hand, the system is operating as well as can be expected. There is likely to always be some specialties where the services find it difficult to maintain adequate manning. However, in the aggregate, the all-volunteer force and its current pay and retirement system appear to have served us well. In terms of direct compensation, the military also benefits from what some might term the "patriot discount." Many in uniform serve not on the basis of some comprehensive compensation structure, but rather due to a sense of duty or calling. While certainly noble, such service does not negate the service component's obligation to compensate those who serve honorably. The key, of course, is designing a scheme that attracts and retains the best Soldiers, Sailors, and Airmen in the world.

Upon closer inspection, it is not clear that the existing system is beyond improvement. Given the persistent fiscal constraints our government is likely to face, the ongoing trends in retirement funding across all public and private sectors, and the influence such forces have on the choices of service members, we might be able to do better. Some factors policy makers might consider in the next round of "reforms," "reviews," or "commissions" on military pay and retirement include improving the correlation of jobs and salaries, considering a move toward a DC retirement plan, and finding any way to reduce (or better integrate) the compounding factors associated with the stay or go decision.

The first factor, and perhaps the most challenging to properly scope and manage, is finding a way to better correlate the nature of the jobs and the associated salaries. By that I mean we need to move away from filling billets solely by rank. Every service member has witnessed countless times where instead of promoting people who we know would do a good job in a particular position, we rotate them out (or promote them on the way out) and find someone already holding the appropriate rank from somewhere else and ask him or her to come learn the nuances of the position. A job is purposely vacated—often for the "needs of the service"—without respect to whether the change will improve or diminish the performance of the organization. While the assignment system serves a larger role in the force-management process, we cannot ignore the negative impacts these scenarios have on our most talented and competent personnel.

The second factor to consider, and one more easily addressed, is to consider the movement to a DC retirement plan. In fact, the military already has a mechanism in place to administer such a DC program. In 1986, the Thrift Savings Plan (TSP) was enacted. TSP is a DC plan—albeit without any contributions by the services. The services could use the TSP as a vehicle to transition from the existing DB plan to a DC plan. Warner outlines several alternatives, including the potential use of the latter.[13] Such a change would allow the services and government to better account for and estimate the associated costs of military retirement. The use of a DC plan will also aid the service member in his or her marginal analysis of whether to stay or separate. A DC plan is portable and moves with the service member should he or she separate. Perhaps surprisingly, the use of a DC plan might even *motivate* individuals to serve in the military. With it, the individual knows exactly what the benefit entails, has more control over that benefit, and the need for estimating probabilities associated with reaching the 20-year point is eliminated. The cliff vesting issues fall away.

As a method to bring this discussion on whether our existing retirement plan should be retired to a close, follow along with the average second-term service member as she contemplates the decision to stay or separate. Assuming this member feels valued by the organization and is optimistic about her ability to navigate the promotion and assignment systems to pursue a career in the military, she is ready to conduct the marginal analysis of the stay versus go decision. The previous statement could, of course, be written entirely differently.[14] No matter how framed, let us walk through the myriad of factors one must address.

First there is the basic salary comparison. What could she make on the outside? Second, there are the indirect cash benefits resulting from the tax-free status of allowances (and perhaps even some pay amounts). Next, one needs to place some value on benefits like access to the fitness center, legal assistance, commissary benefits, and perhaps most importantly, the healthcare component. Even after getting a handle on all those estimates, there are a number of further estimates and assumptions that make the process overwhelming for many.

The first of these additional factors to consider is for the service member to estimate the probability that she will even reach the cliff vesting point of 20 years. While she can calculate the value of a future retirement annuity,[15] she must assign a probability of reaching the 20-year point. Next she must consider whether or not to enroll in the Survivor Benefit Program (SBP).[16] SBP is essentially an insurance program to protect up to 55 percent of one's retirement pay. Without SBP, a retiree's pension annuity ceases when the member dies and his or her spouse/dependents get nothing. So the retiree's net benefit is reduced by the SBP premium. There are no limits to the probability combinations one could make as to whether or not to enroll in SBP, and what is the likelihood the service member will die before the beneficiary.[17]

There are other "costs" of retiring that exist as well. I use "costs" in the sense that many new retirees do not fully appreciate what they must pay for in retirement. Life insurance is but one example. While on active duty, personnel can participate in the Serviceman's Group Life Insurance (SGLI). Upon retirement, veterans can convert to Veterans' Group Life Insurance (VGLI); however, the government subsidy is dramatically reduced. The retiree's net benefit falls again. The same is also true for health care. While the "free" health care provided the member on active duty is a widely touted benefit, it is no longer free upon retirement. Not only is there now a cost associated with health care, finding care can be an issue as well. On active duty, service members know where to go for health care. As a retiree, just because one has TRICARE (the primary health care option open to retirees), not all health care providers accept it as an insurance provider. Using this marginal analysis framework to illustrate the point, we see the number and complexities of the series of actual considerations involved when contemplating separation are remarkable. For those who find just reading about this type of analysis exhausting, you likely are not alone. The default decision in these cases, and certainly for those who are risk averse, is to stay with what you know—and not separate. For service members who understand their value to the organization and compute these marginal benefits and costs as a routine part of their decision making, they often see separating as the best course of action. In short, they will be valued more by organizations outside the military. The services need to determine which type of people we are trying to recruit and retain and act accordingly. However, we should always proceed with caution. With any change to a system, a change in behavior follows. We have had little substantial change in the pay and retirement system over the last 35 years. Maybe that is okay, maybe not.

What Are Policy Makers to Do?

Reconsider Viewpoints

Policy makers should realize many of the service members currently considering the stay or go decision are likely to view compensation and a career of military service much differently than they do. Those thinking about separating today are likely to be significantly younger, aware of the Cold War only from what they studied in history, and unlikely to value their service or oft-marketed benefits in the same manner as do senior leaders. It is not to say they do not value their service, they just do so differently—and across the spectrum of military life. For example, many of today's service members question the resources and emphasis placed on operating commissaries, exchanges, and the club systems at continental US installations.[18] There are of course locales where such services might be needed on a base; however, in many cases I suspect the majority of today's military see those functions as a holdover from an earlier era.

So too with pay and retirement. We should not simply assume the move to a simplified pay system and the use of a DC retirement plan will eliminate any motivation to serve. On the contrary, there may be a significant number of young Americans who would like to serve and may be *more inclined* to do so knowing that they do not have to serve 20 years to earn some type of benefit. Those that stay for one or two terms could transition to other public or private sectors and still have earned a DC benefit. Perhaps more importantly, they would likely leave with a more positive viewpoint on their military experience than if they served 10 or 12 years and separated because they grew tired of trying to navigate the assignment and promotion system. Many of today's service members have goals not associated with attaining a certain rank. They measure their success somewhat differently. We need to consider a pay and retirement system that recognizes those differences.

Develop and Implement a Pilot Program

Policy makers should make a concerted effort to decide whether or not we have an issue—and try *something*. There is no need to commission more reviews or studies to see if there is reasonable rationale to motivate changes to the system. We should not wait to learn of the latest change to guidance for promotion boards or how projected healthcare reform might impact TRICARE. If we do, we will simply add to the list of reviews already conducted and continue to discuss these issues for the sake of discussion.

The services could develop and implement a pilot program for a series of specialties across all the services, or perhaps a narrower program within a single service, where entering personnel could opt for a DC pension plan in lieu of the existing DB plan. Policy makers should select specialties where there exists a civilian need for similar skills. The services could then use TSP as a vehicle to contribute a percentage of the service member's salary, which would be managed by the service member similar to a 401(k) plan for those in the civilian sector. The services could then monitor the continuation rates of those under the traditional DB plan and those in the pilot DC plan within the specialties selected. While the pilot program would have to be carefully crafted and funded, I suggest there are many who would welcome such an opportunity. If the program does not generate the desired results—attracting and retaining the best-qualified Soldiers, Sailors, and Airmen in the world—we can also default back to the existing system (as we did with REDUX). The move to a DC program might actually *attract* a different potential service member, one perhaps not necessarily fully appreciated in the past. Maybe that is okay, maybe not.

Notes

(All notes appear in shortened form. For full details, see the appropriate entry in the bibliography.)

1. Hudson, *Major Military Retirement Reform Proposals*. The Hudson report covers major reform efforts, including the Defense Manpower Commission (1976), the President's Commission on Military Compensation (1978), the Fifth Quadrennial Review of Military Compensation (QRMC) (1984), the Sixth QRMC (1988), and the Defense Advisory Commission on Military Compensation (2006).

2. Warner, *Thinking about Military Retirement*.

3. Murray and Howlett, "Assessing Pay and Benefits."

4. Hudson, *Major Military Retirement Reform Proposals*. He reports that an earlier act (the Department of Defense Authorization Act 1981) had already reduced the benefit from 50 percent of the final base pay to 50 percent of the average of the last 36 months of base pay (Hi-3).

5. Arguden, "There Is No Free Lunch," 529.

6. See Murray and Howlett, "Assessing Pay and Benefits"; Randell, "Job Losses Moderate"; and Warner, *Thinking about Military Retirement* for a detailed treatment on the specifics of the current military retirement system in effect.

7. Murray and Howlett, "Assessing Pay and Benefits."

8. Jennings and Reichenstein, "The Value of Retirement Income Streams."

9. Randell, "Job Losses Moderate."

10. Weisman and Solomon, "Decade of Debt."

11. Warner, *Thinking about Military Retirement*.

12. Ibid.

13. Ibid.

14. Assume our service member contemplating separating is comfortable with the notion that her buddy gets paid more than she does for the same job because the buddy is married. She is also excited about the fact that her new boss knows nothing about her or the organization—but became her new boss simply because it was perceived the old boss had been serving in that role for too long. The old boss did a great job too.

15. Jennings and Reichenstein, "The Value of Retirement Income Streams."

16. Many service members are not aware of the specifics of the DB plan and therefore the potential need to protect a portion of their benefit with the SBP. In short, many simply do not know about the SBP until they have already decided to stay in the military.

17. Extreme examples help illustrate this case. If a retiree enrolls in the SBP and the retiree and the beneficiary die together 20 years after retirement, the member would have paid premiums for 20 years and received no benefit. In contrast, if the retiree lived only one month after retiring and the beneficiary live another 20 years, the beneficiary would have received 55 percent of the retirement pay for 20 years yet only paid one single premium. Any Social Security offset is ignored in this example.

18. However, there is some entertainment value in noting how much value is still placed on having reserved parking spaces at such facilities.

Bibliography

Arguden, R. Yilmaz. "There Is No Free Lunch: Unintended Effects of the New Military Retirement System." *Journal of Policy Analysis and Management* 7, no. 3 (1998): 529.

Boston College. "Pension Coverage of Workers on Current Job, 1992, 2004, and 2007." Center for Retirement Research, www.crr.bc.edu (accessed 25 Nov 2009).

Hudson, Rex. *A Summary of Major Military Retirement Reform Proposals, 1976–2006.* Washington, DC: Federal Research Division, Library of Congress, November 2007.

Jennings, William W., and William Reichenstein. "The Value of Retirement Income Streams: The Value of Military Retirement." *Financial Services Review* 10, no. 1–4 (December 2001): 19.

Murray, Carla T., and Christian Howlett. "Assessing Pay and Benefits for Military Personnel." Economic and Budget Issue Brief, Congressional Budget Office, 15 August 2007.

Randell, Maya J. "Job Losses Moderate, but Unemployment Rate Hits 9.7%." *Wall Street Journal*, 4 September 2009, WSJ.com (accessed 4 Sep 2009).

Warner, John T. *Thinking about Military Retirement.* Alexandria, VA: CNA, January 2006.

Weisman, Jonathan, and Deborah Solomon. "Decade of Debt." *Wall Street Journal*, 28 August 2009, WSJ.com (accessed 4 Sep 2009).

About the Author

Dr. Steve Fraser is currently an assistant professor of finance at Florida Gulf Coast University. He holds a PhD from the University of South Florida and an MBA from the University of Pittsburgh. Dr. Fraser's research interests include individual finance, investment policy, and institutional investors. His research has appeared in academic journals including The Journal of Investing, The Journal of Wealth Management, and Financial Services Review. He previously served on the faculty at the USAF Academy (from which he also graduated in 1990) and retired from the US Air Force in 2007.

APPENDIX

★★★★★★★★★ ★ ★★★★★★★★★

Every man builds his world in his own image. He has the power to choose, but no power to escape the necessity of choice.

—Ayn Rand

A man does what he must—in spite of personal consequences, in spite of obstacles and dangers and pressures—and that is the basis of all human morality.

—Winston Churchill

Be ashamed to die until you have won some victory for humanity.

—Horace Mann

Imagine what might happen if a Rembrandt received a box of sixteen crayons and an average Joe was given a full palette of oil paints, easel and canvas. Which one is more likely to produce a work of art? Though the analogy may not fit exactly, the point is clear—the tools matter less than the talent, training and dedication that create the art. You can't have a masterpiece without a master. I think we forget that sometimes in the realm of warfare.

—The Hon. Ike Skelton
Chairman
House Armed Services Committee

Men occasionally stumble over the truth, but most of them pick themselves up and hurry off as if nothing had happened.

—Winston Churchill

MMX
TIME CAPSULE

As of 1 January 2010

Key Political and Military Persons

President of the United States: Barack H. Obama
Vice President: Joseph R. Biden
Secretary of State: Hillary R. Clinton
Senate Majority Leader: Harry M. Reid
Speaker of the House of Representatives: Nancy Patricia D. Pelosi
Secretary of Defense: Robert M. Gates
Secretary of the Army: John M. McHugh
Secretary of the Navy: Raymond E. Mabus, Jr.
Secretary of the Air Force: Michael B. Donley
Chairman of the Joint Chiefs of Staff: ADM Michael G. Mullen
Chief of Staff of the Army: GEN George W. Casey, Jr.
Chief of Naval Operations: ADM Gary Roughead
Chief of Staff of the Air Force: Gen Norton A. Schwartz
Commandant of the Marine Corps: Gen James T. Conway
US Central Command Commander: GEN David H. Petraeus
Multi-National Force—Iraq Commander: GEN Ray T. Odierno
Commander of US Forces Afghanistan: GEN Stanley A. McChrystal

Force Strength (Active and Reserve)

US Army: 1,000,000 personnel
US Navy: 455,000 personnel, 283 ships, and 3,700 aircraft (11 aircraft carriers)
US Marine Corps: 244,000 personnel
US Air Force: 550,000 personnel, 5,500 aircraft, 180 remotely piloted aircraft, 2,130 air-launched cruise missiles, and 450 intercontinental ballistic missiles

Current Major Military Operations

Iraq—Operation Iraqi Freedom (OIF): began 20 March 2003
Afghanistan—Operation Enduring Freedom (OEF): began 7 October 2001
OIF casualties: 4,358 (as of 8 January 2010)[1]
OEF casualties: 940 (as of 8 January 2010)[2]

Key Economic Statistics

DOD budget: Approx. one-third of US tax receipts ($741 billion/$2.5 trillion)
Military expenditures: 4 percent of gross domestic product (GDP)
Unemployment rate: 10.2 percent
Budget: $3 trillion
Debt: Greater than 60 percent of GDP
Discount rate: 0.5 percent
Inflation rate: 2.7 percent
US GDP: $14.83 trillion
Per capita GDP: $48,000
Average price of a gallon of unleaded gasoline: $2.50
Average price of a hamburger, French fries, and a drink: $5.50
Average price of a movie theater ticket: $8.75

Population

World population: 6.8 billion
US population: 308 million (ranks third behind China and India)

Personal Communication

255 million cell phones
26 billion indexible web pages
223 million internet users
1 trillion unique URLs

Key Military Statistics

Military service: All volunteer force (no draft)
Age requirements: 18 years of age (17 years of age with parental consent)
Maximum enlistment ages:
    Army—42
    Navy—34
    Marines—28
    Air Force—27

Manpower available for military service (ages 16–49):
    Men: 72.7 million
    Women: 71.6 million

Manpower fit for military service (ages 16–49):
    Men: 59.7 million
    Women: 59.4 million

Annual manpower reaching military significant age:
    Men: 2.2 million
    Women: 2.1 million

Emerging Trends

Two press releases—one by the US Census Bureau in 2008 and the other by the United Nations Population Division in 2009—highlight the changing nature of the global population and, more importantly, of that in the United States. The data provided by both organizations suggests that most of the developed world is losing population, the existing population is graying, and, within the United States, a demographic shift is taking place that will see minorities become the majority. Both the international and domestic demographic trends will have a major impact on the US military.

The UN population division and the US Census Bureau have identified the following trends in global and American population:

- The developed world and China are aging to the point where several developed nations—Germany, Japan, Russia, and Italy—will have shrinking populations and all will have significantly aging populations.
- The developing nations will maintain high population growth rates with a youth bulge. In 2008, the median age for Afghanistan was 16.8, for Pakistan 21, and for India 25 (see tables below for the comparison with the developed world).
- The population of the United States will keep growing, and in 2050 it will remain, along with India and China, one of the three largest countries in the world.
- In 2042, minorities will become the majority in the United States, accounting for 54 percent of the country's population.
- The working age population—18 to 64—is expected to become 50 percent minority in 2039, and by 2023 minorities will comprise more than half of all children in the country.
- In 2030, one in five US residents is expected to be 65 years or older.
- Between 2010 and 2050 the United States is expected to receive 1.1 million migrants annually.

The fear in the rest of the developed world is that graying societies will face negative economic and military-strategic consequences. Economically, the lack of a ready base of young people will hurt the productive capability of major developed countries (Japan by 2050 will have a median age of 55 while Germany's will be 51) as aging populations cause a shrinkage in the work force and a consequent drop in national gross domestic products. These countries will also have the burden of caring for a large aging population. Moreover, older societies are likely to be more conservative and less innovative, thus further reducing their economic competitiveness.

Demographic Transition: Select Nonwestern States

Country	2009 (population in millions)	2025 (population in millions)	2050 (population in millions)	2009 (median age)	2050 (median age)
Afghanistan	28.1	44.9	73.9	16.8	23.5
Bangladesh	162.2	195	222	24.1	39.2
China	1.34 billion	1.45 billion	1.41 billion	33.8	45.2
India	1.19 billion	1.43 billion	1.61 billion	24.7	38.4
Iran	74.1	87	96	26.3	41.9
Mexico	109	123	128	27.2	43.9
N. Korea	23.9	25.1	24.5	33.6	41.9
Pakistan	180.8	246.2	335.1	21	32.7

Demographic Transition: Select Developed Countries

Country	2009 (population in millions)	2025 (population in millions)	2050 (population in millions)	2009 (median age)	2050 (median age)
Australia	21.2	24.7	28.7	37.5	42.9
Bulgaria	7.5	6.7	5.3	41.5	49.5
France	62.3	65.7	67.6	39.9	44.8
Germany	82.1	79.2	70.5	43.9	51.7
Japan	127.1	120.7	101.6	44.4	55.1
Poland	38	36.9	32	37.9	51
Russia	140.8	132.3	116	37.9	44
United Kingdom	61.5	66.6	72.3	39.7	42.5
United States	314.6	358.7	403.9	36.5	41.7

From a military-strategic standpoint, aging societies with a small base of young people will be reluctant to commit forces to wars—especially if they risk facing a significant number of casualties.[3]

Implications for the United States

Of all the developed countries, the shifts in demography favor the United States. Its population remains stable albeit getting older. Economically, it will have a larger portion of the developed world's GDP—according to one estimate the United States will control 54 percent of this amount by 2050.[4]

Additionally, the United States will continue to have a large population, and its median age of 41.7 will be lower than that of other developed nations. It will also benefit from the inflow of an additional 1.1 million immigrants every year from 2010 to 2050.

Notes

1. According to http://projects.washingtonpost.com/fallen/.
2. According to http://projects.washingtonpost.com/fallen/.
3. Cited in Amit Gupta, "Australia and Strategic Stability in Asia," in *Strategic Stability in Asia*, edited by Amit Gupta (Aldershot, England: Ashgate, 2008), 157.
4. Richard Jackson and Neil Howe with Rebecca Strauss and Keisuke Nakashima, *The Graying of the Great Powers: Demography and Geopolitics in the 21st Century* (Washington, DC: Center for Strategic and International Studies, 2009), 6.

www.ingramcontent.com/pod-product-compliance
Lightning Source LLC
Chambersburg PA
CBHW060420220326
41598CB00021BA/2233